Lippincott's Illustrated Q&A Review of

Neuroscience

Duane E. Haines, PhD
Professor Emeritus, Department of Anatomy, and
Professor of Neurology and Professor of Neurosurgery
The University of Mississippi Medical Center
Jackson, MS 39216-4505

Special Contributions by

M. Alissa Willis, MD
Department of Neurology,
The University of Mississippi Medical Center

James B. Walker, MD
Department of Neurosurgery,
The University of Mississippi Medical Center

Illustrators

M. P. Schenk, BS, MSMI, CMI, FAMI
W. K. Cunningham, BA, MSMI

Photography

Charles P. Runyan III
G. W. (Bill) Armstrong, RBP

Wolters Kluwer | Lippincott Williams & Wilkins
Health
Philadelphia • Baltimore • New York • London
Buenos Aires • Hong Kong • Sydney • Tokyo

Acquisitions Editor: Crystal Taylor
Product Managers: Kelley A. Squazzo & Catherine A. Noonan
Marketing Manager: Michelle Patterson
Vendor Manager: Alicia Jackson
Manufacturing Manager: Margie Orzech
Design Coordinator: Doug Smock
Compositor: SPi Technologies

First Edition

351 West Camden Street
Baltimore, MD 21201

Two Commerce Square, 2001 Market Street
Philadelphia, PA 19103

Printed in China.

Library of Congress Cataloging-in-Publication Data
Haines, Duane E.
Lippincott's illustrated Q & A review of neuroscience / Duane E. Haines ; special contributions by M. Alissa Willis, James B. Walker ; illustrators, M.P. Schenk, W.K. Cunningham ; photography, Charles P. Runyan III, G.W. (Bill) Armstrong. — 1st ed.
p. ; cm.
Other title: Illustrated Q & A review of neuroscience
Other title: Lipppincott's illustrated Q and A review of neuroscience
Summary: "Lippincott's Illustrated Q&A Review of Neuroscience offers up-to-date, clinically relevant board-style questions-perfect for course review and board prep! Approximately 500 multiple-choice questions with detailed answer explanations cover frequently tested topics in neuroscience. The book is heavily illustrated with photos or clinical images. Online access to the questions and answers provides flexible study options"—Provided by publisher.
ISBN 978-1-60547-822-7 (pbk.)
1. Neurology--Examinations, questions, etc. 2. Nervous system—Diseases—Examinations, questions, etc. I. Willis, M. Alissa (Mary Alissa) II. Walker, James B., M.D. III. Title. IV. Title: Illustrated Q & A review of neuroscience. V. Title: Lipppincott's illustrated Q and A review of neuroscience.
[DNLM: 1. Nervous System Diseases—Examination Questions. 2. Nervous System—Examination Questions. WL 18.2 H153L 2011]
RC343.5.H32 2011
616.80076—dc22

2010024711

DISCLAIMER

Care has been taken to confirm the accuracy of the information present and to describe generally accepted practices. However, the authors, editors, and publisher are not responsible for errors or omissions or for any consequences from application of the information in this book and make no warranty, expressed or implied, with respect to the currency, completeness, or accuracy of the contents of the publication. Application of this information in a particular situation remains the professional responsibility of the practitioner; the clinical treatments described and recommended may not be considered absolute and universal recommendations.

The authors, editors, and publisher have exerted every effort to ensure that drug selection and dosage set forth in this text are in accordance with the current recommendations and practice at the time of publication. However, in view of ongoing research, changes in government regulations, and the constant flow of information relating to drug therapy and drug reactions, the reader is urged to check the package insert for each drug for any change in indications and dosage and for added warnings and precautions. This is particularly important when the recommended agent is a new or infrequently employed drug.

Some drugs and medical devices presented in this publication have Food and Drug Administration (FDA) clearance for limited use in restricted research settings. It is the responsibility of the health care provider to ascertain the FDA status of each drug or device planned for use in their clinical practice.

To purchase additional copies of this book, call our customer service department at (800) 638-3030 or fax orders to (301) 223-2320. International customers should call (301) 223-2300.

Visit Lippincott Williams & Wilkins on the Internet: http://www.lww.com. Lippincott Williams & Wilkins customer service representatives are available from 8:30 am to 6:00 pm, EST.

9 8 7 6 5 4 3 2 1

Preface

Neuroscience is not only an enormously interesting topic on its own but is one that permeates most aspects of modern medicine: central lesions, peripheral nerve disease, trauma, such as nerve root avulsion, and many others. In this respect, an understanding of nervous system structure coupled with knowledge of the corresponding clinical correlates are, simply put, two sides of the same coin. For example, understanding the corticospinal tract is enhanced by the knowledge of its function, and recognizing clinical deficits of this same system requires knowledge of its structure. The nervous system is a challenging and engaging puzzle, the clinical and anatomical pieces of which interlock to create a clear and successful diagnosis of the impaired patient.

Lippincott's Illustrated Q&A Review of Neuroscience offers well-designed questions, the majority of which are clinically oriented and presented in the patient vignette style used by the National Board of Medical Examiners (NBME). These questions cover a wide range of basic science topics, while stressing their use and relevance in a clinical context.

This style of question offers several distinct advantages. First, it allows the users to preview questions and test themselves in a format generally consistent with what they will see on the USMLE Step1. Second, it allows the users to review/test their knowledge of CNS structure and function in a format that mimics what may be seen in the clinical setting. Third, these questions provide excellent practice for any course taught in a professional school, whether it has a distinct clinical leaning or approach.

These questions do not follow one particular book, or source, but reflect content that is most generally addressed in clinically oriented neuroscience/neurobiology courses. While there are what one could call general knowledge questions in a clinical format, there are also challenging questions that will both test and extend the users' knowledge base. The following general approaches, or innovations, have guided this project:

Many questions are in a patient vignette style generally reflecting that used by the NBME. As much as is reasonably possible, information on relevant history/cause, tests/imaging, and essential elements of the examination are used in the stem. The answers are listed in alphabetical order.

Every effort has been made to design brief questions that will test the users' ability to interpret, recall, apply, and integrate information.

Recognizing that most users will be viewing images of the CNS in a clinical orientation or environment (viewer's left is patient's right; viewer's right is patient's left), *all images*, including line drawings, stained sections, brain slices, and brain specimens, are placed in an orientation that reflects how that particular brain part is viewed in MRI and/or CT in the clinical environment. This method automatically imparts to these images the same features that are seen in MRI or CT: the image (line drawing, stained section, brain slice) has right and left sides and the topography of tracts within the image matches exactly that seen in MRI and CT. On the other hand, MRI and CT, by their very nature, have intrinsic direction and laterality. This provides maximum opportunity for the users to practice the essential skill of determining the correlation between side of lesion and laterality of deficit(s), mandatory for the clinical environment. While this is a departure from other question/review books, it reflects an absolute clinical reality: *This reality being that when students view the brain in MRI or CT, their right is the patient's left and their left is the patient's right, and the anatomy in the brainstem is reversed top-to-bottom: the anterior/ventral aspect of the brainstem is up in these images and its posterior/dorsal aspect is down*. Mastering these relationships is *absolutely essential* to successful evaluation, diagnosis, and treatment of the neurologically impaired patient, a skill that should be developed early in the career of all users.

The correct answer is explained, the incorrect answers are also explained, and relevant additional points may be included in the answer.

There are a wide variety of structures within the nervous system that simply need to be learned, remembered, and recalled as part of any diagnostic effort. Chapter 23 provides an opportunity to practice identifying structures, spaces, or clinical defects that may be routinely encountered. While these examples are not designed to be all inclusive, they do provide broad-based samples. In most of the questions in this chapter, the images are of normal brains; three show clinical conditions. The clinical situations posed in all questions are meant to stimulate one's thinking as to *what structures may relate to the deficits in the example*. The items labeled in the image include one or more structures, damage to which would correlate with deficits described in the example. This allows a functional correlation and, at the same time, an opportunity to identify central nervous system structures in a variety of images.

The study of neuroscience is a great adventure with many exciting and interesting twists and turns. In the end, the successful diagnosis brings great satisfaction to the physician and great peace of mind to the patient. It is the goal of this book to impart some of this excitement to the user.

Duane E. Haines
Jackson, MS

Acknowledgments

Many individuals, both directly and indirectly, contribute to a project of this nature, and the author is greatly appreciative of their help, comments, and constructive criticism. I have enjoyed working with them and learning from them.

I am deeply indebted to Dr. Alissa Willis and Dr. James Walker for reading all these questions and answers and for holding my feet to the proverbial fire with regard to "clinical correctness" of the content. They are very knowledgeable and excellent clinicians and very busy, yet they took the time to help with a most essential part of this project. When it comes to a clinical knowledge base, they are the pros and I am the interloper. Alissa and James, thank you very much.

My colleagues in the Department of Anatomy, particularly Dr. James C. Lynch, and including Dr. Kimberly Simpson, Dr. Tony Moore, and Dr. Allen Sinning, were very helpful. My clinical colleagues in the Department of Neurosurgery (Dr. Louis Harkey, Dr. Andy Parent, Dr. Gustavo Luzardo, and Dr. Dale Hoekema [currently at Kadlec Regional Medical Center, Richland, WA]) and in the Department of Neurology (Dr. Hartmut Uschman, Dr. Robert Herndon, and Dr. James Corbett) most graciously put up with my incessant questions about clinical issues, both great and small; I have learned a great deal from them as clinicians and as friends. Dr. Barbara Puder (Samuel Merritt University) also provided critical suggestions that were very useful.

I want to express my sincere appreciation to Dr. Timothy C. McCowan, Chairman of Radiology, for allowing unfettered access to MRI and CT images, both normal and abnormal, and to Mr. W. (Eddie) Herrington, Chief CT/MRI Technologist for his willingness to supply images.

In addition to teaching medical students, another one of my joys is working with residents in neurosurgery and neurology. Of the current group of neurosurgery residents (Drs. Marks, Gaspard, Rey-Dios, Orozco, Johnson, Downes, Stacy, Uribe, and Walker), I want to single out Dr. James B. Walker for his outstanding review of everything and his excellent suggestions and Dr. Jared J. Marks, Dr. Bryan A. Gaspard, and Dr. Ludwig D. Orozco for supplying images and for numerous conversations on correct clinical terminology and interpretations. Of the current group of neurology residents who are on that service (Drs. Murray, Scarff, Hussaini, Shah, Willis, Ali, Bradley, Khan, and Sinclair), I want to single out Dr. Alissa Willis for her exceptional review of everything and numerous excellent suggestions and Dr. Lee Murray and Dr. Syed Hussaini for answering a seemingly endless list of questions and for images. Dr. Spencer Miller, a former neurology resident and currently in the Air Force in Nevada, patiently hunted down dozens of images, some of which are in this work; he did a great job and was very cooperative.

One essential element in the early development of this project was input and suggestions from students on what would make it more useful to their needs. I would like to express my appreciation to the following individuals for their thoughtful comments, insights, and suggestions: Stavros Atsas (University of Louisville); Caroline Botros (University of Medicine and Dentistry of New Jersey, School of Osteopathic Medicine); Rachel Chard (Oregon Health and Science University); Amy Haberman (University of Texas Medical Branch at Galveston); Joseph Hassab (Jefferson Medical College); Jennifer Huang (Mount Sinai School of Medicine); Cameron Jones (Morehouse School of Medicine); Heather Logghe (University of California); Setty Magana (Mayo Clinic College of Medicine, University of Minnesota Medical School); Asheer Singh (University of Otago); Ryckie Wade (University of East Anglia); Anne Williams (Bastyr University Naturopathic Medical School); and Israel Wojnowich (Mount Sinai School of Medicine).

A special thanks is due to my wife, Gretchen, for her laborious reading of every question and every answer. She quoted rules on punctuation that I never knew existed, found every little phrase that could be improved, and astonished me with her ability to catch misspelled words, be they general, basic science, or clinical. I greatly appreciate her outstanding help and support throughout this effort; if there is merit in this work, she is the reason.

Duane E. Haines
Jackson, MS

Contents

Introduction—A Cautionary Tale of Right and Left

Education is both an immediate activity: learn the material and pass the test; and a long-term endeavor: acquire information that will be useful and applicable to the goals of a life-long career. Both students *and* faculty are learning. Students are learning new information and how it applies to solving a clinical problem. Faculty are also learning new information *and* learning how to adapt/adopt new or changing concepts to an ever-evolving medical education environment.

There are two ways of looking at images of the brain. One has a long and distinguished history and, for lack of a better description, can be called the *Anatomical Orientation*. The other recognizes that how the brain is viewed in the clinical setting has been dramatically altered, for all time, by the advent of MRI and CT; this can be called the *Clinical Orientation*. The precept of looking at images in either axial or coronal plane in a *Clinical Orientation reflects contemporary clinical reality and is essential knowledge to master*. Once a criterion for orientation is established, it may be applied to *all* images, be they line drawings, stained sections, or brain slices. Naturally, MRI and CT have a universally recognized orientation.

The *Anatomical Orientation* has been the educational standard for many decades (Figure 1A,B). In this orientation, dorsal/posterior is up in the brainstem image, ventral/anterior is down in the image, but there is *absolutely no standard* of what is right or left. Some argue that right or left depends on whether one is looking at the rostral or the caudal surface of the section: certainly a legitimate point. So, one educator's right is the other educator's left: no consistency. In fact, some tests may use two identical brainstem drawings or stained slices in anatomical orientation with one side labeled 'right' on one drawing/slice and the opposite side labeled 'right' on the other: consistently inconsistent, even within the pages of the same examination! In addition, line drawings, in an anatomical orientation, do not have a 'rostral' or 'caudal' surface; they have only the surface that is seen; and line drawings do not have depth or thickness. Another very important fact is that images in the *Anatomical Orientation*, especially those of the brainstem, do not match MRI or CT images of the same portion of the CNS (Figure 1A–C). In addition, the topographical arrangement of clinically essential tracts of the brainstem, such as the medial lemniscus and spinal trigeminal tract and nucleus, is the reverse of that seen in MRI or CT.

On the other hand, using images[1] in a *Clinical Orientation* automatically solves four very important problems. First, the posterior/dorsal aspect of the brainstem is down in the image, and the anterior/ventral aspect is up in the image *exactly* matching the orientation of the same brain region in MRI/CT (Figure 2A–C). Second, the overall shape of the image exactly matches the corresponding portion of the brain in MRI or CT (Figure 2). So, the positions of internal structures learned in an anatomical image can be transposed directly to the clinical image ***without*** any spatial modification or change; everything matches (Figure 2A–C). Of particular relevance is the fact that the somatotopy of particularly important tracts (such as the medial lemniscus or spinal tract of V) is now correct for the clinical environment. Third, and more importantly, specifying that

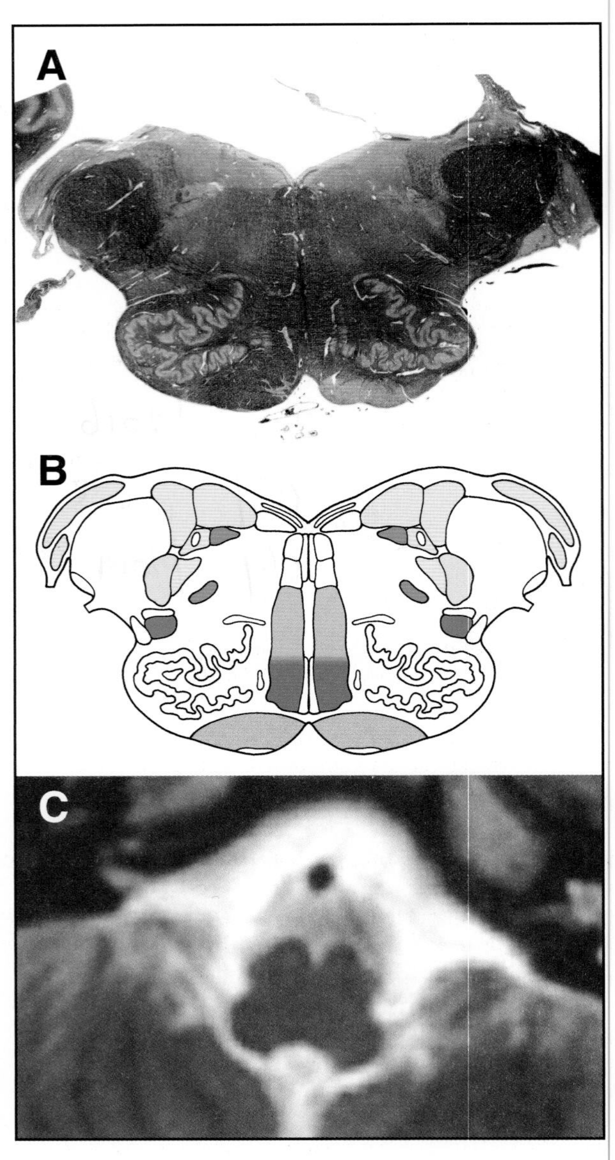

Figure 1. Stained section (**A**) and comparable drawing (**B**), both in Anatomical Orientation, compared to an MRI (**C**) that is in a *Clinical Orientation*.

[1] These images may be line drawings, unstained brain slices, stained sections, or gross brain specimens.

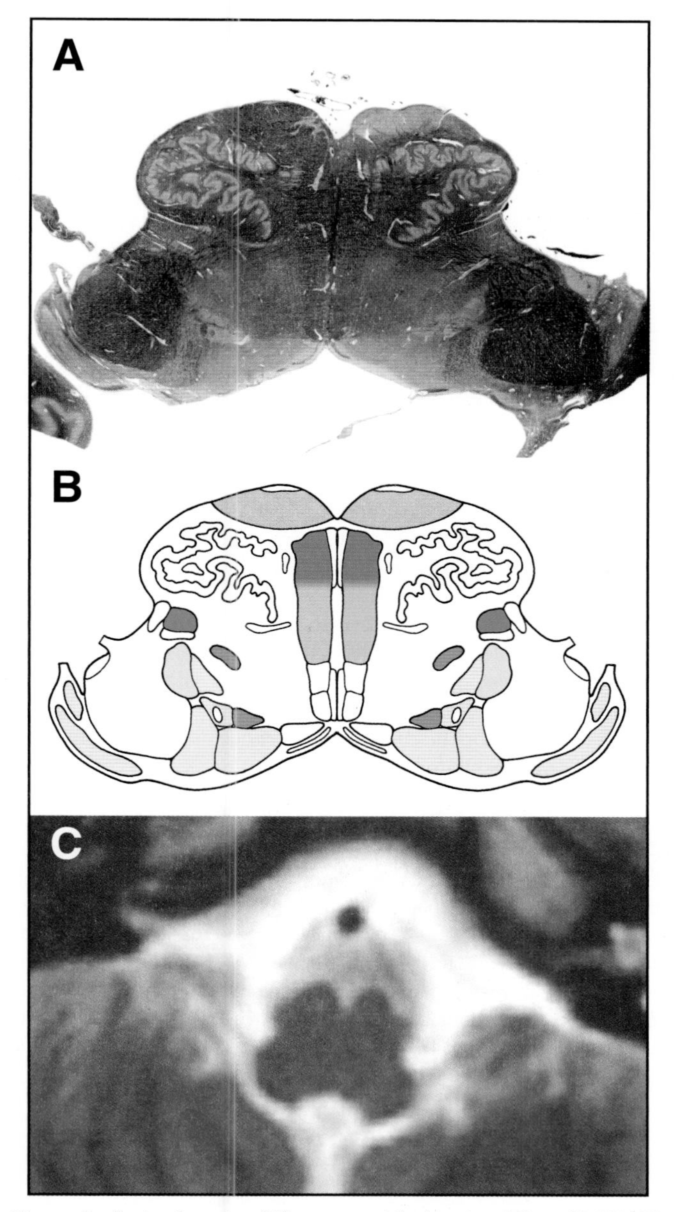

Figure 2. Stained section (**A**), comparable drawing (**B**), and MRI (**C**), all are in a *Clinical Orientation*. The usefulness of the anatomical information, when mastered in this orientation, is directly applicable to the clinical environment.

stained sections, line drawings, or brain slices are being shown in a *Clinical Orientation* automatically solves the right/left dilemma. The sides of the brain in MRI or CT are universally recognized as follows: ***the observer's right is the patient's left, and the observer's left is the patient's right***.[2] A stained section or line drawing in ***the same Clinical Orientation*** matches the MRI or CT orientation. Therefore, the right/left of the MRI transfers to what is right/left on the stained section, brain slice, or line drawing whenever it is clearly established that these latter images are consistently presented in a ***Clinical Orientation***. A criterion is established, and it applies to any and all figures in that orientation. Fourth, the topography of tracts that is essential to a successful diagnosis of the impaired patient now correlates exactly with that seen in MRI and CT.

It is acknowledged that a major test that all medical students will take does not present stained sections or line drawings in a clinical orientation. However, when preparing for any test, the goal of all students is to gain knowledge, understanding, and competency that will benefit them into, through, and well beyond the clinical years. Studying images of the brain in an orientation (anatomical) that may appear in only one major examination, ***and*** in an orientation that will unfortunately not be seen in the clinical setting or in practice, certainly limits the extent and applicability of one's knowledge base. On the other hand, knowledge gained on images in the ***Clinical Orientation*** can be transferred effortlessly and seamlessly into the clinical setting with no modification, change, or confusion. Such information is also essential to long-term professional success. In addition, it is much easier to learn clinical and then go to anatomical (as a specific occasion may require), than to learn anatomical (with its strikingly limited applicability) and then go to the clinical for the clinical years and for an entire career.

Recognizing the value of presenting information in a clinical format, all images in this review (line drawings, stained sections, brain slices, gross brain, and, naturally, MRI and CT), unless otherwise specifically stated in the stem of the question, are presented in a ***Clinical Orientation***, the exact same orientation one sees in MRI/CT. By using this approach, what is right and what is left is automatically established and transferred from image to image. Right and left are not labeled, primarily because right and left are also not labeled on MRI or CT images in the clinical setting; right/left is understood as part of the fundamental knowledge base.

COLOR KEY FOR DRAWINGS

The line drawings used to illustrate brain anatomy, with particular emphasis on the brainstem and spinal cord, are designed to emphasize structures and features that characterize specific levels. While excessive detail is avoided, the important nuclei, tracts, and pathways/systems are shown. These structures are color-coded using the following plan.

Dark blue signifies the gracile fasciculus (fasciculus gracilis) of the spinal cord, the gracile nucleus (nucleus gracilis) of the medulla, and that portion of the medial lemniscus conveying discriminative touch, vibratory sense, and proprioception/position sense from the lower extremity. Lower extremity is commonly abbreviated as LE. The nucleus gracilis is the place where the first-order neurons of this pathway synapse on the cell bodies of the second-order neurons; the axons of these second-order neurons cross the midline, as the sensory decussation, to form the medial lemniscus. The LE fibers in the medial lemniscus terminate in more lateral regions of the ventral posterolateral nucleus of the thalamus and from there are relayed to the LE area of the somatosensory cortex.

Light blue signifies the cuneate fasciculus (fasciculus cuneatus) of the spinal cord, the cuneate nucleus (nucleus cuneatus) of the medulla, and that portion of the medial lemniscus conveying discriminative touch, vibratory sense, and proprioception/position sense from the upper extremity. Upper extremity is

[2]Coronal MRI and CT are viewed as if the physician is looking at the face of the patient: the physician's right is the patient's left and the physician's left is the patient's right. Axial MRI and CT are viewed as if the patient is lying on his/her back and the physician is standing at his/her feet looking up through his/her body; as is the case for coronal images, the physician's right is the patient's left and the physician's left is the patient's right. Consequently, in stained sections, line drawings, or brain slices that are specified as being presented in a *Clinical Orientation*, the physician's right is that patient's left and the physician's left is that patient's right.

commonly abbreviated as UE. The nucleus cuneatus is the place where the first-order neurons of this pathway synapse on the cell bodies of the second-order neurons; the axons of these second-order neurons cross the midline, as the sensory decussation, to form the medial lemniscus. The UE fibers in the medial lemniscus terminate in more medial regions of the ventral posterolateral nucleus of the thalamus and from there are relayed to the UE area of the somatosensory cortex.

Dark green signifies the anterolateral system; this pathway arises from the posterior horn, crosses within about two spinal levels of the cells of origin (these are second-order neurons of the pathway), and coalesces on the opposite side to form the system. The anterolateral system is a composite bundle, conveying pain and thermal sense, which contains a number of important fibers terminating in brainstem nuclei (such as spinoreticular and spinomesencephalic fibers), while the spinothalamic component of the ALS ends in the thalamus.

Light green signifies the sensory root of the trigeminal nerve and the spinal trigeminal tract. These fibers arise from cells in the trigeminal ganglion and are the first-order neurons in the pathway conveying pain and thermal sense, and discriminative touch, from the face and oral cavity. The primary sensory axons conveying pain and thermal information terminate in the spinal trigeminal nucleus pars caudalis. Those fibers concerned with discriminative touch terminate in the principal sensory nucleus. Both of these trigeminal nuclei contain cells that give rise to trigeminothalamic projections. Trigeminothalamic fibers arising in the pars caudalis cross the midline to terminate in the ventral posteromedial nucleus (VPM) of the thalamus; fibers arising in the principal sensory nucleus have ipsilateral and contralateral projections to the VPM.

Dark red specifies motor nuclei that arise from the basal plate of the developing brainstem. The hypoglossal, abducens, trochlear, and oculomotor nuclei innervate muscles that arise from head mesoderm and contain cells that may be classified as Somatic Efferent (SE or general Somatic Efferent). Those muscles that arise from pharyngeal arches are innervated by the trigeminal motor nucleus (arch 1), the facial nucleus (arch 2), and the nucleus ambiguus (arch 3 via the glossopharyngeal nerve, arch 4 via the vagus nerve). The cell bodies of these nuclei contain cells that may be classified as Visceral Efferent (VE or Special Visceral Efferent, SVE). Visceromotor fibers (also called autonomic) arise from the dorsal motor nucleus of the vagus and the inferior salivatory, superior salivatory, and Edinger-Westphal nuclei; these cells may also be classified as Visceral Efferent (VE or general Visceral Efferent, GVE). These visceromotor nuclei contain preganglionic parasympathetic cells that end in peripheral ganglia.

Light red specifics sensory nuclei that arise from the alar plate of the developing brainstem. The cochlear and vestibular nuclei are concerned with hearing and with balance and equilibrium, respectively. These nuclei may be classified as receiving Somatic Afferent (SA or Special Somatic Afferent, SSA) input that may also be referred to as exteroceptive (hearing) or proprioceptive (vestibular). The spinal trigeminal nucleus receives pain and thermal sensations, and the principal sensory nucleus receives discriminative touch, both from the face and oral cavity, which may be classified as Somatic Afferent (SA or general Somatic Afferent, GSA). The solitary tract and nucleus are concerned with viscerosensory input that may be classified as Visceral Afferent (VA); this is subdivided into Special Visceral Afferent (SVA, taste for the rostral solitary nucleus; this part may be called the gustatory nucleus) and General Visceral Afferent (GVA, general sensation for the caudal solitary nucleus; this part may be called the cardiorespiratory nucleus).

Light gray specifies the location of corticospinal fibers. These arise from the somatomotor cortex (area 4), and to lesser degrees from other cortical areas, and enter the posterior limb of the internal capsule. From here, they descend through the crus cerebri and basilar pons to form the pyramid of the medulla. About 85% of corticospinal fibers cross the midline in the motor decussation (sometimes called pyramidal decussation) to form the lateral corticospinal tract and about 15% form the uncrossed anterior corticospinal tract. Corticospinal tract fibers influence somatomotor function on the opposite side of the body.

Chapter 1

General Concepts

QUESTIONS

Select the single best answer.

1 Your patient is a 5-year-old boy who has fallen off of a jungle gym onto a hard surface and has been rendered unconscious. When he arrives at the Emergency Department, he is still unconscious, is bleeding from his nose, but has only minor facial abrasions. Which of the following imaging studies would give you the most immediate and relevant information regarding the condition of this patient?

(A) CT
(B) PET scan
(C) T1-MRI
(D) T2-MRI
(E) X-ray

2 A 31-year-old woman calls your office and says that she has been violently throwing up for several hours. Which of the following would describe this woman's condition at this particular time in the course of her illness?

(A) A deficit
(B) A sign only
(C) A sign + a symptom
(D) A symptom only
(E) Apnea

3 A 69-year-old woman presents at the Emergency Department after losing consciousness at a family picnic. The examination reveals a loss of pinprick sensation on the right upper and lower extremities, a diminution of discriminative touch on the right lower extremity, an uncoordinated gait, and weakness of the facial muscles on the left. Taking all of this woman's deficits into account, which of the following stands out as the best localizing sign in this case?

(A) Sensory loss on both extremities
(B) Sensory loss on lower extremity
(C) Sensory loss on upper extremity
(D) Uncoordinated gait
(E) Weakness of facial muscles

4 A 23-year-old man is brought to the Emergency Department from an accident site. The examination reveals that he is profoundly weak on his right side, has a loss of vibratory sense on the right side of his body, and does not feel pinprick on the left side of his body. The cranial nerve examination is normal, he is oriented as to time and place, and he is well aware of his situation. The lesion in this patient is most likely located in which of the following?

(A) Diencephalon
(B) Metencephalon
(C) Mesencephalon
(D) Myelencephalon
(E) Spinal cord

5 A 6-year-old boy is brought to the Emergency Department with two severe bites on his left hand. The history, provided by the mother, reveals that the boy was playing in his yard and a squirrel approached the boy. When the boy attempted to pet the squirrel, it attacked the boy's hand and viciously bit him. The physician suspects that the squirrel had rabies. If left untreated, which of the following would be a likely progression of this disease?

(A) Anterograde transport of viruses, eventual degradation, only local infection
(B) Anterograde transport of viruses, replication in axon terminals, systemic distribution
(C) No transport of viruses, only a local infection, treatable with topical antibiotics
(D) Retrograde transport of viruses, replication in cell body, systemic distribution
(E) Retrograde transport of viruses, eventual degradation, only local infection

6 A 67-year-old man is brought to the Emergency Department by his wife. The man is lethargic, has motor difficulties, and cannot walk unaided. The history, provided largely by the wife, reveals that the man fell in his barn about 3 or 4 days ago, had a scalp laceration that is healing, and is taking over-the-counter medications for headache. He has deteriorated significantly in the last 24 hours. CT reveals an extraaxial mass in the right hemisphere that has the same texture, gray color, and appearance as the brain. Which of the following would most specifically describe this mass in this man?

(A) Hyperdense
(B) Hypodense
(C) Hyperintense
(D) Isodense
(E) Isointense

7 A 71-year-old man presents to his physician complaining of "numbness" in his hands. The history reveals that this deficit appeared gradually and got progressively worse over several weeks. The examination confirms a loss of pain and thermal sensation over the shoulders, upper extremities, and hands. Further testing reveals that vibratory sense and discriminative touch are normal over upper and lower extremities and there are no cranial nerve deficits. Which of the following most specifically identifies this deficit in this man?

(A) Alternating hemianesthesia
(B) Alternating hemiplegia
(C) Dissociated sensory loss
(D) Paresthesia
(E) Tactile agnosia

8 A 68-year-old woman is brought to the Emergency Department after collapsing at her place of work. The initial examination reveals a slightly overweight female who is somnolent with a profound weakness of her right extremities and a loss of all sensation (pain and thermal, discriminative touch, vibratory sense) also on her right side. Cranial nerve function is normal. Reexamination at 6 hours confirms these results and shows a further reduced level of consciousness. Based on these observations, which of the following represents the most likely location of this lesion?

(A) Forebrain
(B) Medulla
(C) Midbrain
(D) Pons
(E) Spinal cord

9 A 71-year-old man is sitting in his barber's chair when his right upper and lower extremities suddenly become weak; he has no other symptoms. Considering this situation, which of the following is the most likely cause of this man's deficit?

(A) Acute bacterial meningitis
(B) Convexity meningioma
(C) Glioblastoma multiforme
(D) Intracerebral hemorrhage
(E) Subarachnoid hemorrhage

10 Considering the full range of possible treatments (surgery, radiology, chemotherapy), a patient with which of the following tumors would most likely have the longest survival?

(A) Anaplastic astrocytoma
(B) Brainstem glioma
(C) Glioblastoma multiforme
(D) Oligodendroglioma
(E) Meningioma

11 A 36-year-old woman visits her physician with multiple concerns. During the examinations, she reveals that she has double vision (diplopia), has noticed weakness of her right upper and lower extremities, and has a diminution of pain sense on her left hand and forearm. She also reveals that her symptoms have sometimes been worse and then seem to get better for a while. MRI reveals areas of demyelination that correlate with her symptoms. Which of the following explains this woman's deficits?

(A) Damage to the membrane of the axon
(B) Damage to intra-axonal microtubules
(C) Failure of the development of the generator potential
(D) Failure of the development of the end-plate potential
(E) Failure of saltatory conduction along the axon

12 Bipolar cells are characteristically seen in which of the following locations?

(A) Aorticorenal ganglion
(B) Amygdaloid nucleus
(C) Cerebellar cortex
(D) Posterior root ganglion
(E) Vestibular ganglion

13 A 57-year-old man visits his family physician with the complaint of a slight tremor in his left hand. The examination reveals that this tremor basically disappears when the man performs a voluntary movement, but this man also has a slightly shuffling gait and bradykinesia. The results of the evaluation indicate that the man has early Parkinson disease, and the physician decides to start the man on a regimen of levodopa/carbidopa (Sinemet®). Which of the following, if also a component of this man's overall clinical condition, would be an absolute contraindication for using this particular medication in this man?

(A) Anaplastic astrocytoma
(B) Glioblastoma multiforme
(C) Malignant ependymoma
(D) Malignant melanoma
(E) Malignant meningioma

14 A 71-year-old woman is brought to the physician's office by her daughter. The daughter explains that she is concerned that her mother "does not seem right." As part of the evaluation, the physician conducts a mental status examination. Which of the following would be part of this examination?

(A) Ask her to walk heel to toe
(B) Ask the date and her home address
(C) Tap her patellar and biceps tendons
(D) Touch both hands with a sharp pin
(E) Touch both patellae with a tuning fork

15 A 44-year-old man is undergoing a neurological evaluation for a movement disorder. MRI reveals a diminution in the size of the head of the caudate nucleus. The man has two sons who are married but have no children. Which of the following would be an especially important part of the overall examination and evaluation?

(A) Allergies
(B) Family history
(C) Medications
(D) Social history
(E) Surgical history

16 A 17-year-old boy is brought to the Emergency Department from the site of an accident. The history of the current event reveals that the boy was intoxicated and hanging out the car window when he ran into a stop sign, severing his arm just below the shoulder. Which of the following most likely specifies the reaction of lower motor neurons in the cervical spinal cord in this case?

(A) Axonotmesis
(B) Chromatolysis
(C) Neurapraxia
(D) Neurotmesis
(E) Neuroma

17 Pain and thermal information are transmitted by what are commonly called lightly myelinated and unmyelinated nerves that have a slow conduction velocity. Lightly myelinated fibers are also called Aδ fibers and unmyelinated fibers are called C fibers, both having axons of small diameter. Which of the following characterizes the latter of these fiber types?

(A) Axons lying free in the interstitial space/fluid
(B) Axons lying on the surface of Schwann cells in the interstitial space
(C) Axons lying within grooves on the Schwann cells
(D) Axons with only a single layer of formed myelin from Schwann cells
(E) Axons with fewer than three layers of formed myelin from Schwann cells

18 A 64-year-old man is undergoing a neurological examination. The history reveals that this man has not experienced any particular neurological event, but his daughter thinks that his personality is changing. During the examination, the physician asks the man to remember the phrase "dog and cat in the house." The physician proceeds with the exam and about 3 to 4 minutes later asks the man to repeat the phrase back to him. This exercise is most likely testing which of the following?

(A) Anterograde amnesia
(B) Recent memory
(C) Remote memory
(D) Retrograde amnesia
(E) Immediate recall

19 A 72-year-old woman is brought to the Emergency Department by her daughter. The history of the current event, provided by the daughter, reveals that the woman complained of a sudden headache and, after lying down for a period of time, seemed very confused. The examination reveals an otherwise healthy woman who can see but is unable to recognize what she sees; this is probably contributing to her apparent confusion. This woman is most likely suffering which of the following?

(A) Ageusia
(B) Agnosia
(C) Akinesia
(D) Apnea
(E) Apraxia

20 Which of the following cells are essentially the immune cells of the central nervous system and those that react most vociferously to inflammation?

(A) Fibrous astrocytes
(B) Oligodendroglia
(C) Protoplasmic astrocytes
(D) Microglia
(E) Schwann cell

ANSWERS

1 **The answer is A: CT.** CT is fast, is economical, and (most importantly for this case) can immediately identify life-threatening conditions such as skull fracture and/or acute blood in or around the brain. MRI is slower, can be frightening for a young patient (because of the noise), especially if the patient is conscious, and may not clearly identify bone fractures or blood. X-ray would provide information only on bone fractures. PET (Positron Emission Tomography) scan uses intravenous injections of isotopes of fluorine-18-deoxy-glucose to measure relative metabolism within the brain. PET measures metabolic activity but provides little or no anatomic detail, and while it may provide information on certain neurological diseases (neurodegenerative diseases, epilepsy, certain tumors), it is not a standard testing procedure.

2 **The answer is D: A symptom only.** This woman calls and describes a problem that she is experiencing herself (a symptom). You, as the physician, cannot examine, discover, or evaluate this problem on this patient at this time; when you can do so, it then becomes a sign (something that you witness and evaluate). A deficit is generally regarded as a condition resulting from a lesion or disease process, and apnea is the absence of breathing.

3 **The answer is E: Weakness of facial muscles.** Localizing signs are essentially a comparison of all deficits in a particular situation to see which specific deficit may best indicate the probable location of the lesion. The weakness of facial muscles on the left, in combination with long tract signs, localizes the lesion to either the facial nucleus or the root and somewhere in the brainstem close to these structures. In combination with long tract signs, it specifies a lesion in the brainstem (***alternating*** or ***crossed deficits***); in this case, it specifies a lesion in the caudal pons on the left side. Sensory losses on the extremities and an uncoordinated gait may result from lesions at many different levels of the CNS. However, inserting a facial weakness in the equation provides a clear answer. In this case, the gait difficulties may relate to the lower extremity loss of discriminative touch, in which case it is called a sensory ataxia.

4 **The answer is E: Spinal cord.** This case illustrates one general and important concept concerning regional localization of lesions. This patient has alternating sensory losses (one sensory deficit on the right side, another sensory deficit on the left side), a significant motor deficit on one side, and no cranial nerve deficits. This particular combination of problems (alternating sensory + motor + no cranial nerves) is characteristic of a lesion in the spinal cord. Lesions in the diencephalon (or diencephalon + telencephalon) may give rise to motor and all sensory deficits on the same side of the body. Brainstem lesions usually present as alternating, or crossed, deficits: a cranial nerve deficit on one side and a long tract deficit (frequently the corticospinal) on the opposite side.

5 **The answer is D: Retrograde transport of viruses, replication in cell body, systemic distribution.** Rabies viruses use the method of retrograde transport: the active movement/transport of substances or particles from the distal part of the neuron (the axon terminals) proximally toward the cell body of the neuron. Once in the cell body, these viruses replicate, are shed by the neuron, and are subsequently taken up by other axons that terminate on the cell bodies or in the immediate vicinity of the shed viruses. By this mechanism, the disease is spread in the body. Anterograde transport is the movement of substances from the cell body of the neuron distally to the axon terminals of the neuron. This type of transport does not take place in rabies.

6 **The answer is D: Isodense.** The terms isodense, hyperdense, and hypodense are terms that refer to the appearance of a lesion/mass in comparison to the appearance of brain tissue in CT. A lesion that is essentially indistinguishable from brain (same color, texture) is isodense to brain (of equal density). Hyperdense refers to a shift toward the appearance of bone (more white), while hypodense refers to a shift toward the appearance of cerebrospinal fluid (more black). If the lesion in CT is hyperdense, it is whiter than the surrounding brain, while a lesion that is hypodense is darker than the surrounding brain. The terms hyperintense and isointense are used in T1 and T2 MRI; hyperintense is a shift toward more white, and isointense is used when the mass and the brain are essentially the same in appearance.

7 **The answer is C: Dissociated sensory loss.** A loss of pain and thermal sensation, with sparing of posterior column modalities, constitutes a dissociated sensory loss. Such deficits are most frequently associated with spinal cord lesions and may be the opposite of that described here (loss of posterior column modalities, preservation of pain/thermal sensations). This patient has a syringomyelia in cervical levels of the spinal cord; this is a classic sensory loss in a "cape distribution." The loss of a sensation on the body with sparing of cranial nerves also localizes the lesion to the spinal cord. Alternating hemianesthesia is the loss of the same sensory modality on one side of the body and on the opposite side of the face, and an alternating hemiplegia is a corticospinal deficit (motor) on one side of the body and a cranial nerve motor deficit on the opposite side. Paresthesia refers to the spontaneous occurrence of prickling, burning, or other abnormal sensations in the absence of actual stimulation; tactile agnosia is the loss of the ability to recognize objects by touch when peripheral nerves are intact.

8 **The answer is A: Forebrain.** Lesions in the forebrain are generally characterized by motor and sensory deficits (all modalities) on the same side of the body in the absence of cranial nerve deficits. This is one of the main examples of the principle of general localization of CNS lesions. Such lesions may be located in the caudal thalamus, posterior limb of the internal capsule, or over expansive areas of the cerebral cortex. Also, large hemisphere lesions may, but not always, have some facial weakness. However, the latter may be quite large and result in an instantaneously unconscious patient. Naturally, there may be variations on this theme, namely, some cranial nerve involvement if the lesion extends into the genu of the internal capsule and if the patient can be tested to confirm this. Lesions in the brainstem (choices B, C, and D) characteristically present with alternating (or crossed) deficits: long tract (such as corticospinal) signs on one side and cranial nerve signs on the other side. In addition, brainstem lesions, especially those in the pons and medulla, rarely produce somnolence. Spinal cord lesions are usually characterized by alternating motor and sensory deficits or by deficits that have motor and/or sensory levels, but with no cranial nerve signs.

9 **The answer is D: Intracerebral hemorrhage.** The "sudden onset" is a hallmark of a hemorrhagic event, especially in the absence of other symptoms. Recall that the substance of the brain has no pain endings, so the event is signaled by the deficit created, not specifically by pain. Subarachnoid hemorrhage (SAH) is also sudden onset. However, it is accompanied by sudden and excruciating headache (blood vessels do have pain fibers, i.e., a source of the pain), and there may, or may not, be instantaneous focal signs. A variety of signs and symptoms, including focal ones, may appear within hours in SAH. Acute bacterial meningitis develops rapidly but takes a few days to become symptomatic, and glioblastoma multiforme, while a rapidly growing tumor, may take months to produce signs and symptoms. Meningiomas, at any location, almost always develop very slowly and may take years to become symptomatic.

10 **The answer is E: Meningioma.** Meningiomas are generally slow growing, extraaxial (located outside the brain substance), and therefore noninvasive and are frequently benign (they do not metastasize to other locations). The most common treatment is surgical resection, and if the entire tumor can be resected, the cure may be complete. If total resection is not possible, survival may still be measured in many years due to their slow growth. The 5-year survival rate is over 90%. Brainstem gliomas are commonly seen in children or young adults (75%+ are less than 20 years of age), are difficult to treat surgically, and usually do not respond to chemotherapy but may respond to radiation, and the patient may survive 6 to 15 months after diagnosis. Both anaplastic astrocytoma (AA) and glioblastoma multiforme (GBM) are malignant, invasive to the brain, and somewhat responsive to radiation therapy and chemotherapy. These tumors are not curable; survival for AA may be up to 18 to 24 months and for GBM in the range of 6 to 12 months. Oligodendrogliomas are usually adult tumors, most frequently found in the white matter, and the choice of treatment is chemotherapy followed by surgery in some instances. Survival rates are in the general range of 2 to 4.5 years but may be longer in some cases.

11 **The answer is E: Failure of saltatory conduction along the axon.** This patient is suffering from multiple sclerosis, a demyelinating disease. This disease is characterized by the fact that the deficits may wax and wane, multiple systems may be involved, and MRI is a preferred diagnostic tool. The loss of myelin results in a failure of saltatory conduction due to the fact that the paranodal potassium channels oppose the sodium current at the node and the action potential degrades rather than reaching threshold. There is no damage to the axon membrane or to intracellular microtubules. The generator potential is a local depolarization at a sensory ending that is proportionate to the stimulus strength; if this potential reaches threshold, an action potential is produced. The end-plate potential is the local potential that results from an opening of acetylcholine channels and the sudden inward movement of sodium ions. Neither of these potentials is directly affected.

12 **The answer is E: Vestibular ganglion.** Bipolar cells are characteristically found in association with cranial nerves that conduct special senses, for example, in the retina, olfactory epithelium, and the peripheral ganglia of the vestibular and cochlear nerves. The functional component associated with these bipolar cells is special somatic afferent or special visceral afferent. The aorticorenal ganglion, amygdaloid nucleus, and the cerebellar cortex have multipolar cells that may assume a variety of sizes and shapes, but they all fall into the multipolar category as broadly defined. The posterior root ganglion characteristically contains unipolar or pseudounipolar cell bodies. The functional component of pseudounipolar cell bodies is general somatic afferent, exteroceptive and proprioceptive.

13 **The answer is D: Malignant melanoma.** Dopamine, an essential part of this medication regimen, is a precursor in the synthesis of melanin. Melanin is found in the cells of the substantia nigra pars compacta, which are part of the nigrostriatal dopaminergic pathway. The administration of a precursor to melanin in a patient with malignant melanoma will most likely stimulate the growth of this tumor. Malignant melanoma can be found in the brain. The administration of levodopa would not have any significant untoward effects if any of the other choices (anaplastic astrocytoma, glioblastoma, malignant ependymoma, and malignant meningioma) were concurrently present in this man, unless they contained pigmented cells. Also, ependymomas are usually a lesion of younger persons and malignant meningiomas are infrequently seen.

14 **The answer is B: Ask the date and her home address.** The mental status examination is an absolutely necessary part of all neurological examinations, especially in the elderly. It includes questions related to orientation (time, place, and person), registration (the physician names several objects, and the patient repeats the same objects in the same sequence), language (the patient is asked to identify common objects or follow a series of directions), attention and calculation (spell a word backward, add a set of numbers), and recall (repeating something done earlier in the examination). This examination provides a broad range of data on the general mental state of the patient. Asking her to walk heel-to-toe is a test of cerebellar function, and tapping the patellar and triceps tendons is part of a motor examination (reflexes). The two major sensory pathways are tested by touching the hand with a pin (pain/thermal sense; anterolateral system) and touching the patellae (knees) with a tuning fork (vibratory sense; posterior column–medial lemniscus system).

15 **The answer is B: Family history.** A reduction in the size of the head of the caudate nucleus is characteristic of an individual with Huntington disease. This is an inherited neurodegenerative disease with prominent motor deficits, dementia, and eventual death from the disease. This disease is not treatable or curable. Since this is a disorder that is inherited, the most important part of the overall evaluation is to do a detailed investigation of the family history. Such information may not be freely offered. In this man's case, he has two sons of childbearing age, and it would be an option to do genetic testing to see if they carry evidence of the trinucleotide repeat on chromosome 4. Allergies are not relevant; this is not a curable disease, so medications may shed little light; a surgical history would also provide little insight. A social history may provide confirmatory data based on interaction within his community.

16 **The answer is B: Chromatolysis.** The lower motor neurons whose peripheral processes were severed in this case undergo chromatolysis: the cell body swells, the Nissl substance is dispersed to the periphery of the cell body, and the nucleus assumes an eccentric location within the swollen soma. Depending on a number of factors (the degree of the trauma [clean cut vs. ragged tear], distance from the cell body), chromatolytic cells may eventually recover or may continue to degenerate until cell death. Axonotmesis is damage to a nerve, usually a crush-type injury, which disrupts the axon but not its connective tissue coverings or the myelin sheath. These injured axons may degenerate but are also likely to regenerate with time. Neurapraxia is a mild type of lesion that does not significantly damage the axon or its myelin sheath and does not result in degeneration. Temporary compression of a peripheral nerve with numbness and tingling of an extremity, but with rapid recovery, is an example. Neurotmesis is a severe injury in which the nerve is damaged to the point where the axon, myelin, and connective tissue coverings are damaged with resultant degeneration distal to the point of the injury. Neuroma designates a tumor of neural origin; depending on their location, they may have a more specific term.

17 **The answer is C: Axons lying within grooves on the Schwann cells.** Recall that the conduction of an action potential along an axon depends on insulation of some type (a separation of the axon from interstitial fluid) that provides for exposed segments of axons at either end of the insulated segment, which has sodium channels that are necessary for salutatory conduction. Axons that lie within grooves on the Schwann cells are surrounded by the cytoplasm of the Schwann cell; although there is no myelin, the Schwann cell serves an insulation function and provides a short internodal distance. Slowly conducting fibers have small-diameter axons, short internodal distances, and little (lightly myelinated) or no (unmyelinated) myelin coverings. Axons lying free in the interstitial space or on Schwann cells within the interstitial space have no structural mechanism for saltatory conduction. Axons with a myelin coat of any thickness are myelinated (unmyelinated refers to no myelin).

18 **The answer is B: Recent memory.** Recent memory is the process of turning newly learned material into long-term memory; it is the recall of an event or information that happened just a few minutes ago, but with the anticipation that it may be available in memory in the future. Immediate recall is when the physician asks the patient to immediately repeat (with no intervening interval) a series of numbers or words. Remote memory is the ability of the patient to recall events that may be months or years old. Anterograde and retrograde amnesia are most frequently applied to memory deficits that follow a traumatic event or stroke, anterograde referring to an impaired memory immediately after the disorder and retrograde referring to memory deficits immediately before the disorder.

19 **The answer is B: Agnosia (visual agnosia).** Visual agnosia is an inability to recognize objects or persons even though the patient can see the object or person; he or she just cannot figure out what the object, or who the person, is. In a comprehensive view, agnosia is a general term that indicates difficulty recognizing or understanding a variety of sensory inputs; a patient may have visual agnosia, auditory agnosia, tactile agnosia, and so on. Ageusia is loss of the sense of taste; it may be to a particular tastant or to all tastants and may be an inherited disorder, resultant to disorders of the taste buds, or rarely a CNS disorder. Akinesia is a movement disorder resulting from disease or lesions of the basal nuclei; the patient has great difficulty initiating, or is unable to initiate, a voluntary movement. Apnea is an inability to breathe. Apraxia is basically an inability to execute purposeful and useful movements: the patient understands what is to be done, has the muscle strength, has normal sensory function, but has difficulty actually performing the movement.

20 **The answer is D: Microglia.** Microglial cells are first identified within the CNS with the invasion of blood vessels and blood cells; this is evidence of their hemopoietic origin. These glial cells are found in all parts of the CNS, are small in their resting state, but become large and engage in phagocytic activity when stimulated by inflammation or CNS damage. Active microglial cells are frequently described as lipid-containing; this is due to the fact that they ingest myelin fragments. Microglial cells in their active phagocytic state are also called gitter cells. Astrocytes have a myriad of functions that include modulating the neuronal environment, secretion of cytokines and growth factors, reaction to injury (contribution to glial scar formation), and the metabolism of neurotransmitter substances, particularly glutamate. Fibrous (also called fibrillary) astrocytes are located predominately in the white matter, while protoplasmic astrocytes are numerous in the gray matter. Oligodendroglial cells are the myelin-forming cells in the CNS, while the Schwann cell does the same function in the PNS.

Chapter 2

Development and Developmental Defects

QUESTIONS

Select the single best answer.

1 Which of the following combinations of nuclei originate from the basal plate of the developing brain of the embryo?
(A) Abducens + spinal trigeminal
(B) Cochlear + oculomotor
(C) Hypoglossal + nucleus ambiguus
(D) Hypoglossal + vestibular
(E) Mesencephalic of trigeminal + oculomotor

2 A 7-year-old boy is brought to the pediatrician by his mother. The history reveals that the boy has had progressive difficulty seeing the assignments posted by his teacher. The examination reveals a boy who is noticeably small for his age, has a partial loss of vision in both eyes, but otherwise appears normal. An MRI shows a craniopharyngioma (Rathke pouch tumor), which is an aberrant outpocketing of the stomodeum. Which of the following is most directly involved in this developmental defect in this boy?
(A) Adenohypophysis
(B) Neurohypophysis
(C) Optic chiasm
(D) Stalk of the pituitary
(E) Lamina terminalis

3 The CT of a 4-day-old baby reveals cerebral hemispheres that are essentially devoid of gyri and sulci and display enlarged ventricles. The pediatric neurologist explains to the family that one contributing factor was the lack of proper cell migration within the cerebral cortex in early developmental stages. Which of the following most likely describes the appearance of the brain in this neonate?
(A) Anencephaly
(B) Lissencephaly
(C) Microgyria
(D) Pachygyria
(E) Schizencephaly

4 A child is born with unusually close-set eyes (hypotelorism). Suspecting additional developmental defects, the physician orders an MRI. The results of this procedure reveal that the interhemispheric fissure is incomplete, the falx cerebri is incomplete, the lateral ventricles are well formed, but there may be a partial fusion of either frontal or occipital lobes. Which of the following specifies this defect?
(A) Alobar holoprosencephaly
(B) Frontal encephalocele
(C) Lissencephaly
(D) Semilobar holoprosencephaly
(E) Occipital encephalocele

5 A 26-year-old woman develops abdominal pain over a several hour period and then has a bloody vaginal discharge. She is transported to the Emergency Department. The examination reveals a slightly obese woman who is pregnant (she claims that she was unaware of this fact), having considerable lower abdominal discomfort and increasing discharge. A few hours later, she has a miscarriage. Examination of the embryo reveals that neurulation did not take place. Exposure to which of the following would explain this occurrence?
(A) Colchicine
(B) Folic acid
(C) Glutamate
(D) Thiamine
(E) Vitamin A

6 A 31-year-old woman has a sonogram at the end of the first trimester of her second pregnancy. This study reveals a fetus with a small malformation in the lumbosacral area suggestive of a failure of the posture neuropore to close at the appropriate time and the probability of this baby being born with a spina bifida. Normally, this portion of the neural tube is closed by about which of the following times?
(A) 20 days
(B) 22 days
(C) 26 days
(D) 30 days
(E) 34 days

7 In addition to contributing to the formation of the nucleus pulposus of the adult, which of the following is also an important function of the notochord?
(A) Formation of a congenital dermal sinus
(B) Formation of the somite
(C) Induction of ectoderm to form the neural plate
(D) Induction of mesoderm to form muscle
(E) Induction of neural plate to form the neural crest

8 A female child is born with a single midline eye (cyclopia), small head, and a small misshapen nose. CT reveals a rim of brain (with rudimentary gyri) in the skull, a large single ventricle, no evidence of a corpus callosum, and little evidence of basal nuclei or thalami. This significant developmental defect is best characterized as which of the following?

(A) Agenesis of the corpus callosum
(B) Anencephaly
(C) Alobar holoprosencephaly
(D) Schizencephaly
(E) Semilobar holoprosencephaly

9 A 5-year-old boy is brought to the family physician by his mother. Her concern is that she thinks he is "slower" than his two brothers were at about the same age. The examination reveals a normal-appearing active boy who interacts appropriately with the physician, his nurse, and the mother. Further evaluation suggests a very modest developmental delay, and an MRI is ordered. While these images show a fundamentally normal brain, they do reveal continuous layers of gray matter located in the white matter internal to the cortex. These layers are separated from the cortex by obvious ribbons of white matter but are isointense to the overlying cortex. The physician concludes that these are layers of misplaced cortical neurons. Which of the following most likely specifies this developmental defect?

(A) Cortical agenesis
(B) Cranioschisis
(C) Heterotopia
(D) Hypertelorism
(E) Myeloschisis

10 A 27-year-old woman has a 3-year-old son who was born with spina bifida aperta. She and her husband want to have another child and are seeking advice on what can be done to decrease the probability of a recurrence of this event in her second pregnancy. Dietary supplementation with which of the following would be a likely strategy?

(A) Carbamazepine
(B) Folic acid
(C) Vitamin A
(D) Vitamin C
(E) Warfarin

11 The developing hindbrain (rhombencephalon) becomes organized into segments designated r1–r8 soon after closure of the neural tube: these are rhombomeres. Which of the following rhombomeres is associated with the location of neurons that are developing in relation to the facial nerve?

(A) r1 only
(B) r2 + r3
(C) r4 + r5
(D) r6 + r7
(E) r8

12 The MRI of a 4-day-old female newborn reveals cerebral hemispheres that have some large broad gyri interspersed with areas containing very small irregular gyri; these are pachygyri and microgyri, respectively. Which of the following events during pregnancy could explain this outcome in this baby?

(A) Agenesis of the corpus callosum
(B) Dietary deficiency of vitamin A
(C) Failure of microglia to invade the CNS
(D) High intake of dietary folic acid
(E) Improper neuronal migration

13 A 9-year-old girl is brought to the pediatrician by her mother. The girl thinks that there is something wrong with her eyes because she is having a lot of trouble playing goalie on her soccer team. Testing, which includes an ophthalmoscopic examination, reveals that vision in her right eye is essentially normal, but when her right eye is covered, visual acuity is extremely low in the left eye, depth perception is almost nonexistent, and she is unable to perceive shapes in a random-dot stereogram. Which of the following is a likely cause of this problem in this girl?

(A) Amblyopia
(B) Cataract
(C) Myopia
(D) Nyctalopia
(E) Presbyopia

14 The correct orientation of Purkinje cell dendrites in the adult cerebellum is dependent on the orientation and developmental patterns of which of the following?

(A) Climbing fibers
(B) Golgi cells
(C) Granule cells
(D) Mossy fibers
(E) Oligodendrocytes

15 A 2-day-old female baby presents with a small pigmented area at the midline on her lower back. Close examination reveals a small tuft of delicate hairs associated with a small cutaneous dimple; clear fluid issues from what appears to be a small opening in the dimple when palpated. Which of the following most closely signifies this defect?

(A) Myeloschisis
(B) Encephalocele
(C) Spina bifida aperta
(D) Spina bifida cystica
(E) Spina bifida occulta

16 A male baby is born with a saccular structure approximately 6 cm in diameter located in the lumbosacral region. As one step in the assessment of possible treatment options, CT and MRI are conducted to clarify the bony relationships and soft tissue contents of this lesion. These images reveal that this structure contains fluid, portions of the spinal cord, and nerve roots. Which of the following would most correctly describe this lesion?

(A) Encephalocele
(B) Meningocele
(C) Meningoencephalocele
(D) Meningomyelocele
(E) Spinal bifida occulta

17 The sonogram of a 23-year-old pregnant woman, 2 months prior to term, reveals that her fetus has a growth or defect in the frontal area of the skull. The history reveals that the woman has a freewheeling lifestyle and has not been attentive to proper diet, including regularly taking her vitamin supplements. The child is born with a large mass on the forehead, a partially cleft nose, and ocular hypertelorism. CT and MRI on day 3 reveal a defect in the frontal bone and a mass containing fluid, portions of the frontal lobe, and a cavity that is continuous with the anterior horn of the lateral ventricle. Which of the following is specifically descriptive of this defect?

(A) Meningocele
(B) Meningoencephalocele
(C) Meningohydroencephalocele
(D) Meningomyelocele
(E) Microencephaly

18 A 32-year-old mother from a poor rural county gives birth to a male child. Based on her circumstances, she sought almost no prenatal care, visiting a physician only once when she first thought she was pregnant. This child is born with a catastrophic developmental defect associated with a failure of the anterior neuropore to close. Which of the following most specifically designates this developmental failure?

(A) Anencephaly
(B) Dermal sinus
(C) Encephalocele
(D) Meningocele
(E) Meningomyelocele

19 A 26-year-old woman gives birth to a male child with spina bifida and a bifid cranium. The mother's nutritional status and general health are unremarkable. She has been taking multivitamin supplements daily since her teens, and there is no evidence of drug use or family history of developmental defects. The only significant feature of the mother's medical history is that she is an epileptic and is on long-term therapy to control a generalized seizure disorder. Given these facts, which of the following is the most likely cause of this child's birth defects?

(A) Hypertension
(B) Ingestion of folic acid
(C) Ingestion of vitamin D
(D) Vitamin C toxicity
(E) Valproate toxicity

20 A failure of cell migration during development of the central nervous system may result in minor defects that result in little or no deficits or may result in catastrophic defects that are not compatible with life. Thus, the successful migration of developing neurons is absolutely essential in the creation of a normal nervous system. Which of the following is a necessary element in the success of this process?

(A) Fibrous astroglia
(B) Microglia
(C) Oligodendroglia
(D) Protoplasmic astroglia
(E) Radial glia

ANSWERS

1 **The answer is C: Hypoglossal + nucleus ambiguus.** Motor nuclei of the brainstem arise from the basal plate and sensory nuclei from the alar plate. Both nuclei in (C) are motor nuclei; all other choices are a combination of one motor and one sensory nucleus, thereby indicating that both the alar and the basal plates contribute. The basal plate of the brainstem and its derivatives are located medial to the sulcus limitans of the fourth ventricle. The alar plate and its derivatives are located lateral to the sulcus limitans.

2 **The answer is A: Adenohypophysis.** The pituitary is formed by the apposition of an outpocketing from the roof of the stomodeum (this outpocketing is called the Rathke pouch) that comes together with a downward outgrowth from the neuroectoderm. The anterior lobe (adenohypophysis) originates from the stomodeum and the posterior lobe (neurohypophysis, or pars nervosa) originates from the neuroectoderm. The optic chiasm is related to the neuroectoderm; the stalk of the pituitary also arises from the same area as the pars nervosa. The lamina terminalis, the anterior wall of the third ventricle, originates from the neural plate and represents the position of the closed anterior neuropore.

3 **The answer is B: Lissencephaly.** Lissencephaly, also called agyria, is a situation in which the gyri and sulci are poorly developed or completely lacking. This is the result of improper cell migration of maturing neurons on radial glia. A patient with lissencephaly may also have other developmental defects. Anencephaly is a failure of the anterior neuropore to close, resulting in a catastrophic defect (much of the forebrain is absent, calvaria absent) that is not compatible with life. Microgyria is the presence of many unusually small gyri (exuberant cell migration), and pachygria is the presence of fewer but larger gyri (slower aberrant migration, but not absent migration). Schizencephaly is the presence of unilateral or bilateral clefts of varying size that extend from the surface of the brain into the ventricular space.

4 **The answer is D: Semilobar holoprosencephaly.** This defect represents a failure of the embryonic forebrain to properly separate into the two hemispheres of the normal adult. Semilobar holoprosencephaly is the intermediate stage between alobar and lobar holoprosencephaly and, as its name implies, the hemispheres are only partially (semi) formed as separate entities. Alobar is the most severe form and is characterized by a single ventricle, fused frontal lobes, with other developmental defects usually present. Encephaloceles are defects in the skull with resultant protrusion of meninges and brain through the defect; these may be present anywhere but are more common in occipital and frontal areas. Lissencephaly is the failure of gyri and sulci to form properly as a sequel to aberrant cell migration in the cortex.

5 **The answer is A: Colchicine.** Colchicine, an alkaloid used to treat gout, disrupts microtubule formation and function. Because microtubules (and microfilaments) are essential to the folding process of the neural tube, exposure to colchicine will disrupt this process and result in a failure of neurulation. Deficiency of folic acid is associated with an increased risk of certain neural tube defects (anencephaly, spina bifida) but not a complete failure of neurulation. Low levels of vitamin A and thiamine are associated with certain clinical conditions (but not neurulation failure), and these may be taken in therapeutic doses for some diseases or conditions. Glutamate is an excitatory neurotransmitter in wide areas of the central nervous system.

6 **The answer is C: 26 days.** The anterior neuropore is first to close, at about 24 days, and the posterior neuropore is last to close, at about 26 days (just 2 days later). The principal defect associated with failure of the anterior neuropore to close is anencephaly, which is a catastrophic defect that is not compatible with life. The deficits associated with the failure of the posterior neuropore to close are in the category of spinal bifida as broadly defined and may present as spina bifida cystica or spina bifida occulta. Important developmental events may take place at the other days, but these are not related to closure of the posterior neuropore.

7 **The answer is C: Induction of ectoderm to form the neural plate.** Substances (secreted factors and transcription factors) within the notochord, the inducing tissue, stimulate the overlying ectoderm, the responding tissue, to form the neural plate from which the nervous system will arise. Formation of the vertebrae, which will eventually surround the spinal cord, requires inductive signals from the notochord, to form the vertebral bodies, and from the neural plate and neural tube, to form the vertebral arches. Remnants of the notochord will form portions of the nucleus pulposus in the adult.

8 **The answer is C: Alobar holoprosencephaly.** This combination of severe deficits is the first, and worst, of the holoprosencephaly triad that consists of alobar, semilobar, and lobar holoprosencephaly. These defects in development of the brain are usually accompanied by other defects of other parts of the body, such as polydactyly, omphalocele, renal malformations, and others. The prognosis for a neonate with this condition is poor. Agenesis of the corpus callosum may be found, sometimes incidentally, in an otherwise normal individual. Anencephaly is a failure of the anterior neuropore to close; much of the forebrain and skull is absent, and these individuals do not survive. Schizencephaly is an opening (unilateral, bilateral, may be small or large) in the hemisphere that may communicate between the ventricles and the subarachnoid space. In semilobar holoprosencephaly, hemispheres are largely present (they may be partially joined), ventricles are present, and a partial interhemispheric fissure and falx cerebri are present.

9 **The answer is C: Heterotopia.** In this boy, neurons that were migrating toward the cerebral cortex on radial glia stopped migrating before reaching the cortex. These neurons are essentially normal but have taken up final positions in the subcortical white matter. Heterotopias can be band-like (frequently called band heterotopias), nodular, diffuse, or focal, and may take up positions anywhere between the ventricular surface and the cortex. Cortical agenesis refers to a failure of the cortex to develop (not the case in this boy), and hypertelorism refers to paired organs/structures that are spaced abnormally far apart. For example, ocular hypertelorism refers to an abnormally large distance between the eyes. Cranioschisis and myeloschisis describe, respectively, a developmental defect in which the skull is not completely closed (usually accompanied by brain malformations) and an incomplete closure of the neural folds of the spinal cord (usually accompanied by spina bifida in some form).

10 **The answer is B: Folic acid.** The results of an MRC Vitamin Study showed that women who had previously given birth to an infant with a dysraphic defect could reduce the probability of further such incidences by 70% by supplementing their diet with folic acid. Carbamazepine is an anticonvulsant medication and warfarin is used as a blood thinner in antithrombotic therapy. Adequate dietary levels of vitamin A contribute to healthy vision; deficits may result in night blindness due to an interruption of the production and resynthesis of rhodopsin. Vitamin C is a reducing agent and antioxidant; deficiencies of this substance may result in scurvy, although this is infrequently seen in civilized countries. Linus Pauling, the American chemist who won two Nobel Prizes (Chemistry, Peace), maintained that large doses of vitamin C might prevent, or reduce the symptoms of, the common cold.

11 **The answer is C: Rhombomeres (r) 4+5.** Neurons that will associate with the facial nerve develop in relation to rhombomeres r4 + r5 with the root of the nerve lying at level r4. Rhombomere r1 is related to neurons of the trochlear nerve, r2 + r3 to the trigeminal nerve, and r6 + r7 to the glossopharyngeal nerve. Rhombomere r8 is related to the neurons of the hypoglossal nerve.

12 **The answer is E: Improper neuronal migration.** Variation from the normal pattern of cortical gyri and sulci reflects abnormal patterns of neuronal migration on radial glia. In the case of microgyri, there is exuberant migration and smaller, more numerous patterns are created; in the case of pachygyri, migration is aberrant and slow with the creation of fewer larger gyri. Agenesis of the corpus callosum is sometimes seen in individuals with no frank neurological deficits and is even occasionally discovered as an incidental finding. Visual deficits may result from dietary deficiency of vitamin A (due to its involvement in rhodopsin production), and high levels of folic acid have no untoward effects and may even protect against dysraphic defects. Microglia invade the CNS at the same time as blood vessels; a failure in this process would most likely correlate with aberrant vascular development.

13 **The answer is A: Amblyopia.** Amblyopia is a developmental condition that results when incoming visual information from one eye is basically "ignored" by the visual cortex because of wiring problems within the cortex. If discovered early, it can be treated by wearing a patch over the good eye to force the visual projections mediated through the bad eye to remodel up to, and including, the visual cortex. By about 10 years of age, this option may not be available. Cataract (opacity of the lens), myopia (near sightedness), and presbyopia (loss of accommodation) are all visual problems that are almost exclusively seen in much older patients. Nyctalopia is a lessened ability to see in reduced illumination; this may be seen in vitamin A deficiency, since this type of vision is a function of the rods.

14 **The answer is C: Granule cells.** The precursors of granule cells differentiate in the external germinal layer of the early cerebellar anlage. At the same time, precursors of the Purkinje cells are proliferating in the internal germinal layer of the ventricular zone. The developing cells that will become Purkinje cells migrate outward on radial glia to take up their position in the molecular layer, while the precursors to granule cells migrate inward on radial glia (these are Bergmann fibers/Golgi epithelial cells of the adult) to take up their positions in the granule layer. As the dendritic tree of the Purkinje cell grows outward, it encounters the inwardly migrating granule cell. Axonal sprouts of the granule cells encounter the Purkinje cell dendrites and make primitive synaptic contacts. As the migration of the granule cell proceeds, these axonal processes are elongated due to their apposition to the dendrites of the Purkinje cells, and the granule cell continues inward to reach the granule layer where it sprouts its short dendrites. These two patterns are intimately related one to the other. Climbing and mossy fibers establish their synaptic relationships later; Golgi cells arise from the internal germinal layer. Oligodendrocytes are glial elements.

15 **The answer is E: Spina bifida occulta.** Neural tube defects (NTDs) may occur rostrally where the anterior neuropore closes (anencephaly) or caudally where the posterior neuropore closes (spina bifida in its several forms). In spina bifida occulta, the skin is essentially closed, but there may be cutaneous evidence of the defect, such as pigmentation or reddish spot, hairs, dimple, possibly a dermal sinus tract, or other signs. In spina bifida aperta, the skin is not closed over the defect. As the name implies, spina bifida cystica is characterized by a sac-like deformation that may contain only meninges and cerebrospinal fluid (CSF) (meningocele) or meninges, CSF, and spinal cord (meningomyelocele). Myeloschisis is a term that signifies a general failure of the spinal neural folds to close. Encephalocele is just like a meningocele but is associated with the head (occipital, frontal).

16 **The answer is D: Meningomyelocele.** This structure contains meninges, cerebrospinal fluid (CSF), and neural structures; hence, it is a meningomyelocele: a hernia or tumor (-cele) containing spinal cord (myelo-) and meninges. Almost all of these saccular lesions contain at least meninges and CSF; the addition of portions of the spinal cord increases the level of complexity and treatment. An encephalocele and a meningoencephalocele are quite similar. The encephalocele is described as a cranial defect (bifid cranium) with herniation of brain but not necessarily also containing meninges. The meningoencephalocele is the defect in the skull with extrusion of meninges and brain through the defect. Meningocele is a spinal defect, again most likely in the lumbosacral area, consisting of a sac containing only meninges and CSF, while a spina bifida occulta has no associated saccular structure but is a defect of a considerably less catastrophic nature.

17 **The answer is C: Meningohydroencephalocele.** This lesion is a frontal encephalocele that, based on its structure, is a meningohydroencephalocele: a hernia or tumor (-cele), containing brain tissue (encephalo-), with a fluid-containing cavity (hydro-), and surrounded by meninges (meningio-) and cerebrospinal fluid (CSF). The term meningocele usually applies to spinal lesions and is a sac containing only meninges and CSF. A meningoencephalocele is a saccular defect associated with the skull that contains meninges, brain, and CSF, but no part of the ventricular system; a meningomyelocele is a spinal defect containing meninges, CSF, and portions of the spinal cord. Microencephaly refers to an unusually small head that consequently contains a small brain.

18 **The answer is A: Anencephaly.** The lamina terminalis of the adult is the structure representing the general location of the closed anterior neuropore. Failure of the anterior neuropore to close (anencephaly) results in a combination of profound defects: much of the forebrain is absent or severely malformed, the calvaria is largely missing, and there are associated facial abnormalities. Anencephaly is seen in 0.03% to 0.70% of births; most of these (95%) occur in families with histories of neural tube malformations. Affected infants are usually stillborn or die within a few days; a very few may survive for 2 to 3 months. Meningocele and meningomyelocele are defects associated with the spinal cord, and encephalocele is a skull defect through which brain may protrude; all of these are survivable defects. A dermal sinus (a tract extending from the skin to the meninges or subarachnoid space) may be found in the lumbosacral or occipital areas and is a minor problem that is easily (surgically) corrected.

19 **The answer is E: Valproate toxicity.** The antiseizure medications valproate and carbamazepine, among others, can cause dysraphic defects. This is through an interference with the folding process of the neural plate. Hypertension is not mentioned in this woman's case and has not been associated with developmental defects. Ingestion of folic acid and vitamin D is actually good since supplemental doses of folic acid are associated with prevention of dysraphic defects (especially in patients with a known risk), and vitamin D promotes growth through proper metabolism of calcium (teeth and bone growth). Vitamin C, taken in high doses, is not known to be harmful, mainly because it is water soluble. In fact, high doses of vitamin C have been suggested to combat the common cold.

20 **The answer is E: Radial glia.** The cell bodies of radial glia are located near the luminal surface of the developing neural tube, and their apical processes extend to the abluminal surface, the external surface of the developing embryo. In this respect, the radial glia are unique in that their processes extend the entire width of the developing neural tube and, by doing so, they provide a framework for maturing neurons to migrate from their origin in the ventricular zone to their final location. Radial glia are also the first glial cells to appear since they are essential for future stages of development. All other glial cells appear at later stages. Microglia are seen in the developing nervous system after it has been invaded by blood vessels, giving rise to the accepted view that they originate from hematopoietic cells of the bone marrow. Fibrous astroglia (also called fibrillary astroglia) and oligodendroglia are found predominately in white matter, and protoplasmic astroglia are found mainly in the gray matter. A commonly used alternative term for astroglia is astrocyte.

Chapter 3

The Meninges

Recall the Cautionary Tale: The images in this chapter are presented in a Clinical Orientation; this approach emphasizes how structure and function correlate with deficits in a clinical setting. MRI and CT have a universally recognized orientation and laterality. Line drawings, stained sections, brain slices, or gross brain specimens in the Clinical Orientation are viewed in the identical manner that one views an MRI: your right, as the physician-observer, is the patient's left, and the observer's left is the patient's right. The laterality of deficits and reference to connections follow accordingly.

QUESTIONS

Select the single best answer.

1 A 41-year-old woman presents to her family physician. The history reveals that she has never had headaches until several months ago, but she has not only started having them, but they seem to be getting more frequent. She states that she has been using over-the-counter (OTC) medications to treat her headaches. MRI reveals a 2-cm meningioma at the location indicated by the red circle in the image below. Which of the following represents the most likely structure of origin of this tumor?

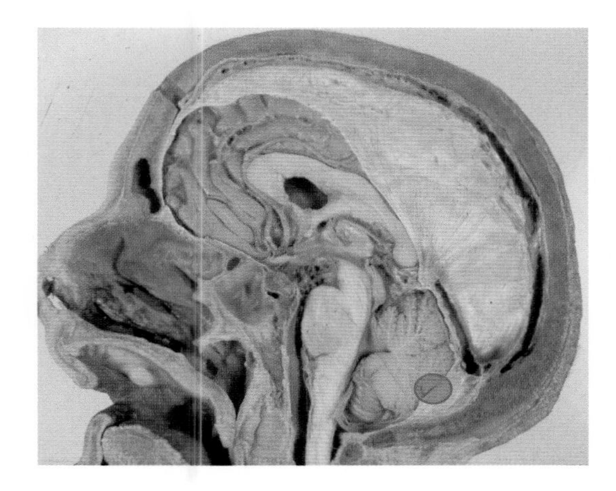

(A) Clivus
(B) Diaphragma sellae
(C) Falx cerebelli
(D) Falx cerebri
(E) Tentorium cerebelli

2 You are in the Emergency Department when a 58-year-old man is brought in accompanied by his wife. The wife explains that her husband had an explosive headache (the husband described it like a bomb going off in his head) and lost consciousness. She also said that, as they waited for the emergency vehicle, he awakened and was nauseated. The examination reveals an acutely ill man with meningismus and photophobia. This patient most likely suffered which of the following?

(A) Aneurysm rupture
(B) Bacterial meningitis
(C) Epidural hemorrhage
(D) Leakage from an AVM
(E) Subdural hemorrhage

3 A 39-year-old woman presents with a complaint of headache that is refractory to OTC medications. The history reveals that she became aware of her persistent headache "a long time ago" but cannot specify a time frame in weeks or months. An MRI reveals the lesion shown below. Recognizing the appearance of this lesion, it is most likely which of the following?

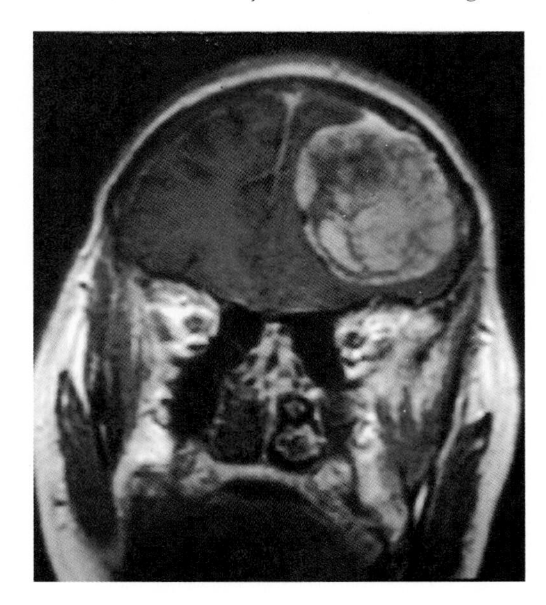

(A) Convexity meningioma
(B) Glioblastoma multiforme
(C) Parasagittal meningioma
(D) Arteriovenous malformation
(E) Suprasellar meningioma

4 A 41-year-old construction worker falls from a scaffolding at work, strikes his head, and is rendered unconscious. He is transported to the Emergency Department (ED) and has regained consciousness by the time he arrives. The examination reveals that he seems to be completely oriented (time, place, profession, date, name of wife), has no motor or sensory deficits, and is ambulatory. He has a scalp laceration stitched and is sent home. Six hours later, he is brought back to the ED in a somnolent state, becomes progressively more unresponsive, and dies. Which of the following is the most likely cause of this man's clinical course?

(A) Convexity meningioma
(B) Epidural hemorrhage
(C) Subarachnoid hemorrhage
(D) Subdural hemorrhage
(E) Rupture of an AVM

5 The mother of a 3-month-old boy expresses concern about a small red dimple associated with an equally small pigmented spot in the lumbosacral area of her son's back. At examination, the physician notes that the boy is in no discomfort and has no other signs or symptoms, but manipulation expresses a small amount of purulent fluid from the dimple; culture and Gram staining reveal bacteria. Surgery is recommended to which the mother agrees. If left untreated, which of the following would most likely be an outcome in this case?

(A) Cauda equina syndrome
(B) Chronically increased intracranial pressure
(C) Enlarged ventricles on MRI
(D) Recurrent meningitis
(E) Tethered cord syndrome

6 A 37-year-old man is transported to the Emergency Department from a rural community by EMS; they report that he was initially unconscious but was revived at the scene. The history reveals that the man is a farmer, was painting the roof of a two-story chicken house, and slipped and fell off the roof. During the examination, the man was relatively alert and oriented but complained of headache; a small scalp laceration is sutured. A CT of the man's head is below. Which of the following would be the most appropriate clinical plan in this man's case?

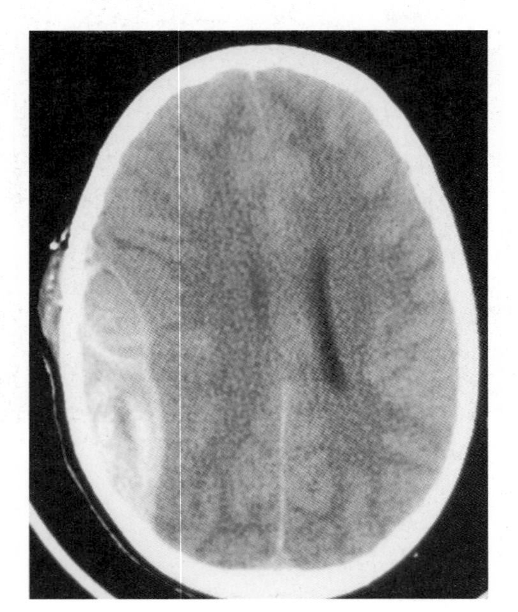

(A) Observe the patient, lesion will resorb
(B) Observe the patient, repeat CT only if focal symptoms appear
(C) Observe the patient, surgery after 5 days
(D) Observe the patient, surgery at earliest opportunity
(E) Observe the patient, surgery only if signs of herniation appear

7 A 48-year-old man presents with a complaint of diplopia. The examination reveals no other cranial nerve deficits and no long tract signs. The MRI shows a large fusiform aneurysm completely filling the ambient cistern. Which of the following would be most specifically affected by this vascular defect?

(A) Abducens nerve
(B) Oculomotor nerve
(C) Trigeminal nerve
(D) Trochlear nerve
(E) Vein of Rosenthal

8 A 39-year-old male physician becomes acutely ill within 12 hours of returning from a mission trip. The history reveals that he started having a headache and general malaise about 4 hours before his return; he thinks his illness is about 18 to 24 hours old. The examination reveals fever, a decreased level of awareness, and meningismus. Analysis of cerebrospinal fluid obtained via lumbar puncture confirms meningitis. A contrasted MRI of this patient, as seen below, provides further clarity of his medical condition. Which of the following is the most likely source of his disease?

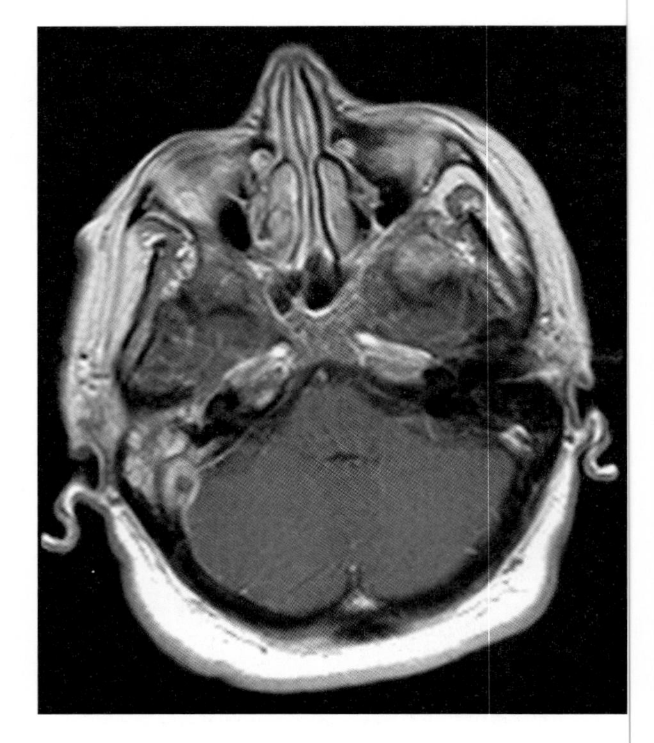

(A) Mastoiditis
(B) Otitis media
(C) Pharyngitis
(D) Rhinosinusitis
(E) Vasculitis

9 In the developing human embryo and fetus, the meninges originate from the endomeninx and the ectomeninx. A failure of the ectomeninx to develop properly would most directly affect which of the following?

(A) Arachnoid mater
(B) Dura mater
(C) Denticulate ligament
(D) Filum terminale internum
(E) Pia mater

10 A 47-year-old man presents with a complaint of persistent headache that is largely refractory to OTC medications. The history and examination reveal that he has experienced progressive weakness and sensory losses over both lower extremities; he acknowledges that these symptoms have gotten more noticeable over the "last year or so." MRI reveals one 2.5 × 3.0-cm tumor. Which of the following would most likely specify this lesion?

(A) Convexity meningioma
(B) Falcine meningioma
(C) Parasagittal meningioma
(D) Suprasellar meningioma
(E) Tentorial meningioma

11 A 69-year-old man is brought to the Emergency Department (ED) of a local hospital. His wife explains that he had a sudden "explosive" headache, following which he became nauseated and vomited several times. The examination at the ED reveals that the man is acutely ill, has meningismus, and is photophobic. A CT was done. Which of the following specifies the appearance of the subarachnoid space in relation to the brain substance in this man, as seen in the image below?

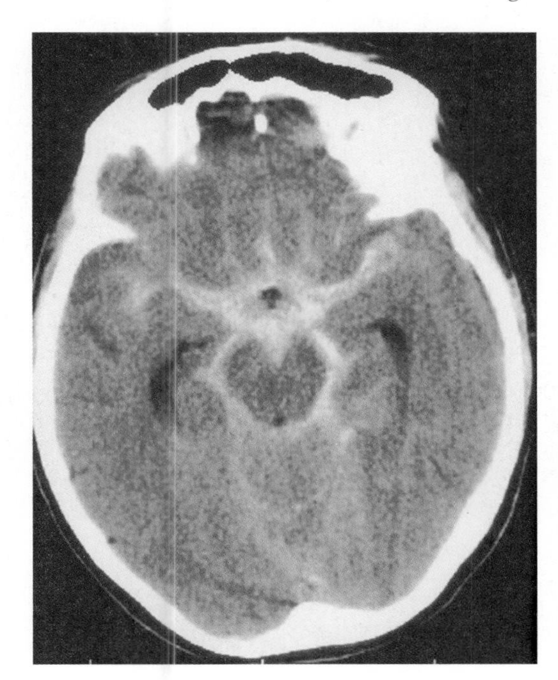

(A) Hyperdense
(B) Hyperintense
(C) Hypodense
(D) Hypointense
(E) Isodense

12 A 54-year-old woman who has a history of neoplastic disease is admitted to the hospital through the Emergency Department. The history reveals that the woman is being treated for breast cancer. The examination reveals that she is experiencing refractory headaches and persistent nausea. MRI and CT reveal several small lesions that may be metastatic from the primary site, including one in the junction of the falx cerebri with the tentorium cerebelli. Which of the following is the likely location of this lesion and might represent an obstruction to venous flow?

(A) Inferior sagittal sinus
(B) Sigmoid sinus
(C) Straight sinus
(D) Superior petrosal sinus
(E) Transverse sinus

13 A homeless man is brought to a local Emergency Department by the local police department. He claims to be about 60 years old, but this cannot be confirmed. The history indicates that he was found lying on the curb of a busy street either sleeping or unconscious. The examination reveals that this man has a small abrasion on his left forehead (cause unknown), is quite lethargic, is essentially not cooperative, and complains that he is not allowed to sleep. This man's CT is shown below. Which of the following most likely specifies this man's lesion?

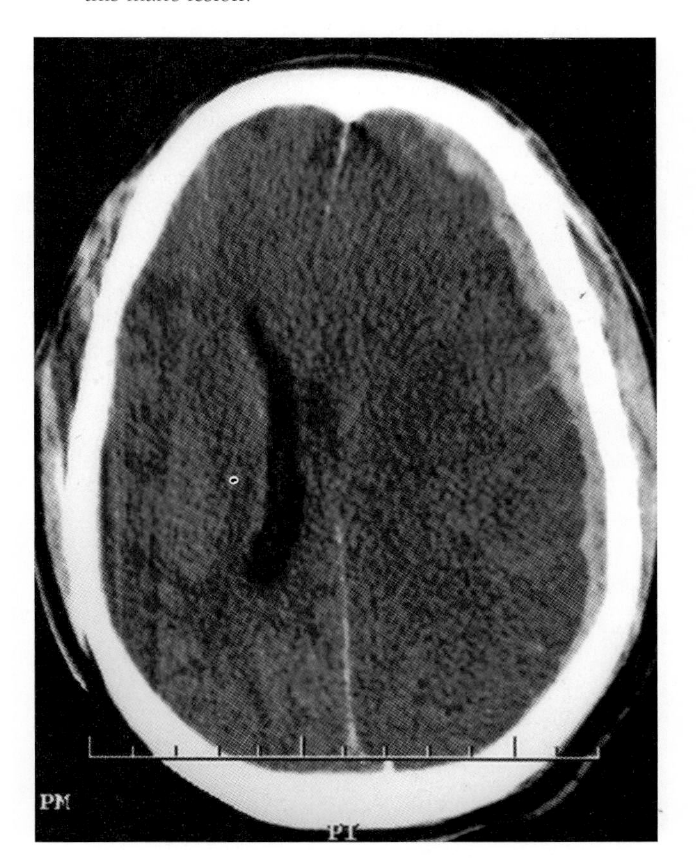

(A) Cerebral hemorrhage
(B) Epidural hemorrhage
(C) Subarachnoid hemorrhage
(D) Subdural hemorrhage
(E) Subgaleal hemorrhage

14 A 39-year-old woman becomes acutely ill with headache and vomiting; she is taken to the Emergency Department by her husband. The examination reveals a decreased level of consciousness, confusion, and a high fever. Suspecting bacterial meningitis, the physician orders a lumbar tap, which reveals gram-positive bacteria. Which of the following is the most likely causative agent in this patient?

(A) *Escherichia coli*
(B) Group B streptococci
(C) *Haemophilus influenzae*
(D) *Neisseria meningitidis*
(E) *Streptococcus pneumoniae*

15 Which of the following layers of the meninges characteristically has numerous tight (occluding) junctions between its cells?

(A) Arachnoid mater
(B) Dura + arachnoid
(C) Meningeal dura
(D) Periosteal dura
(E) Pia mater

16 A 39-year-old woman presents to her physician with a complaint of some awkwardness while walking. The history reveals that she noticed a minor problem several months ago and tried to compensate by increasing her exercise program. Her symptoms got slowly, and progressively, worse. The examination reveals an unsteady gait, dysmetria, and difficulty with the heel-to-shin test. An MRI with contrast is shown below. This woman is most likely suffering which of the following?

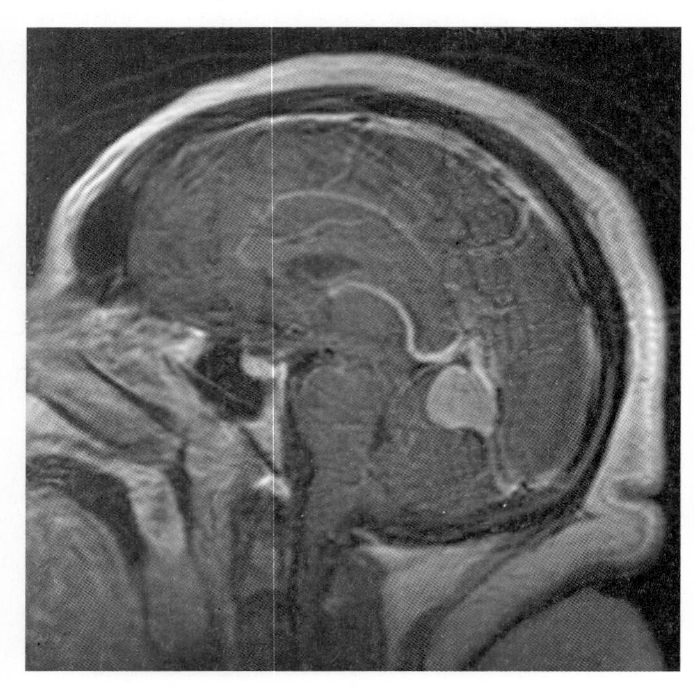

(A) Cerebellar hemorrhage
(B) Cerebellar astrocytoma
(C) Falx cerebelli meningioma
(D) Falx cerebri meningioma
(E) Tentorium cerebelli meningioma

17 In the normal adult, the brain does not "float" but rather is suspended by the arachnoid trabeculae within the cerebrospinal fluid (CSF) of the subarachnoid space. Which of the following represents the approximate weight of a 1,350-g brain when suspended in CSF?

(A) 160 to 200 g
(B) 100 to 140 g
(C) 40 to 70 g
(D) 10 to 35 g
(E) 0 g

Questions 18 and 19 are based on the image below.

A 33-year-old woman is brought to the Emergency Department from her home. The history of the precipitating event, provided by her neighbor and the woman's 11-year-old son, indicated the following. The woman was working in her flowerbed and suddenly fell over and, according to her son, "started jerking and twitching and bubbles came out of her mouth." The examination revealed that the woman was coherent, was oriented, and had no obvious neurological deficits. She did say that she has had recurring headaches over the last year or so but has never had a "spell" like this before. T2-weighted MRI reveals the lesion shown below.

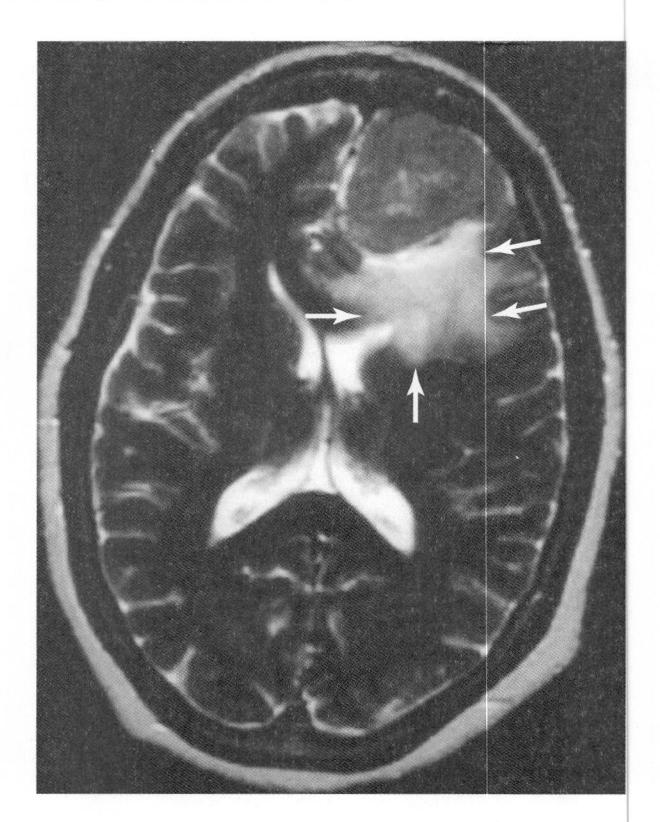

18 Which of the following most likely specifies the lesion in this woman?

(A) Frontal lobe astrocytoma
(B) Frontal lobe arteriovenous malformation
(C) Frontal lobe hemorrhage
(D) Frontal lobe meningioma
(E) Frontal lobe oligodendroglioma

19 Which of the following specifies the appearance of the lighter area, indicated by the white arrows, in the image of this woman's brain?

(A) Hyperdense
(B) Hyperintense
(C) Hypodense
(D) Hypointense
(E) Isointense

20 A 41-year-old man presents to his family physician with a complaint of difficulty walking. The history reveals that these symptoms have become slowly more noticeable over many months; the man cannot remember when they first started. The examination reveals that the man is unable to perform the heel-to-shin test with his right lower extremity, has a right intention tremor, and has a visual loss, essentially a left homonymous hemianopsia. Recognizing this as a somewhat unusual combination of deficits, the physician orders an MRI that shows a large tumor impinging on the occipital lobe and on the cerebellar hemisphere. There is no effacement of the midline. Which of the following represents the most likely origin/location of this lesion?

(A) Diaphragma sellae
(B) Falx cerebelli
(C) Falx cerebri
(D) Tentorial incisure
(E) Tentorium cerebelli

21 Which of the following structures is identified by the arrows in the image below?

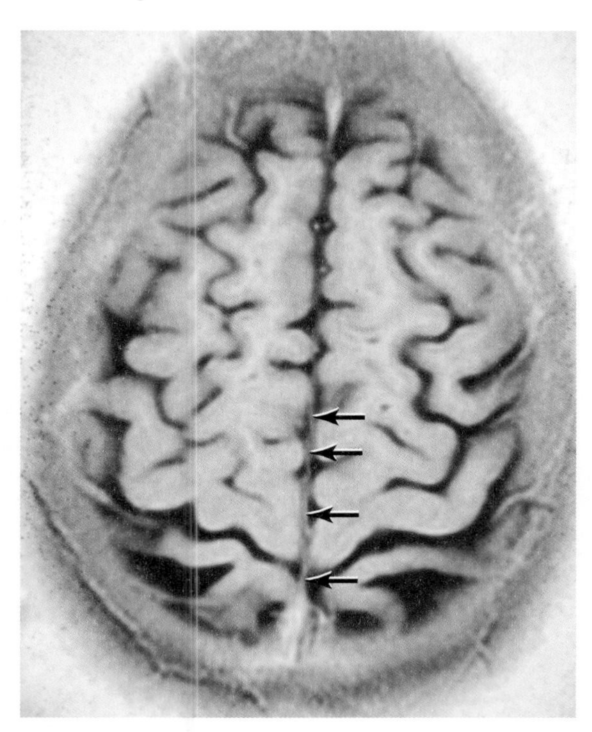

(A) Anterior cerebral arteries
(B) Falx cerebelli
(C) Falx cerebri
(D) Inferior sagittal sinus
(E) Superior sagittal sinus

22 A 14-year-old girl is brought to the Emergency Department by her mother. The history, provided by the mother, reveals that the girl became suddenly ill over the last day or two and was just completing a treatment regimen for a sinus infection. The examination reveals an acutely ill girl with fever, nausea, vomiting, and severe headache. A lumbar puncture confirms the physician's initial suspicion of bacterial meningitis (bacteria identified from a sample of cerebrospinal fluid). Which of the following is most likely involved in this girl's disease?

(A) Arachnoid only
(B) Arachnoid + pia
(C) Dura + arachnoid + pia
(D) Meningeal dura only
(E) Pia only

23 A 56-year-old man presents to his physician with a complaint of "something is wrong with my eyes." The history reveals that this symptom has had an insidious onset. The examination reveals that the man has lost most, but not all, voluntary movements of the right eye, he has diplopia, the eyelid droops slightly, and the pupil is dilated. He has no other signs or symptoms. MRI reveals an aneurysm within the subarachnoid space. This lesion is most likely located in which of the following?

(A) Ambient cistern
(B) Chiasmatic cistern
(C) Interpeduncular cistern
(D) Prepontine cistern
(E) Superior cistern

24 The MRI of a 46-year-old man reveals a meningioma in the location shown in the image below. Based on its location, which of the following clinical conditions or deficits would most likely be seen?

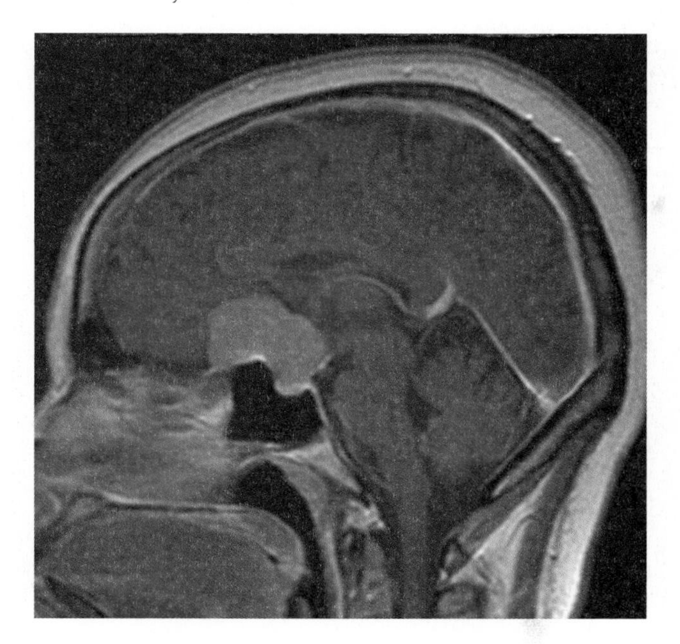

(A) Anosmia
(B) Diplopia
(C) Hemianesthesia
(D) Hemiplegia
(E) Visual deficits

25 An 85-year-old man is brought to your clinic by his daughter. The history, largely supplied by the daughter, reveals that he has had a persistent headache for 2 days, and his daughter feels that he has also been somewhat confused over the same period of time. When pressed for details, he did say that he tripped on a curb but did not fall, 3 days prior this visit. His blood pressure is within normal ranges. MRI reveals a characteristic hemorrhagic lesion. This man has had no frank brain trauma, no abnormalities of the cerebral vessels, but slight brain atrophy. Which of the following might you expect to see in this man's MRI?

(A) Epidural hematoma
(B) Intracerebral hemorrhage
(C) Intraventricular hemorrhage
(D) Subarachnoid hemorrhage
(E) Subdural hematoma

26 Which of the following meningeal layers is present on the convexity of the hemisphere but is not present in the dural reflections, such as the falx cerebri?

(A) Arachnoid barrier cell layer
(B) Arachnoid trabeculae
(C) Dural border cell layer
(D) Meningeal dura layer
(E) Periosteal dura layer

27 A 20-year-old woman is brought to the Emergency Department by her husband. The history, largely offered by the husband, reveals that the woman became ill over about the last 36 hours. The examination reveals that she has no evidence of trauma, but she does have fever and meningismus and does not realize that she is in a hospital. The enhanced MRI below shows changes within the brain or on its surface. This patient most likely suffered which of the following?

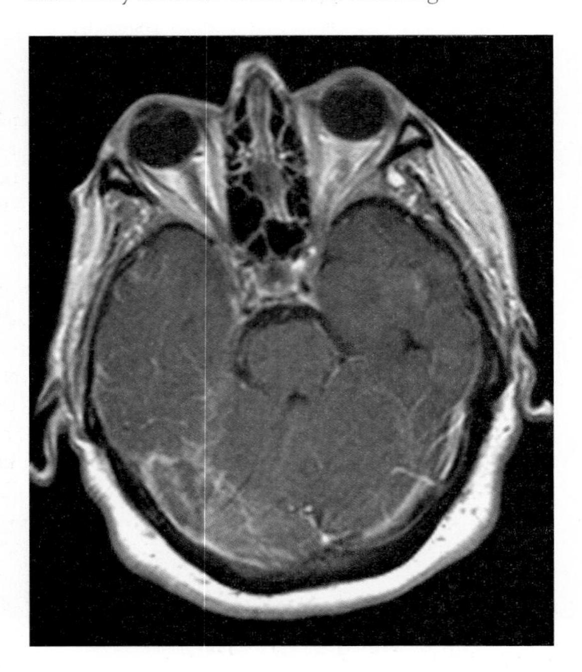

(A) Aneurysm rupture
(B) Bacterial meningitis
(C) Epidural hematoma
(D) Leak from an AVM
(E) Subarachnoid hemorrhage

28 Which of the following structures is indicated by the arrows in the image below?

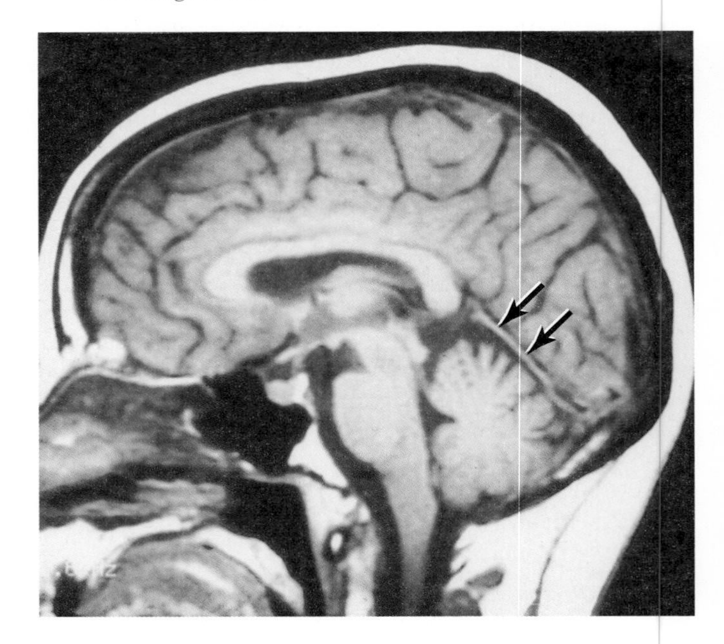

(A) Falx cerebelli
(B) Internal cerebral vein
(C) Superior petrosal sinus
(D) Tentorium cerebelli
(E) Vein of Galen

29 Which of the following features do the larger venous sinuses of the meninges share with larger veins in the body?

(A) An occasional valve
(B) Elastic fibers
(C) Endothelial lining
(D) Sparse circular muscle cells
(E) Sparse longitudinal muscle cells

ANSWERS

1 **The answer is C: Falx cerebelli.** This lesion represented here is internal to the occipital bone, inferior to the straight sinus (which is obvious in this image), and overlapping the falx cerebelli. In this location, it is also inferior to, but separated from, the tentorium cerebelli and, consequently, is in an infratentorial position. The falx cerebelli is the only dural reflection within the infratentorial compartment. The clivus is the base of the skull consisting of the sphenoid body and the basilar part of the occipital bone extending from the dorsum sellae caudally to the foramen magnum. The diaphragma sellae is the small dural reflection that forms a cuff through which the stalk of the pituitary passes. The falx cerebri is the large sickle-shaped dural reflection (seen in this image) located between the cerebral hemispheres; it attaches to the skull at the position of the superior sagittal sinus and attaches to the superior surface of the tentorium at the position of the straight sinus. The falx cerebri separates the supratentorial area into right and left compartments. The tentorium cerebelli is the large dural reflection insinuated in between the cerebellum and the occipital lobes of the cerebral hemispheres.

2 **The answer is A: Aneurysm rupture.** The sudden severe headache with nausea and usually vomiting is characteristic of aneurysm rupture; this is the second most common cause of blood in the subarachnoid space (the first being trauma). These patients experience syncope, usually have meningismus (neck pain from meningeal irritation), and are light sensitive (photophobia) to varying degrees. About one third die before, or soon after, reaching a medical facility; about one third survive but have permanent disabilities; about one third survive with minimal neurological deficits. Although it may have a sudden onset (measured in hours or days), bacterial meningitis does not appear this suddenly and is accompanied with fever. Epidural and subdural hemorrhages are most commonly seen consequent to some sort of trauma; this may be quite minor in the elderly such that the patient may not even remember the event. Leakage from an AVM may be relatively silent or cause headache of a more insidious type.

3 **The answer is A: Convexity meningioma.** A careful survey of this image reveals a large tumor that is extraaxial (inside the skull but outside the brain) but has caused little or no effacement of the midline despite its significant size. These are general characteristics of meningioma. These tumors represent about 15% to 18% of intracranial tumors, are usually benign, are very slow growing, and arise from arachnoid mater. The tumor in this patient originates at the convexity of the frontal lobe, hence its name. Glioblastoma multiforme (GBM) is intraaxial, is highly invasive, and would significantly efface the midline, if this large. Patients with GBM have a survival rate of less than 1 year after diagnosis, while the 5-year survival rate for meningioma after treatment is about 92%. Parasagittal and suprasellar meningiomas are at different locations (adjacent to the superior sagittal sinus and at the sella turcica, respectively), while arteriovenous malformations (AVMs) are usually located within the brain substance and will have many flow voids.

4 **The answer is B: Epidural hemorrhage.** This sequence of events is a potential outcome of an untreated epidural hemorrhage. The patient is rendered unconscious, regains consciousness (has a lucid interval), deteriorates rapidly, and dies; this is called "talk and die." The most common cause of epidural hemorrhage is skull fracture with concomitant damage to the middle meningeal artery. However, epidural lesions may occur, and may be fatal, in the absence of skull fracture. Meningiomas may present with a variety of deficits, but sudden death is not one. Indeed, they may be present for years without any symptoms or signs. Subarachnoid and subdural hemorrhages and arteriovenous malformation (AVM) rupture may result from trauma. These patients will also have neurological issues to varying degrees, but a sudden rapid deterioration would be rare.

5 **The answer is D: Recurrent meningitis.** The occurrence of a red dimple and pigmented spot in this body area in a very young patient points to the presence of the opening of a congenital sinus (dermal sinus tract). Dark hairs may also be present at the site. This small defect at this location is indicative of the opening of a small epithelial-lined channel (the dermal sinus tract) that extends from the surface of the body to become continuous with the spinal meninges. These young patients may have repeated bouts of bacterial meningitis. The treatment of choice is surgical removal of the sinus tract. Cauda equina syndrome results from degenerative disc disease or disc injury (and is an adult condition); tethered cord may occur in children but is accompanied by many obvious problems (weakness, pain, bladder dysfunction, scoliosis). The other choices (B and C) are mainly seen in cases of excessive cerebrospinal fluid (CSF) production or a failure to be reabsorbed into the venous system. Cephalomegaly is not a feature in this patient, so absorption/overproduction of CSF is not an issue in his clinical course.

6 **The answer is D: Observe the patient, surgery at earliest opportunity.** This man has suffered an epidural hematoma, sometimes also called an extradural hematoma. The location (immediately internal to the skull), shape (lenticular shaped, short and fat), and the characteristic appearance specify that this lesion does not cross suture lines; all clearly indicate epidural hematoma. In addition, this lesion is about 2 cm at its widest point, and there is some effacement of the midline. This type of lesion may present with the triad of initial unconsciousness, followed by a "lucid interval" (patient is alert and oriented to time and place), potentially followed by signs/symptoms of herniation resulting in death. In general, rapid surgical intervention is indicated if (1) the patient initially presents with symptoms, (2) the lesion is greater than 1 cm thick at its widest point, and (3) the patient is asymptomatic but has a thick lesion. Large epidural hematomas are, on the one hand, very unlikely to resorb in a timely manner or at all and, on the other hand, are likely to result in herniation, sometimes within hours. This type of lesion increases intracranial pressure, which is the potential source of an eventual herniation. Small asymptomatic lesions, less than 1 cm wide, may be monitored medically with surgery always as an option.

7 **The answer is D: Trochlear nerve.** Diplopia results from damage to cranial nerves that innervate the extraocular muscles. Of the three cranial nerves that innervate these muscles, only the trochlear nerve traverses the ambient cistern; the oculomotor passes through the interpeduncular cistern, and the abducens passes through the prepontine cistern. All three of these cranial nerves do course through the cavernous sinus. Damage to the trigeminal does not result in diplopia, and this nerve or its three major branches are not within the ambient cistern. The trigeminal nerve is motor to the muscles of mastication and sensory for the face, lining of the oral cavity, and teeth. Although the vein of Rosenthal is in the ambient cistern, injury to this structure, in isolation, will not result in diplopia.

8 **The answer is A: Mastoiditis.** A comparison of the right mastoid area with the left in this contrasted MRI clearly reveals significant amounts of purulent material in the mastoid process on the right side. The infection spread from the mastoid process into the nervous system, resulting in inflammation of the sigmoid sinus on the right side (compare with the left). The thin white lines extending rostromedially and caudally from the sigmoid sinus indicate the purulent exudate in the subarachnoid space adjacent to the cerebellum. In addition, note the thin irregular white lines in the temporal lobe on the right; these are exudate within the sulci of the temporal lobe. This MRI contains portions of the nasal pharynx and shows some internal areas of the petrous portion of the temporal bone and some of the nasal conchae. There is no evidence of inflammation in these areas, ruling out rhinosinusitis, otitis media, and pharyngitis as potential sources of this man's disease. In addition, other than the obvious involvement of the sigmoid sinus, there is no evidence of vasculitis (arteries or veins) in the brain or head.

9 **The answer is B: Dura mater.** The ectomeninx is the main source of the dura mater; a failure in this developmental sequence would most adversely affect the dura of the adult, or structures derived therefrom. The endomeninx gives rise to the arachnoid and pia and to structures that are derived from these portions of the meninges. The denticulate ligament, arachnoid trabeculae, and the filum terminale internum arise from the pia and are, consequently, derived from the endomeninx.

10 **The answer is B: Falcine meningioma.** Meningiomas are slow growing, usually benign, tumors that may attain significant size with little, or no, displacement of the brain. They may present in subtle ways that belie their size or extent and with deficits that may progress slowly over time. A meningioma that causes bilateral lower extremity deficits would be located in the falx cerebri. In this position, it impinges on the anterior and/or posterior paracentral gyri, these being the primary somatomotor and somatosensory cortices for the lower extremity, respectively. A convexity meningioma is found on the convexity of the hemisphere, and a parasagittal meningioma is located immediately adjacent to the superior sagittal sinus. Suprasellar meningiomas are found in the immediate vicinity of the sella turcica, and a tentorial meningioma is in the tentorium cerebelli and may extend into either the supratentorial or the infratentorial compartments.

11 **The answer is A: Hyperdense.** The terms hyperdense, hypodense, and isodense apply to CT. Hyperdense refers to a shift toward the appearance of bone, which is white in CT, and hypodense refers to a shift toward the appearance of cerebrospinal fluid (CSF), which is black in CT. In this man, the blood in the subarachnoid space is hyperdense; it is whiter than the surrounding brain, a shift toward the whiteness of bone. Isodense refers to a lesion/condition in CT that appears the same as brain, no lighter or no darker; it is isoequal density. The terms hyperintense and hypointense are terms used in relation to MRI. In T1/MRI, hyperintense refers to a shift toward the appearance of fat, which is white, and in T2, it refers to a shift toward the appearance of CSF, which is also white. Hypointense in both T1 and T2 refers to the shift toward the appearance of bone, which is more black.

12 **The answer is C: Straight sinus.** The straight sinus is located at the point where the falx cerebri attaches to the tentorium cerebelli; it is a major channel for venous drainage from the internal areas of the hemispheres (internal cerebral veins), medial temporal lobe, and superior cerebellar surface. Although some patients have a true sinus confluens, in many individuals, the superior sagittal sinus becomes the right transverse sinus, and the straight sinus becomes the left transverse sinus. The inferior sagittal sinus is very small and is located in the free edge of the falx cerebri; the superior petrosal sinus travels from the cavernous sinus along the edge of the petrous portion of the temporal bone to end in the transverse sinus. The transverse sinus extends from the sinus confluens along the inner aspect of the skull, then turns downward toward the jugular foramen as the sigmoid sinus.

13 **The answer is D: Subdural hemorrhage.** This lesion has the following characteristics: it is long and thin, obviously does not respect suture lines, extends from the frontal pole to the parietal area, and appears hyperdense (a shift toward more white, toward the appearance of the skull). This lesion is a subdural hemorrhage. Cerebral hemorrhage refers to extravasated blood being located within the cerebral hemisphere; subgaleal hemorrhage refers to hemorrhage within the scalp and external to the skull. Neither of these is apparent in this man. Epidural hemorrhage results in lesions that are characteristically lens/lenticular shaped, hence short and fat, this reflecting the fact that they do not cross suture lines, also not a feature in this man. Subarachnoid hemorrhage (SAH) specifies blood within the subarachnoid space. In SAH, the blood within the subarachnoid space frequently outlines gyri by filling the intervening sulci and may also fill the larger extent of the Sylvian sulcus, both not apparent in this man.

14 **The answer is E: *Streptococcus pneumoniae*.** While acute bacterial meningitis may be caused by several different agents, there is a propensity for certain agents to be associated with this disease in patients of different ages. *Streptococcus pneumoniae* (followed by *Neisseria meningitidis*) is most commonly seen in adults. Both *Escherichia coli* and group B streptococci are found in newborns and very young children, while *Haemophilus influenzae* is present in older children. *Neisseria meningitidis* may also be the causative agent in older children and young adults.

15 **The answer is A: Arachnoid mater.** The arachnoid mater (also called the arachnoid barrier cell layer) is composed of only a few cell layers and has no extracellular space, and the cells have many tight (occluding) junctions between them. This structural feature serves to prohibit (under normal conditions) the movement of fluid (CSF) from the subarachnoid space outward. In this respect, the construction of the arachnoid serves as a "barrier" to fluid movement. The meningeal and periosteal dura are characterized by long, slender fibroblasts interlaced with extracellular collagen; this feature gives these layers great strength. The pia mater is frequently a single layer of cells located on the surface of the brain and spinal cord; the pia mater also extends into the sulci on the cerebral cortex and into the fissures of the cerebellum.

16 **The answer is E: Tentorium cerebelli meningioma.** This lesion is clearly extraaxial (located outside the substance of the nervous system), is of very slow onset (history), and is clearly located in an infratentorial position (less than 10% of meningiomas are located in the posterior fossa). In addition, this tumor is clearly attached to the inferior surface of the tentorium at the caudal edge of the tentorial incisure and is impinging on the superior aspect of the cerebellum. Note that even though this tumor is large for its location, with the exception of the superior aspect of the cerebellum, the other contents of the posterior fossa are in their normal positions. Since this is an extraaxial lesion, the possibility of hemorrhage into the cerebellum or a cerebellar astrocytoma is ruled out; both of these would be intraaxial. The falx cerebri attaches to the superior surface of the tentorium above the straight sinus (clearly visible in this image); consequently, this lesion is in the wrong position to involve the falx cerebri. The falx cerebelli is located on the inner surface of the skull at the midline below the tentorium; this tumor is also in the wrong position to involve this meningeal reflection.

17 **The answer is C: Approximately 40 to 70 g.** The brain loses about 95% to 97% of its total weight when suspended in CSF; it does not lose all of its weight. When the brain is removed from the skull during an autopsy and placed in a fluid fixative, it will not float but will slowly sink to the bottom of the container. This is why the failure to suspend the brain with a ligature, usually tied around the basilar artery, will result in the brain settling to the bottom of the container and becoming slightly flat on one side.

18 **The answer is D: Frontal lobe meningioma.** This left-sided lesion is clearly extraaxial (it does not invade the brain), appears to originate from the frontal convexity (note the space and falx cerebri clearly visible at the midline), and has not effaced the midline to any significant degree. In addition, the lack of effacement of the midline, or of any significant displacement of brain structures, clearly suggests that this lesion grew very slowly, which is a characteristic of meningioma. This lesion does not have the flow voids that are characteristic of arteriovenous malformations (AVMs), and AVMs are located within the brain substance. Also excluded are frontal lobe astrocytoma and frontal lobe hemorrhage, since their appearance is quite different; they are intraaxial lesions, and both cause more brain displacement than is seen in this case. Oligodendrogliomas are frequently found in the frontal lobes (45% to 60% of cases) but are intraaxial and are located predominately within the white matter. In addition, oligodendroglioma may contain cysts or calcium deposits and may present with seizure (also seen in other CNS tumors and AVM).

19 **The answer is B: Hyperintense.** The very light area within the left hemisphere immediately caudal to the tumor in this T2-weighted MRI in this woman's brain is edema. This represents a compromise of the blood-brain barrier with consequent movement of fluid into the brain tissue adjacent to this lesion. In this T2-weighted image, the presence of excessive fluid in the brain appears as a shift toward the appearance of cerebrospinal fluid; this is a shift toward more white; this is hyperintense. Hypointense in both T1 and T2 is a shift toward the appearance of air or bone in a normal patient, which is black; hypointense specifies a shift toward more black. Isointense is a situation where the lesion and the surrounding brain are of equal intensities. Hyperdense and hypodense are terms used in CT. Hyperdense refers to a shift toward the appearance of bone in CT, which is white; hyperdense is a shift toward more white. Hypodense refers to a shift toward the appearance of fluid or air in CT, which is black; hypodense is a shift toward more black.

20 **The answer is E: Tentorium cerebelli.** This is a meningioma (slow growing—symptoms appeared over many months, no midline effacement) that has arisen from the tentorium cerebelli and has simultaneously impinged on the occipital lobe (above the tent, visual deficits) and the cerebellar hemisphere (below the tent, cerebellar deficits). Meningiomas that arise from the diaphragma sellae (sellar, suprasellar) frequently impinge on optic structures and the hypothalamus; those that arise from the falx (falcine meningiomas) may impinge on the paracentral gyri (bilateral motor and/or sensory deficits on lower extremity). If these falcine lesions impinge on more rostral portions of the medial aspect of the frontal lobe, they may remain "silent" for some time and may become quite large. The falx cerebelli is quite small, or absent in some individuals, and rarely gives rise to tumors. The tentorial incisure is the opening within the tentorium (and therefore not a structure) that communicates between supratentorial and infratentorial compartments. Meningioma may occur at the rim of the tentorial incisure.

21 **The answer is C: Falx cerebri.** This axial MRI is located at a more superior plane through the hemisphere, below the level of the superior sagittal sinus, but above the level of the corpus callosum. The falx cerebri is the dural reflection located at the midline between the cerebral hemispheres, dividing this part of the cranial cavity into right and left compartments, attached to the skull at the midline position of the superior sagittal sinus, and caudally attached to the superior surface of the tentorium cerebelli. The inferior sagittal sinus is located in the free edge of the falx cerebri. Rostrally, the falx cerebri attaches to the crista galli of the ethmoid bone. The falx cerebelli is located on the midline, inferior to the tentorium cerebelli and in between the cerebellar hemispheres; it may be poorly developed in some individuals.

22 **The answer is B: Arachnoid + pia.** Meningitis, of bacterial origin, is a disease that characteristically presents as an acute illness that can be rapidly fatal if not treated. The bacterial organisms are in the subarachnoid space (this is how they can be retrieved via lumbar puncture), and the inflammation involves the membranes lining this meningeal space, the arachnoid on the external aspect and the pia on the internal surface. This also explains why this disease is commonly called leptomeningitis: the pia and arachnoid collectively form the leptomeninges, or leptomeninx. Pachymeningitis, involvement of the dura mater, may be seen in a limited number of cases but only if the inflammation accesses the dura. However, pachymeningitis may also arise from other sources such as inflammation originating from the skull (especially its inner table) or as a complication of syphilis (hypertrophic cervical pachymeningitis).

23 **The answer is C: Interpeduncular cistern.** The combination of a loss of most eye movement, a dilated pupil, and drooping eyelid points to involvement of the oculomotor nerve, which traverses the interpeduncular cistern. Indeed, aneurysms of the basilar bifurcation commonly impinge on the oculomotor root (unilaterally or bilaterally) in the interpeduncular cistern. A lesion in the ambient cistern or the superior cistern (involving the root of the trochlear nerve) or in the prepontine cistern (involving the abducens root) may result in diplopia but will not result in a dilated pupil or drooping eyelid. In fact, damage to cranial nerve IV and VI will result in a loss of specific eye movements but not a loss of most eye movements on the involved side. The chiasmatic cistern is immediately adjacent to the optic chiasm, and this patient has no visual deficits. Recall that cisterns are simply enlarged areas of the subarachnoid space.

24 **The answer is E: Visual deficits.** This tumor is located on the midline (you can tell the plane is at the midline because the cerebral aqueduct is clearly seen) and is specifically located at the tuberculum sellae; this is a tuberculum sellae meningioma. At this location, the two major structures that could be damaged are the pituitary and its stalk and the optic chiasm and the proximal parts of the optic nerves before they enter the optic canals. In the absence of any comment on hypothalamic dysfunction, the only choice is visual deficits reflecting chiasm or proximal optic nerve damage. These visual deficits are usually asymmetric but most frequently involve both visual fields. Anosmia is seen in meningioma of the olfactory groove; the lesion in this image is too caudal to involve, to any significant degree, olfactory structures. Diplopia may be seen with damage to any of cranial nerves III, IV, and VI. These nerves travel through the cavernous sinus and exit the skull via the superior orbital fissure; this fissure is slightly lateral and superior to the optic canal which transmits the optic nerve. These cranial nerves are out of the territory commonly affected by this type of meningioma. Hemiplegia and hemianesthesia are motor and sensory deficits usually reflecting damage in the cerebral hemisphere or brainstem.

25 **The answer is E: Subdural hematoma.** In a normal individual, the dura mater is attached to the skull; the dura is attached to the arachnoid mater (at the dura border cell layer); the arachnoid trabeculae are attached to the inner surface of the arachnoid and extend across the subarachnoid space to attach to the pia mater, which, in turn, is attached to the surface of the brain. The brain is suspended within its narrow envelope of cerebrospinal fluid-containing subarachnoid space by this continuity of connections; it does not "float"; it is tethered. Recognizing this anatomical relationship, in older patients as the brain atrophies and becomes smaller, it pulls on the arachnoid mater via the trabeculae. The arachnoid mater is tough: it has many continuous tight junctions and does not tear. On the other hand, the dural cell layer, which is immediately external to the arachnoid, is a structurally weak layer at the dura-arachnoid interface (few cell junctions, comparatively large extracellular spaces, no collagen). A seemingly innocuous event, particularly in an elderly patient (such as tripping on a curb), may jolt the brain just enough to shear open the dural border cell layer creating an artifactual space at the dura-arachnoid interface in which blood collects; this is the subdural hematoma. What is commonly called a "subdural hematoma" is actually a collection of blood within the sheared opened dural border cell layer. The other choices would require vascular pathology or significant trauma; neither is a factor in this man's case.

26 **The answer is E: Periosteal dura layer.** The periosteal layer of the dura mater is the outermost part of the dura, tightly adherent to the inner surface of the skull; characteristically it consists of fibroblasts and copious amounts of collagen. The periosteal dura is especially adherent to the inner table at suture lines. Since this part of the dura is continuous over all aspects of the inner surface of the skull, and is continuous externally to the dural sinuses, it does not participate in the formation of the dural reflections; these reflections originate from the location of dural sinuses. From without to within, the meningeal dura, dural border cell layer, arachnoid barrier layer, and arachnoid trabeculae all follow with the dural reflections. These relationships are most obvious with the larger reflections (falx cerebri, tentorium cerebelli) and not so distinct with the smaller reflections (diaphragma sellae). Naturally, the subarachnoid space also follows the dural reflections.

27 **The answer is B: Bacterial meningitis.** The comparatively sudden onset, signs, and symptoms (fever, neck stiffness [meningismus], and confusion or decreased awareness) all point to bacterial meningitis. The MRI reveals lacey-appearing white elements on the cerebellum, thin white tendrils that are most obvious in the temporal lobe, and a thin white layer on the surface of the cerebellum and temporal lobe; all are more noticeable on the right side. This white material is the purulent exudate characteristic of acute bacterial meningitis (also called acute pyogenic meningitis) that is enhancing within the leptomeningeal space. The delicate white lines that appear to be within the right temporal lobe are actually this exudate extending into the sulci of the temporal lobe. Recall that the subarachnoid space (leptomeningeal space) is present on the brain surface but also extends into the sulci since the pia mater intimately follows all the details of the brain surface. This image reveals no indication of an AVM: the patient had no trauma, she did not initially present with headache (insidious or sudden), and she has no focal signs. These features, along with the MRI, rule out the other choices.

28 **The answer is D: Tentorium cerebelli.** This image is very close to the midline, based on the observation that the cerebral aqueduct is visible within the midbrain. The tentorium cerebelli is located between the cerebellum and the occipital lobes, attaches to the inner surface of the occipital bone, and also attaches along the upper edge of the petrous portion of the temporal bone. The superior petrosal sinus is located at this attachment to the temporal bone. The shape of the tentorium cerebelli is like that of a "tent" over the cerebellum. The internal cerebral vein is located along the superior and medial edge of the dorsal thalamus and empties caudally into the great cerebral vein of Galen. In turn, the great cerebral vein arches sharply up and caudally to enter the straight sinus. The straight sinus is located in the junction of the falx cerebri and the tentorium cerebelli and ends at the sinus confluens.

29 **The answer is C: Endothelial lining.** Venous sinuses of the meninges are lined by an endothelium that is basically the same as the endothelial lining of veins. In other respects, venous sinuses are uniquely different. They do not have elastic fibers in their walls; instead, they have fibroblasts and large amounts of collagen. Venous sinuses do not have smooth muscle cells (circular or longitudinal) in their walls, and they do not have valves. It is this lack of valves that may expedite the passage of infectious agents, under certain circumstances, into and through veins and venous sinuses within the skull. In addition, venous sinuses have rigid walls and they do not collapse.

Chapter 4

The Ventricles and Cerebrospinal Fluid

Recall the Cautionary Tale: The images in this chapter are presented in a Clinical Orientation; this approach emphasizes how structure and function correlate with deficits in a clinical setting. MRI and CT have a universally recognized orientation and laterality. Line drawings, stained sections, or brain slices in the Clinical Orientation are viewed in the identical manner that one views an MRI: your right, as the physician-observer, is the patient's left, and the observer's left is the patient's right. The laterality of deficits and reference to connections follow accordingly.

QUESTIONS

Select the single best answer.

1 Which of the following most specifically identifies the white area indicated by the tip of the pointer in the MRI below?

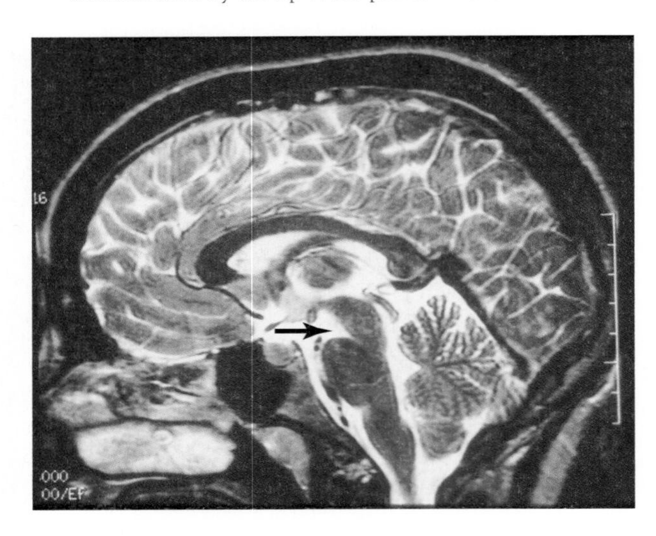

(A) Ambient cistern
(B) Crural cistern
(C) Interpeduncular cistern
(D) Subarachnoid space
(E) Third ventricle

2 A 12-year-old boy is brought to your office with the complaint of headache, fever, chills, and somnolence. A history from the mother reveals that he became ill rather quickly, as she remembers, in only a day or two. A sample of cerebrospinal fluid taken from this boy is cloudy and contains 12,800 white blood cells (mostly polymorphonuclear leukocytes) per ml. Which of the following would most likely explain these observations in this boy?

(A) Acute bacterial meningitis
(B) Acute subarachnoid hemorrhage
(C) Acute viral meningitis
(D) Glioblastoma multiforme
(E) Multiple sclerosis

3 An 89-year-old man is brought to the family physician by his daughter who is concerned that her father is getting forgetful. The history reveals that the man has been quite active all his life and currently is working part-time at a garden supply store. The examination reveals a healthy, ambulatory man with no overt neurological signs and whose cognitive functions are within an appropriate range for his age. An MRI shows enlarged ventricles, but no lesions in the brain or ventricles. These observations suggest that that this man's condition is most likely which of the following?

(A) Communicating hydrocephalus
(B) Congenital hydrocephalus
(C) Hydrocephalus ex vacuo
(D) Obstructive hydrocephalus
(E) Posthemorrhagic hydrocephalus

4 A 58-year-old man is suffering from a neurodegenerative disease characterized by numerous CAG (cytosine-adenine-guanine) repeats on chromosome 4. His MRI reveals bilaterally enlarged anterior horns of the lateral ventricles. Which of the following would be most obviously diminished in this man to result in his enlarged ventricles?

(A) Amygdaloid nucleus
(B) Anterior nucleus of thalamus
(C) Body of the caudate nucleus
(D) Head of the caudate nucleus
(E) Hippocampus proper

5 A 26-year-old man is brought to the Emergency Department from the site of a motor vehicle collision. The man is unconscious, and the initial examination reveals that he has fractures of his left hand, a compound fracture of his left humerus, and multiple facial and scalp injuries. A CT reveals a fractured skull and a traumatic brain injury (TBI). Which of the following labeled areas in the image below identifies blood within a part of the ventricular system in this man?

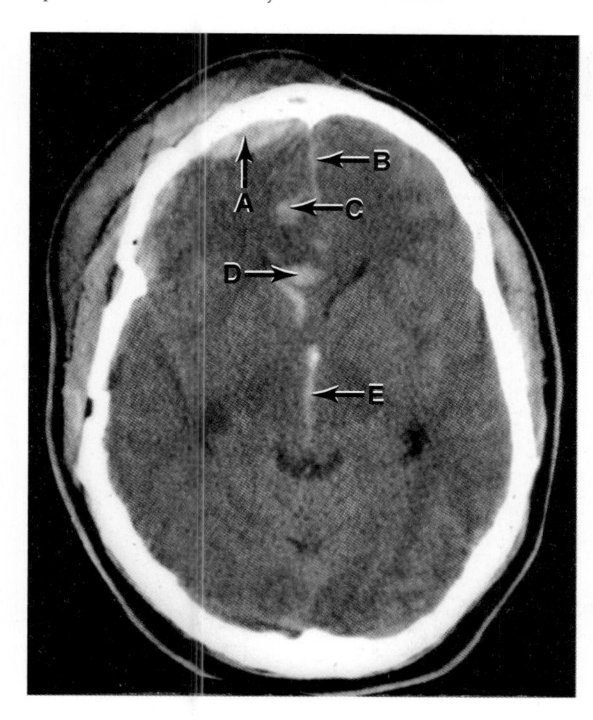

(A) A
(B) B
(C) C
(D) D
(E) E

6 Which of the following specifies the space indicated by the white arrow in the brain specimen below?

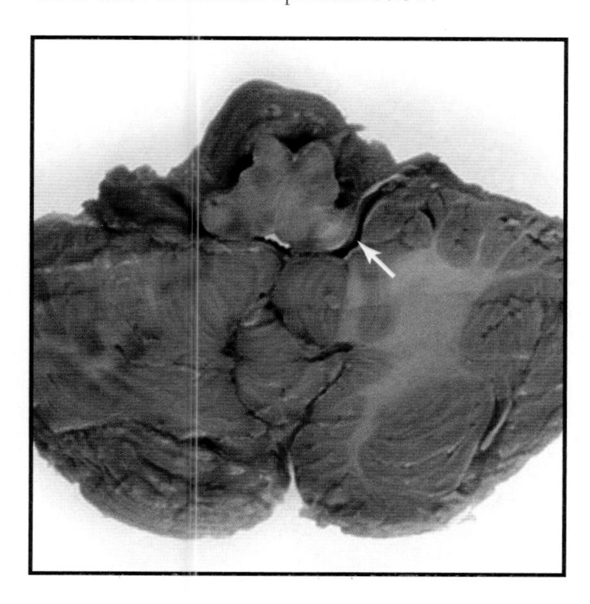

(A) Ambient cistern
(B) Cerebellomedullary cistern
(C) Cerebral aqueduct
(D) Lateral recess
(E) Subarachnoid space

7 A 62-year-old woman presents at the Emergency Department with the complaint of sudden onset of a headache. CT reveals a small amount of subrachnoid blood and, as an incidental finding, a calcified glomus choroideum. This structure is usually found in which of the following locations?
(A) Anterior horn, lateral ventricle
(B) Atrium of the lateral ventricle
(C) Fourth ventricle
(D) Interventricular foramen
(E) Third ventricle

8 The sonogram of a 31-year-old woman who is 6 months pregnant reveals enlarged ventricles in the fetus. This observation suggests that the foramina allowing egress of cerebrospinal fluid (CSF) from the ventricular system into the subarachnoid space have failed to develop properly or have developed improperly. Which of the following represents the approximate time at which these openings usually become patent?
(A) 1 month of gestation
(B) 1.5 months of gestation
(C) 2 weeks of gestation
(D) End of first trimester
(E) End of second trimester

9 A 21-year-old male medical student goes to the Student Health Center; he complains of progressive weakness. The history reveals that this student had an intestinal virus with diarrhea and vomiting that resolved over 2 days. About 10 days later, he noticed tingling sensations and numbness in his feet and hands, followed by progressive weakness. Over the last 8 days it has gotten so bad that he is having trouble standing and walking, and is unable to continue attending class. A CSF sample obtained via lumbar puncture is clear, contains four monocytes per mm^3 and 760 mg protein including increased immunoglobulins; the opening pressure is 90 mm H_2O. This student is most likely suffering from which of the following?
(A) Acute viral meningitis
(B) Benedikt syndrome
(C) Gubler syndrome
(D) Guillain-Barré syndrome
(E) Myasthenia gravis

10 Production of cerebrospinal fluid in a healthy normal individual is an active process that proceeds against a pressure gradient. Which of the following is especially numerous in cells subserving this function?
(A) Endoplasmic reticulum
(B) Lysosomes
(C) Mitochondria
(D) Nucleus
(E) Synaptic vesicles

11 A 67-year-old man is transported to the Emergency Department from his home. The history of the current event was provided by his daughter, a surgeon. He had a sudden severe headache, began vomiting, and had a significantly reduced level of consciousness. CT reveals the pattern of intracranial blood shown in the image below. Which of the following is the reason for the interface of the white area with the black area at the position of the white arrow?

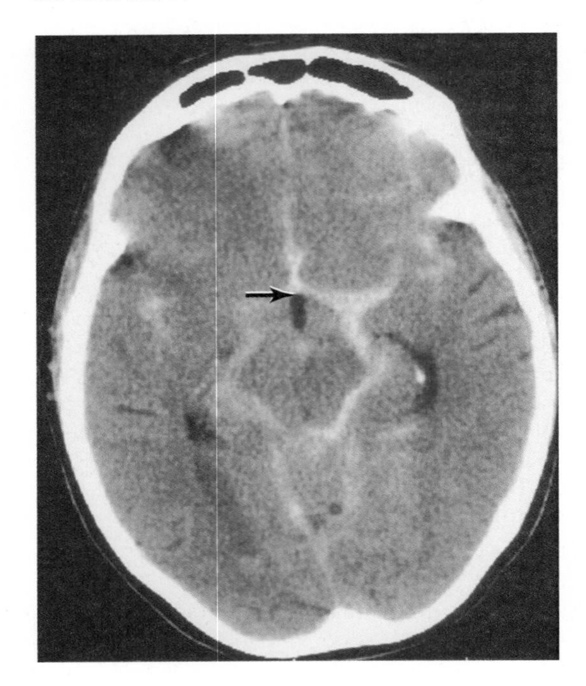

(A) Presence of subarachnoid clot
(B) Presence of a leptomeningeal cyst
(C) Presence of lamina terminalis
(D) Presence of stria medullaris thalami
(E) Presence of stria terminalis

12 The epithelium of the choroid plexus rests on a basal lamina; the fenestrated epithelium that lines the capillaries within the choroid plexus also rests on a basal lamina. These two basal laminae are separated from each other by a small tissue space. Which of the following are usually located within this space?
(A) Elastic fibers only
(B) Fibroblasts and collagen
(C) Fibroblasts and elastic fibers
(D) Many red blood cells
(E) Many white blood cells

13 The MRI of an 11-year-old girl reveals a highly vascular tumor that appears to originate from the rostral portions of the choroid plexus located within the temporal horn of the lateral ventricle. Which of the following is the most likely candidate for the vascular supply to this tumor?
(A) Anterior choroidal artery
(B) Anterior inferior cerebellar artery
(C) Lateral posterior choroidal artery
(D) Medial posterior choroidal artery
(E) Medial striate artery

14 The mother of a 20-month-old boy brings her son to his pediatrician. She explains that over the last several weeks he has become progressively more irritable, complains more often that his "head hurts," has been sick, and frequently vomits. The examination confirms the mother's impressions and reveals that the boy also has cephalomegaly. MRI reveals the lesion shown in the image below. This large lesion is located primarily in which of the following?

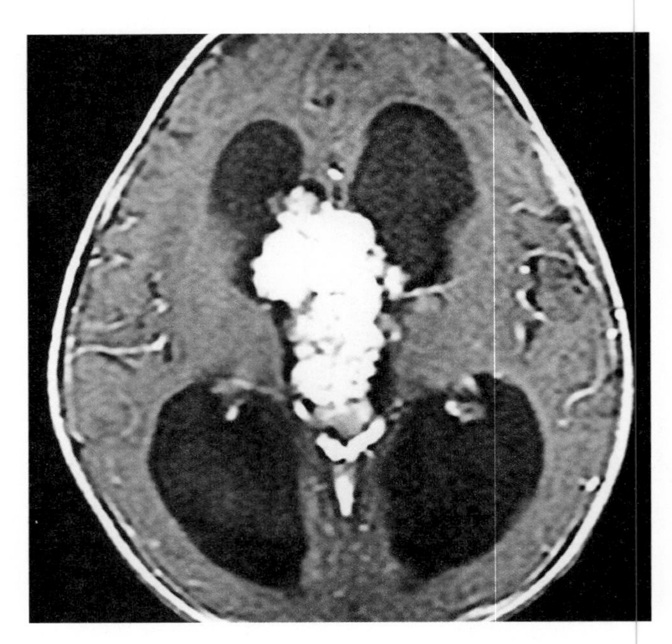

(A) Anterior horn, lateral ventricle
(B) Atrium, lateral ventricle
(C) Dorsal thalamus
(D) Fourth ventricle
(E) Third ventricle

15 A 4-year-old boy is brought to the pediatrician by his mother. The history reveals that the boy has had headaches off and on for several months, but over the last 3 weeks they have gotten persistently worse. Over this same recent period, he has also become progressively more lethargic and experienced nausea and vomiting on several occasions. MRI reveals a tumor within the third ventricle that appears to arise from the choroid plexus. Which of the following is the most likely major blood supply to this tumor?
(A) Anterior choroidal artery
(B) Anterior inferior cerebellar artery
(C) Lateral posterior choroidal artery
(D) Medial posterior choroidal artery
(E) Posterior inferior cerebellar artery

16 A 64-year-old man presents with gait disturbance, urinary incontinence, and dementia (primarily memory and thinking slowly). The history reveals that the man had been treated for bacterial (streptococcal) meningitis. MRI reveals that the ventricles are somewhat, and uniformly, enlarged, and lumbar puncture indicates a slightly elevated pressure within the subarachnoid space. CSF is normal. This man is most likely suffering which of the following?

(A) Aqueductal stenosis
(B) Communicating hydrocephalus
(C) Idiopathic hydrocephalus
(D) Normal pressure hydrocephalus
(E) Obstructive hydrocephalus

17 A 36-year-old woman presents with internuclear ophthalmoplegia, a slight left inferior quadrantanopia, and some loss of discriminative touch on the left upper extremity. To a certain extent, these symptoms seem to wax and wane over time. Analysis of a CSF sample obtained via lumbar puncture reveals a slightly elevated white cell count and clearly elevated immunoglobulin levels when compared to other proteins, but no other constituents of note. This woman is most likely suffering from which of the following?

(A) Acute bacterial meningitis
(B) Glioblastoma multiforme
(C) Intracerebral hemorrhage
(D) Multiple sclerosis
(E) Myasthenia gravis

18 A 43-year-old man is brought to the Emergency Department by his wife. She explains that he just returned from a trip to a foreign country and 2 days later became ill. The examination reveals that the man is somewhat lethargic, has a slight fever, and has sore muscles. Suspecting a rapidly involving CNS condition, the physician orders a lumbar puncture. The opening pressure at the puncture is 105 mm H_2O. Three tubes of CSF are collected; the first is red, the second is less red, and the third is very light pink-to-clear. When allowed to set for a few minutes, the supernatant in the first tube is clear; small clots appear in tube one, but not in the others. Subsequent testing reveals only a very slight elevation in CSF protein levels. These observations in this man suggest which of the following?

(A) Aneurysm rupture
(B) Arteriovenous malformation bleed
(C) Intracerebral hemorrhage
(D) Subarachnoid hemorrhage
(E) Traumatic tap

19 A 57-year-old woman visits her family physician with the complaint of headache. The history and examination reveal an otherwise healthy, slender woman who claims to have an active life style. She indicates that she has never had headaches, but started having them about 2 months ago, and that they are getting progressively worse. OTC medications help, but not always. MRI reveals an aneurysm on the proximal part of the right A2 segment. If this aneurysm begins to leak blood will most immediately occupy which of the spaces indicated in the given image?

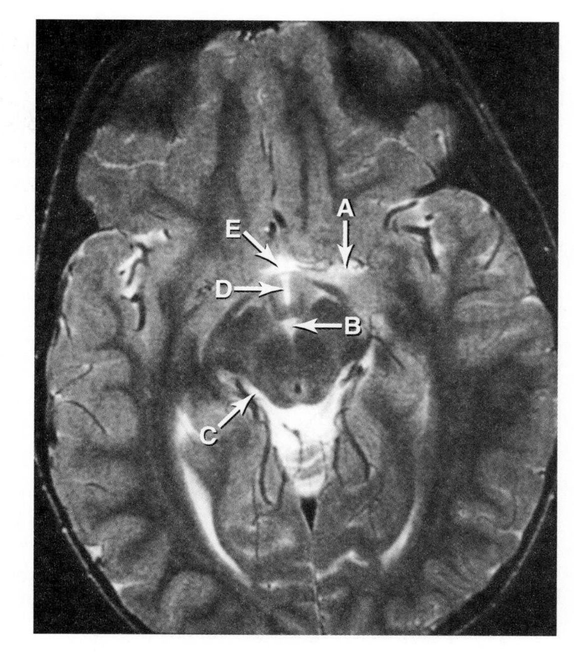

(A) A
(B) B
(C) C
(D) D
(E) E

20 The location of structures relative to the ventricular spaces is important in the diagnosis of a broad variety of clinical conditions. Which of the following structures is characteristically located in the lateral wall of the lateral ventricle?

(A) Amygdaloid nucleus
(B) Caudate nucleus
(C) Corpus callosum
(D) Dorsal thalamus
(E) Hippocampus

21 A 6-year-old girl is brought to the Emergency Department by her parents. The history reveals that this is the second visit in 3 weeks; on the first visit she was treated for bacterial meningitis. The current examination reveals an acutely ill child who is lethargic and has headache and fever. Suspecting a recurrence of the meningitis the physician orders a lumbar puncture and 2 to 3 mL of CSF is removed for analysis. The mother is concerned about the procedure but the physician assures her that the daily production rate of the fluid removed (CSF) will quickly offset the small amount removed. What is this approximate production rate, per 24-hour period, in an adult?

(A) 100 to 175 mL/day
(B) 175 to 300 mL/day
(C) 275 to 375 mL/day
(D) 425 to 500 mL/day
(E) 600 to 900 mL/day

22 A 32-year-old morbidly obese woman presents to her physician with the complaint of a persistent, and recurring, headache that she has been treating with OTC medications. The examination reveals that the woman has partial loss of vision in both eyes; she was not aware of this, and she has papilledema. Lumbar puncture reveals elevated opening pressure (270 mm H_2O) but normal CSF, and MRI reveals normal ventricles and no effacement of sulci. This woman is most likely suffering which of the following?

(A) Aqueductal stenosis
(B) Hydrocephalus ex vacuo
(C) Idiopathic intracranial hypertension
(D) Normal pressure hydrocephalus
(E) Obstructive hydrocephalus.

23 A 23-year-old man presents with nausea, vomiting, and lethargy, cardinal signs of increased intracranial pressure. The history reveals that his symptoms have been slowly developing over several weeks but have become suddenly worse in the last 4 days. The CT below reveals a tumor of the choroid plexus in the atrium of the lateral ventricle with recent bleeding into the atrium, posterior horn, and temporal horn of the left lateral ventricle. In addition to blood in this ventricle, which of the following is particularly obvious in this patient?

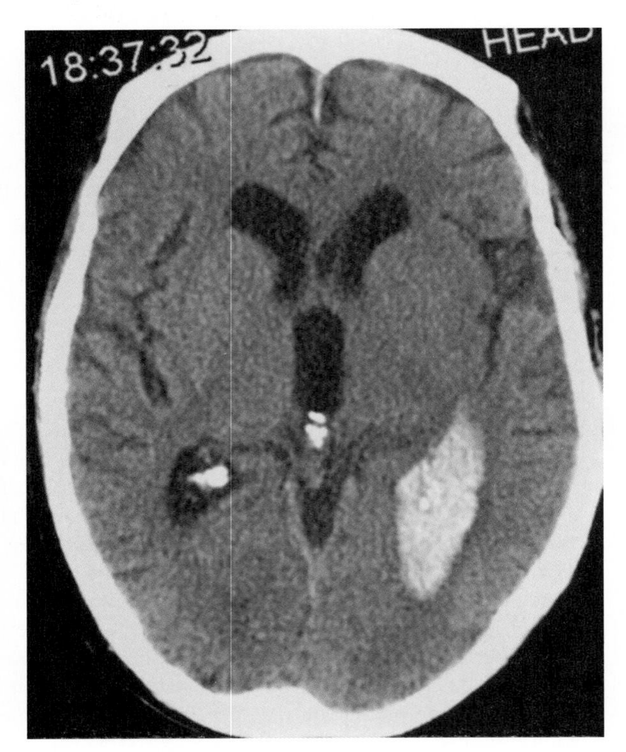

(A) Enlarged anterior horn, lateral ventricle
(B) Enlarged atrium, lateral ventricle
(C) Enlarged superior cistern
(D) Enlarged Sylvian cistern
(E) Enlarged third ventricle

24 A 4-year-old girl is brought to her pediatrician by her mother. The mother explains that her daughter has complained of headache, has become more lethargic, and frequently vomits. These symptoms have become progressively worse over the last 6 to 8 weeks. MRI reveals a tumor within the fourth ventricle. The tumor is removed and frozen sections are prepared. Histologically, this tumor is characterized by the presence of perivascular pseudorosettes. This girl most likely has which of the following?

(A) Anaplastic astrocytoma
(B) Choroid plexus papilloma
(C) Ependymoma
(D) Glioblastoma multiforme
(E) Oligodendroglioma

25 Which of the following structures would be regarded as being located in the medial wall of the lateral ventricle?

(A) Amygdaloid nucleus
(B) Caudate nucleus, head, body, tail
(C) Hippocampus and fornix
(D) Stria terminalis, terminal vein
(E) Ventricular surface of dorsal thalamus

26 A 19-year-old man is brought to the Emergency Department from the site of a motorcycle collision. He has compound fractures of the radius and ulna, a dislocated right shoulder, and head injuries and facial abrasions. CT, shown below, reveals traumatic brain injury (TBI) including blood at a number of locations within the cranial cavity. The enlarged inferior horn of the lateral ventricles correlates with blockage of cerebrospinal fluid (CSF) flow at which of the following locations?

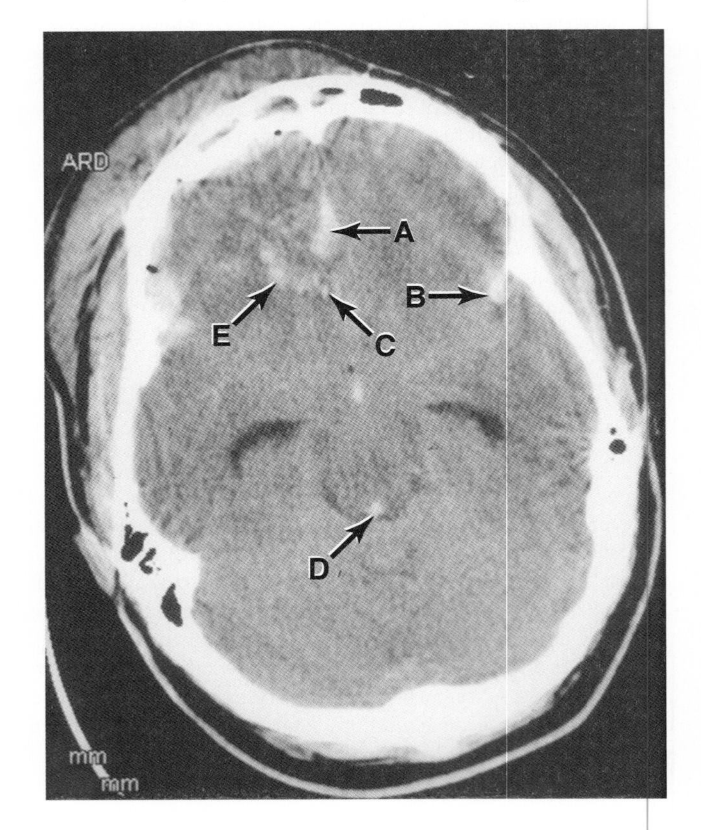

(A) A
(B) B
(C) C
(D) D
(E) E

27 A 17-year-old boy is transported to the Emergency Department by his parents. His mother indicates that he suddenly became ill over the last 3 days with vomiting and fever. The examination confirms the mother's observations and further reveals that the boy is confused and has meningismus. MRI is inconclusive. Lumbar puncture (the opening pressure is above the normal range) retrieves a sample of cerebrospinal fluid (CSF) that is cloudy, contains 17,000 WBCs per cubic mm (primarily polymorphonuclear leukocytes [PMNs]) and elevated protein (450 mg%). Based on this history and laboratory results, this boy is most likely suffering which of the following?

(A) Acute bacterial meningitis
(B) Acute viral meningitis
(C) Guillain-Barré syndrome
(D) Multiple sclerosis
(E) Subarachnoid hemorrhage

28 Which of the following characterize the epithelium covering the choroid plexus within the ventricles of the brain?

(A) Columnar-shaped, stereocilia, few mitochondria, tight junctions
(B) Columnar-shaped, microvilli, many mitochondria, no tight junctions
(C) Cuboidal-shaped, microvilli, few mitochondria, no tight junctions
(D) Cuboidal-shaped, microvilli, many mitochondria, tight junctions
(E) Cuboidal-shaped, stereocilia, many mitochondria, tight junctions

29 A 20-month-old boy is evaluated for vomiting and lethargy that has gotten progressively worse over the last 10 weeks. The mother indicates that her son has been quite fussy over the last several weeks to the point where she sometimes cannot console him. MRI reveals enlarged ventricles as shown in the image below. Assuming that this is a nonsecreting tumor, which of the following is the most likely cause of this ventricular enlargement in this boy?

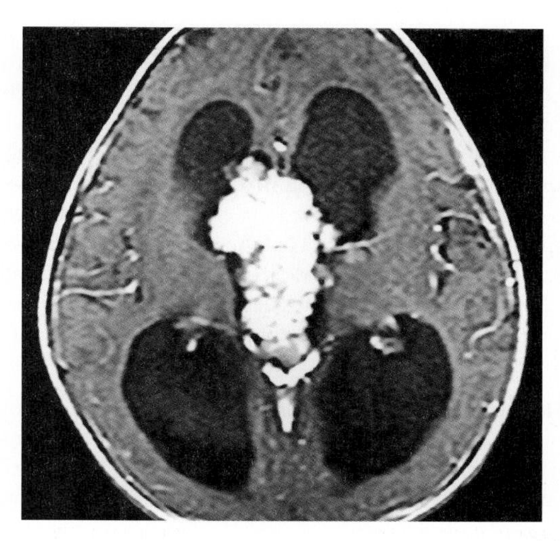

(A) Alobar holoprosencephaly
(B) Communicating hydrocephalus
(C) Hydrocephalus ex vacuo
(D) Idiopathic intracranial hypertension
(E) Obstructive hydrocephalus

ANSWERS

1 **The answer is C: Interpeduncular cistern.** The space caudal to the mammillary body, rostral to the basilar pons, and located on the midline between the two basis pedunculi is the interpeduncular fossa. This portion of the subarachnoid space contains many important structures such as the bifurcation of the basilar artery, oculomotor root, the P_1 segment, many small perforating arteries, and small veins. The ambient cistern is located on the lateral aspect of the midbrain and contains the trochlear nerve and several important arteries, and the crural cistern is located on the surface of the crus cerebri and contains the anterior choroidal artery and the basal vein. The subarachnoid space is a very general term that describes *any space* on the surface of the brain or spinal cord between the pia mater and the arachnoid mater. Cisterns are enlarged, and specifically named, regions of the overall subarachnoid space. The third ventricle is inside the brain and communicates rostrally with the lateral ventricles and caudally with the fourth ventricle via the cerebral acqueduct.

2 **The answer is A: Acute bacterial meningitis.** The combination of sudden onset, fever/chills, a decreased level of consciousness, and a CSF sample with high white blood cell count is characteristic for acute bacterial meningitis. Acute subarachnoid hemorrhage is also a sudden onset situation, but the CSF sample would contain many RBCs, not WBCs. Viral meningitis is slower onset and the sample would not have significantly elevated WBCs, while the sample from multiple sclerosis may have slightly elevated lymphocytes or monocytes and elevated immunoglobulin G. Glioblastoma multiforme is a rapidly growing tumor (weeks to months for symptoms to appear); a CSF sample would show elevated proteins and very possibly malignant cells shed by the tumor.

3 **The answer is C: Hydrocephalus ex vacuo.** This man is elderly, has an age-appropriate mental condition (a general forgetfulness), but no other general or focal neurological problems. His enlarged ventricles are due to general brain atrophy, an age-related condition, especially recognizing that he has no neurological deficits. The other choices are not correct because of age (congenital hydrocephalus is seen in the very young) and the lack of lesions in the brain or ventricles: obstructive, communicating, or posthemorrhagic hydrocephalus would be seen in cases with lesions, blood, or other impediments to CSF flow in the subarachnoid space, all lacking in this man.

4 **The answer is D: Head of the caudate nucleus.** Huntington disease is an incurable neurodegenerative disorder that affects several brain regions, but is most noticeable by the loss of the head of the caudate nucleus in the lateral wall of the anterior horn of the lateral ventricle. While the body of the caudate is also affected, the overwhelmingly obvious loss is of the caudate head. A CT of a Huntington patient may also show movement artifacts within the image due to tremor during the procedure. The greater the number of CAG repeats, the more rapid the onset and more severe the disease. Fewer numbers of CAG repeats usually correlate with a slower onset and a less severe disease. In either situation, the disease eventually results in the death of the patient. The anterior nucleus of the thalamus may be affected in Wernicke syndrome. The amygdaloid nucleus and the hippocampus are in the anterior and medial walls of the temporal horn of the lateral ventricle; they may be diminished in cases of dementia and some syndromes of the temporal lobe that have dementia and behavioral components. These temporal lobe structures are also susceptible to trauma in certain injuries.

5 **The answer is E (Third ventricle).** This man has clear soft tissue damage to his head, a skull fracture, and intracranial injuries. This CT shows acute blood in the third ventricle (choice E), in the medial longitudinal fissure between the frontal lobes (choice B), and immediately internal to the inner table of the skull as a small epidural hemorrhage (choice A). In addition, this man has hemorrhagic lesions in his right frontal lobe (choices C and D). The hemorrhagic lesion at C is in the medial area of the lobe adjacent to the falx cerebri, while that at D is in the brain substance immediately rostral to the corpus callosum. This patient also has blood in the anterior horn of the lateral ventricle on the right, but no blood in the anterior horn on the left. Not only is the left anterior horn devoid of blood, but so are the slightly enlarged temporal horns (bilaterally), and there is no blood in the superior cistern (these all appear black in this image).

6 **The answer is D: Lateral recess.** The lateral recesses (i.e., one on each side) are the lateral extensions of the fourth ventricle. They form a sleevelike structure that arches around the lateral aspect of the medulla at the pons-medulla junction to open, as the foramen of Luschka, into the subarachnoid space at the cerebellopontine angle. A small part of the choroid plexus of the fourth ventricle extends out of the foramen of Luschka and into the subarachnoid space. The ambient cistern is located on the side of the midbrain and contains the trochlear nerve and several major arteries and veins. The cerebellomedullary cistern consists of a dorsal portion (the cisterna magna) and a lateral portion on the lateral and anterior aspect of the medulla; this is sometimes called the inferior cerebellopontine cistern. The cerebral aqueduct is the portion of the ventricular system in the midbrain, and the subarachnoid space is extensive, covering all portions of the brain and spinal cord. Cisterns are enlarged areas of the subarachnoid space.

7 **The answer is B: Atrium of the lateral ventricle.** The large tuft of choroid plexus that is located in the atrium of the lateral ventricle is the glomus choroideum; this is frequently calcified to varying degrees in older patients. The atrium is at the junction of the temporal horn, posterior horn, and body of the lateral ventricle. The choroid plexus of the lateral ventricle extends along the roof of the temporal horn, through the atrium, along the floor of the body of the ventricle, through the interventricular foramen, and along the roof of the third ventricle. The anterior horn of the lateral ventricle does not contain choroid plexus.

8 **The answer is D: End of first trimester.** The only route for CSF to leave the ventricular system in the normally developing fetus is through the two lateral foramina (Luschka) and the single medial foramen (of Magendie); both are openings of the lateral recess and in the caudal roof of the fourth ventricle, respectively. These openings are normally established at about 10 to 12 weeks. A failure of these openings to develop may result in enlarged ventricles since the choroid plexus is forming and producing CSF by this time.

9 **The answer is D: Guillain-Barré syndrome.** The advent of sensory deficits followed by a rapid progression of bilateral weakness of the extremities is a characteristic feature of Guillain-Barré syndrome. These patients usually develop trouble breathing and may require artificial ventilation. This syndrome is a self-limiting autoimmune disease that may follow viral or bacterial infections, may result in life-threatening deficits, and whose treatment is complicated by the severity and course of the disease. Acute viral meningitis is characterized by rapid onset, headache, fever, and lethargy, but no sensory or motor deficits; these patients usually recover with supportive care. The Benedikt syndrome is a lesion of the central and medial midbrain that involves the crus cerebri, oculomotor root, red nucleus, and cerebellothalamic fibers. The Gubler syndrome is a brainstem syndrome that involves corticospinal fibers and the facial root producing an alternating (or crossed) hemiplegia. Myasthenia gravis (MG), like Guillain-Barré, is an acquired autoimmune disease involving the neuromuscular junction. In contrast to Guillainn-Barré, muscular weakness in MG patients usually appears first in ocular and facial musculature; waxing and waning of the weakness and muscle fatigability are hallmarks of MG, while sensory deficits are not.

10 **The Answer is C: Mitochondria.** This particular organelle is the source of energy for the cell, and is especially numerous in the epithelial cells that form the choroid plexus. All cells have a nucleus; they are essential to cell function, but not a source of cell energy. Lysosomes are membrane-bound vesicles that consolidate exogenous material; they do not produce energy. Synaptic vesicles are located in the distal terminals of axons (except in rare situations), and the endoplasmic reticulum may be rough (has ribosomes) or smooth (does not have ribosomes). Ribosomes synthesize proteins, and the endoplasmic reticulum serves to transport this material to distal regions of the cell.

11 **The answer is C: Presence of lamina terminalis.** In this axial CT, the distinct oval-shaped black area on the midline is the lower portion of the third ventricle, and the white area immediately rostral to this dark area is the cistern of the lamina terminalis; the structure at this interface is the lamina terminalis. The sudden onset and all signs and symptoms experienced by this man, all point to a subarachnoid hemorrhage. This axial CT is through the rostral midbrain and inferior portions of the diencephalon (specifically the hypothalamus); the positions of the hypothalamus, interpeduncular cistern, and other cisterns of the midbrain are clearly seen. A subarachnoid clot would not present with a sharp flat surface, would be irregular in size and shape, and could be anywhere in the subarachnoid space. Leptomeningeal cysts (also called arachnoid cysts) are usually large fluid-filled structures, are developmental in origin, and 75% are found in children; none of these features is seen in this case. The stria terminalis is a bundle of efferents from the amygdaloid nucleus that is located in the groove between the dorsal thalamus and caudate, arching from the amygdale toward the septal area. The stria medullaris thalami extends along the upper edge of the dorsal thalamus from the area of the septum and interventricular foramen caudally to the habenular nucleus.

12 **The answer is B: Fibroblasts and collagen.** This narrow tissue space between the basal lamina of the choroid epithelium and the basal lamina of the capillary epithelium of the choroid vasculature contains fibroblasts, collagen, and, of course, fluid that will eventually become cerebrospinal fluid. In the normal, and healthy, individual there are no red or white blood cells in this space and there are no elastic fibers. These would appear resultant to some type of trauma or ongoing pathologic process.

13 **The answer is A: Anterior choroidal artery.** The anterior choroidal artery originates from the cerebral part of the internal carotid artery and passes caudolaterally along the course of the optic tract, but generally lateral to this important structure. It serves part of the genu, the optic tract, inferior portions of the posterior limb of the internal capsule and the lenticular nucleus, portions of the hippocampus and amygdaloid nucleus, and, of course, the choroid plexus (CP) within the temporal horn. The anterior inferior cerebellar artery serves the CP sticking out of the foramen of Luschka, the medial posterior choroidal artery serves the CP of third ventricle, and the lateral posterior choroidal artery enters the lateral ventricle in the general area of the glomus of the CP of the lateral ventricle. This vessel may also contribute to the CP in the temporal horn. Branches of M_1 serve only structures within the hemisphere.

14 **The answer is E: Third ventricle.** This tumor is a choroid plexus papilloma; about 70% of these lesions are found in children. In this axial view, the location of the tumor is clearly in a significantly enlarged third ventricle. The two dorsal thalami are evident on either side of the third ventricle, and are separated from the tumor mass by portions of the ventricular space. The two interventricular foramina are enlarged by portions of the tumor, but this lesion does not invade the anterior horn (or the atrium). The anterior horn, atrium, and posterior horn of the lateral ventricle are significantly enlarged and, by logical extension, so are the temporal horns of the lateral ventricle. The fourth ventricle is not involved. The treatment of choice is surgical removal.

15 **The answer is D: Medial posterior choroidal artery.** The medial posterior choroidal artery originates from the P_2 segment of the posterior cerebral artery in the interpeduncular cistern, passes around the midbrain, and enters the caudal aspect of the third ventricle via the velum interpositum to serve the choroid plexus of the third ventricle. The anterior inferior cerebellar artery arises from the basilar artery and, en route to the cerebellum, serves the tuft of choroid plexus sticking out of the foramen of Luschka; the posterior inferior cerebellar artery arises from the vertebral artery, and provides blood to the choroid plexus within the fourth ventricle. The choroid plexus of the lateral ventricle is served by the lateral posterior choroidal (body + atrium) and the anterior choroidal (temporal horn to atrium) arteries.

16 **The answer is D: Normal pressure hydrocephalus.** Classically, normal pressure hydrocephalus presents as the triad of gait problems, dementia, and urinary incontinence. This condition may follow trauma, subarachnoid hemorrhage, meningitis, tumor, or other causes. All portions of the ventricular system may be enlarged to varying degrees. In general, ventriculoperitoneal shunt is the treatment of choice. Aqueductal stenosis results from an obstruction of the cerebral aqueduct with consequent enlargement, if left untreated, of the third and lateral ventricles. Communicating hydrocephalus results from an obstruction of CSF movement within the subarachnoid space, and obstructive hydrocephalus is exactly what it sounds like, an obstruction to CSF flow at some point within the ventricular system. Idiopathic hydrocephalus is seen in obese women of child-bearing age or in individuals with chronic renal disease.

17 **The answer is D: Multiple sclerosis.** The combination of diffuse deficits, symptoms that wax and wane, and increased immunoglobulins in the CSF points to multiple sclerosis. The deficits are not likely be vascular related, since they represent multiple vascular territories. It is highly unlikely that a CSF sample would be taken from a patient suspected of having an intracerebral tumor (glioblastoma) due to the significant risk of herniation and possible sudden death. Acute bacterial meningitis is rapid onset (not the case here) and can be rapidly fatal if not treated; bacteria would be seen in the CSF. Intracerebral hemorrhage is usually characterized by long tract (such as corticospinal) and cranial nerve deficits that have a sudden onset; these symptoms may get better, worse, or remain constant, but they do not wax and wane to any degree. Myasthenia gravis is a neurotransmitter disease that is usually characterized by weakness of ocular, facial, and oropharyngeal muscle initially, followed by weakness of the limbs.

18 **The answer is E: Traumatic tap.** In this man, the opening pressure is within normal ranges (about 80 to 140 mm H_2O), there is a precipitous decrease in the red color as the tap proceeds, the supernatant is clear, and small clots appear only in the first tube. All of these are characteristic of a traumatic tap. Aneurysm rupture (the second most common cause of blood in the subarachnoid space), bleeding from an AV malformation, and subarachnoid hemorrhage (traumatic—not the case in this man, nontraumatic) all result in an increased opening pressure and xanthochromia of the supernatant; neither is a factor in this man. In addition, subarachnoid hemorrhage resulting from aneurysm rupture (nontraumatic) would be signaled by a sudden severe headache; this man has no such complaint. Intracerebral hemorrhage will increase intracranial pressure, may result in focal neurologic deficits reflecting the location of the lesion (not a complaint in this patient), and, if the bleeding accesses the subarachnoid space, the appearance of CSF would be similar to that in subarachnoid hemorrhage.

19 **The answer is E (Cistern of the lamina terminalis).** The A_2 segment of the anterior cerebral artery extends from the anterior communicating artery rostrally to the point where it begins to curve around the genu of the corpus callosum where it becomes A_3. The proximal part of A_2 is located in the cistern of the lamina terminalis; bleeding from this vessel is most frequently (and initially) found in this cistern. In fact, hemorrhage in this location may also invade the inferior and medial portions of the frontal lobe and/or the anterior horn of the lateral ventricle. Choice A is part of the Sylvian cistern which contains M_1 and portions of the superficial and deep middle cerebral veins. Choice B is the interpeduncular cistern which contains P_1, a number of other arterial branches, and the oculomotor root. Choice C is the ambient cistern; this portion of the subarachnoid space contains P_{2-3}, the trochlear nerve, and several other important arteries such as the superior cerebellar and quadrigeminal arteries. Choice D identifies the more inferior parts of the third ventricle; this is not part of the subarachnoid space.

20 **The answer is B: Caudate nucleus.** The caudate nucleus, in all of its parts, is found in the lateral wall of the lateral ventricle (head-anterior horn, body-body of ventricle, tail-inferior/temporal horn). Huntington disease is characterized, in MRI or CT, by the loss of the shape of the head of the caudate nucleus protruding into the anterior horn. The amygdaloid nucleus is located in the anterior end of the temporal horn, and the hippocampus is found in what is regarded as the medial wall of the temporal horn. The corpus callosum forms the anterior wall (internal surface of the genu) of the anterior horn, and the body of the corpus callosum forms the roof of the body of the ventricle. The dorsal thalamus makes up much of the floor of the body of the lateral ventricle.

21 **The answer is D: Approximately 425 to 500 mL/day.** At any given moment, there is about 125 to 150 mL of CSF within the ventricles and subarachnoid space of an adult. This may vary slightly based on the size of the individual: lower for small individuals, higher for large persons. Recognizing that CSF is reabsorbed about three times in every 24-hour period, this puts the production rate in the general range of 450+ for the same time frame. Excessively low or high production rates may correlate with specific clinical signs or symptoms. Also, recognizing that the total intracranial volume in an adult is about 1,400 mL, it is easy to understand why the sudden blockage of CSF egress or reabsorption may be catastrophic within a very short time frame. In other words, the daily production of 450 mL of fluid will completely fill the cranial cavity with fluid in about 3 days, leaving no space for the brain!

22 **The answer is C: Idiopathic intracranial hypertension.** Idiopathic intracranial hypertension, also called pseudotumor cerebri, is commonly seen in obese women of child-bearing age, and is characterized by headache, a variety of visual disturbances that may eventually include blindness if left untreated, and papilledema. These patients may also have diplopia and retrobulbar pain when moving their eyes. Treatment includes weight loss, possibly a shunt procedure, or fenestration of the optic sheath to relieve pressure on the optic nerve. Aqueductal stenosis is an abnormally small or constricted cerebral aqueduct that, in cases of hemorrhage within the ventricle, may become occluded. Hydrocephalus ex vacuo is a condition of the elderly that is related to brain atrophy; the ventricles may appear enlarged, but there are no abnormalities or neurological deficits. Normal pressure hydrocephalus is also seen in the elderly, but it is usually characterized by the triad of dementia, impaired gait, and urinary incontinence or urgency. Obstructive hydrocephalus is simply an obstruction to CSF flow within the ventricular system or subarachnoid space.

23 **The answer is E: Enlarged third ventricle.** In this axial view, the third ventricle in this patient is about five times wider than that seen in a normal patient of comparable age. It is clearly and dramatically enlarged compared to the normal third ventricle and when compared to the other ventricular spaces in this image. The anterior horn of the lateral ventricle is somewhat enlarged, about twice the size that one would expect to see in a normal patient, and the atrium is also somewhat enlarged. The superior cistern (also called quadrigeminal cistern) is within normal size ranges of a normal brain, as is that portion of the Sylvian cistern located on the insular cortex.

24 **The answer is C: Ependymoma.** This type of tumor is frequently seen in children, frequently (70%) found in the fourth ventricle, and not encapsulated, but well defined from surrounding tissue. While the histological appearance may vary even within a single tumor, the presence of perivascular pseudorosettes, or of true rosettes, is characteristic of this tumor. The former are polygonal or columnar cells arranged around a small vessel; the latter are similar types of cells encircling a small lumen. Anaplastic astrocytomas and glioblastoma multiforme (GBM) both arise from astrocytes, are found within brain tissue, and are characterized by pleomorphic cells reflecting their origin. In addition, GBM are rapidly growing, infiltrate the surrounding brain, and may contain mitotic spindles indicative of their rapid growth patterns. Choroid plexus papillomas are comparatively rare and characterized by elongated cuboidal, or columnar, epithelium, and a thickened stroma within the leaves of the tumor. Oligodendroglioma are tumors of the white matter, and are most frequently found in supratentorial locations.

25 **The answer is C: Hippocampus and fornix.** While the shape of the temporal horn of the lateral ventricle may vary somewhat among individuals, the hippocampus is considered to be located in its medial wall within the temporal horn. The fornix arises from the hippocampus as a flattened part called the crus; this fiber bundle arches toward the midline, becomes rounded, and is medially located in relation to the body of the lateral ventricle. Rostrally, the body of the fornix is located adjacent to its opposite counterpart, becomes the column of the fornix, lies immediately behind the anterior commissure, and passes through the hypothalamus to end in the mammillary nuclei. The amygdala is located in the rostral end of the temporal horn, and the caudate, in all of its parts, is located in the lateral wall of the lateral ventricle (anterior horn, body, temporal horn). The ventricular surface of the dorsal thalamus is immediately inferior to the body of the lateral ventricle, and the stria terminalis and terminal vein are in the groove formed by the apposition of the body of the caudate and the dorsal thalamus.

26 **The answer is D (Cerebral aqueduct).** Cerebrospinal fluid (CSF) is produced by the choroid plexuses of the lateral (about 70%), third (about 5%), and fourth (about 5%) ventricles. About 20% of the CSF is generated from the ependymal lining of the ventricles, from small vessels located in the sleeves of spinal nerve roots, from cerebral interstitial spaces, and from other similar sources. CSF produced in the lateral ventricles flows into the third ventricle, and from the third into the fourth ventricle. The cerebral aqueduct (choice D) is small (about 2 mm in diameter), and is the prime point at which the flow of CSF can be comprised. Blood within the ventricles may block CSF flow through packing of blood cells at the opening of, or within, the cerebral aqueduct, and/or through small blood clots that occur at these two locations. The blood at choice A is between the frontal lobes and appears to be dissecting into the inferior and medial portion of the frontal cortex. The blood at choices C and E is located within the substance of the brain at the inferior (orbital) aspect of the frontal lobe. While this generally resembles the shape of the anterior horn of the lateral ventricle, this blood is not within the ventricular space. The lower portion of the left Sylvian cistern (a part of the subarachnoid space) also contains blood (choice B).

27 **The answer is A: Acute bacterial meningitis.** The sudden onset (this may be over one to several days), fever, vomiting, and CSF with significantly elevated polymorphonuclear leukocytes (PMNs) are characteristic of acute bacterial meningitis (also called acute pyogenic meningitis). In a normal individual, CSF is colorless, contains no PMNs, and is about 15 to 45 mg% protein. Viral meningitis may also have an acute onset, but the CSF appears normal; the cells are monocytes and are seen in a few hundreds (not thousands), and the protein at about 40 to 100 mg% is barely above normal. Guillain-Barré syndrome (GBS) also has a comparatively rapid onset (over 4 to 6 days), has normal appearing CSF with normal cells (no PMNs, 0–5 monocytes), but has elevated proteins. GB has motor symptoms that are not a feature in this boy. Multiple sclerosis has a slow onset, waxes and wanes, and has CSF that appears normal but has increased monocytes and protein. In the case of rupture of an intracranial aneurysm (resulting in subarachnoid blood/subarachnoid hemorrhage), the onset is sudden, accompanied by severe headache, and the CSF appears red with a xanthochromic supernatant) with increased red blood cells.

28 **The answer is D: Cuboidal-shaped, microvilli, many mitochondria, tight junctions.** The epithelium of the choroid plexus is composed of cuboidal-shaped cells with dome-shaped apices facing the lumen of the ventricle. These domes are covered by a thick mat of microvilli which greatly increases the surface area of the cells and, thereby, greatly enhances their secretory ability. The cuboidal cells are attached to each other by continuous tight junctions located between the cells near their luminal surfaces. These cells have numerous mitochondria, a well-developed rough endoplasmic reticulum, and intracellular vesicles, all correlating with their significant secretory function. The nucleus of the cell is generally located near the base of the cell, and the cells sit on a basal lamina. Internal to this basal lamina are fibroblasts and collagen of the connective tissue core of the villus.

29 **The answer is E: Obstructive hydrocephalus.** The most obvious feature in this image, in addition to the significantly enlarged ventricles, is the large choroid plexus tumor in the third ventricle. This tumor occupies both interventricular foramina, thus blocking flow of CSF from the lateral ventricles into the third ventricle; this hydrocephalus results from this obstruction to flow. Alobar holoprosencephaly is a severe developmental defect in which the hemispheres do not develop, and there is a single midline ventricle. In communicating hydrocephalus, there is normal flow of CSF through the ventricular system and subarachnoid space with blockage being at the arachnoid villi by, as examples, blood cells (subarachnoid hemorrhage) or tumor proteins. Hydrocephalus ex vacuo is a condition commonly seen in the elderly; the enlarged ventricles are related to atrophy of the brain. However, hydrocephalus ex vacuo may also be seen in younger patients under certain specific conditions, for example, in a nonelderly alcoholic individual. Idiopathic intracranial hypertension (also called pseudotumor cerebri) is a condition most commonly seen in obese women of child-bearing age; in these patients there is increased intracranial pressure, but the ventricles are usually not enlarged.

Chapter 5

An Overview of the Cerebrovascular System

Recall the Cautionary Tale: The images in this chapter are presented in a Clinical Orientation; this approach emphasizes how structure and function correlate with deficits in a clinical setting. MRI and CT have a universally recognized orientation and laterality. Line drawings and stained sections in the Clinical Orientation are viewed in the identical manner that one views an MRI: your right, as the physician-observer, is the patient's left, and the observer's left is the patient's right. The laterality of deficits and reference to connections follow accordingly.

QUESTIONS

Select the single best answer.

1 A 21-year-old man is brought to the Emergency Department from the site of a motorcycle collision. CT reveals that his C3 and C4 vertebrae are fractured and the integrity of the transverse foramen of both is compromised. Which of the following portions of the vertebral artery is most likely damaged in this patient?

(A) V_1
(B) V_2
(C) V_3
(D) V_4

2 Which of the following is a branch of the vascular structure indicated at the tip of the arrow in the given image?

(A) Anterior communicating artery
(B) Great cerebral vein
(C) Middle cerebral vein
(D) Ophthalmic artery
(E) Quadrigeminal artery

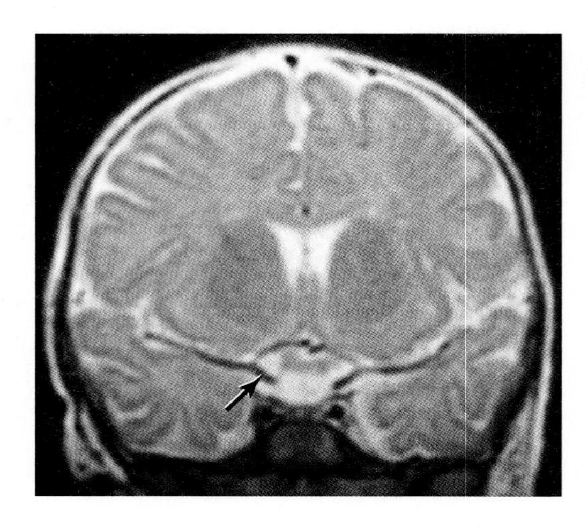

3 The normal healthy human brain weighs about 2% of the total body weight but consumes about what percent of the total oxygen used by the body?

(A) About 5%
(B) About 10%
(C) About 20%
(D) About 30%
(E) About 40%

4 After surgery for a saccular aneurysm, a 63-year-old woman does not awaken. A detailed review of the procedure and the postoperative images suggest that the aneurysm clip inadvertently occluded an especially important branch of P_1. Which of the following vessels was most likely involved?

(A) Anterior choroidal artery
(B) Lateral posterior choroidal artery
(C) Medial posterior choroidal artery
(D) Thalamogeniculate artery
(E) Thalamoperforating artery

5 A 59-year-old man is transported to the Emergency Department following the sudden onset of neurological deficits. The history reveals that the man is morbidly obese and hypertensive; he is refractory to taking his medications as prescribed. CT reveals a clear-cut defect in the cerebral hemisphere as shown in the image below. Which of the following neurological deficits is this man most likely experiencing?

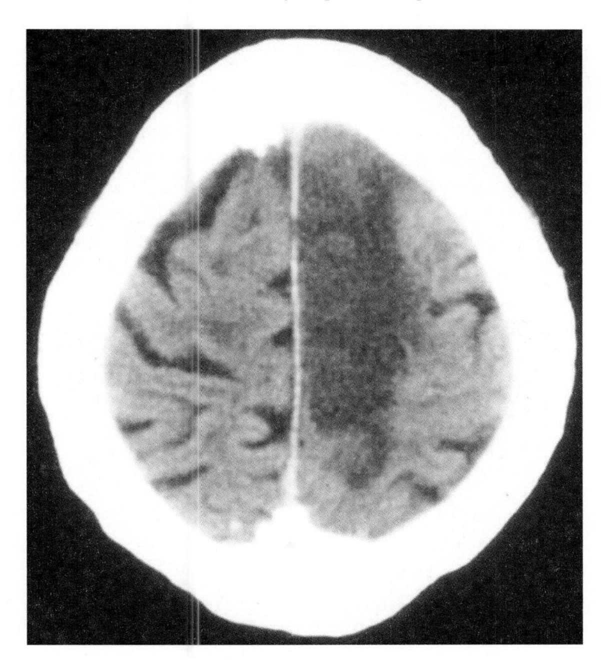

(A) Left lower extremity weakness
(B) Left homonymous hemianopia
(C) Right lower extremity weakness
(D) Right homonymous hemianopia
(E) Right upper extremity weakness

6 The neurologist, who is treating a 43-year-old woman for intractable headache, orders an MRI. While this MRI shows no gross abnormalities of the forebrain or brainstem, it does reveal that this woman has an unusually small (in diameter) right vertebral artery. This part of this vessel is usually located in which of the following cisterns?
(A) Ambient
(B) Dorsal cerebellomedullary
(C) Interpeduncular
(D) Lateral cerebellomedullary
(E) Prepontine

7 A 53-year-old woman presents with a complaint of persistent headache and double vision. The MRI reveals what appears to be a vascular lesion within the cavernous sinus, and angiogram confirms that it is an aneurysm of the intracavernous portion of the internal carotid artery. Which of the following structures would most likely be the first structure to be affected by this expanding lesion?
(A) The abducens nerve
(B) The mandibular nerve
(C) The maxillary nerve
(D) The oculomotor nerve
(E) The trochlear nerve

8 A 69-year-old man presents at the Emergency Department with the complaint of a sudden onset of difficulty walking. The examination reveals that he is ataxic, is unable to do the heel-to-shin test, and has dysmetria. An MRI (T2 weighted) reveals the lesion shown in the image below. Based on this image, occlusion of which of the following most likely resulted in this man's deficits?

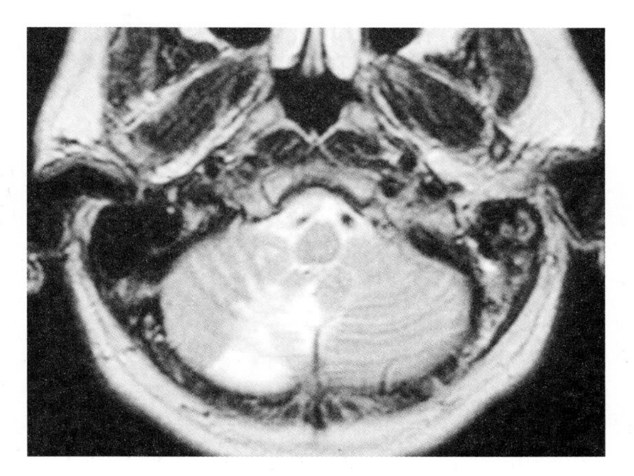

(A) Anterior inferior cerebellar artery
(B) Distal posterior inferior cerebellar artery
(C) Medial posterior choroidal artery
(D) Proximal posterior inferior cerebellar artery
(E) Superior cerebellar artery

9 Which of the following vessels, identified in the image below, is the origin of the thalamogeniculate artery and the medial and lateral posterior choroidal arteries?

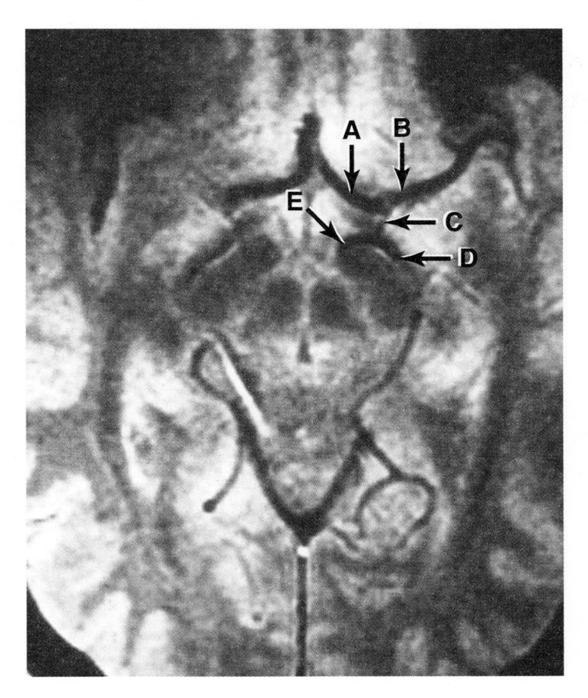

(A) A
(B) B
(C) C
(D) D
(E) E

10 A 3-week-old infant presents with bulging fontanels, prominent veins in the scalp, and hydrocephalus. Angiogram and MRI reveal that this patient has a vein of Galen malformation (these may sometimes be called a vein of Galen aneurysm), which is a type of arteriovenous malformation. Based on their various locations, which of the following could be the most likely feeding artery to this lesion?

(A) Angular branches, M_4
(B) Anterior choroidal
(C) Medial choroidal
(D) Pericallosal, A_5
(E) Temporal branches, P_3

11 An 11-year-old boy is brought to the Emergency Department by his mother with a serious infection on his face. The history reveals that he fell while visiting his grandfather's farm and scraped the right side of his face hard enough that it bled. The examination reveals an infection with purulent discharge and an acutely ill boy with headache, fever, somnolence, and a stiff neck. Suspecting bacterial meningitis, a lumbar puncture is ordered. Which of the following is the most likely route through which this infection could access the central nervous system (CNS)?

(A) Anterior cerebral vein(s)
(B) Deep middle cerebral vein
(C) Internal cerebral veins
(D) Ophthalmic vein(s)
(E) Superficial middle cerebral vein

12 Which of the following structures is located at the position of the red dot in the figure below?

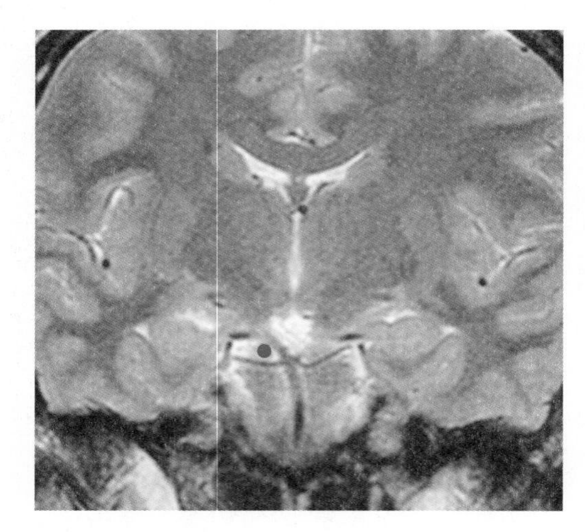

(A) Abducens nerve
(B) Facial nerve
(C) Oculomotor nerve
(D) Trigeminal nerve
(E) Trochlear nerve

13 A 47-year-old man presents to his family physician with the complaint that he sees "two of everything." The history reveals that this condition has been getting worse over the last several weeks to the point where the man is now having serious problems at his job (accountant). The examination shows weakness of the right superior oblique muscle and a slight down-and-out rotation of that eye. MRI reveals a mass in the ambient cistern presumably compressing the trochlear nerve. In addition to this nerve, which of the following vascular structures would most likely be compromised by this mass?

(A) Anterior cerebral vein
(B) Basal vein of Rosenthal
(C) Deep middle cerebral vein
(D) Internal cerebral vein
(E) Vein of Galen

14 A 3-year-old girl is brought to the pediatrician by her mother, who explained that the girl complained of a headache for about 2 days and then had a rather sudden onset of fever, vomiting, and lethargy. The examination reveals a middle ear infection that has spread to the mastoid air cells. MRI shows an apparent venous thrombosis and probable meningitis. Which of the following is the most likely venous structure involved in the spread of this patient's infection?

(A) Basal vein
(B) Cavernous sinus
(C) Deep middle cerebral vein
(D) Sigmoid sinus
(E) Straight sinus

15 A 72-year-old man has a sudden onset of motor and sensory deficits at his home, and he is transported to the Emergency Department. The examination reveals a profound paralysis and loss of all sensation (pain, thermal, discriminative touch, vibratory sense) on his left side. CT shows a hemorrhagic lesion in the region of the brain indicated in the image below. This hemorrhagic stroke most likely originates from which of the following?

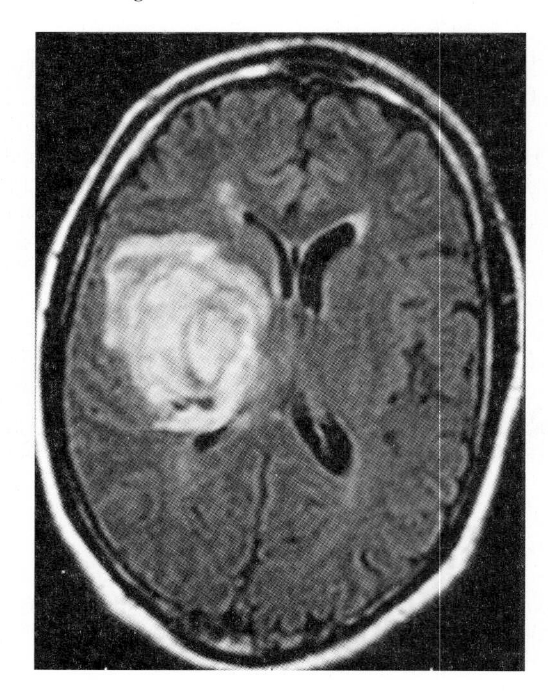

(A) Central/Rolandic branch of M_4
(B) Superior trunk of M_2
(C) Lateral striate artery/arteries
(D) Medial striate artery/arteries
(E) Thalamogeniculate artery

16 The CT of a 76-year-old man reveals a hemorrhagic stroke within the right hemisphere involving mainly the putamen, globus pallidus, and the adjacent portions of (the lateral parts of) the posterior limb of the internal capsule. The neurologist concludes that this is the result of a vascular event in the territory of the lenticulostriate arteries. Which of the following is the principal site of origin of these vessels?

(A) A_1
(B) A_2
(C) M_1
(D) M_2
(E) P_1

17 A 62-year-old man presents with the complaint of a sudden headache that abated after over-the-counter (OTC) medication. The history reveals that the man has been taking medication for many years for high blood pressure and that his father died of nontraumatic (spontaneous) subarachnoid hemorrhage. Magnetic resonance angiography (MRA) shows a saccular aneurysm on the internal carotid arterial tree. Which of the following is the most common location of aneurysms on this part of the cerebral vascular system?

(A) Anterior cerebral, A_3
(B) Anterior communicating artery
(C) Basilar bifurcation
(D) Middle cerebral, M_2
(E) Thalamoperforating artery

18 Which of the following structures is indicated by the white arrow in the T2-weighted MRI below?

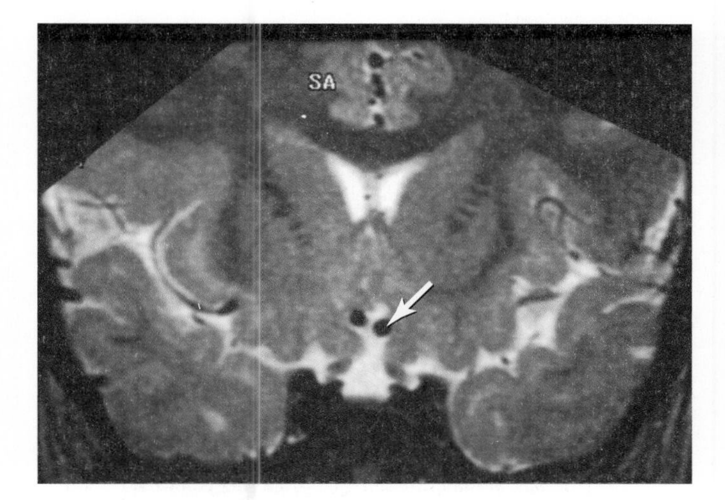

(A) Left A_1
(B) Left A_2
(C) Left A_3
(D) Left M_1
(E) Left P_1

19 A 47-year-old overweight woman presents with the complaint of difficulty seeing. The examination reveals a significant loss of vision in her left eye, and the results of the funduscopic examination suggest that the central artery of the retina has been occluded. Which of the following arteries is the most likely source of this small, but very important, vessel?

(A) A_1 segment
(B) Anterior choroidal
(C) Internal carotid
(D) M_1 segment
(E) Ophthalmic

20 A 54-year-old man complains of nonspecific visual deficits. A visual field examination by confrontation is conducted for both eyes by the physician in her office. The results clearly suggest a right homonymous hemianopsia. An MRI is ordered and reveals a hemorrhagic cortical lesion. An occlusion of which of the following vessels would most likely lead to the deficits experienced by this patient?

(A) A_5
(B) M_4
(C) P_1
(D) P_2
(E) P_4

21 A 20-year-old man is brought to the Emergency Department following a generalized seizure that occurred while he was watching a football game on TV. The history reveals that the man had several of what he called "spells" while at work. In addition, the man also indicated that he has had a number of persistent headaches over the last 2 months that he treated with OTC medications. MRI reveals the lesion, an arteriovenous malformation (AVM), shown in the image below. Based on its location, which of the following most likely represents the feeding artery and the draining venous structure?

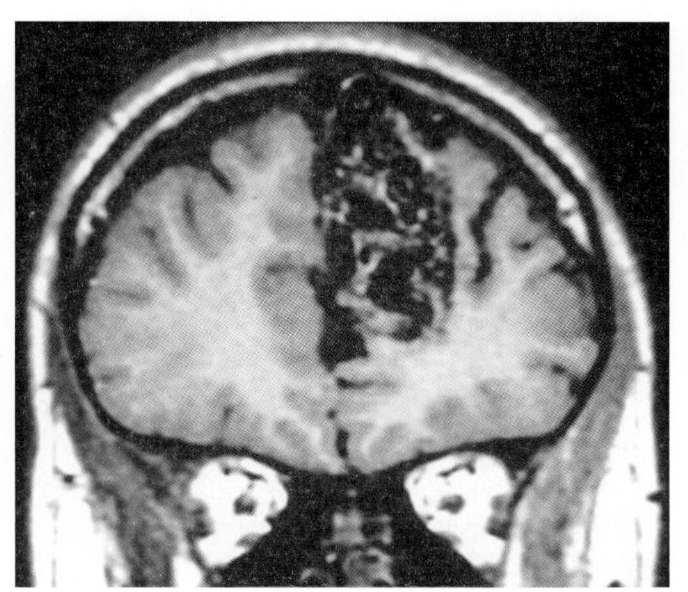

(A) A_1 branches—cavernous sinus
(B) A_2 branches—deep middle cerebral vein
(C) $A_{3,4}$ branches—superior sagittal sinus
(D) $M_{3,4}$ branches—superior sagittal sinus
(E) M_4 branches—superficial middle cerebral vein

22 A 27-year-old graduate student presents at the University Health Center with persistent headache. The history reveals that these headaches have gotten worse over the last several months. No frank neurological deficits are discovered in the neurological examination, although the student states that he thinks he has become more irritable over the last semester. An MRI reveals a fusiform aneurysm in the groove between the corpus callosum and the cingulate gyrus, which is impinging on both of these structures. This aneurysm is most likely located on which of the following?

(A) Angular artery
(B) Callosomarginal artery
(C) Frontopolar branches
(D) Orbitofrontal artery
(E) Pericallosal artery

23 A 21-year-old man is brought to the Emergency Department from the site of a motorcycle collision. He has extensive cuts, lacerations, a compound fracture of his left humerus, and is bleeding from his left ear. CT reveals a basal skull fracture that passes through the jugular foramen. In addition to the sigmoid sinus, which of the following vascular structures is also most likely damaged in this man?

(A) Confluence of sinuses
(B) Inferior petrosal sinus
(C) Straight sinus
(D) Superior petrosal sinus
(E) Transverse sinus

24 Which of the following represents the significance of the vascular loop identified by the white arrow in the image below?

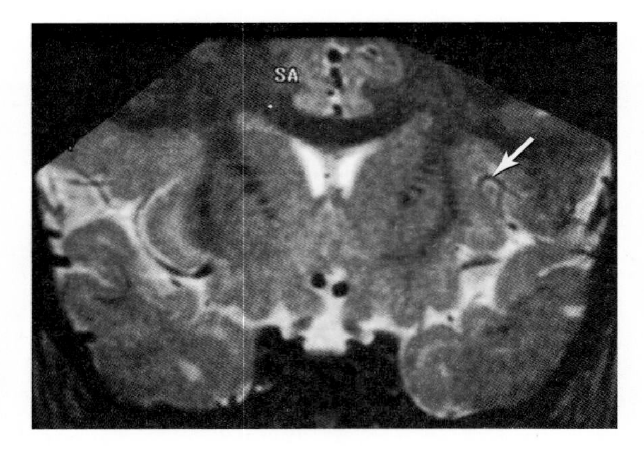

(A) Transition from A_1 to A_2
(B) Transition from A_2 to A_3
(C) Transition from M_1 to M_2
(D) Transition from M_2 to M_3
(E) Transition from P_1 to P_2

25 A 59-year-old man presents with persistent headache and nausea; both have become progressively worse over the last 6 to 8 weeks. The angiogram of this man shows a clearly abnormal location and shape of the venous angle. MRI reveals that this important venous landmark is displaced by a grape-sized tumor. Which of the following meet to form this venous structure?

(A) Basal vein (Rosenthal) and great cerebral vein (Galen)
(B) Longitudinal caudate vein and transverse caudate veins
(C) Superficial middle cerebral vein and vein of Trolard
(D) Thalamostriate vein and internal cerebral vein
(E) Transverse sinus and sigmoid sinus

26 Which of the following joins with the vein of Trolard and the vein of Labbé on the lateral surface of the cerebral hemisphere to form an obvious tripod-shaped vascular structure?

(A) Anterior cerebral vein
(B) Basal vein of Rosenthal
(C) Deep middle cerebral vein
(D) Superficial middle cerebral vein
(E) Superior cerebral veins

27 A 31-year-old woman experiences a small vascular hemorrhage, confirmed in CT, in her left thalamus that extends into her hypothalamus on the same side. Following this event, she experiences repeated bouts of systemic hypotension that require medical intervention. If not resolved, this woman may suffer which of the following?

(A) Advent of arteriovenous malformation
(B) Aneurysm rupture
(C) Subarachnoid hemorrhage
(D) Subdural hemorrhage
(E) Watershed infarct

28 Which of the following structures receive a blood supply from small penetrating branches that arise from the vessel indicated by the arrow in the image below?

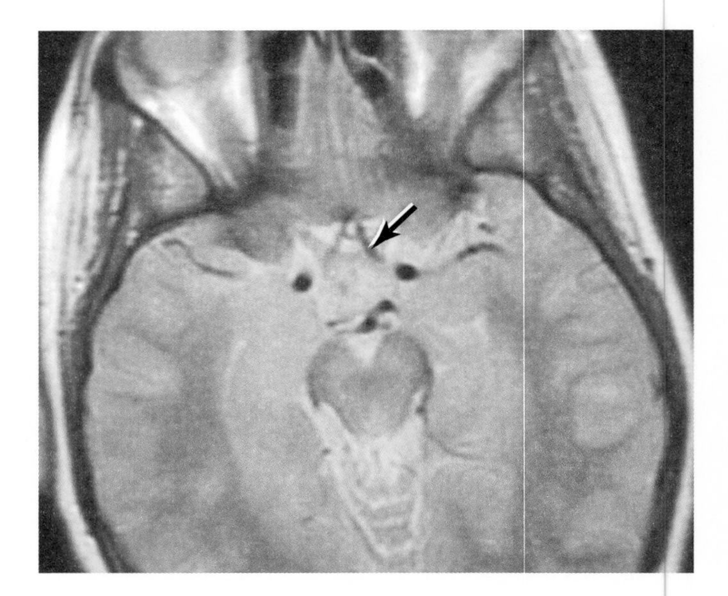

(A) Dorsomedial, ventromedial, arcuate nuclei
(B) Medial and lateral segments of globus pallidus, putamen
(C) Optic tract, adjacent internal capsule
(D) Preoptic, supraoptic, paraventricular, suprachiasmatic nuclei
(E) Posterior nucleus, nuclei of the mammillary body

29 A 29-year-old woman complains to her family physician that she believes her sense of smell has changed. During the history and examination, the physician learns that the woman's sense of smell has diminished over the last several months to the point where she sometimes does not smell her infant's soiled diaper. By her own estimation, she has become sometimes uncharacteristically short tempered in recent weeks. MRI reveals a large AVM in the medial portion of the orbital aspect of the frontal lobe on the left, including the gyrus rectus, medial orbital gyri, and olfactory structures. Based on the location of this vascular defect, which of the following is most likely the principal draining vein?

(A) Anterior cerebral vein
(B) Inferior cerebral veins
(C) Internal cerebral vein
(D) Superior cerebral veins
(E) Superficial middle cerebral vein

30 A 71-year-old man is transported to the Emergency Department after collapsing at his home. The examination reveals an unusual combination of deficits: right-sided paralysis and right-sided loss of somatosensory information, all on the upper and lower extremities, and a right homonymous hemianopsia. MRI shows a lesion resulting from occlusion of a small vessel. Which of the following vessels is most likely obstructed in this case?

(A) Anterior choroidal artery
(B) Lateral striate artery
(C) Medial striate artery
(D) Lateral posterior choroidal artery
(E) Thalamoperforating artery

ANSWERS

1 **The answer is B: V_2.** The V_2 segment of the vertebral artery passes through the transverse foramina from C6-C2. The other segments of the vertebral artery are as follows: V_1 is that part of the vertebral artery which is between its origin from the subclavian artery and where it enters the foramen at C6; V_3 is that part which exits C2, courses through the foramina of C1, arches caudally around the lateral mass of C1, and penetrates the atlanto-occipital membrane; V_4 is that part which penetrates the dura and is located between the dura and the point where the two vertebral arteries join to form the basilar artery.

2 **The answer is D: Ophthalmic artery.** The pointer identifies the right internal carotid artery (cerebral part) that clearly divides, in this image, into the middle and anterior cerebral arteries. The most consistent larger branches of this portion of this vessel are the ophthalmic, posterior communicating, and anterior choroidal arteries. The anterior communicating artery connects the two anterior cerebral arteries (ACAs) at the junction of the A_1 and A_2 segments of the ACA. The quadrigeminal is a branch of P_1 and arches around the midbrain to serve the colliculi. The great cerebral vein (of Galen) is associated with the superior cistern, empties into the straight sinus and the middle cerebral vein, and may be either deep (on the insular cortex) or superficial (in the lateral sulcus).

3 **The answer is C: About 20%.** The brain requires a constant flow of oxygenated blood and is a voracious consumer of oxygen. Even though the brain represents a small portion of total body weight (about 2%), it consumes a disproportionately large amount of the total oxygen (about 20% is needed by the brain) required by the body. Interruption of the supply of oxygenated blood to the brain will result in syncope (after 5 to 15 seconds) and, within a few minutes (2 to 4), permanent brain damage and disability and/or death.

4 **The answer is E: Thalamoperforating artery.** There are two ways to look at this question. First, which of these vessels is a branch of P_1, and the answer is thalamoperforating. The second is to consider which of these vessels serves the portions of the thalamus that are involved in cortical activation/arousal, and again the answer is thalamoperforating. In about 8% of individuals, the thalamoperforating artery arises as a single trunk from one P_1 and gives rise to branches that serve both sides of the thalamus. If this single trunk is inadvertently trapped in an aneurysm clip when approaching a basilar tip aneurysm, bilateral thalamic infarcts may result and interrupt the cortical arousal pathway. The anterior choroidal artery is a branch of the internal carotid, and the thalamogeniculate, medial, and lateral posterior choroidal arteries are branches of P_2. None of the other choices serves the thalamic areas involved in cortical arousal.

5 **The answer is C: Right lower extremity weakness.** In this axial CT, the infarcted area is clearly in the territory of the anterior cerebral artery on the left side. On the right side of the image, the central sulcus is obvious and is bordered by a long gyrus located immediately rostral (precentral gyrus) and caudal (postcentral gyrus) to the central sulcus. Using these landmarks as reference points, this lesion involves the hip area and the lower extremity area (anterior paracentral gyrus) of the left hemisphere, resulting in weakness of the right lower extremity. If this lesion also involved the posterior paracentral gyrus, the patient would also have a sensory loss of all modalities on the right lower extremity. Left motor deficits are on the wrong side, and the right and left visual deficits are not choices since the anterior cerebral artery does not serve any portion of the visual cortex.

6 **The answer is D: Lateral cerebellomedullary.** This cistern contains the intracranial segment of the vertebral artery, the roots of cranial nerves IX, X, XI, and XII, and small veins that drain the medulla. Variations in the vertebral artery (VA) include right hypoplasia (about 10% of cases), left hypoplasia (about 5%), failure to communicate with the basilar artery (2% on right, 3% on left), and the left VA may (about 4%) arise directly from the arch of the aorta. The ambient cistern contains many important structures, including the trochlear nerve, segments P_2 and P_3 of the posterior cerebral artery, the basal vein, and the quadrigeminal artery. Distal portions of the posterior inferior cerebellar artery are located in the dorsal cerebellomedullary cistern (also called cisterna magna); the basilar artery (other small arteries and veins) and the abducens nerve are in the prepontine cistern. The interpeduncular cistern contains many important structures, including the oculomotor nerve, upper basilar artery and bifurcation and the P_1 segment, thalamoperforating artery, and the origin of the quadrigeminal artery.

7 **The answer is A: The abducens nerve.** In this particular situation, there are three structures located in the cavernous sinus that, if damaged, could lead to diplopia: the abducens, oculomotor, and trochlear nerves. However, the position of the abducens nerve, immediately adjacent to the internal carotid artery, makes it the most likely of these three nerves to be initially damaged. The oculomotor and trochlear nerves, along with the maxillary nerve, are located in the lateral wall of the cavernous sinus and may be injured later as the aneurysm enlarges. Damage to the maxillary nerve may result in somatic sensory loss over the cheek area and upper teeth and palate on the same side. The mandibular nerve does not traverse this sinus but exits the skull via the foramen ovale.

8 **The answer is B: Distal posterior inferior cerebellar artery.** The posterior inferior cerebellar artery (PICA), a branch of the vertebral artery, courses around the medulla to access and serve the medial portion of the inferior aspect of the cerebellar hemisphere. En route around the medulla, PICA provides important branches to the lateral portions of this region of the brainstem; these branches arise from the proximal portion of PICA. In this particular case, the occlusion is distal to the branches serving the medulla (note that the medulla is spared) but affects those branches serving the cerebellar cortex; this lesion involves the distal PICA. The anterior inferior cerebellar artery, a branch of the basilar artery, serves the lateral portion of the inferior surface of the cerebellar hemisphere; this region is spared in this case. The superior cerebellar artery, a branch of the basilar artery, serves the superior (tentorial) surface of the cerebellum and the cerebellar nuclei. The medial posterior choroidal artery arises from the P_2 segment and serves the choroid plexus in the third ventricle (and adjacent structures en route).

9 **The answer is D (P_2 segment).** The P_2 segment of the posterior cerebral artery (PCA) is located immediately distal to the intersection of the PCA/P_1 and the posterior communicating artery (intersection of choice E with choice C); this segment gives rise to the thalamogeniculate artery and to the medial and lateral posterior choroidal arteries. The appearance of the first temporal branch of the PCA indicates the beginning of the P_3 segment. The A_1 segment of the anterior cerebral artery (choice A) gives rise to penetrating branches that serve parts of the optic chiasm, anterior regions of the hypothalamus, and adjacent structures. The M_1 segment of the middle cerebral artery (MCA) (choice B) gives rise to the lenticulostriate arteries and to small arteries that serve the temporal pole. Penetrating branches of the posterior communicating artery (choice C) serve the hypothalamus, and the P_1 segment (choice E) is the origin of the thalamoperforating and quadrigeminal arteries.

10 **The answer is C: Medial choroidal artery.** The medial choroidal artery (also called the posterior medial choroidal artery) originates from the P_2 segment, wraps around the brainstem in the area of the midbrain-thalamus junction, and enters the third ventricle to serve the choroid plexus. In this trajectory, it gives rise to small branches that serve brain structures it passes. In this location, branches of the medial posterior choroidal artery are in a position to contribute arterial blood to this vascular malformation. The angular branches serve the lateral hemisphere in the general area of the angular gyrus, the temporal branches of the posterior cerebral artery serve primarily the inferior surface of the temporal lobe, and the medial striate serves the head of the caudate nucleus and part of the anterior limb of the internal capsule. The optic tract, inferior portions of the internal capsule, and the choroid plexus of the temporal horn (plus immediately adjacent structures) are served by the anterior choroidal artery.

11 **The answer is D: Ophthalmic vein(s).** An infection on the face, especially in the general vicinity of the eye, maxilla, or forehead, has direct access to the CNS. Bacteria may enter and move retrogradely through the superficial facial veins, then the superior and inferior ophthalmic veins, and then enter the cavernous sinus. The anterior cerebral veins are located on the inferior aspect of the frontal lobe and, along with the deep middle cerebral vein (from the insular cortex), join the basal vein of Rosenthal. The internal cerebral veins drain the interior of the cerebral hemisphere and then continue into the great cerebral vein. The superficial middle cerebral vein is located on the surface of the hemisphere in the lateral sulcus.

12 **The answer is C: Oculomotor nerve.** The oculomotor nerve exits the medial edge of the cerebral peduncle and immediately passes between the superior cerebellar artery (the smaller vessel located immediately below the red dot in this image) and the P_1 segment of the posterior cerebral artery (the larger vessel located immediately above the dot in this image). This is a characteristic and predictable relationship that explains the myriad of oculomotor deficits seen in vascular lesions, mainly aneurysm, at this location. The abducens nerve exits at the caudal aspect of the pons, generally in-line with the preolivary fissure; the facial nerve exits at the caudal aspect of the pons, generally in-line with the postolivary sulcus. The exit of the trigeminal nerve on the lateral aspect of the pons specifies the junction of the basilar pons with the middle cerebellar peduncle. The trochlear nerve is the only cranial nerve to exit the dorsal aspect of the brainstem; it exits immediately caudal to the inferior colliculus.

13 **The answer is B: Basal vein of Rosenthal.** The basal vein of Rosenthal receives venous blood from the anterior cerebral vein, which is located on the orbital surface of the frontal lobe, and from the deep middle cerebral vein, which is located on the cortex of the insular lobe. After receiving these important tributaries, the basal vein passes through the ambient cistern along with several important arteries (P_2 and P_3, quadrigeminal artery, origin of the lateral posterior choroidal arteries, superior cerebellar artery) and the trochlear nerve. The internal cerebral vein is located close to the superior and medial edge of the dorsal thalamus in the cistern of the velum interpositum, and the vein of Galen receives several important veins, including the basal vein, and is located in the superior (quadrigeminal) cistern.

14 **The answer is D: Sigmoid sinus.** The sigmoid sinus is located on the internal aspect of the skull immediately posterior to the petrous portion of the temporal bone and close to the mastoid air cells and, therefore, close to the source of this infection in the mastoid process. The sigmoid sinus descends from the anterior end of the transverse sinus, along the caudal aspect of the petrous bone, to end at the jugular foramen where it is continuous with the external jugular vein. The deep middle cerebral vein is on the insular cortex, the basal vein (of Rosenthal) is located at the medial aspect of the temporal lobe, and the cavernous sinuses are laterally adjacent to the sella turcica at the base of the brain. The straight sinus is found where the falx cerebri attaches to the tentorium cerebelli on the midline. None of these sinuses (sigmoid excepted) represents a conduit into the CNS from the mastoid infection.

15 **The answer is C: Lateral striate artery/arteries.** The lateral striate arteries are branches of M_1, are commonly called the lenticulostriate branches of M_1, and serve the large central area of the hemisphere, including the lenticular nucleus, the posterior limb, and parts of the genu and anterior limb, all parts of the internal capsule. While this lesion extends beyond the strict territory of the lateral striate (lenticulostriate) branches, it is common for these large vascular lesions to recruit some of the surrounding tissue into the lesion. Recall the penumbra in strokes. Note that the medial portions of the thalamus on the right side appear largely intact. The cerebral cortex lateral to the lesion is unaffected; this clearly indicates that the cortical branches are still perfused, eliminating choice A. Also, the superior trunk of M_2 serves the insular cortex, gives rise to M_3 and M_4 branches, and does not serve significant internal areas of the hemisphere. The medial striate artery serves the head of the caudate nucleus and parts of the anterior limb of the internal capsule; the head of the caudate on the right is unaffected. The thalamogeniculate artery arises from P_2 and serves about the caudal half of the thalamus, a territory not consistent with the size, extent, and location of this lesion.

16 **The answer is C: M_1.** The various segments of the Anterior, Middle, and Posterior cerebral arteries are designated by the letters A, M, and P; the subscript numbers designate particular segments (1 = proximal, higher numbers more distal). These letters and numbers are commonly used in clinical settings. The lenticulostriate arteries (also called lateral striate arteries) arise from M_1. The A_1 segment is located between the internal carotid and the anterior communicating artery; its branches serve mainly the anterior hypothalamus. A_2 (infracallosal segment) is located between the anterior communicator and the genu of the corpus callosum and serves medial parts of the frontal lobe. The M_2 (insular part of MCA) is located on, and serves, the insular cortex. The P_1 segment is found between the bifurcation of the basilar artery and the intersection of the PCA with the posterior communicating artery; its branches serve the caudal hypothalamus and the medial areas of the rostral midbrain; it is the origin of the thalamoperforating artery.

17 **The answer is B: Anterior communicating artery.** The anterior communicating artery (ACom), or the junction of ACom with the anterior cerebral artery, is the most common site of aneurysm on the internal carotid system. These lesions may arise from the ACom, from the larger vessel in the case of duplicate ACom or fenestrated ACom, or from the junction between A_2 and the ACom. When these lesions rupture, they may invade the frontal lobe, the anterior horn of the lateral ventricle, and eventually the third ventricle. Aneurysms may occur on the other choices, or at branch points of small vessels arising from these larger vessels, but their occurrence is much less than that seen in relation to the ACom.

18 **The answer is B: Left A_2.** This coronal image reveals a number of vascular structures that are emphasized by the white cerebrospinal fluid. The plane of the image is through rostral portions of the cerebral hemispheres (note the head of the caudate, its continuation with the putamen, and the anterior limb of the internal capsule), rostral to the level of the optic chiasm but caudal to the level of the genu of the corpus callosum. Along the midline, between the hemispheres, the two black structures are the A_2 segments (infracallosal) of the anterior cerebral artery (ACA). The A_1 segment (precommunicating) is the portion of the ACA between its origin at the branching of the internal carotid artery (ICA) into the ACA and the middle cerebral artery and the anterior communicating artery (ACom). The A_2 segment extends from the ACom rostrally to just inferior to the genu where it continues by arching around the genu as A_3 (precallosal segment). M_1 is the larger (middle cerebral artery) of the terminal branches of the ICA that passes laterally through the Sylvian cistern. P_1 is the part of the posterior cerebral artery between the basilar bifurcation and its junction with the posterior communicating artery.

19 **The answer is E: Ophthalmic artery.** The ophthalmic artery is usually the first branch of the cerebral part (also called the supraclinoid part) of the internal carotid artery. The ophthalmic artery courses inferior to the optic nerve, through the optic foramen, and gives rise to the central retinal artery once in the orbit. Aneurysms in the immediate area of the origin of the ophthalmic artery that impinge on this vessel may result in visual deficits that may be transient or long term. The A_1 segment gives rise to small penetrating vessels that serve the anterior hypothalamus and sometimes to the medial striate artery. The anterior choroidal artery is usually the last branch of the internal carotid; it passes along the course of the optic tract and serves this structure and others in the immediate vicinity. The major branches of the M_1 segment are the lenticulostriate arteries.

20 **The answer is E: P_4: A segment of the posterior cerebral artery.** The right homonymous hemianopsia reveals three important bits of information. First, the lesion is in the territory of the posterior cerebral artery, since this is the vessel that serves the visual cortex. Second, it is in the distal portion of this vessel (choice E; this eliminates choices C and D, since these latter segments have branches that serve diencephalic targets and, if occluded, would also infarct these targets). There is no mention, in the stem, of deficits even intimating that proximal parts of the posterior cerebral artery may be involved. Third, being a deficit of a larger area of the visual field, choice B is eliminated, since much more than the macular vision (central vision) is lost. Distal branches of the anterior cerebral territory do not serve the visual cortex. Point to remember: lesions of optic pathways that are caudal to the optic chiasm result in deficits in the opposite hemifield or quadrant.

21 **The answer is C: $A_{3,4}$ branches—superior sagittal sinus.** This AVM is clearly in the territory of the anterior cerebral artery (ACA) and is located immediately adjacent to the superior sagittal sinus. At this coronal plane, this malformation is in the A_3 (precallosal segment) and rostral A_4 (supracallosal segment) territories of the ACA. Therefore, the primary contributing artery is likely via the branches of the A_3 and A_4 segments, and the draining venous structures are most probably superficial cerebral veins and the superior sagittal sinus. The A_1 and A_2 territories are devoid of any part of the AVM, and the cavernous sinus and deep middle cerebral vein are not located in the area of this lesion. In similar manner, the lateral aspect of the hemisphere contains no part of the malformation, and the superficial middle cerebral vein is on the most lateral aspect of the cerebral hemisphere and not close to this malformation.

22 **The answer is E: Pericallosal artery.** The pericallosal artery generally originates in the area immediately rostral to the genu of the corpus callosum where the main trunk of the anterior cerebral artery divides into the pericallosal and callosomarginal arteries. The former is located in the callosal sulcus (located between the corpus callosum and the cingulate sulcus) and the latter is found in the area of the cingulate sulcus (located between the cingulate gyrus and the medial aspect of the superior frontal gyrus). The angular artery is an M_4 branch that serves the general area of the angular gyrus (a part of the Wernicke area), and orbitofrontal arteries are also M_4 branches that serve the lateral aspect of the frontal lobe. Frontopolar branches originate from the anterior cerebral artery and serve the medial aspects of the rostral portions of the frontal lobe.

23 **The answer is B: Inferior petrosal sinus.** The inferior petrosal sinus lies in a shallow groove that passes from the cavernous sinus caudally to enter the anteromedial portion of the jugular foramen; the petrooccipital suture forms the base of this groove. The inferior petrosal sinus joins the internal jugular vein at the jugular foramen. This foramen also contains three cranial nerves and small arteries. The confluence of sinuses is located at the junction of the superior sagittal, straight, and both transverse sinuses on the inner aspect of the occipital bone. The straight sinus is located at the attachment of the falx cerebri and tentorium cerebelli and is rostrally continuous with the great cerebral vein and caudally with the confluence (or the left transverse sinus). The superior petrosal sinus passes along the upper edge of the petrous portion of the temporal bone from the cavernous sinus to the point where the transverse sinus turns to become the sigmoid sinus. The transverse sinuses are located on the inner aspect of the occipital bone on either side between the confluence of sinuses and the sigmoid sinuses.

24 **The answer is D: Transition from M_2 to M_3.** The M_1 segment of the middle cerebral artery (MCA) begins at the bifurcation of the internal carotid artery into the M_1 segment of the MCA and the A_1 segment of the anterior cerebral artery. The M_1 segment proceeds laterally through the Sylvian cistern for about 1.5 to 2.0 cm (about 5/8 to 3/4 in) before branching at the entrance to the insula (limen insulae) into superior and inferior trunks. These trunks branch over the insular cortex as the M_2 segments; their small penetrating branches serve this cortical area. As the branches constituting the M_2 segment approach the edges of the insula, and the position of the circular sulcus of the insula, they form sharp loops (shown in this MRI) that leave the insula and join the inner surface of the opercula (parietal, frontal, temporal); this loop represents the M_2 to M_3 transition. These vessels on the inner aspect of the opercula are the M_3 segment. The other transitions are not found close to this location and are not represented in this MRI.

25 **The answer is D: Thalamostriate vein and internal cerebral vein.** The venous angle is the sharp curve, variously described as U- or V-shaped, formed by the junction of the thalamostriate vein with the internal cerebral vein, which takes place at the position of the interventricular foramen (of Monro). This is also the immediate location of the column of the fornix and the anterior tubercle (and anterior nucleus) of the thalamus. The thalamostriate vein (also called the superior thalamostriate vein at this particular location) is located in the groove between the caudate and the dorsal thalamus; this is a continuation of the vein located adjacent to the tail of the caudate in the temporal lobe (also called the inferior thalamostriate vein at this location). The other junctions of vascular structures may have important implications in evaluation and diagnosis, but these junctions do not have specific names.

26 **The answer is D: Superficial middle cerebral vein.** The superficial middle cerebral vein is a large venous structure located on the surface of the cerebral hemisphere, usually at the position of the lateral sulcus. In the caudal area of the lateral sulcus, it connects with the superior anastomotic vein of Trolard (which also drains into the superior sagittal sinus) and with the inferior anastomotic vein of Labbé (which also drains into the transverse sinus). From this location, the superficial middle cerebral vein arches around the temporal lobe at the position of the lateral sulcus to enter the lateral aspect of the cavernous sinus. The superior cerebral veins are small, are located on the cerebral cortex, and drain into the superior sagittal sinus. The anterior cerebral vein is on the orbital surface of the frontal lobe and joins the basal vein of Rosenthal that travels along the most medial aspect of the temporal lobe to end in the vein of Galen. The deep middle cerebral vein is found on the insular cortex and also joins the basal vein.

27 **The answer is E: Watershed infarct.** Systemic hypotension, a sudden decrease in blood pressure throughout the vascular bed, will result in decreased flow, particularly in the distal parts of the vascular bed. A blood pressure (BP) below 85/45, or a mean arterial pressure (MAP) of 60 or below, represents a dangerously low BP and is a condition under which a watershed infarct may occur. The overlap of the distal territories of major vessels, especially the cerebral and cerebellar arteries, is especially vulnerable with a sudden decrease in the flow of oxygenated blood. When there is a sudden drop in systemic BP, blood retreats from the distal regions first, and these overlapping territories (watershed zones) are vulnerable to a watershed infarct. Systemic hypotension may be seen in cases such as this example or in cases of significant blood loss. AVMs are usually developmental defects that the patient is born with. A decrease in BP would not precipitate aneurysm rupture (no evidence that she has one) or subarachnoid hemorrhage (the first and second causes of which are trauma and aneurysm rupture). A precipitating event of some type would precede subdural hemorrhage, especially in a young woman.

28 **The answer is D: Preoptic, supraoptic, paraventricular, suprachiasmatic nuclei.** The arrow is identifying the left A_1 segment of the anterior cerebral artery. The left M_1 segment of the middle cerebral artery, the internal carotid arteries bilaterally, the anterior communicating artery, and the right A_1 are also clearly seen in this image. These anteriorly located nuclei of the hypothalamus are served by penetrating branches arising from the A_1 segment. Small branches of the anterior communicating artery serve the anterior commissure, septum, and may also send branches to the suprachiasmatic nucleus. The dorsomedial, ventromedial, and arcuate nuclei are in the tuberal region of the medial hypothalamic zone and receive their blood supply from small penetrators from the internal carotid and rostral portions of the posterior communicating artery (PCom). The posterior hypothalamic nucleus and mammillary nuclei receive arterial blood from the caudal portions of PCom and, to a small degree, from the P_1 segment of the posterior communicating artery. Optic tract and the immediately adjacent internal capsule are served by the anterior choroidal artery, while the lenticular nuclei (globus pallidus and putamen) receive their blood supply from the lenticulostriate branches (also called lateral striate arteries) of M_1.

29 **The answer is A: Anterior cerebral vein.** The arterial supply to this malformation is from the orbital branches of the A_2 segment of the anterior cerebral artery and possibly from orbitofrontal branches from the M_4 segment of the middle cerebral artery. The anterior cerebral vein drains medial aspects of the orbital frontal cortex and is joined, at about the position of the anterior perforated substance, by the deep middle cerebral vein to form the basal vein (of Rosenthal). The basal vein, in turn, courses around the medial aspect of the temporal lobe to end in the great cerebral vein (of Galen). The inferior cerebral veins are small cortical veins that drain lower cortical gyri on the convexity into the transverse sinus or larger cortical veins (middle cerebral, Trolard, Labbé), while the superior cerebral veins are equally small veins that drain the upper cortical gyri on the convexity into the superior sagittal sinus. The superficial middle cerebral vein is one of these large cortical veins; this vein drains into the cavernous sinus. The internal cerebral vein arises at the venous angle as a continuation of the superior thalamostriate vein and ends in the great cerebral vein; it drains large central regions of the hemisphere.

30 **The answer is A: Anterior choroidal artery.** The anterior choroidal artery arises from the cerebral part (also called the supraclinoid part) of the internal carotid artery and courses caudolateral along the trajectory of the optic tract. En route, it gives off branches to the optic tract and branches that serve the inferior portion of the internal capsule just as it is condensing to enter the crus cerebri; this vessel continues to the medial temporal lobe. The damage to the optic tract is self-explanatory. Those branches to the inferior portion of the internal capsule serve corticospinal fibers and thalamocortical projections that arise in the ventral posteromedial and ventral posterolateral nuclei, enter the internal capsule, and course to the overlying somatosensory cortex. The lateral striate arteries are branches of M_1 that serve the lenticular nucleus and immediately adjacent parts of the posterior limb of the internal capsule, while the medial striate artery is a branch of the anterior cerebral artery (distal A_1/proximal A_2) that serves the head of the caudate and medial parts of the anterior limb of the internal capsule. The lateral posterior choroidal artery arises from P_2 and provides blood to the choroid plexus of the lateral ventricle; the thalamoperforating artery arises from P_1 to serve rostral regions of the thalamus.

Chapter 6

The Spinal Cord

Recall the Cautionary Tale: The images in this chapter are presented in a Clinical Orientation; this approach emphasizes how structure and function correlate with deficits in a clinical setting. MRI and CT have a universally recognized orientation and laterality. Line drawings and stained sections in the Clinical Orientation are viewed here in the identical manner that one views an MRI: your right, as the physician-observer, is the patient's left, and the observer's left is the patient's right. The laterality of deficits and reference to connections follow accordingly.

QUESTIONS

Select the single best answer.

1 A 30-year-old man falls from the second story of his barn and strikes his neck on the edge of a wooden gate. Subsequent examination will reveal that the man fractured the atlas and axis, with severe damage to the spinal cord at C1 and C2 in the area shown in the image below. Which of the following would most likely be the primary and immediate medical concern for this patient?

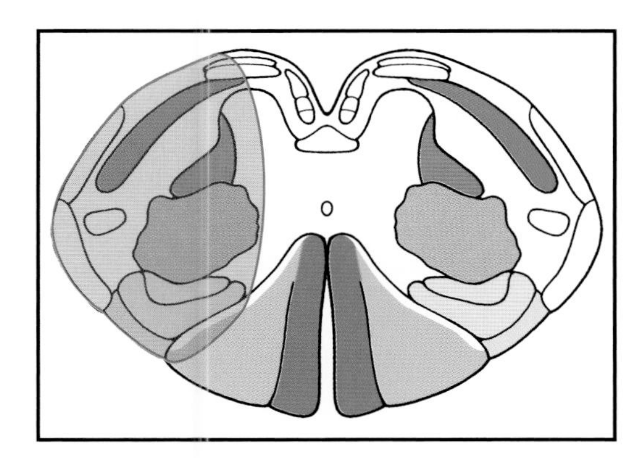

(A) Alternating hemianesthesia
(B) Alternating hemiplegia
(C) Difficulty breathing
(D) Difficulty swallowing
(E) Dysdiadochokinesia

2 During a routine neurological examination on a 35-year-old woman, the physician gently taps on the patellar tendon to activate the knee jerk, or quadriceps, reflex. The resulting reflex is within a normal range. Which of the following receptors is activated in response to this stimulus?

(A) Golgi tendon organ
(B) Merkel cell complex
(C) Muscle spindle
(D) Pacinian corpuscle
(E) Ruffini complex

3 Based on his signs and symptoms, a 41-year-old man is suspected of potentially having an autoimmune demyelinating disease. A lumbar puncture is ordered to retrieve a sample of CSF for analysis. When you perform this procedure, which of the following intervertebral spaces would be most appropriate to use?

(A) L1-2
(B) L2-3
(C) L4-5
(D) L5-S1
(E) T12-L1

4 A 23-year-old man is brought to the Emergency Department from a construction site. The history of the current event is that the man fell from a two-story building onto his back. The examination reveals that the man has abrasions and bruises on his back, on the back of his head, and on his right elbow. CT shows that the laminae on two adjacent vertebrae are shattered with bone spicules damaging the spinal cord in the shaded area in the image below. Which of the following would most likely be compromised in this patient?

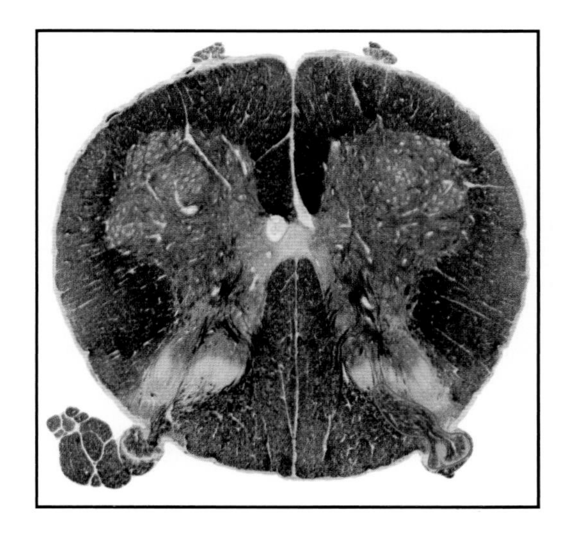

(A) Ankle jerk reflex
(B) Biceps reflex
(C) Crossed extension reflex
(D) Nociceptive reflex
(E) Patellar reflex

5 A 27-year-old construction worker falls about 20 ft at a work site and is transported to the Emergency Department. The examination reveals profound weakness of his right upper and lower extremities and a loss of pain and thermal sensation on his left upper and lower extremities. CT confirms a spinal fracture in cervical levels with bone fragments impinging on the spinal cord. Damage to which of the shaded areas in image below would account for the deficits seen in this patient?

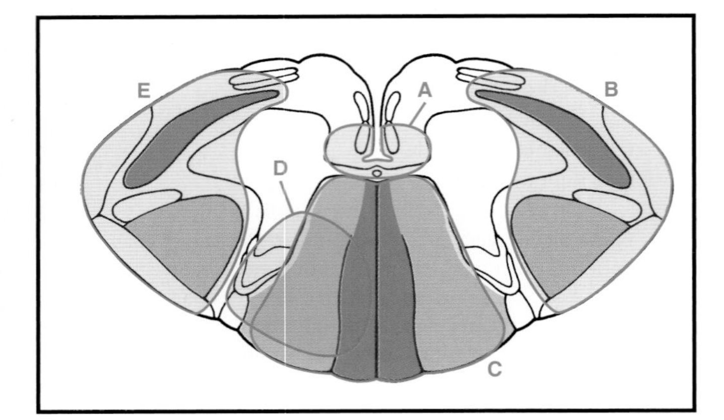

(A) A
(B) B
(C) C
(D) D
(E) E

6 Which of the following would best designate/describe a reflex?
(A) A chronic discomfort experienced by the patient
(B) A clearly specified localizing sign
(C) A poorly localized sensory input
(D) A voluntary response to a specific sensory input
(E) An involuntary response to specific sensory input

7 Which of the following laminae of the spinal cord gray matter (the Rexed laminae) contains the cell bodies of the intermediolateral cell column?
(A) IV
(B) V
(C) VI
(D) VII
(E) VIII

8 The light structure at the tip of the arrows in the image below is essential to the transmission of pain and thermal sensations. Which of the following specifies this structure?

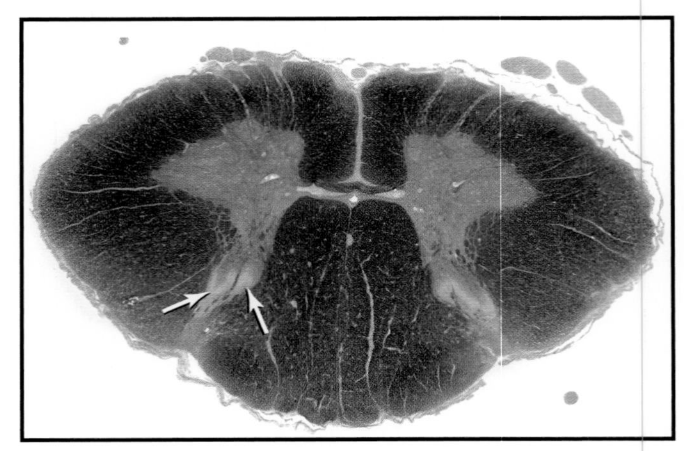

(A) Nucleus ambiguus
(B) Nucleus proprius
(C) Intermediolateral nucleus
(D) Posteromarginal nucleus
(E) Substantia gelatinosa

9 A 41-year-old man comes to your clinic and describes the following symptoms. About 2 days ago, while at his construction job, he experienced sudden lower back pain. He has noticed a gradual loss of sensation over his buttocks and inner thighs, difficulty urinating, and what he describes as "just getting weaker." Based on these symptoms, you suspect this man is suffering which of the following?
(A) Cauda equina syndrome
(B) Central cervical disc herniation
(C) Lateral lumbar disc herniation on the left
(D) Lateral lumbar disc herniation on the right
(E) Syringomyelia at lumbar levels

10 A 47-year-old man is recovering from a serious back injury that involved spinal cord damage. The motor examination at this point in his recovery reveals a full range of movement (he can flex and extend), but only if the influence of gravity is removed (his physician holds his arm while he attempts movements). When recorded on the chart, the neurologist will record this observation as which of the following grades?
(A) 0
(B) 1
(C) 2
(D) 3
(E) 4
(F) 5

11 A 25-year-old man is transported to the Emergency Department from the site of a motorcycle collision. The examination reveals no signs indicative of damage to either the forebrain or brainstem. He does have motor and sensory signs indicative of damage to the cervical spinal cord. In addition, he has a Horner syndrome. Damage to fibers in which of the areas labeled in the given image would most likely result in this visceral deficit?

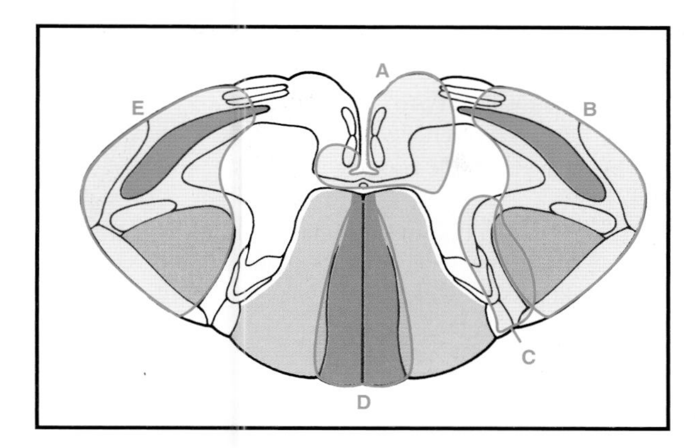

(A) A
(B) B
(C) C
(D) D
(E) E

12 Which of the following ascending fiber populations arises from the structure indicated by the arrow in the image below?

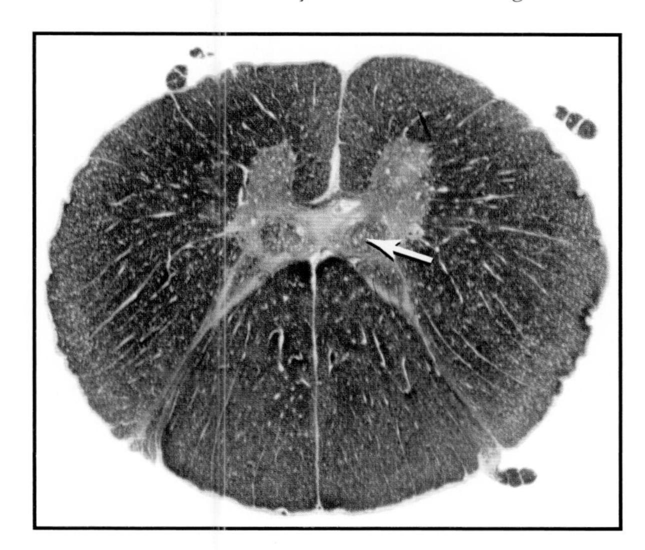

(A) Anterior spinocerebellar
(B) Anterolateral system
(C) Posterior spinocerebellar
(D) Spinohypothalamic
(E) Spinoreticular

13 The examination and evaluation of a 39-year-old man clearly suggest that a lumbar puncture is necessary to retrieve a sample of CSF for analysis. Which of the following would be a contraindication for this procedure?
(A) Suspicion of elevated protein in CSF (neurosyphilis)
(B) Suspicion of immunoglobulins in CSF (in multiple sclerosis)
(C) Suspicion of increased white blood cells in CSF (bacterial meningitis)
(D) Suspicion of mass lesion and metastatic cells in CSF (glioblastoma)
(E) Suspicion of subarachnoid blood in CSF (ruptured aneurysm)

14 A 34-year-old construction worker presents with the complaint that he has persistent pain in his lower back. The history reveals that this discomfort started suddenly about a week ago during heavy lifting. The examination reveals a healthy male who complains that he has a painful "burning sensation shooting down my leg" during certain movements. This man is most likely suffering which of the following?
(A) Brain tumor
(B) Mononeuropathy
(C) Nerve root avulsion
(D) Polyneuropathy
(E) Radiculopathy

15 A 17-year-old boy is brought to the Emergency Department by the police. The history of the immediate event reveals that this boy was in an altercation and stabbed in the back. The results of a CT suggest that the knife passed between adjacent thoracic vertebrae (T6 and T7), damaged the cord, and impaled the vertebral body. The results of the neurological examination suggest that the area of the spinal cord shaded in the image below was transected. Based on the extent of his injury, which of the following deficits would this boy most likely experience?

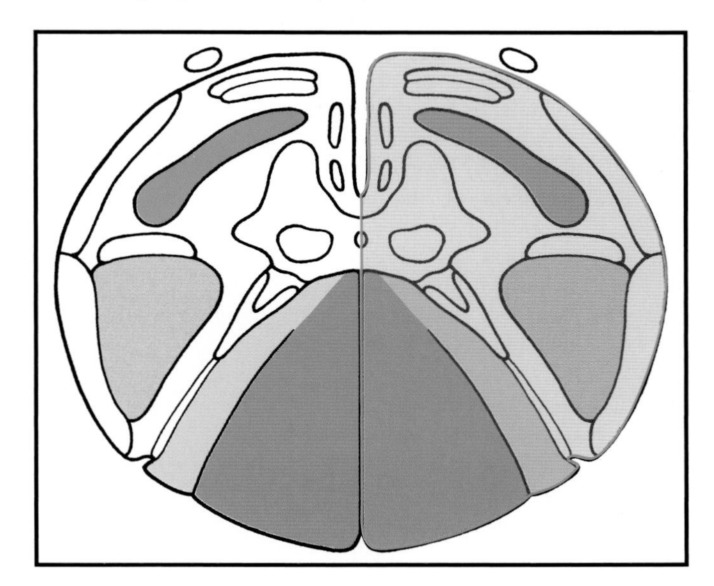

(A) Horner syndrome on the right
(B) Left-sided pain and thermal loss
(C) Left-sided hemiplegia
(D) Right-sided discriminative touch loss
(E) Right-sided hemiplegia

16 Which of the following neurotransmitters is common to both lower motor (α motor) neurons and to preganglionic parasympathetic neuron cell bodies located in the intermediolateral cell column?
(A) Acetylcholine
(B) Aspartate
(C) Dopamine
(D) γ-Aminobutyric acid
(E) Glutamate

17 A 37-year-old woman visits her family physician with the complaint of clumsiness and difficulty ambulating. The history reveals that this woman's symptoms have been getting progressively worse over several months. MRI reveals an extraaxial lesion (black in the image below) impinging on the spinal cord. Radiological evidence, as well as the symptoms experienced by this woman, indicates that this lesion initially impinged on the light red area, then progressed to also involve the dark red area in the image below. Which of the following most specifically describes the progression of this woman's symptoms?

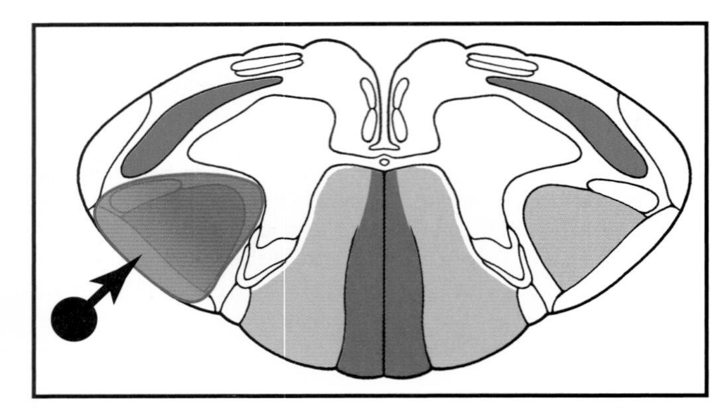

(A) Loss of pain and thermal sense, left upper extremity followed by lower extremity
(B) Loss of pain and thermal sense, right upper extremity followed by lower extremity
(C) Weakness of the extremities, left lower extremity followed by the upper extremity
(D) Weakness of the extremities, right lower extremity followed by the upper extremity
(E) Weakness of the extremities, right upper extremity followed by the lower extremity

18 The conus medullaris is anchored to the inner aspect of the spinal dural sac by which of the following?
(A) Arachnoid trabeculae
(B) Coccygeal ligament
(C) Denticulate ligament
(D) Filum terminale externum
(E) Filum terminale internum

19 A 46-year-old construction worker is brought to the Emergency Department following a fall at his workplace. The examination reveals a compound fracture of the humerus, and cuts and lacerations across his back and at the base of his neck. Muscle stretch reflexes are within normal ranges. The man has a bilateral loss of pain and thermal sensations over his upper extremities; this is frequently called a "cape distribution" sensory loss. Which of the following is most likely damaged to result in this deficit?
(A) Anterior white commissure
(B) Anterolateral system on the left
(C) Anterolateral system on the right
(D) Bilateral posterior columns
(E) Bilateral gracile fascicule

20 The neurons within the area identified by the white arrow in the image below innervate muscles fibers of the sternocleidomastoid muscle. Which of the following specifically identifies this cell group?

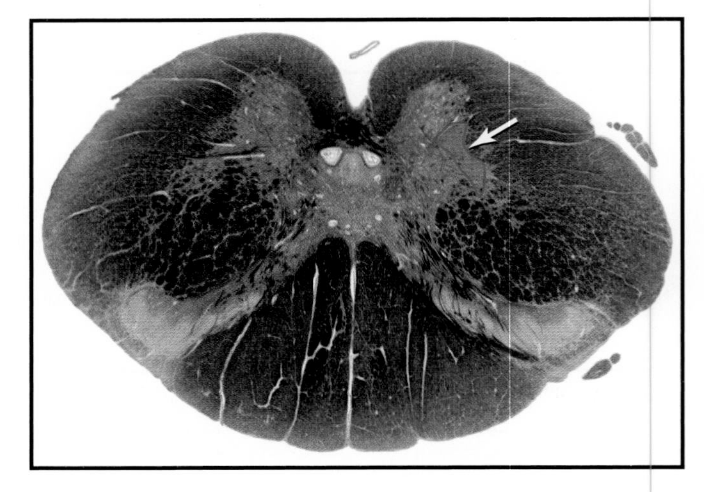

(A) Accessory nucleus
(B) Lateral spinal motor neurons
(C) Medial spinal motor neurons
(D) Nucleus ambiguus
(E) Phrenic nucleus

21 Although it is infrequently seen in some nations, polio is a disease that directly affects lower motor neurons or affects their ability to act. This may affect muscles of the body (spinal form) or of cranial nerves (bulbar form). Recognizing the effect of this disease on the body, cells in which of the following spinal laminae would be most affected?
(A) I
(B) II
(C) V
(D) VIII
(E) IX

22 Sympathetic innervation of the heart and lungs is essential to maintaining a neurophysiological balance within the cardiovascular system. Which of the following nuclei contains cells that participate in this pathway?
(A) Accessory nucleus
(B) Dorsal nucleus of Clarke
(C) Intermediolateral nucleus
(D) Nucleus proprius
(E) Posteromarginal nucleus

23 The fibers found within the area outlined by red in the image below have their cell bodies of origin in which of the following locations?

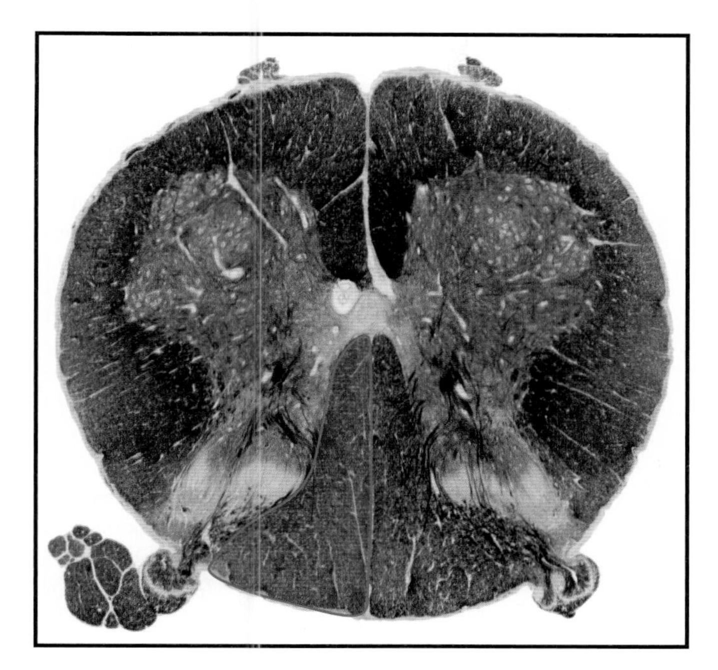

(A) Left posterior root ganglia above T6
(B) Left posterior root ganglia below T6
(C) Right posterior horn below T6
(D) Right posterior root ganglia above T6
(E) Right posterior root ganglia below T6

24 A 19-year-old factory worker is walking through the metal shop and steps on a sharp object that penetrates his right shoe and pierces his foot. He reflexively withdraws his right foot but continues to maintain his posture and balance. Which of the following contains the circuits that account for this response in this man?
(A) Crossed extension reflex
(B) Inverse myotatic reflex
(C) Muscle stretch reflex
(D) Nociceptive reflex
(E) Withdrawal reflex

25 A 32-year-old woman presents to her family physician with the complaint of general weakness. The history reveals that her weakness is worse toward the end of the day and when she tries to exercise. The examination shows that she is generally weak, her left eyelid droops slightly compared to the right, and her smile is not symmetrical. Her physician concludes that she may have a lower motor neuron disease. The function of which of the following neurotransmitters is most likely compromised in this woman?
(A) Acetylcholine
(B) Aspartate
(C) Dopamine
(D) Glutamate
(E) Glycine

26 The dark bundles identified by the black arrows in the image below constitute which of the following functional categories of afferent fibers?

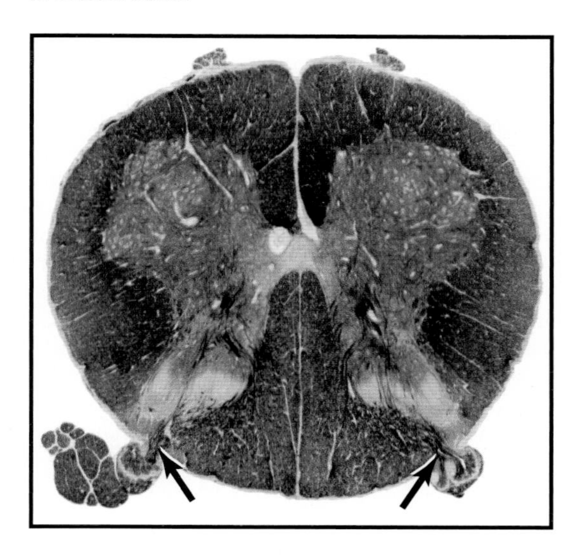

(A) Heavily myelinated, rapidly conducting, conveying pain and thermal sense
(B) Heavily myelinated, slowly conducting, conveying pain and thermal sense
(C) Heavily myelinated, rapidly conducting, conveying discriminative touch
(D) Lightly myelinated, slowly conducting, conveying pain and thermal sense
(E) Lightly myelinated, rapidly conducting, conveying discriminative touch

27 A 23-year-old man sustains a spinal cord injury and initially experiences difficulty breathing. Damage to which of the following cell groups, or nuclei, would most directly contribute to this deficit in this man?
(A) Accessory nucleus
(B) Dorsal nucleus of Clarke
(C) Lateral anterior horn neurons
(D) Medial anterior horn neurons
(E) Phrenic nucleus

28 Fibers of the lateral division of the posterior root form which of the following distinct fiber tracts/bundles in the spinal cord before they synapse in the posterior horn?
(A) Anterior white commissure
(B) Fasciculus proprius
(C) Posterolateral tract
(D) Postsynaptic posterior column fibers
(E) Spinospinal fibers

ANSWERS

1 **The answer is C: Difficulty breathing.** Severe damage to the spinal cord at C1 and C2 will result in sudden and significant breathing difficulty. The anatomical basis, for what can be a medical emergency, is the interruption of fibers arising in the lateral medulla (above the lesion) that descend to terminate in the phrenic nucleus (located below the lesion at about C3-7). These medullary reticulospinal fibers supply rhythmic impulses to the phrenic nucleus which, in turn, innervates that half of the diaphragm causing it to contract. Remove those fibers, and that half of the diaphragm does not function. This man would have an alternating hemianesthesia (this is not a medical emergency), but would not have an alternating hemiplegia; he would have a right-sided Upper Extremity (UE) and Lower Extremity (LE) paralysis. Even though the posterior and anterior spinocerebellar tracts are involved, dysdiadochokinesia would not be seen because the extremities are paralyzed on that side. Also, the nucleus ambiguus is not involved, so it is highly unlikely that this man would have difficulty swallowing. Alternating hemianesthesia refers to a sensory loss on one side of the body and the opposite side of the face. Alternating hemiplegia refers to corticospinal tract injury accompanied by a cranial nerve deficit (usually motor). Both of these may also be called crossed deficits.

2 **The answer is C: Muscle spindle.** The afferent limb of the reflex originates from the spindle(s) in the muscle stretched by the tendon tap. The efferent limb originates from lower motor neurons that are monosynaptically activated and, in turn, innervate extrafusal muscles containing the activated spindle, while the activity of antagonistic muscles is decreased via inhibitory interneurons. Hyperreflexia is an excessive reaction, hyporeflexia is a slow/diminished reaction, and areflexia is an absent reflex. While all may be indicative of some type of neurological disorder, it should be kept in mind that some patients may have increased or decreased reflex activity and no pathology. The reflex activity must be viewed in the larger context of the complete neurological examination. The Golgi tendon organ is the receptor for the inverse myotatic reflex, which is a measure of muscle tension. Merkel cell complexes, Pacinian corpuscles, and Ruffini complexes are sensory receptors that measure tactile discrimination, vibratory sense, and proprioception, respectively.

3 **The answer is C: L4-5.** The point of choice for a lumbar puncture is classically at the L4-5 interspace followed by the L3-4 interspace. Level T12-L1 is clearly too high, since the spinal cord is present at this point. Introducing a needle at this interspace would damage the spinal cord. The end of the spinal cord (the conus medullaris) is generally located at L1-2. Introducing the needle at this level, or at L2-3, could be dangerously close to the cord or damage the cord. The dural sac (and therefore the lumbar cistern) ends at about S1-2; consequently, introduction of a needle at the L5-S1 level could possibly miss the cistern. However, some clinicians prefer the L5-S1 and use it with consistent success.

4 **The answer is E: Patellar reflex.** The characteristics of the spinal cord in this image (round shape, large anterior and posterior horns, proportionately smaller amount of white matter, and obvious clusters of large neurons in the anterior horn) clearly specify it as a lumbar level. Correctly identifying the spinal cord level and the specific part of the cord damaged are key to answering this question. The lesion involves the medial division fibers (heavily myelinated, rapidly conducting, proprioceptive) of the posterior root and clearly avoids the posterolateral tract that receives lateral division fibers (lightly and unmyelinated, slowly conducting, pain and thermal sense). These medial division fibers convey information related to discriminative touch, vibratory sense, and position sense/proprioception, and are the afferent fibers for the muscle stretch reflex. Being a lumbar level, the only choice is the patellar reflex. The ankle jerk reflex is mediated through sacral levels and the biceps reflex is mediated through cervical levels; this is the wrong level for these reflexes. Afferent fibers for the nociceptive reflex traverse the lateral division of the posterior root and posterolateral tract; both are not involved and this modality is not conveyed by medial division fibers. The crossed extension reflex is the contralateral part of the nociceptive reflex.

5 **The answer is E (Area of lateral corticospinal fibers and anterolateral system).** The characteristics of this level (oval shape, large amount of white matter, large anterior and posterior horns, large posterior columns) clearly specify a cervical level. The damage to the right corticospinal tract (gray) accounts for the weakness of the right Upper Extremity (UE) and Lower Extremity (LE), and damage to the right anterolateral system (green) explains the left-sided loss of pain and thermal sense. The primary deficit in lesion A would be a bilateral loss of pain and thermal sense (damage to anterior white commissure) reflecting the spinal levels involved. Damage at B would produce a left weakness UE + LE and right-sided pain and thermal loss on the same extremities, and damage at C would result in bilateral (UE + LE) loss of discriminative touch and vibratory sensation. The deficit in lesion D would be a weakness and a loss of much discriminative touch and vibratory sense, on the UE + LE, and both on the right side.

6 **The answer is E: An involuntary response to specific sensory input.** Reflexes are the result of activation of a specialized sensory ending (such as a muscle spindle, Pacinian corpuscle, or naked pain ending) and the resulting, and totally involuntary, motor responses. In many situations, the response may take place before the patient is even aware of the stimulus. None of the other choices represents an involuntary response, and by what they describe, they are excluded.

7 **The answer is D: Lamina VII.** Lamina VII contains, in thoracic levels, the intermediolateral cell column (laterally placed) and the nucleus of Clarke (medially placed). It contains sympathetic preganglionic cell bodies. Laminae IV-VI are parts of the posterior horn and contribute to ascending tracts conveying sensory information; lamina VIII is located in the anterior horn and contains interneurons and cell bodies that contribute to spinospinal fibers.

8 **The answer is E: Substantia gelatinosa.** The characteristics of this level (oval shape, large amount of white matter, large anterior and posterior horns, large posterior columns) clearly specify a cervical level. The substantia gelatinosa (spinal lamina II) receives input from fibers conveying pain and thermal information (these are C and A-delta fibers) and gives rise to both tract cells and spinal interneurons that convey this information to higher levels. The nucleus ambiguus is located in the medulla, not the spinal cord, and serves muscles of the pharynx and larynx. The nucleus proprius (spinal lamina III + IV) is largely populated by spinal interneurons; the intermediolateral nucleus (spinal lamina VII) contains sympathetic preganglionic neurons. The posteromarginal nucleus (spinal lamina I) is not much more than about a single layer of neurons forming a cap over the gelatinosa; in this respect, its location is well known (in relation to the gelatinosa), but it cannot be seen in an overview. The axons of many lamina I cells enter the contralateral anterolateral system.

9 **The answer is A: Cauda equina syndrome.** In this case, the sudden onset and low back pain both point to a disc problem. The subsequent bilateral loss of sensation over the buttocks and thighs, weakness, and difficulty urinating indicate bilateral root damage; this is characteristic of impingement on the cauda equina which is made up of roots distributing to both sides. A central cervical disc could result in upper extremity weakness and/or long tract signs. A lumbar disc herniation on either side will usually result in a radiculopathy with its characteristic unilateral signs and symptoms; syringomyelia is usually characterized by a bilateral loss of pain and thermal sense over restricted dermatomal levels that correlate with the cord levels of the syrinx.

10 **The answer is C: 2.** These standards for the grading of spinal cord injury were established by the American Spinal Injury Association and are internationally recognized. The grades are as follows: 0 = complete paralysis; 1 = contractions that can be palpated or visible, but without other movement; 2 = full range of movement with gravity removed; 3 = full range of movement against gravity; 4 = full range of movement against gravity and moderate resistance; and 5 = normal full range of movement against full resistance. This man's deficit would also be graded 2, or 2/5, using the Royal Medical Research Council scale. This grading system may also be used to indicate the motor abilities of patients with corticospinal damage at levels above the motor decussation.

11 **The answer is C (Area of white matter adjacent to the posterior horn).** The characteristics of this level (oval shape, large amount of white matter, large anterior and posterior horns, large posterior columns) clearly specify a cervical level. The hypothalamus sends important descending fibers to the intermediolateral cell column on the ipsilateral side; these hypothalamospinal fibers are sympathetic in function. These fibers are located in the lateral funiculus between the posterior horn and the lateral corticospinal tract and are only in the area of C. Damage to these descending fibers removes this input to the intermediolateral cell column resulting in an ipsilateral Horner syndrome (ptosis, miosis, anhidrosis). Enophthalmos may be mentioned as part of a Horner, but this is doubted by some neurologists, and facial vasodilation (a sense of warmth) may be apparent in some patients. The main symptoms from injury in area A would be bilateral sensory deficits (anterior white commissure) and weakness of axial muscles on the left (medial nuclei of anterior horn). A lesion at B would result in a right-sided loss of pain and thermal sense, and weakness on the left side, both below the lesion level; at D the deficit would be a bilateral loss of discriminative touch and vibratory sense on the lower extremities. The lesion at E is a loss of pain and thermal sense on the left side, and weakness on the right side, both below the level of the lesion: the mirror of B.

12 **The answer is C: Posterior spinocerebellar.** The characteristics of this level (round shape, relatively large amount of white matter, small anterior and posterior horns, lateral horn) clearly specify a thoracic level. The structure labeled is the dorsal nucleus of Clarke (also called the posterior thoracic nucleus), also a prime feature of a thoracic level. Cell bodies of this nucleus send their axons to form the ipsilateral posterior spinocerebellar tract; they ascend to enter the cerebellum via the restiform body and terminate in the cerebellar nuclei and cortex, as mossy fibers, on the same side. Anterior spinocerebellar fibers arise from cells in the anterior horn, particularly spinal border cells, course through the anterior spinocerebellar tract, and enter the cerebellum by coursing over the surface of the superior cerebellar peduncle. Fibers comprising the anterolateral system (ALS) originate from neurons located in the posterior horn of the spinal cord gray matter, course through the anterior white commissure, and form the ALS; spinoreticular and spinohypothalamic fibers are part of the ALS.

13 **The answer is D: Suspicion of mass and metastatic cells in CSF (glioblastoma).** When there is a suspicion of a mass lesion, and the naturally increased intracranial pressure that almost always accompanies a mass lesion, doing a lumbar puncture could well have a catastrophic outcome. The introduction of the needle into the dural sac and the sudden pressure shift downward may cause tonsillar herniation with the consequent pressure on the medulla and potential compression of medullary respiratory and cardiac centers. In all of the other choices, unless there is clear evidence of increased intracranial pressure, a lumbar could be done safely.

14 **The answer is E: Radiculopathy.** The characteristics of shooting pains that may be exacerbated by movement or position, and that generally follow a dermatomal distribution, are seen in intervertebral disc disease or extrusion; these pains typify root compression or irritation. A mononeuropathy is a lesion of a specific peripheral nerve with consequent sensory and motor deficits, the most common being carpal tunnel syndrome. Polyneuropathy is a disease process affecting both sensory and motor nerves. A common scenario is diabetic neuropathy in which the loss is in a distal-to-proximal stocking-glove pattern. Root avulsion is most commonly seen at cervical levels.

15 **The answer is C: Left-sided hemiplegia.** This patient has suffered a traumatic spinal cord hemisection on the left side. This results in the classic signs and symptoms that are seen in Brown-Sequard syndrome. In this case, with a lesion on the left side, these are (1) a left-sided hemiplegia: in this case, only the lower extremity is involved because the lesion is at midthoracic levels; (2) a left-sided loss of discriminative touch and vibratory sense below the lesion level; and (3) a right-sided loss of pain and thermal sense beginning about two segments below the lesion, in this case beginning at about T9. Hypothalamospinal fibers to the intermediolateral cell column end by about the T3-4 levels, so this patient will not have a Horner.

16 **The answer is A: Acetylcholine.** Acetylcholine is associated with the neuromuscular junction (a special type of synapse on skeletal muscles), with neurons of the intermediolateral cell column (sympathetic) and their synapse on postganglionic neurons, and with postganglionic terminations on sweat glands, and the smooth muscle of blood vessels located in skeletal muscles. Aspartate and glutamate are excitatory neurotransmitters located at many CNS locations; GABA is an inhibitory neurotransmitter also found at diverse locations. Dopamine has a restricted distribution, the substantia nigra being one well-known location.

17 **The answer is D: Weakness of the extremities, right lower extremity followed by the upper extremity.** The characteristics of this level (oval shape, large amount of white matter, large anterior and posterior horns, large posterior columns) clearly specify a cervical level. This lesion impinges on the lateral corticospinal tract on the right side, first involving the more lateral portions of this tract, then involving the lateral plus the medial portions. The somatotopy of corticospinal fibers in cervical levels is such that fibers passing to the lower cord levels (lumbosacral, for innervation of the lower extremity) are located laterally in the tract, and fibers passing to upper levels (cervical, for innervation of the upper extremity) are located medially. Corticospinal fibers to thoracic levels occupy a location between the lower extremity and upper extremity fibers. As this extraaxial lesion enlarges, it first impinges on the lateral part of the right lateral corticospinal tract, resulting in weakness of the right lower extremity; it then further impinges on the medial part of the tract to result in weakness of the right upper extremity. Both extremities on the right are eventually involved. This may be called an ascending motor deficit, since it starts with the lower extremity and progresses to involve the upper. Even though this lesion also involves the posterior spinocerebellar tract, motor deficits of the cerebellar type are not seen; one cannot observe movement disorders in a paralyzed extremity. Pain and thermal sense pathways are also not involved.

18 **The answer is E: Filum terminale internum.** The filum terminale internum is also called the pial part of the filum terminale, since it is composed primarily of pia mater. As this structure descends from the conus to insert at the caudal end of the dural sac, it passes through the center of the cauda equina. The filum terminale externum (also called the coccygeal ligament) attaches the dural sac to the coccyx and, therefore, can be called the dural part of the filum terminale. The denticulate ligaments are found on the lateral aspect of the spinal cord midway between the posterior and anterior roots; these sheets of pia pass from the cord laterally to attach to the inner surface of the dura. Arachnoid trabeculae are present only around the brain.

19 **The answer is A: Anterior white commissure.** Nerve cells in the posterior horn conveying pain and thermal (P/T) sense send their axons through the anterior white commissure to enter the anterolateral system (ALS) on the contralateral side. Consequently, this commissure contains fibers conveying these modalities from both sides of the body; damage to the commissure causes bilateral P/T loss. The fact that the loss is over the upper extremities indicates damage to this structure within the cord from about C4 to about T1. The ALS conveys P/T sense from the opposite side of the body; lesions of the ALS result in a loss on the opposite side, not bilaterally. The posterior columns convey discriminative touch, vibratory sense, and proprioception, not P/T; the gracile fasciculus carries comparable information from the lower extremity.

20 **The answer is A: Accessory nucleus.** The accessory nucleus is located at spinal levels C1 to about C5/C6 in the lateral portion of the anterior horn. Cells located in the rostral portion of this nucleus (C1-2) innervate the ipsilateral sternocleidomastoid muscle, while those in its more caudal regions (C3-C5/C6) innervate the ipsilateral trapezius. Damage to this nucleus, or the accessory nerve, results in a paralysis of these muscles, most frequently seen as an inability to perform a movement "against resistance." Lateral spinal motor neurons are found in the lateral portion of the anterior horn at cervical and lumbosacral enlargements and serve appendicular musculature. Medial spinal motor neurons are found in the medial portion of the anterior horn at all levels and innervate the epaxial (deep back) muscles. The nucleus ambiguus (medulla) serves muscles of the larynx and pharynx, and the phrenic nucleus (C3-C6/C7) innervates the diaphragm.

21 **The answer is E: Lamina IX.** Lower motor neurons, the large α motor neurons that innervate skeletal muscles, are found in lamina IX of the spinal cord and in the motor nuclei of cranial nerves III-VII and IX-XII that innervate skeletal muscle. Lamina I, II, and V are in the posterior horn and are predominately related to the transmission of somatosensory input. Lamina VIII consists mainly of spinal interneurons.

22 **The answer is C: Intermediolateral nucleus.** The intermediolateral nucleus (spinal lamina VII) is commonly referred to as the intermediolateral cell column, since it forms a continuous columnar nucleus from about T1 to about L1-2. These cells are sympathetic preganglionic cells that exit the anterior roots, enter the sympathetic chain ganglia via the white ramus, and terminate on postganglionic cell bodies whose axons distribute to the heart as cervical and thoracic cardiac nerves. The accessory nucleus is in the cervical levels of the spinal cord and innervates the trapezius and sternocleidomastoid muscles, while the dorsal nucleus of Clarke gives rise to spinocerebellar fibers. The posteromarginal nucleus (lamina I) contains mainly tract cells, and the nucleus proprius (laminae III, IV) contains both tract cells and spinal interneurons.

23 **The answer is E: Right posterior root ganglia below T6.** The characteristics of this level (round shape, small amount of white matter, proportionately very large anterior and posterior horns) clearly specify a lumbar level. The outlined area is the right fasciculus gracilis in this lumbar level. This spinal cord tract is made up of the central processes of posterior root ganglion cells that are located on the right (ipsilateral) side and below T6; they ascend uncrossed to terminate in the gracile nucleus on the same side. Both posterior column tracts are uncrossed within the spinal cord. Posterior root ganglia above T6 contribute their central processes to the cuneate fasciculus; these fibers ascend to end in the cuneate nucleus. The right posterior horn (above and below T6) projects mainly into the left anterolateral system via the anterior white commissure.

24 **The answer is A: Crossed extension reflex.** The circuits for the crossed extension reflex mediate flexion of the injured extremity (excitation of flexor muscles, inhibition of extensors) and extension of the opposite extremity (excitation of extensor muscles, inhibition of flexor muscles). This reflex is in play in the walking, standing, or running patient. The withdrawal reflex (also called the flexor reflex or the nociceptive reflex) is essentially the ipsilateral part of the crossed extension reflex; it does not include the opposite leg unless the person is ambulatory. The inverse myotatic reflex (also called autogenic inhibition) is involved in the inhibition of muscles that are put under high tension. The muscle stretch is a classic reflex; tapping on a tendon stretches the muscle spindles in the muscle attached to that tendon generating an action potential on the Ia fiber. This Ia fiber innervates α motor neurons that innervate the muscle containing the stretched spindle; the muscle contracts.

25 **The answer is A: Acetylcholine.** This woman's signs and symptoms (fluctuating weakness, affecting spinal and cranial motor neurons, worse with exertion or at day's end) are classic for myasthenia gravis. This disease is caused by circulating antibodies to nicotinic acetylcholine receptors at the neuromuscular junction; this is the terminal of the axon on the lower motor neuron innervating skeletal muscle. Consequently, the function of acetylcholine is significantly compromised. Aspartate and glutamate are excitatory neurotransmitters found at many CNS locations; dopamine is largely found in the nigrostriatal pathway; glycine is found in spinal interneurons. With the exception of dopamine, neurotransmitter disease is not associated with these substances.

26 **The answer is C: Heavily myelinated, rapidly conducting, conveying discriminative touch.** The characteristics of this level (round shape, small amount of white matter, proportionately very large anterior and posterior horns) clearly specify a lumbar level. The arrows are pointing to the medial division of the posterior root. These fibers are the central processes of heavily myelinated fibers that are rapidly conducting (these axons have long internodal distances that correlate with their rapid conduction velocities) and that convey discriminative touch and vibratory sensations. Sometimes these are collectively called proprioceptive fibers. In addition to their thick coat of myelin, the axons of these fibers are of large diameter. Lightly myelinated fibers are slowly conducting, have small diameter axons, have a comparatively thin myelin coat with short internodal distances, and conduct information related to the perception of pain and thermal sensations.

27 **The answer is E: Phrenic nucleus.** The phrenic nucleus is located in central portions of the anterior horn extending from about C3-7 and gives rise to axons that innervate the musculature of the diaphragm. The accessory nucleus is also located in the cervical region (C1 to about C5/6); this nucleus is the origin of axons that ascend through the foramen magnum and exit via the jugular foramen and innervate the trapezius and sternocleidomastoid muscles. The dorsal nucleus of Clarke (also called the posterior thoracic nucleus) is the origin of axons that form the posterior spinocerebellar tract. Lateral neurons in the anterior horn are found in the cervical and lumbosacral enlargements and innervate the muscles of the extremities. Medial neurons in the anterior horn are essentially found at all levels, and innervate axial and deep back muscles and, in thoracic levels, intercostal muscles. While it is true that the intercostal muscles serve a supportive role in respiration, the diaphragm is "the essential respiratory muscle."

28 **The answer is C: Posterolateral tract.** Lateral division fibers are thinly myelinated, slowly conducting fibers that convey pain and thermal sensations. The central processes of these fibers enter the spinal cord and immediately form the posterolateral tract (also called the dorsolateral tract, or tract of Lissauer). This tract caps the posterior horn, lying immediately external to spinal lamina I on the surface of the cord. These central processes ascend and descend varying distances within the posterolateral tract depending on the particular information they convey. The anterior white commissure conveys second-order neurons in the pain/temperature pathway that cross the midline. The fasciculus proprius is located in the posterior, lateral, and anterior funiculi at the interface of the gray and white matter; spinospinal fibers are located within the fasciculus proprius. The postsynaptic posterior column system consists of cells located in the posterior horn that enter the fasciculi gracilis and cuneatus and ascend to terminate in the posterior column nuclei.

Chapter 7

The Medulla Oblongata

Recall the Cautionary Tale: The images in this chapter are presented in a Clinical Orientation; this approach emphasizes how structure and function correlate with deficits in a clinical setting. MRI and CT have a universally recognized orientation and laterality. Line drawings, stained sections, and gross brain images in the Clinical Orientation are viewed here in the identical manner that one views an MRI: your right, as the physician-observer, is the patient's left, and the observer's left is the patient's right. The laterality of deficits and reference to connections follow accordingly.

QUESTIONS

Select the single best answer.

1 An 81-year-old man experiences a sudden sensory loss on his right upper extremity. The examination reveals a profound right-sided loss of discriminative touch and vibratory sense, but pain and thermal sensations on all extremities and on the face are within normal ranges. MRI reveals a small medullary infarct. Damage to which of the following structures, identified below, would most likely result in this man's deficit?

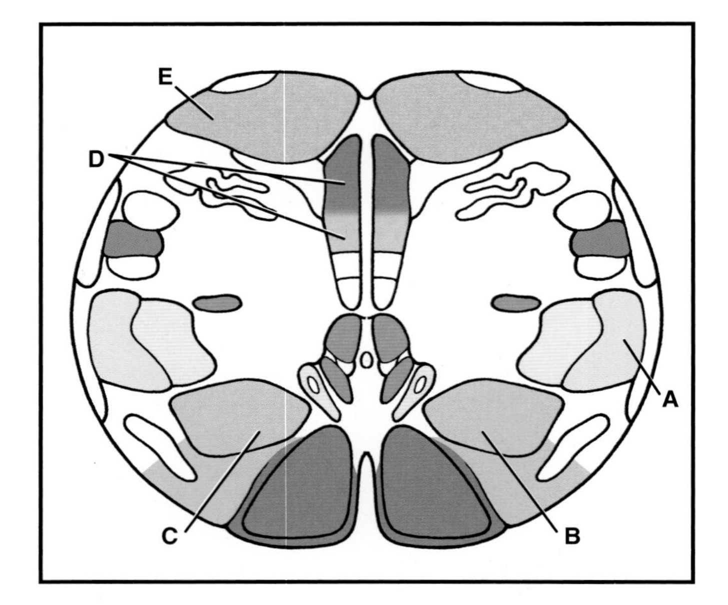

(A) A
(B) B
(C) C
(D) D
(E) E

2 The MRI of a 71-year-old woman reveals a defect in the territory served by the posterior inferior cerebellar artery. Which of the following combinations of structures is located within the territory of this vessel?

(A) Anterolateral system + medial lemniscus
(B) Hypoglossal nerve root + spinal trigeminal nucleus
(C) Pyramid + anterolateral system
(D) Spinal trigeminal tract + nucleus ambiguus
(E) Vestibular nuclei + medial longitudinal fasciculus

3 A 76-year-old man presents to the Emergency Department with the complaint of persistent headache. The examination reveals evidence of increased intracranial pressure (lethargy, nausea, headache). During the examination, the man loses consciousness, his blood pressure suddenly and dramatically increases and then precipitously drops, he has bradycardia, and then his heart rate ceases with respiration following a similar course, all within about 6 minutes. Which of the following regions of the brain contains centers that, if rapidly compromised, would give rise to this man's sudden symptoms and catastrophic decline?

(A) Diencephalon/thalamus
(B) Mesencephalon/midbrain
(C) Metencephalon/pons
(D) Myelencephalon/medulla
(E) Telencephalon/cerebrum

4 A 69-year-old man presents with the complaint of a sudden loss of sensory function. The examination reveals a significantly overweight man who suffers from sleep apnea. CT reveals a small vascular lesion within the territory of the posterior spinal artery in the medulla. Which of the following structures would be found within this vascular region?

(A) Anterior spinocerebellar tract
(B) Anterolateral system
(C) Gracile and cuneate nuclei
(D) Medial lemniscus
(E) Principal olivary nucleus

5 Which of the following structures is derived from the alar plate of the developing brainstem?
(A) Ambiguus nucleus
(B) Dorsal nucleus of the vagus
(C) Oculomotor nucleus
(D) Spinal trigeminal nucleus
(E) Trochlear nucleus

6 A 59-year-old woman presents with rapidly progressing weakness. The results of the neurologic examination clearly suggest that she has involvement of corticospinal fibers within the medulla, which is shown on the image below. Which of the following indicates the location of these specific fibers?

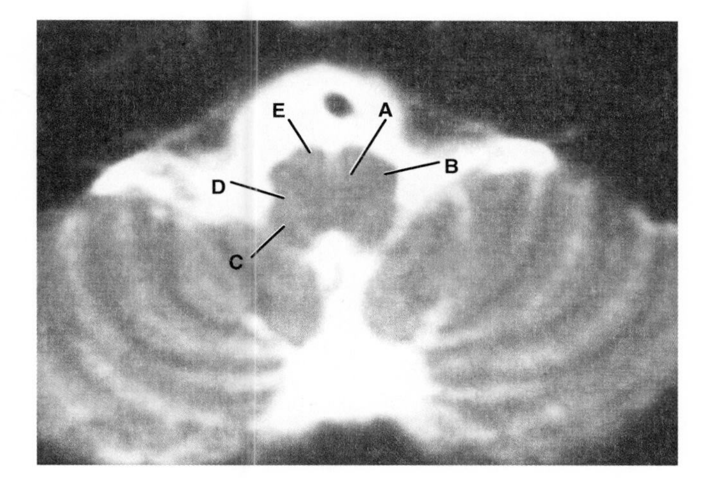

(A) A
(B) B
(C) C
(D) D
(E) E

7 A 71-year-old man presents to his physician's office with the complaint that his "...left leg feels funny...." The examination reveals a loss of vibratory sense and discriminative touch on his left lower extremity; pain and thermal sense are intact, and the man's leg is not weak. The upper extremities have normal sensation and strength. CT reveals a small vascular defect. Damage to which of the following would explain this deficit in this patient?
(A) Accessory nucleus
(B) Accessory cuneate nucleus
(C) Cuneate nucleus
(D) Gracile nucleus
(E) Spinal trigeminal nucleus

8 A 56-year-old woman is brought to the Emergency Department by her daughter. The woman complains that she suddenly had difficulty walking. The examination reveals that strength in all extremities is within a normal range but that she has a loss of proprioception and discriminative touch on both lower extremities. MRI reveals a small, well-localized medullary lesion, presumably of vascular origin. Damage in which of the following areas in the image below, each representing a lesion, would most likely explain this woman's deficits?

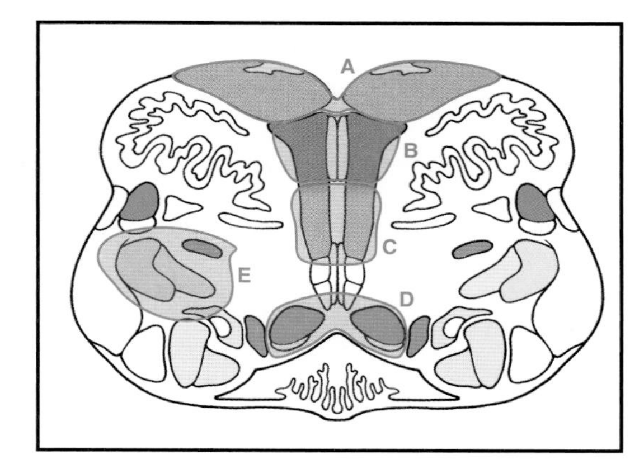

(A) A
(B) B
(C) C
(D) D
(E) E

9 Which of the following nuclei gives rise to axons that project to the cerebellar cortex on the contralateral side where they end as climbing fibers?
(A) Accessory cuneate nucleus
(B) Inferior olivary nucleus
(C) Lateral reticular nucleus
(D) Solitary nucleus
(E) Vestibular nucleus

10 A previously healthy 7-year-old boy suddenly develops episodic emesis. In between bouts of vomiting, he feels fine and engages in age-appropriate activities. He has no other signs or symptoms. MRI reveals a small tumor in the caudal portion of the fourth ventricle. Impingement on which of the following would be the most likely cause of this patient's symptoms?
(A) Area postrema
(B) Hypoglossal trigone
(C) Striae medullares
(D) Vagal trigone
(E) Vestibular area

11 Which of the following most specifically characterizes the black fibers identified by the white arrows in the figure below?

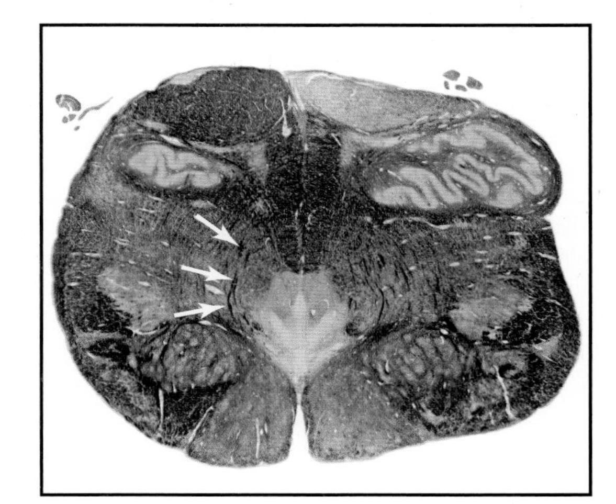

(A) Convey pain and thermal sense from the right side of the body
(B) Convey pain and thermal sense from the left side of the body
(C) Convey proprioceptive information from the right side of the face
(D) Convey proprioceptive information from the right side of the body
(E) Convey proprioceptive information from the left side of the body

12 A 67-year-old man presents with a loss of pain and thermal sensation on the left side of his face and on the right side of his body. In addition, he has difficulty swallowing, hoarseness of speech, and an unsteady gait. MRI reveals that these deficits resulted from a brainstem stroke. Bleeding in which of the following arterial territories would result in these deficits?
(A) Anterior inferior cerebellar
(B) Anterior spinal
(C) Paramedian branches
(D) Posterior inferior cerebellar
(E) Posterior spinal

13 The dark fiber bundle indicated by the arrow in the image below is composed of the central process of primary sensory fibers that convey a specific type, or types, of information. Which of the following most specifically identifies the information conveyed by this bundle?

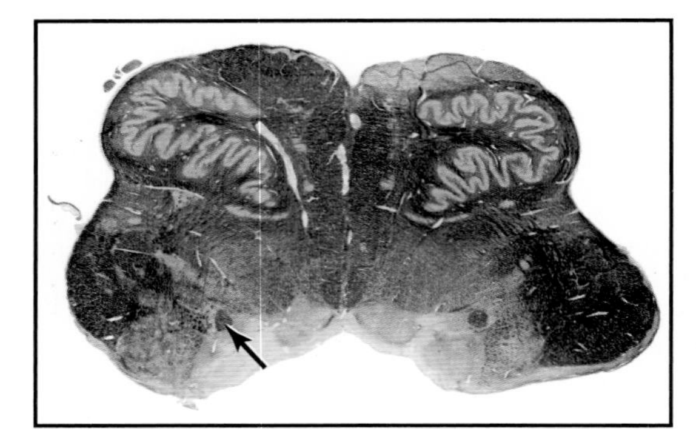

(A) General somatic sensations from the face and oral cavity
(B) General visceral sensations only from the face and head
(C) General and special visceral sensations from the head
(D) Special visceral sensations only from the face and head
(E) Special somatic sensations from the head

14 A 34-year-old man is brought to the Emergency Department from the scene of a motor vehicle collision. He has head and neck injuries and is taken for a CT that reveals damage to his brainstem and cerebellum. There is concern of possible tonsillar herniation with rapid cessation of respiratory function. Which of the following medullary nuclei is most intimately involved in this function?
(A) Cuneate
(B) Dorsal motor vagal
(C) Hypoglossal
(D) Inferior olivary
(E) Reticular nuclei

15 An 80-year-old woman presents with a sudden onset of hoarseness, dysarthria, and dysphagia. MRI reveals a small defect, presumably vascular, in the medulla. Damage to which of the following structures would explain these deficits in this woman?
(A) Dorsal motor vagal nucleus
(B) Hypoglossal nucleus
(C) Nucleus ambiguus
(D) Solitary nucleus
(E) Spinal trigeminal nucleus

16 Which of the following cranial nerves exits the medulla from the groove identified by the white arrow in the image below?

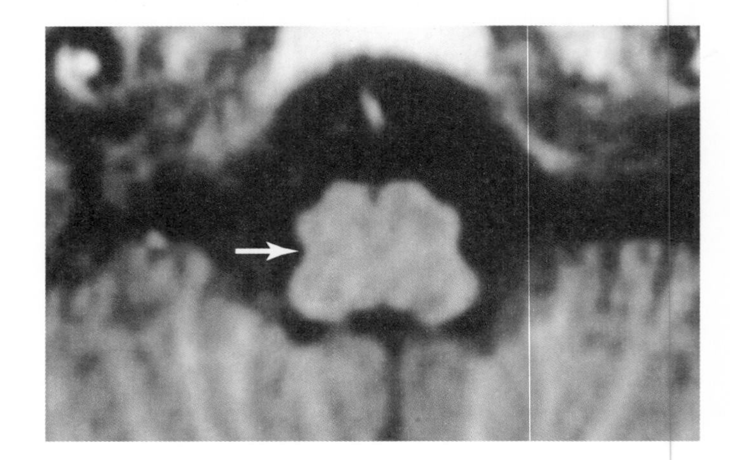

(A) Abducens
(B) Facial
(C) Hypoglossal
(D) Vagus
(E) Vestibulocochlear

17 Which of the following parts of the trigeminal complex is most specifically concerned with the receipt of, and transmission of, pain and thermal sensations?
(A) Mesencephalic nucleus
(B) Principal sensory nucleus
(C) Spinal trigeminal nucleus, pars caudalis
(D) Spinal trigeminal nucleus, pars interpolaris
(E) Spinal trigeminal nucleus, pars oralis

18 A 67-year-old man presents with a loss of discriminative touch and profound weakness, both deficits on the right side of his body, and with a deviation of his tongue to the left on attempted protrusion. The CT reveals that this is the result of a vascular incident. Which of the following vascular territories is most likely involved?
(A) Anterior inferior spinal artery
(B) Anterior spinal artery
(C) Posterior inferior cerebellar artery
(D) Posterior spinal artery
(E) Vertebral artery

19 Which of the following nuclei within the medulla, indicated in the image below, contains neurons that, after a synapse, will influence the activity of visceral tissues (cardiac and/or smooth muscle, glandular epithelium)?

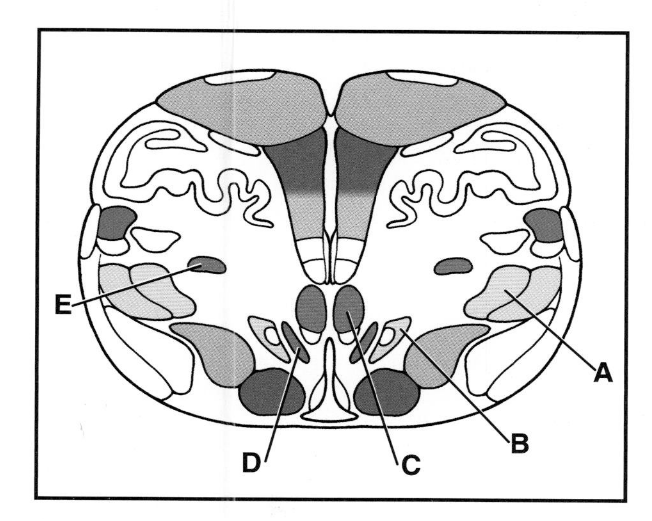

(A) A
(B) B
(C) C
(D) D
(E) E

20 A 32-year-old man presents at the Emergency Department with the complaint of a loss of sensation on his forehead. The history reveals that the man is an abuser of several illegal drugs. The examination confirms the loss of sensation on one side of his forehead, and an MRI reveals a cerebral lacuna within the medulla. Damage to which of the following areas, each representing a somatotopic region of the body, would explain this man's deficit?

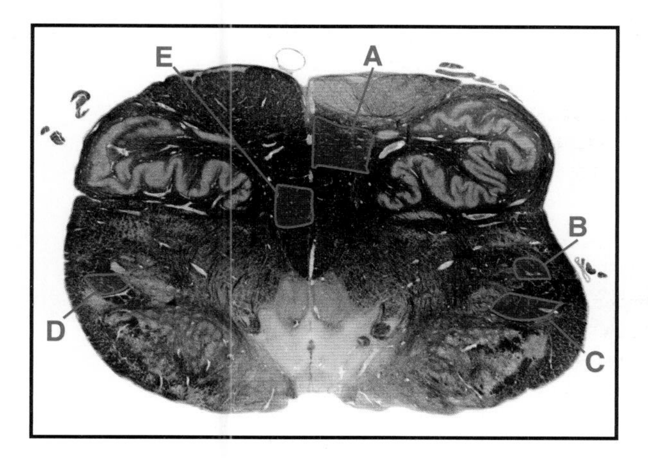

(A) A
(B) B
(C) C
(D) D
(E) E

21 A 45-year-old man presents with tinnitus and vertigo. An enhanced MRI reveals a tumor in the cerebellopontine angle that probably involves two cranial nerves. Which of the following cranial nerves are most likely involved with this tumor?

(A) Abducens and facial
(B) Abducens and hypoglossal
(C) Glossopharyngeal and vagus
(D) Vestibulocochlear and facial
(E) Vestibulocochlear and vagus

22 A 23-year-old woman is transported to the Emergency Department with burns on her hands and forearms. The EMS personnel indicate that an accidental fire started in the woman's kitchen, when her 10-month-old daughter was in a highchair a few feet away; the woman aggressively grabbed the burning material and threw it out the back door to protect her daughter. Pain transmission from the woman's hands was suppressed during this emergency. Which of the following is the source of these fibers?

(A) Nucleus raphe dorsalis
(B) Nucleus raphe magnus
(C) Nucleus raphe pontis
(D) Solitary nucleus
(E) Spinal trigeminal nucleus

23 The axons of the neuron cell bodies located in the nucleus, indicated by the black arrow in the image below, terminate in which of the following?

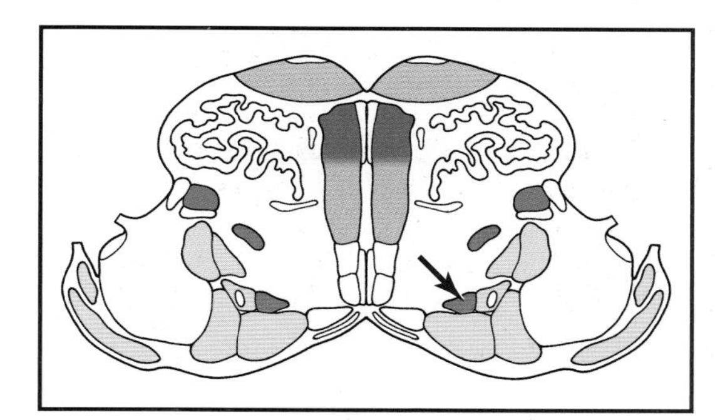

(A) Geniculate ganglion
(B) Intramural ganglion
(C) Otic ganglion
(D) Pterygopalatine ganglion
(E) Submandibular ganglion

24 A 67-year-old man is diagnosed with a lateral medullary syndrome, commonly called a posterior inferior cerebellar artery (PICA) syndrome. One of the features of this syndrome is ataxia that may relate to involvement of the vestibular nuclei; these nuclei are in the PICA territory. Damage to which of the following structures, also in the PICA territory, may also contribute to this deficit?

(A) Anterior spinocerebellar tract fibers
(B) Anterolateral system fibers
(C) Restiform body fibers
(D) Reticulospinal fibers
(E) Solitary tract fibers

25 Which of the following structures is located at the irregularity indicated by the black arrow in the fissure shown in the image below?

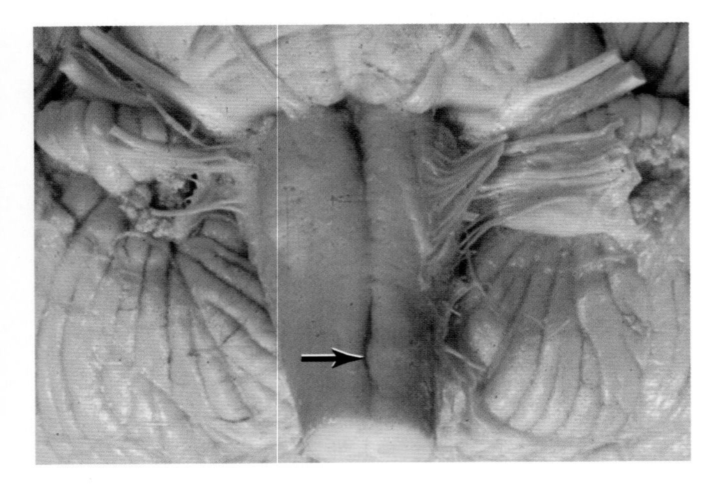

(A) Decussation of trigeminothalamic fibers
(B) Exit of accessory nerve
(C) Exit of phrenic nerve
(D) Level of motor decussation
(E) Level of sensory decussation

26 During a routine physical examination of a 53-year-old woman, the physician tests several cranial nerve reflexes. He notices that the stylopharyngeus muscle does not respond, even though sensation in the pharynx appears intact. Which of the following contains neurons that innervate this structure?
(A) Dorsal motor vagal nucleus
(B) Inferior salivatory nucleus
(C) Nucleus ambiguus
(D) Nucleus raphe magnus
(E) Trigeminal motor nucleus

27 A 69-year-old man is brought to the Emergency Department from his farm at the request of his wife. The history of the current event, as provided by the man, indicated that he was replacing the siding on his barn and kept repeatedly looking up (extending his neck) to assess his progress. His complaint is that the right side of his face suddenly became "numb" and he became unsteady; the examination confirms these symptoms and also reveals that he has a loss of pain and thermal sense on his left upper and lower extremities. MRI reveals an infarct. Damage to which of the following structures would explain this man's facial deficit?
(A) Anterolateral system
(B) Nucleus ambiguus
(C) Solitary tract
(D) Spinal trigeminal tract
(E) Trigeminal mesencephalic tract

28 The results of a series of diagnostic tests, including MRI with and without contrast, strongly suggest that a 5-year-old boy has an ependymoma within the fourth ventricle. During removal, the neurosurgeon notes that this tumor has made a noticeable dimple in the floor of the medullary portion of this ventricle lateral to the sulcus limitans. Which of the following nuclei would most likely be initially affected by the impingement of this tumor?
(A) Abducens and facial motor
(B) Dorsal motor vagal and hypoglossal
(C) Hypoglossal and solitary
(D) Spinal trigeminal and ambiguus
(E) Spinal and medial vestibular

29 An 11-year-old girl is taken to the pediatrician by her mother. According to the mother, the girl has developed progressive difficulties swallowing and talking. The examination reveals that the tongue protrudes symmetrically, the girl's smile is symmetrical, but her palate droops slightly on the left and the uvula deviates to the right when she says "Ah." MRI reveals a small extraaxial tumor in the posterior fossa. Which of the following is the most likely location of this tumor?
(A) At the motor decussation
(B) At the preolivary sulcus
(C) At the postolivary sulcus
(D) Laterally at the caudal pons
(E) Medially at the caudal pons

30 Which of the following structures contains motor fibers that innervate striated muscles that arise from head mesoderm and, consequently, have a general somatic efferent functional component?

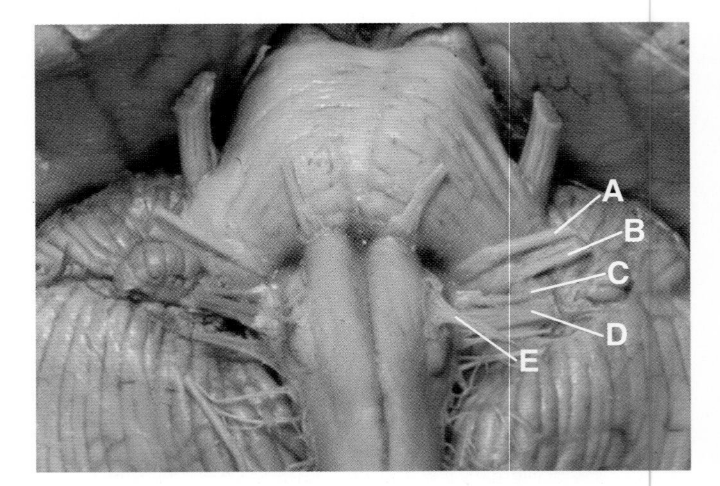

(A) A
(B) B
(C) C
(D) D
(E) E

ANSWERS

1 **The answer is C: Cuneate nucleus on the right.** The cuneate nucleus on the right (C) receives discriminative touch and vibratory sense information from the right upper extremity (UE); the gracile nuclei (dark blue) receive comparable input from the lower extremities (LE) on the same side of the body. This ascending information is uncrossed in the spinal cord and terminates in the cuneate nucleus on the same side. The spinal trigeminal tract (A) is conveying pain and thermal sensations from the left side of the face and left side of the oral cavity, and the left cuneate nucleus (B) is relaying discriminative touch from the left side of the upper extremity. The medial lemniscus on the right (D) is conveying vibratory sense and discriminative touch from UE and LE on the left side; the medullary pyramid (E) contains corticospinal fibers that will influence the left UE and LE.

2 **The answer is D: Spinal trigeminal tract + nucleus ambiguus.** The main structures located within the territory of the posterior inferior cerebellar artery (PICA) are (1) the spinal trigeminal tract and nucleus, (2) nucleus ambiguus, (3) anterolateral system, and (4) the vestibular nuclei. There are several other smaller structures also found in the PICA territory. The hypoglossal nerve root, medial lemniscus, pyramid, and medial longitudinal fasciculus are in the anterior spinal artery territory. Deficits resulting from a PICA lesion, or from occlusion of penetrating branches of the anterior spinal artery, reflect damage characteristic of the structures located within these vascular territories.

3 **The answer is D: Myelencephalon/medulla.** The lateral portions of the reticular formation of the medulla oblongata (commonly called medulla) contain groups of cells that maintain heart rate and that project to the phrenic nucleus of the cervical spinal cord to maintain respiration; the phrenic nucleus innervates the diaphragm. These medullary centers are especially susceptible to tonsillar herniation with sudden and catastrophic results. The only other part of the brain that may influence heart rate and respiration is the diencephalon via connections of the hypothalamus; damage to the hypothalamus is rarely suddenly catastrophic. It is true that central herniation may damage the midbrain and pons, but the cessation of breathing or cardiac function in these cases is measured in days, and there is a multitude of other characteristic deficits.

4 **The answer is C: Gracile and cuneate nuclei.** The posterior spinal artery serves the gracile and cuneate nuclei, and their respective nuclei, within the medulla. The anterolateral system (conveying pain and thermal sense) and the anterior spinocerebellar tract are in the territory of branches of the posterior inferior cerebellar artery. Penetrating branches that arise from the vertebral artery serve the principal olivary nucleus, and branches from the anterior spinal artery serve the medial lemniscus within the medulla.

5 **The answer is D: Spinal trigeminal nucleus.** Of all the choices in this question, D, the spinal trigeminal nucleus, is the only one that is a sensory nucleus within the brainstem. This nucleus receives pain and thermal sense from the ipsilateral side of the face via the spinal trigeminal tract. The ambiguus nucleus supplies motor fibers to the IXth and Xth cranial nerves, and the dorsal nucleus of the vagus (also called the dorsal motor nucleus of the vagus) gives rise to preganglionic parasympathetic fibers that distribute to most of the viscera of the thorax and abdomen via the vagus nerve. Oculomotor and trochlear nuclei are motor to all of the major extraocular muscles except the lateral rectus muscles.

6 **The answer is E (Pyramid).** Corticospinal fibers are located within the pyramid at the level of the medulla oblongata. These fibers (E) will cross at the motor decussation to influence lower motor neurons on the contralateral (left in this example) side of the spinal cord. The noticeable elevation on the lower and lateral aspect of the medulla (A) is the inferior olivary nucleus, and the large bulge on the dorsolateral aspect of the medulla (C) is the restiform body; the latter is the larger part of the inferior cerebellar peduncle. The positions of the medial lemniscus and the anterolateral system are identified at B and D, respectively. These latter two are sensory pathways conveying discriminative touch/vibratory sense and pain/thermal sense, respectively, from the opposite side of the body.

7 **The answer is D: Gracile nucleus.** The gracile nucleus receives discriminative touch, vibratory sense, and proprioception from the lower extremity on the same side of the body. The fibers that terminate in this nucleus are the central processes of primary sensory fibers. Axons arising in this nucleus cross in the sensory decussation to enter the medial lemniscus. The accessory nucleus provides motor fibers to the trapezius and sternocleidomastoid muscles. The accessory cuneate nucleus sends proprioceptive information to the cerebellum via the restiform body; the cuneate nucleus sends comparable information to the contralateral thalamus (for conscious perception of two-point discrimination and vibratory sense) via the medial lemniscus. The spinal trigeminal, especially the pars caudalis, is concerned with the transmission of pain and thermal sense from the ipsilateral side of the face.

8 **The answer is B (Lower portions of the medial lemniscus bilaterally).** This woman has an interesting problem: she perceives that she is weak (difficulty walking), but her muscles are not weak when actually examined for strength. A bilateral lesion of those portions of the medial lemnisci conveying information from the lower extremities (LEs) removes the proprioception, discriminative touch, and vibratory sense from reaching higher levels. This results in a sensory ataxia. Receipt of this sensory input from the LEs is essential for successful, effective, and coordinated walking. A bilateral lesion at A (pyramids) would result in quadriplegia and at C (bilateral upper extremity [UE] areas of medial lemnisci) would result in deficits of the UEs that are essentially the same as seen in the LEs. Bilateral injury of the hypoglossal nuclei (D) would result in total paralysis of the tongue, and the unilateral lesion at E results in a right-sided loss of pain and thermal sense on the face, dysarthria, and dysphagia due to its involvement of the spinal trigeminal tract and nucleus and the nucleus ambiguus.

9 **The answer is B: Inferior olivary nuclei.** The only source of cerebellar afferents that terminates as climbing fibers in the molecular layer of the cerebellar cortex is the inferior olivary nuclei; these fibers are exclusively crossed and send collaterals into the cerebellar nuclei. The principal olivary nucleus projects to the lateral cortex, and the dorsal and medial accessory olivary nuclei project to the vermis and intermediate cortex. The main targets of climbing fibers are Purkinje cell dendrites that they excite. Fibers arising from the accessory cuneate nucleus, the lateral reticular nucleus, and the vestibular nuclei terminate predominately as mossy fibers in the granular layer. The solitary nucleus is a visceral afferent center in the medulla that has insignificant cerebellar connections.

10 **The answer is A: Area postrema.** The area postrema is a small area in the lateral wall of the caudal area of the fourth ventricle that has an unusually rich vascular supply, glia population, and small nerve fibers. Irritation of the area postrema results in emesis. The following trigones are elevations over their respective nuclei: hypoglossal trigone—hypoglossal nucleus; and vagal trigone—dorsal motor nucleus of the vagal nerve. The vestibular nuclei are found internal to the vestibular area, which is lateral to the sulcus limitans (recall that motor nuclei are medial to the sulcus limitans). While irritation of the vestibular area may result in vomiting, there would also be other deficits (gait problems, nystagmus). The striae medullares are fibers traversing the floor of the fourth ventricle from the midline laterally and consisting of axons from caudal pontine and arcuate nuclei.

11 **The answer is D: Convey proprioceptive information from the right side of the body.** The arrows identify the internal arcuate fibers on the right side of the medulla. These fibers originate from neuron cell bodies in the cuneate and gracile nuclei and arch through the caudal medulla toward the midline where they cross at the sensory decussation to enter and form the medial lemniscus on the contralateral (in this case, the left) side. The right cuneate and gracile nuclei receive primary sensory input from the right side of the body. These fibers are conveying proprioception, discriminative touch, and vibratory sense that will cross at the sensory decussation to become part of the left medial lemniscus. These fibers do not convey pain and thermal sense from the body, and proprioceptive input from the face traverses the trigeminal nerve.

12 **The answer is D: Posterior inferior cerebellar artery.** The posterior inferior cerebellar artery serves the lateral medulla, which includes the spinal trigeminal tract and nucleus, anterolateral system, nucleus ambiguus, vestibular nuclei, and the restiform body. Damage to these structures explains the deficits experienced by this man. The anterior inferior cerebellar artery provides no significant blood supply to the medulla; paramedian branches, which arise from the basilar artery, serve medial areas of the pons. The anterior spinal artery serves the medial area of the medulla (medial lemniscus, corticospinal fibers, hypoglossal root); the posterior spinal artery serves caudal and dorsal regions of the medulla (mainly the gracile and cuneate nuclei) caudal to the obex.

13 **The answer is C: General and special visceral sensations from the head.** The structure labeled by the arrow is the solitary tract, and the immediately surrounding cell bodies collectively constitute the solitary nucleus. The solitary tract is composed of the central processes of general visceral afferent fibers that convey sensations from areas of the pharynx and larynx and from the salivary glands; we all have experienced that tingling sensation in the salivary glands when something especially tasty is placed in the oral cavity. This portion of the solitary tract and nucleus is more caudally located in the medulla and is called the cardiorespiratory nucleus. The solitary tract also contains the central processes of special visceral afferent fibers conveying taste; this is its more rostral part and is called the gustatory nucleus. Cranial nerves VII, IX, and X contain fibers that enter the solitary tract and terminate in the surrounding nucleus. In this respect, this tract and nucleus are the visceral afferent center of the brainstem. General somatic sensation enters the brainstem on the trigeminal nerve, traveling in the spinal trigeminal tract, to end in the spinal trigeminal nucleus. Special somatic sensations are conveyed on the VIIIth cranial nerve.

14 **The answer is E: Reticular nuclei.** The reticular nuclei of the medulla, located in its ventrolateral region in mid-to-caudal levels, contain neurons that project to the phrenic nucleus, which innervates the diaphragm, and to spinal motor neurons that innervate intercostal muscles. The diaphragm is the most important muscle for successful breathing. Damage to these reticular nuclei may result in respiratory arrest, which constitutes a medical emergency. Even if the projection to motor neurons innervating the intercostal muscles survives, this projection, on its own, is not adequate to sustain breathing. The gracile nucleus receives proprioceptive input from the ipsilateral upper extremity; the dorsal motor nucleus of the vagus consists of preganglionic parasympathetic neurons. The hypoglossal nucleus is motor to the ipsilateral genioglossus muscle, and the inferior olivary nucleus sends a projection to the cerebellar cortex that ends as climbing fibers.

15 **The answer is C: Nucleus ambiguus.** The nucleus ambiguus, through the glossopharyngeal nerve, innervates the stylopharyngeus muscle and, through the vagus nerve, innervates the palate, constrictor muscles of the throat, and the vocalis muscle. Injury to this nucleus results in a paralysis of these muscles on the ipsilateral side. This can be a serious problem since the patient may aspirate food or fluids into the lungs when swallowing. The dorsal motor vagal nucleus contains parasympathetic preganglionic neurons; the hypoglossal nucleus innervates the ipsilateral genioglossus muscle. The solitary nucleus receives general and special visceral senses (such as taste); the spinal trigeminal nucleus receives pain and thermal sensations from the ipsilateral face and oral cavity.

16 **The answer is D: Vagus.** The arrow identifies the postolivary sulcus (also called the retro-olivary sulcus or groove). The two cranial nerves that exit from the postolivary sulcus are the glossopharyngeal nerve, usually as a small single root close to the pons-medulla junction, and the vagus nerve, as a row of rootlets exiting caudal to the glossopharyngeal. The abducens nerve exits at the caudal border of the pons generally in-line with the preolivary sulcus, and the facial nerve exits the caudal border of the pons generally in-line with the postolivary sulcus; neither the abducens nerve nor the facial nerve exits *from* these respective sulci. The hypoglossal nerve exits the preolivary sulcus, which is located between the pyramid and the inferior olivary eminence; the vestibulocochlear nerve is located at the posterolateral aspect of the brainstem at the medulla-pons junction directly adjacent to the restiform body.

17 **The answer is C: Spinal trigeminal nucleus, pars caudalis.** The pars caudalis of the spinal trigeminal nucleus extends from the spinal cord–medulla junction rostrally to the level of the obex. It receives pain and thermal input from the ipsilateral side of the face and oral cavity via the spinal trigeminal tract; this tract is made up of the central processes of divisions V_1, V_2, and V_3. The tract also receives general sensory input via the VIIth, IXth, and Xth cranial nerves, from the ear, external auditory meatus, and surface of the eardrum. The pars interpolaris extends from the obex to the rostral aspect of the hypoglossal nucleus, and the pars oralis extends from the rostral end of the hypoglossal nucleus to the level of the principal sensory nucleus. The principal sensory nucleus is located at midpontine levels and receives discriminative touch information; the mesencephalic nucleus, located in the rostral pons and extending into the mesencephalon, receives proprioceptive input and is the nucleus for the muscle stretch reflexes mediated via the trigeminal nerve.

18 **The answer is B: Anterior spinal artery.** Penetrating branches of the anterior spinal artery enter the medial portions of the medulla and serve the medial lemniscus (discriminative touch loss), corticospinal fibers in the pyramid (weakness), and the exiting root of the hypoglossal nerve (tongue deviation on protrusion). Although there is only a single midline anterior spinal artery, the penetrating branches (called sulcal arteries) of this vessel alternate, one to the right, one to the left, etc. Occlusion of anterior spinal branches results in the Déjèrine syndrome. The anterior inferior cerebellar artery largely serves inferior and lateral aspects of the cerebellum and gives rise to the labyrinthine artery. The posterior inferior cerebellar artery (PICA) serves the lateral medulla and the inferior and medial aspects of the cerebellum. An occlusion of this vessel is known as the PICA syndrome. The posterior spinal artery is usually a branch of PICA that serves the posterior column nuclei. Penetrating branches from the vertebral artery serve the general area of the inferior olivary nuclei.

19 **The answer is D (Dorsal motor nucleus of the vagus).** The dorsal motor nucleus of the vagus nerve is located immediately lateral to the hypoglossal nucleus, but medial to the solitary nucleus and tract, and medial to the position of the sulcus limitans (when this sulcus is visible). The general visceral efferent preganglionic parasympathetic neurons of this nucleus project, via the vagus nerve, to terminal (also called intermural) ganglia of organs in the thorax and abdomen (to the level of the splenic flexure of the colon). After a synapse, the postganglionic neurons send their axons to innervate smooth muscle and glandular epithelium and, if going to the heart, cardiac muscle. The spinal trigeminal nucleus (A) receives pain and thermal sensations from the ipsilateral face and oral cavity and projects to the contralateral thalamus; the solitary nucleus (B) receives visceral input (both general and special) via cranial nerves VII, IX, and X. The hypoglossal nucleus (C) provides motor innervation to the ipsilateral half of the tongue; the nucleus ambiguus (E) innervates laryngeal and pharyngeal musculature on the ipsilateral side.

20 **The answer is B (The lower portions, about the lower one-third, of the spinal trigeminal tract).** In this *clinical orientation*, the face is somatotopically represented right-side-up in the spinal trigeminal tract and in the medially adjacent nucleus. The ophthalmic division (V_1) is represented most ventrally, the maxillary division (V_2) is represented in the middle, and the mandibular division (V_3) is represented most dorsally. A lacuna is a small, well-circumscribed lesion that most frequently results from the occlusion of a small penetrating vessel; these may be located anywhere in the central nervous system. This patient has a lacuna in the area of B. In the supratentorial location, these may be called cerebral lacunae. The lower extremity is somatotopically represented by A, and the upper extremity is represented at E; in both cases, the modality is discriminative touch, vibratory sense, and proprioception. The mandibular division and maxillary division of the trigeminal nerve are represented at C and D, respectively.

21 **The answer is D: Vestibulocochlear and facial.** The deficits experienced by this patient, and the location of the lesion, clearly point to the likelihood that this is a vestibular schwannoma. This particular lesion is commonly called a cerebellopontine angle tumor. This particular lesion is characteristically located at, or in, the internal acoustic meatus; the two nerves that traverse this opening are the vestibulocochlear and facial nerves. Deficits indicating VIIIth nerve damage appear early and in a high percentage of cases, while deficits indicating VIIth nerve injury appear late and in a lesser percentage of cases. The abducens root is located at the pons-medulla junction, but close to the midline, roughly in-line with the preolivary fissure. The hypoglossal nerve exits directly from the preolivary sulcus. The glossopharyngeal and vagus nerves exit from the postolivary fissure.

22 **The answer is B: Nucleus raphe magnus.** The nucleus raphe magnus (NRM) contains serotonergic cells that project bilaterally to the posterior horn of the spinal cord as raphespinal fibers; these presynaptically inhibit, through interneurons, the central processes of A-delta and C fibers conveying pain and thermal sensations. The NRM receives descending projections from the periaqueductal gray of the brainstem and also inhibits pain transmission in the spinal trigeminal nucleus, pars caudalis, through the same synaptic mechanism. The nucleus raphe dorsalis and nucleus raphe pontis give rise to serotonergic projections to a variety of forebrain and brainstem areas, but they do not contribute to the raphespinal pathway. The solitary nucleus receives special visceral sense (taste) and general visceral sense in rostral and caudal portions of the nucleus, respectively. Pain and thermal sensations from the face and oral cavity end in the spinal trigeminal nucleus, particularly its pars caudalis.

23 **The answer is C: Otic ganglion.** The inferior salivatory nucleus, indicated by the black arrow, is located directly medial to the rostral end of the solitary tract and nuclei and inferior to the medial vestibular nucleus. This visceromotor nucleus sends its preganglionic parasympathetic fibers to the otic ganglion via the glossopharyngeal nerve. The otic ganglion, in turn, sends postganglionic fibers to the parotid gland where they innervate smooth muscle and glandular epithelium. The geniculate ganglion is located on the facial nerve and contains sensory cells conveying general and special senses (taste) to the spinal trigeminal tract/nucleus and solitary tract/nucleus, respectively. Intramural ganglia are located in the gut wall and receive their preganglionic input from the dorsal motor nucleus of the vagus nerve. The pterygopalatine and submandibular ganglia receive their preganglionic input from the superior salivatory nucleus, which is the visceromotor nucleus of the facial nerve.

24 **The answer is C: Restiform body fibers.** The restiform body, the larger portion of the inferior cerebellar peduncle, contains a wide variety of fibers that are conveying afferent information to the cerebellum. These include, but are not limited to, dorsal spinocerebellar fibers, reticulocerebellar fibers, olivocerebellar fibers, cuneocerebellar fibers, and others. Broadly speaking, they function to ensure muscle coordination, synergy, and integration of movement. A lesion of the restiform body disrupts this function. Damage to anterior spinocerebellar fibers, on their own, does not cause ataxia; anterolateral system fibers convey pain and thermal sense. Solitary tract fibers are related to visceral (special and general) sense; reticulospinal fibers innervate spinal motor neurons, some of which influence extensor muscles.

25 **The answer is D: Level of motor decussation.** This small irregularity in the anterior median fissure of the medulla is the superficial indicator of the location of the decussation of corticospinal fibers, commonly called the motor decussation (a functional designation) or sometimes called the pyramidal decussation (not a functional designation). About 85% to 90% of corticospinal fibers cross within, and on the surface of, the brainstem at the cervical cord–medulla junction. The decussation of trigeminothalamic fibers takes place within the medulla over a widespread area and not at a particular medullary level. The rootlets of the accessory nerve exit the lateral aspect of the cervical spinal cord, ascend through the foramen magnum, briefly join with the caudal rootlets of the vagus, and then exit the skull via the jugular foramen. The phrenic nerve arises from the ventral rami of C4 with smaller contributions from C3 and C5 and descends to innervate the ipsilateral half of the diaphragm. The level of the sensory decussation, the crossing of fibers of the posterior column–medial lemniscus system, is within the medulla immediately rostral to the motor decussation.

26 **The answer is C: Nucleus ambiguus.** The cells of the nucleus ambiguus provide somatic motor fibers for the IXth (glossopharyngeal) and Xth (vagus) nerves; the striated muscles served by these nerves originate from the third and fourth pharyngeal arches and, therefore, have a special visceral efferent functional component. The stylopharyngeus muscle is innervated by the IXth nerve, and the throat constrictor muscles and the vocalis are innervated by the Xth nerve. The stylopharyngeus elevates the pharynx during the gag reflex. The dorsal motor vagal and the inferior salivatory nuclei contain parasympathetic preganglionic cells (visceral motor) that project to peripheral ganglia. The nucleus raphe magnus gives rise to raphespinal projections that are involved in the descending inhibition of pain, and the trigeminal motor nucleus innervates the masticatory muscles (including the pterygoids), tensor tympani, tensor veli palatine, mylohyoid, and the anterior belly of the digastric.

27 **The answer is D: Spinal trigeminal tract.** This man has had an infarct in the territory of the posterior inferior cerebellar artery (PICA); choices A to D are in this vascular territory. It is believed that repeated extension of the neck, especially hyperextension, may contribute to this type of lesion through several mechanisms, some being damage to the V_3 segment of the vertebral artery. The spinal trigeminal tract is composed of primary sensory fibers that convey pain and thermal sensations for the same side of the face and oral cavity; these fibers terminate in the spinal trigeminal nucleus. The anterolateral system conveys pain and thermal sense from the contralateral side of the body; the solitary tract contains visceral fibers (general and special [taste]) that enter the brainstem on cranial nerves VII, IX, and X. Laryngeal and pharyngeal muscles, and the vocalis muscle, are innervated by neurons of the nucleus ambiguus. The mesencephalic nucleus is not in the PICA territory and relays proprioceptive information; this man has no such deficits.

28 **The answer is E: Spinal and medial vestibular nuclei.** The spinal and medial vestibular nuclei are located lateral to the sulcus limitans and are superficially located in the floor of the medullary portion of the fourth ventricle. In general, they function in the coordination of eye movement and in the innervation of axial/postural muscles. An impingement in this area, enough to produce a visible dimple in the floor of the ventricle, would affect the function of these nuclei. Ependymomas constitute about 9% to 10% of all brain tumors in pediatric patients. The hypoglossal and dorsal motor vagal nuclei are located in the medial floor of the medullary portion of the fourth ventricle, and the abducens nucleus is located immediately internal to the facial colliculus in the caudal portion of the pontine part of the ventricle; all of these nuclei are motor and are located medial to the sulcus limitans. The spinal trigeminal and solitary nuclei are located lateral to the sulcus limitans but are deeper within the medulla than the vestibular nuclei. The facial nucleus and the nucleus ambiguus are medial to the sulcus limitans, but they are also located deep within the medulla.

29 **The answer is C: At the postolivary sulcus.** The palate, which is weak in this girl, is innervated by the vagus nerve, as are the muscles of the larynx and pharynx that are involved in vocalization and swallowing. This is an extraaxial tumor (outside the substance of the brain) that is located in the postolivary sulcus and is impinging on the root of the vagus nerve. A tumor at the motor decussation would potentially result in a weakness of all extremities, and a similar problem at the preolivary sulcus would result in weakness of one side of the tongue with consequent deviation toward that side on attempted protrusion. The medial caudal pons is the location of the exit of the abducens nerve, and the point where the corticospinal fibers exit the lateral caudal pons is the location of the positions of the facial and vestibulocochlear nerves. The only deficits described for this patient clearly relate to the root of the vagus nerve.

30 **The answer is E (Hypoglossal nerve).** Muscles of the tongue (innervated by the hypoglossal nerve), and the extraocular muscles (innervated by cranial nerves III, IV, and VI), all arise from general head mesoderm and, consequently, all share a common functional component (General Somatic Efferent, GSE). The facial nerve (choice A), glossopharyngeal nerve (choice C), and the vagus nerve (choice D) all innervate striated muscle that originates from pharyngeal arches (arches 2, 3, and 4, respectively). Due to the fact that these particular muscles arise from mesoderm within the pharyngeal arches, the functional component associated with these fibers is Special Visceral Efferent (SVE). The facial motor nucleus and the nucleus ambiguus (for IX and X) are the source of these SVE fibers. Cranial nerves VII, IX, and X also contain visceromotor (autonomic) fibers that are parasympathetic preganglionic and that originate from the superior salivatory nucleus (VII), the inferior salivatory nucleus (IX), and the dorsal motor nucleus of the vagus (X). The vestibulocochlear nerve (choice B) is sensory; its functional component is Special Somatic Afferent (SSA) due to the fact that some of the VIIIth nerve sensory apparatus arises from the surface of the head.

Chapter 8

The Pons

Recall the Cautionary Tale: The images in this chapter are presented in a Clinical Orientation; this approach emphasizes how structure and function correlate with deficits in a clinical setting. MRI and CT have a universally recognized orientation and laterality. Line drawings, stained sections, and gross brain images in the Clinical Orientation are viewed here in the identical manner that one views an MRI: your right, as the physician-observer, is the patient's left, and the observer's left is the patient's right. The laterality of deficits and reference to connections follow accordingly.

QUESTIONS

Select the single best answer.

1 The CT of an 81-year-old woman reveals that she has had a small stroke in the territory of the paramedian branches of the basilar artery. Your examination reveals that this patient has diplopia. Damage to which of the following would explain this deficit in this woman?

(A) Abducens nucleus/nerve
(B) Corticonuclear fibers
(C) Lateral lemniscus
(D) Oculomotor nucleus/nerve
(E) Trochlear nucleus/nerve

Questions 2 and 3 are based on the following case:

A 51-year-old woman presents with the complaint that she is having difficulty seeing; she says that she seems to see "two of everything." The general history is not remarkable. The examination reveals that this woman is unable to perform a voluntary conjugate eye movement in the horizontal plane to the right with either eye and, when attempting a similar voluntary movement to the left, only the left eye looks lateral; the right eye does not move. The CT reveals a small brainstem lesion.

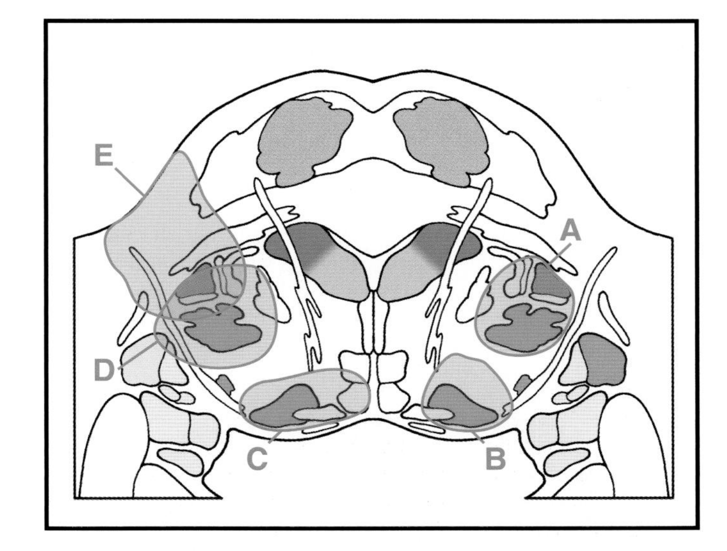

2 Which of the areas in the image above, each representing a potential lesion, represents the most likely location of this lesion?

(A) A
(B) B
(C) C
(D) D
(E) E

3 Recognizing the probable location of this lesion, which of the following deficits might you also expect to see in this patient?

(A) Left internuclear ophthalmoplegia only
(B) Loss of pain/thermal sense on left side
(C) Loss of pain/thermal sense on right side
(D) Weakness of facial muscles on the left
(E) Weakness of facial muscles on the right

4 While reviewing a series of axial MRI through the brainstem, the radiologist needs to establish a landmark for describing whether the small area of demyelination is in the middle cerebellar peduncle or in the basilar pons. Which of the following represents the landmark separating these two structures?

(A) Exit of the abducens nerve
(B) Exit of the facial nerve
(C) Exit of the trigeminal nerve
(D) Position of the trochlear nerve
(E) Position of the glossopharyngeal nerve

5 Which of the following cranial nerves conveys parasympathetic preganglionic fibers to the submandibular ganglion?

(A) Accessory
(B) Facial
(C) Glossopharyngeal
(D) Oculomotor
(E) Vagus

6 A 37-year-old woman visits her family physician. She complains that she periodically stumbles when she jogs, an exercise regimen that she has maintained for many years. She has no other concerns. MRI reveals an area of demyelination in the position of the white oval in the image below. This defect is most likely located in which of the following?

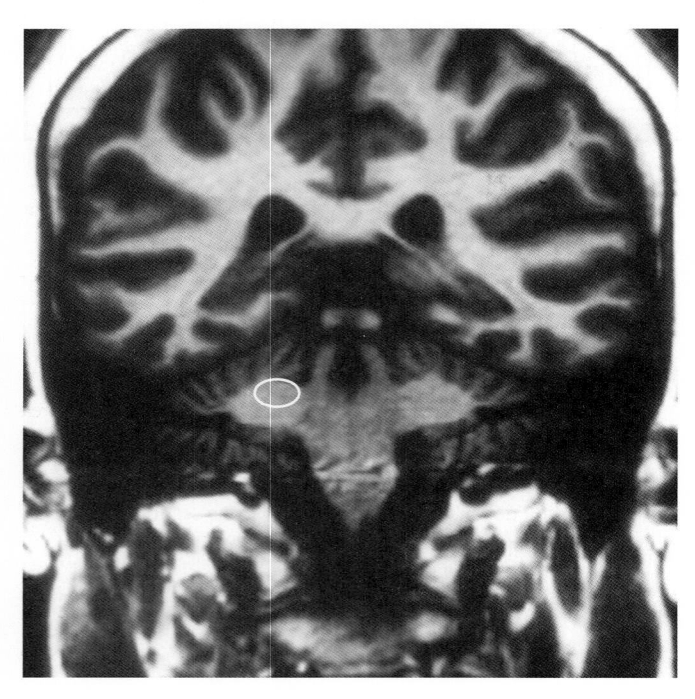

(A) Brachium conjunctivum on the right
(B) Brachium pontis on the left
(C) Brachium pontis on the right
(D) Juxtarestiform body on the right
(E) Restiform body on the right

7 The MRI of a 41-year-old man is shown below. In this T2-weighted image, the arrow is identifying a round black structure. Which of the following most specifically identifies the location of this structure?

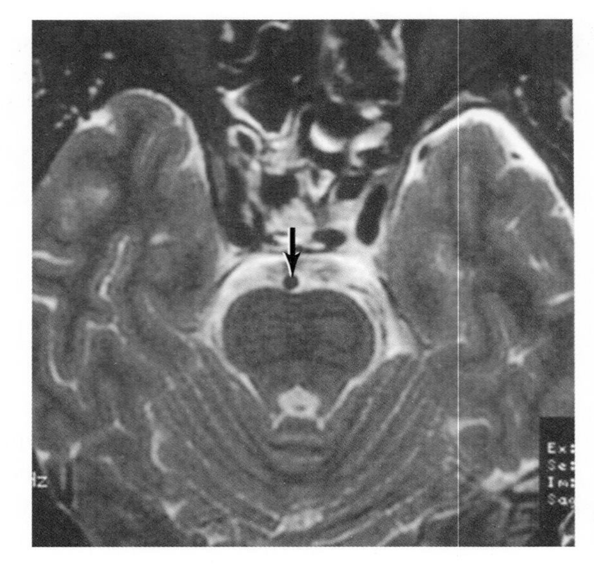

(A) Cisterna magna
(B) Interpeduncular cistern
(C) Lateral cerebellomedullary cistern
(D) Premedullary cistern
(E) Prepontine cistern

Questions 8, 9, and 10 are based on the following case:

A 71-year-old man is brought to the Emergency Department from his home. He complains that his hand suddenly became numb; the physician notices that the man's speech seems slightly slurred. The examination reveals weakness of the facial muscles on upper and lower aspects of his face on the left and a loss of pain and thermal sense on the left side of his face and right side of his body. Masticatory function is normal, and his tongue does not deviate on attempted protrusion.

8 The loss of pain and thermal sense on this man's body indicates damage to which of the following?

(A) Anterolateral system
(B) Anterior white commissure
(C) Medial lemniscus
(D) Ventral posterolateral nucleus
(E) Posterior limb, internal capsule

9 The loss of pain and thermal sense on this man's face indicates damage to which of the following?

(A) Spinal trigeminal nucleus, pars interpolaris
(B) Spinal trigeminal nucleus, pars oralis
(C) Spinal trigeminal tract
(D) Trigeminal nerve
(E) Ventral trigeminothalamic fibers

10 Assuming that this is a vascular lesion (sudden onset), which of the following vessels would most likely be involved?

(A) Anterior spinal artery
(B) Circumferential branches of the basilar artery
(C) Paramedian branches of the basilar artery
(D) Posterior inferior cerebellar artery
(E) Superior cerebellar artery

11 If the elevation identified by the black arrow in the image below were to be damaged by a vascular lesion, tumor, or trauma, significant and demonstrable deficits would result. Which of the following specifies this elevation?

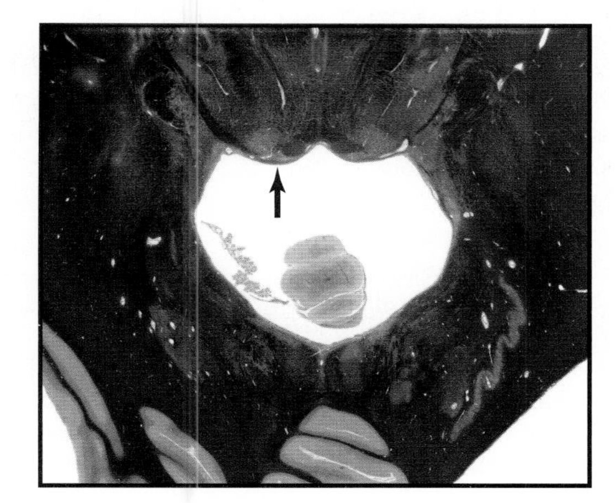

(A) Arcuate eminence
(B) Facial colliculus
(C) Hypoglossal trigone
(D) Vagal trigone
(E) Vestibular area

12 The visceromotor (autonomic) preganglionic fibers that distribute with peripheral branches of the facial nerve originate in which of the following?
(A) Edinger-Westphal nucleus
(B) Facial motor nucleus
(C) Inferior salivatory nucleus
(D) Nucleus ambiguus
(E) Superior salivatory nucleus

13 A 57-year-old woman visits her family physician complaining of a sensory problem. The history reveals that the woman has sensory deficits that have slowly been getting more noticeable. MRI reveals a small pontine lesion in the area of the white oval in the image below; this lesion damages a distinct dark line (at the arrows) that traverses the substance of the lesion. Which of the following most likely represents this distinct dark line in this image?

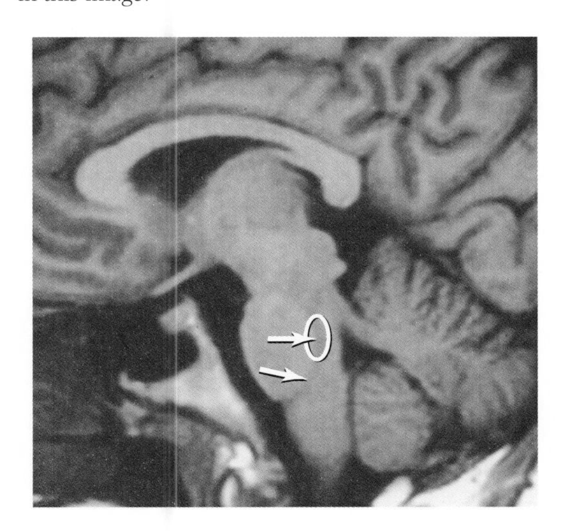

(A) Anterolateral system
(B) Lateral lemniscus
(C) Medial lemniscus
(D) Medial longitudinal fasciculus
(E) Spinal trigeminal tract

14 A 37-year-old man presents to his family physician with the complaint of "difficulty seeing." The history reveals that this started several months ago, and, in the words of the man, it "gets better, then worse, then better." The examination reveals that the man has no visual field deficits, has no somatomotor or somatosensory deficits, but does have a gaze deficit in the horizontal plane. When he looks to the left, his left eye abducts and his right eye adducts. When he looks to the right, his right eye abducts but his left eye does not adduct. Damage to which of the following would explain this man's deficit?
(A) Abducens nucleus on the left
(B) Abducens nucleus on the right
(C) Medial longitudinal fasciculus bilaterally
(D) Medial longitudinal fasciculus on the left
(E) Medial longitudinal fasciculus on the right

15 The MRI of a 61-year-old woman reveals a vascular lesion in the caudal basilar pons in the territory of the paramedian branches of the basilar artery. The woman has a profound weakness of her left upper and lower extremities. Recognizing the location of this lesion, which of the following would this patient most likely also experience?
(A) Aphasia
(B) Diplopia
(C) Dysarthria
(D) Dysphagia
(E) Hoarseness

16 A 71-year-old man presents with the complaint of numbness on one side of his face. The examination reveals that the man has a loss of pain and thermal sense on the right side of his face and weakness of his masticatory muscles on the same side. An MRI reveals a small, well-localized pontine lesion. Which of the outlined areas in the image below, each representing a lesion, most likely indicates the position of the area damaged?

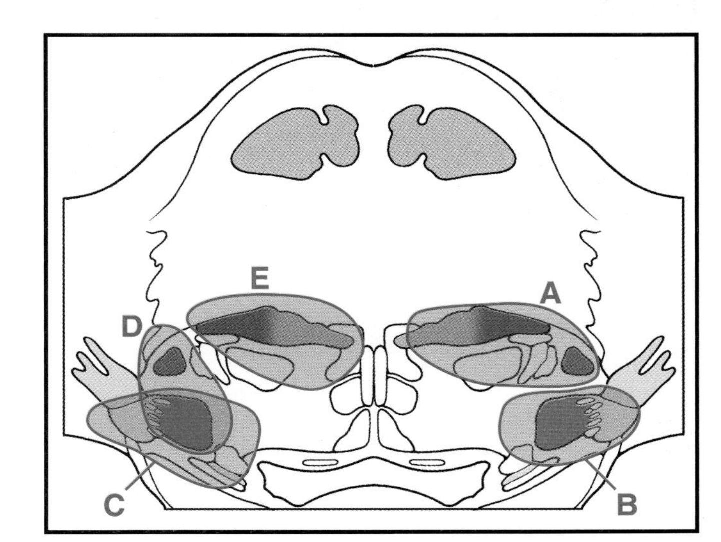

(A) A
(B) B
(C) C
(D) D
(E) E

17 A newborn male baby presents with signs that clearly suggest that the muscles arising from the second pharyngeal arch failed to develop. Which of the following would be most noticeable in this baby and would lead the physician to this conclusion?

(A) Difficulty breathing
(B) Difficulty moving eyes
(C) Loss of facial movements
(D) Loss of mastication
(E) Loss of tongue movements

18 A 45-year-old woman presents to her family physician with a complaint of anxiety. The history is unremarkable. The examination reveals no frank motor or sensory deficits, and cranial nerve function is within normal ranges. An MRI (T2 weighted) reveals a dark structure in the posterior fossa shown at the black pointer in the image below. Which of the following most likely represents this structure?

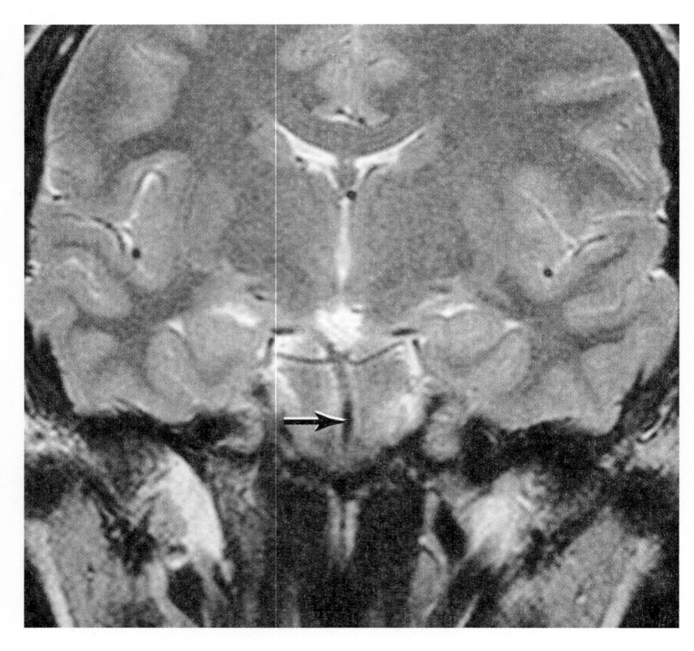

(A) Clivus meningioma
(B) Basilar artery
(C) Posterior cerebral artery
(D) Vein of Galen
(E) Vertebral artery

19 A 71-year-old woman is brought to the Emergency Department by EMS personnel. The history reveals a sudden onset of profound weakness. The examination reveals a right hemiplegia and inability to abduct the left eye, this resulting in diplopia. CT shows a localized hemorrhagic lesion. Which of the following vascular territories would most likely be involved to produce these deficits?

(A) Anterior spinal artery of medulla
(B) Long circumferential basilar branches
(C) Paramedial basilar branches at caudal levels
(D) Paramedian basilar branches at rostral levels
(E) Paramedian branches of basilar bifurcation

20 Which of the following receives input from muscle spindles in the temporalis muscle that will eventually influence a muscle stretch reflex?

(A) Edinger-Westphal nucleus
(B) Facial nucleus
(C) Mesencephalic nucleus
(D) Principal sensory nucleus
(E) Spinal trigeminal nucleus

21 A 41-year-old man presents to his family physician with what he calls "trouble seeing." The examination reveals that he has internuclear ophthalmoplegia. MRI reveals an area of demyelination suggestive of multiple sclerosis. Which of the structures indicated in the image below would explain this man's deficits if it were the site of this lesion?

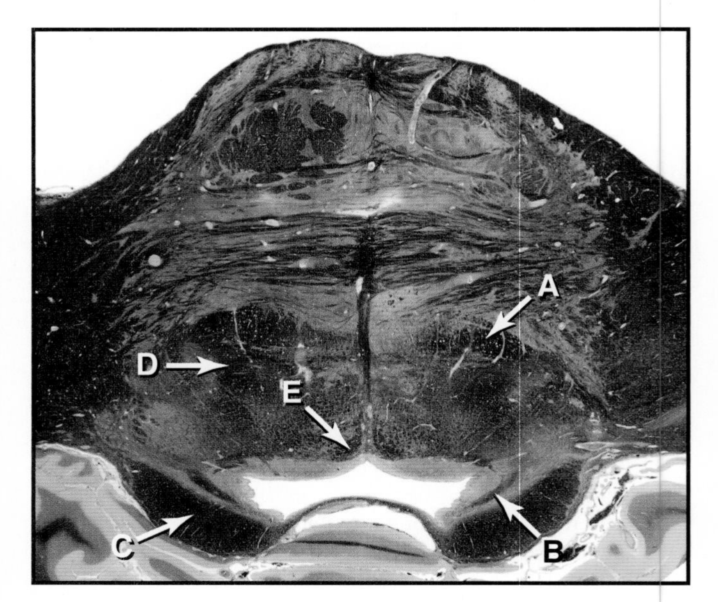

(A) A
(B) B
(C) C
(D) D
(E) E

22 A 63-year-old man visits his physician for a routine checkup. While he does not have any specific complaints, his physician notices that the man has some vague somatosensory losses on his right lower extremity; he also has some clear difficulties localizing sound in space, but he is not deaf in either ear. An MRI reveals a small lesion in the pontine tegmentum. Damage to which of the following could explain the hearing deficits?

(A) Central tegmental tract
(B) Cochlear nuclei
(C) Inferior olivary nuclei
(D) Lateral lemniscus
(E) Medial lemniscus

23 A 32-year-old woman presents with signs and symptoms suggestive of a demyelinating disease. MRI reveals a plaque, a well-defined area of myelin loss, at the location indicated by the white arrow in the image below. Which of the following represents the most likely location of this lesion?

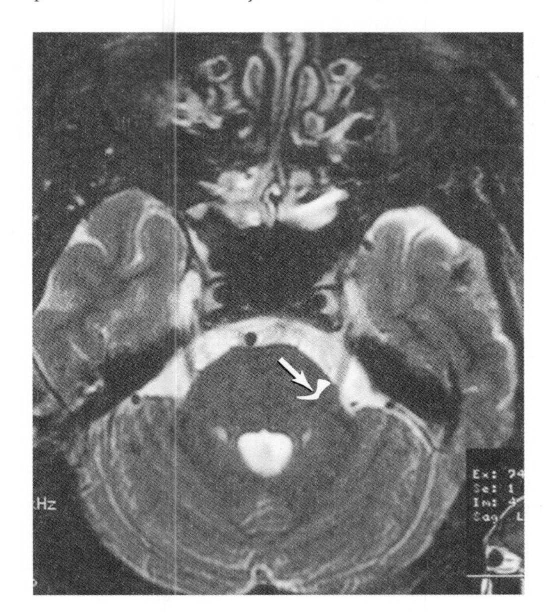

(A) Basilar pons
(B) Inferior cerebellar peduncle
(C) Middle cerebellar peduncle
(D) Pontine tegmentum
(E) Superior cerebellar peduncle

24 The masseter, temporalis, and pterygoid muscles originate embryologically from which of the following?
(A) Pharyngeal arch 1
(B) Pharyngeal arch 2
(C) Pharyngeal arch 3
(D) Pharyngeal arch 4
(E) Pharyngeal arch 6

25 The image below is through the caudal pons just rostral to the medulla-pons junction. Internal structures of the brainstem are undergoing a transition from their positions in the medulla to their characteristic appearance and location in the caudal pons. Which of the following outlined areas includes structures involved only in the conveyance of somatosensory information from the body and movements of the facial muscles?

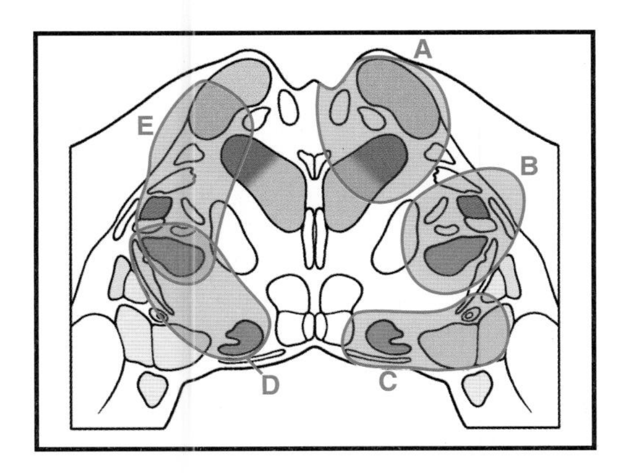

(A) A
(B) B
(C) C
(D) D
(E) E

26 A 70-year-old man presents with sudden-onset weakness and diplopia. The history reveals that the man has had poorly controlled hypertension for about 20 to 25 years and is also being treated for diabetes. The examination reveals an obese man who has a profound weakness of his left upper and lower extremities and an inability to abduct his right eye. All other eye movements are within normal ranges; there are no other cranial nerve deficits and no sensory losses. This man is most likely suffering which of the following?
(A) Benedikt syndrome
(B) Claude syndrome
(C) Collet-Sicard syndrome
(D) Foville syndrome
(E) Gubler syndrome

27 A 49-year-old woman presents with the sudden onset of difficulty walking and tremor. The examination reveals that she has signs and symptoms that clearly suggest that she has suffered a lesion of cerebellar efferent fibers that traverse the pons to eventually influence the function of the motor cortex. CT shows a circumscribed hemorrhagic lesion. Damage to which of the areas indicated in the image below would most likely correlate with this woman's deficit?

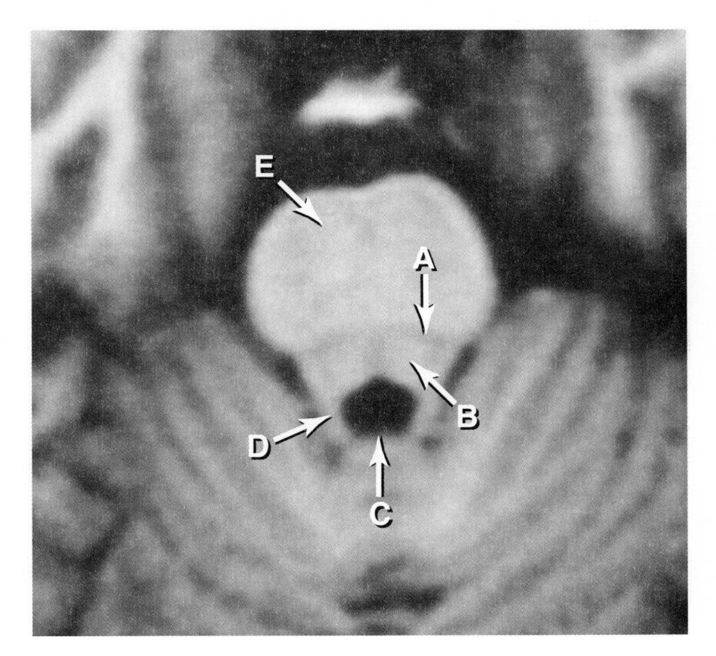

(A) A
(B) B
(C) C
(D) D
(E) E

28 A 5-year-old girl is brought to her pediatrician by her mother. The history provided by the mother indicates that the girl has been irritable, has not been eating well, and complains that "her head hurts." The examination reveals that the girl is unable to abduct her left eye, has a weakness of muscles around her eye and mouth on the left, and is indeed fussy. MRI shows a tumor in the fourth ventricle. Which of the following would explain these deficits in this patient?

(A) Damage from a clivus meningioma
(B) Damage to the facial colliculus
(C) Damage to the hypoglossal trigone
(D) Damage to the vagal trigone
(E) Damage to the vestibular area

29 The red nucleus of the midbrain sends important projections to the inferior olivary nucleus of the medulla on the same side. These rubroolivary fibers traverse the pons en route to the olivary nuclei of the medulla. These fibers descend through the pons as constituent fibers of which of the following?

(A) Central tegmental tract
(B) Dorsal longitudinal fasciculus
(C) Lateral lemniscus
(D) Medial lemniscus
(E) Medial longitudinal fasciculus

ANSWERS

1 **The answer is A: Abducens nucleus/nerve.** The abducens nucleus and nerve is in the territory of the paramedian branches of the basilar artery; damage to either will result in diplopia on lateral gaze to the weak side. Lesions of the oculomotor and trochlear nuclei/nerves will also cause diplopia, but none of these structures is located in the specified vascular territory. Damage to corticonuclear fibers affects motor functions of cranial nerves but does not directly influence eye movement; the lateral lemniscus is a sensory tract for auditory input.

2 **The answer is C (Abducens nucleus, medial longitudinal fasciculus).** This lesion involves the right facial colliculus, the floor of the fourth ventricle affecting the abducens nucleus, and the medially adjacent medial longitudinal fasciculus (MLF). This lesion damages the right abducens motor neurons (serving the right lateral rectus) and the interneurons that project (through the left MLF) to oculomotor neurons serving the left medial rectus. It also damages the right MLF, interrupting axons of interneurons in the left abducens that cross to the right to serve the right medial rectus; the left lateral rectus is intact. Consequently, on attempted lateral gaze to the right and left, the only intact movement is the left eye abducting to the left. This combination of deficits is the one-and-a-half syndrome. The lesion in B affects the left abducens nucleus, but the left MLF is not affected. The other lesions are not compatible with the deficits presented.

3 **The answer is E: Weakness of facial muscles on the right.** A lesion that involves the right abducens nucleus and adjacent structures, as shown here, is located in the facial colliculus. In this position, there is most likely damage to the internal genu of the facial nerve as it wraps around the VIth nucleus. This would result in a profound weakness of the facial muscles on the right side of the face. Other lesions on the right side (D, E) may result in a facial weakness on that side, but not in combination with eye movement disorders. Lesion B will also likely have a facial weakness, but it is on the wrong side when considered in light of the eye movement problems.

4 **The answer is C: Exit of the trigeminal nerve.** The border between the middle cerebellar peduncle and the basilar pons is represented by the exit/entrance of the root of the trigeminal nerve; the middle peduncle is superior to this point, and the basilar pons is inferior to this point. Pontocerebellar fibers originating in the basilar pons become a part of the middle peduncle as they pass the point of the root of the trigeminal nerve. The exit of the abducens nerve is at the caudal border of the pons in line with the preolivary sulcus; the exit of the facial nerve is at the caudal border of the pons in line with the postolivary sulcus. The trochlear nerve is adjacent to the rostral edge of the middle peduncle, momentarily, as it passes through the ambient cistern; the glossopharyngeal nerve exits via the postolivary sulcus.

5 **The answer is B: Facial nerve.** The superior salivatory nucleus contains general visceral efferent (GVE) cell bodies that send their axons via the facial nerve (and through the chorda tympani) to the submandibular ganglion; these are preganglionic parasympathetic fibers. GVE fibers are part of the autonomic nervous system (commonly called the visceromotor system) and are either preganglionic or postganglionic fibers; the latter innervate smooth muscle, cardiac muscle, glandular epithelium, or a combination of these three tissues. The glossopharyngeal (inferior salivatory nucleus), oculomotor (Edinger-Westphal nucleus), and vagus (dorsal motor nucleus) nerves contain GVE preganglionic fibers that distribute, respectively, the otic, ciliary, and intramural ganglia. The accessory nerve does not contain GVE fibers.

6 **The answer is C: Brachium pontis on the right.** In a coronal MRI (this is a T1-weighted image), the hemisphere is seen in cross section, a plane perpendicular to its long axis, while the brainstem is seen in an acutely oblique angle that is almost parallel to its long axis. For example, in this MRI, one can identify the inferior colliculus and portions of the medulla; envision the plane that would be necessary on the midsagittal MRI to include these two rather divergent structures. The white oval, representing demyelination, is located in the brachium pontis (middle cerebellar peduncle, MCP) on the right; the plane through the brainstem is superior to the exit of the trigeminal nerve. The right brachium conjunctivum is clear in this image but is not in the area of the lesion. The juxtarestiform body on the right is at the caudomedial aspect of the MCP, basically opposite the location of the white oval; it only contains fibers passing between vestibular and cerebellar structures. The restiform body is the large bundle of fibers located on the posterolateral aspect of the medulla; it contains a variety of cerebellar afferent fibers and will form the inferior cerebellar peduncle when it joins with the juxtarestiform body in the base of the cerebellum.

7 **The answer is E: Prepontine cistern.** The dark structure at the tip of the black arrow is the basilar artery; it is, along with the origin of the anterior inferior cerebellar artery, small perforating arteries, small veins, and the abducens nerve, one of the most prominent structures within the prepontine cistern. The basilar artery appears dark, not because its walls are imaged but because of what is called a flow void (or signal void). This is due to the fact that blood flowing through this vessel emits no signal because there are no activated protons in blood. The flowing blood appears black. The cisterna magna is posterior to the medulla and between it and the cerebellum; the interpeduncular cistern is at the interpeduncular fossa. The lateral cerebellomedullary cistern is located at the lateral aspect of the medulla; the premedullary cistern is at the medial and inferior aspect of the medulla.

8 **The answer is A: Anterolateral system.** This patient has a loss of pain and thermal sense on the left side of his face and right side of his body; this is an alternating hemianesthesia, also called a crossed sensory deficit. The anterolateral system conveys this information from the spinal cord, through the brainstem, and into the thalamus. The anterior white commissure contains these fibers as they cross the midline; however, a lesion of this structure results in a bilateral sensory deficit restricted to the levels involved. The medial lemniscus transmits proprioception, discriminative touch, and vibratory sense. The ventral posterolateral nucleus of the thalamus conveys all somatosensory information, not just pain and temperature; the posterior limb also conveys all sensation plus corticospinal fibers. These combinations of deficits are lacking in this patient.

9 **The answer is C: Spinal trigeminal tract.** A loss of pain and thermal sense from the face could suggest a trigeminal root lesion or a spinal trigeminal tract lesion; the latter are the central processes of sensory neurons entering the brainstem on the trigeminal nerve. However, when considered in light of the sensory losses on the body, the paralysis of facial muscles, and the intact masticatory function, the lesion is localized to the tract in the caudal pons; the trigeminal nerve is excluded. The spinal trigeminal nucleus pars oralis and pars interpolaris do not convey pain and thermal sense. Ventral trigeminothalamic fibers cross in the medulla and, once crossed, take up a position adjacent to the medial lemniscus. In this respect, they are not located in the area of this patient's lesion, and this patient has no discriminative touch and vibratory sense deficits. Damage to ventral trigeminothalamic tract fibers would, by nature of their position and combined blood supply, most likely also involve the medial lemniscus.

10 **The answer is B: Circumferential branches of the basilar artery.** The structures involved in the lesion in this patient (anterolateral system, spinal trigeminal tract, and motor facial nucleus) are all located in close proximity to each other in caudal regions of the lateral pontine tegmentum. This area of the pontine tegmentum is served by short and long circumferential branches that arise from the basilar artery, pass around the pons, and penetrate the lateral surface of the pons. The anterior spinal artery gives rise to alternating branches to the right and left that serve the medial medulla. Paramedian branches of the basilar artery serve medial regions of the pons; the superior cerebellar artery provides blood supply to the superior aspect of the cerebellar cortex and to most or all of the cerebellar nuclei. The posterior inferior cerebellar artery serves the lateral region of the medulla and the inferior and medial region of the inferior cerebellar surface.

11 **The answer is B: Facial colliculus.** The facial colliculus is in the caudal floor of the fourth ventricle immediately adjacent to the median fissure of the fourth ventricle and rostral to the striae medullares of the fourth ventricle. The facial colliculus is associated with motor functions of cranial nerves and is located medial to the sulcus limitans. Internal to the facial colliculus is the abducens nucleus and the internal genu of the facial nerve. The latter forms a caudal-to-rostral loop around the abducens nucleus. The arcuate eminence is a bony elevation of the anterior/superior aspect of the petrous portion of the temporal bone, indicating the location of the superior semicircular canal. The hypoglossal and vagal trigones are elevations in the caudal floor of the fourth ventricle, signifying the locations of the hypoglossal nucleus and the dorsal motor nucleus of the vagus, respectively. These elevations are caudal to the striae medullares of the fourth ventricle. While the vestibular area and some of its underlying nuclei are visible in this image, they are lateral to the sulcus limitans.

12 **The answer is E: Superior salivatory nucleus.** The superior salivatory nucleus (SSNu) gives rise to General Visceral Efferent (GVE) fibers that join fascicles of the facial nerve lateral to the internal genu of the facial nerve; these preganglionic fibers do not arch through the internal genu. Cell bodies in the SSNu send their axons to the pterygopalatine ganglion, which, in turn, sends postganglionic fibers to the nasal, palatine, and lacrimal glandular epithelium, and to the submandibular gland, which sends postganglionic fibers to sublingual and submandibular salivary glands. The facial motor nucleus and the nucleus ambiguus supply motor (Special Visceral Efferent, SVE) innervation to the muscles of facial expression (via VII) and to pharyngeal and laryngeal muscles (via IX and X). The Edinger-Westphal nucleus sends GVE preganglionic fibers to the ciliary ganglion, which sends postganglionic fibers to the sphincter pupillae muscle. The inferior salivatory nucleus sends its GVE preganglionic axons on the glossopharyngeal nerve to the otic ganglion, which, in turn, innervates the parotid salivary gland.

13 **The answer is C: Medial lemniscus.** This sagittal MRI is slightly lateral to the midline; the main features that indicate this fact are the lack of the cerebral aqueduct, the attenuation of the medulla, and the prominence of the superior and inferior colliculi. The medial lemniscus is the rostrocaudally oriented bundle of fibers located at the junction of the basilar pons and the pontine tegmentum, the latter two of which are clearly evident in this image. The medial lemniscus is part of the posterior column–medial lemniscus system and conveys discriminative touch, vibratory sense, and proprioception/position sense from the body. The anterolateral system (pain and thermal sense from the body) and the lateral lemniscus (auditory) are both sensory pathways but are located in the lateral-most parts of the pontine tegmentum and are not in this sagittal plane. The medial longitudinal fasciculus is located immediately adjacent to the midline and would be seen in a sagittal image that would be close enough to the midline to include the cerebral aqueduct, which is not seen here. The spinal trigeminal tract (pain and thermal sense from the face) is also located in the lateral part of the pontine tegmentum and in the lateral medulla.

14 **The answer is D: Medial longitudinal fasciculus on the left.** This man has a left internuclear ophthalmoplegia (commonly called INO)—left because the left medial rectus muscle is weak on attempted horizontal gaze movements. The medial longitudinal fasciculus (MLF) contains the axons of internuclear neurons that innervate medial rectus motor neurons on the ipsilateral side and whose cell bodies are located in the contralateral abducens nucleus. In this man, the left weak medial rectus specifies a lesion in the left MLF. If the right medial rectus were weak, the lesion would be on the right. Lesions of the abducens nucleus would concomitantly result in weakness of the lateral rectus on the ipsilateral side and medial rectus on the contralateral side. Damage to the MLF bilaterally would result in bilateral INOs; both medial recti are weak on horizontal gaze in either direction. This man has multiple sclerosis.

15 **The answer is B: Diplopia.** The corticospinal tract (weakness of the extremities) and the exiting roots of the abducens nerve are both in the territory of the paramedian branches of the basilar artery in the caudal portions of the pons. Damage to the abducens root results in weakness of the lateral rectus on the ipsilateral side; that eye deviates slightly inward (and will not abduct), thus producing the diplopia. This diplopia is more severe on attempted lateral gaze to the side of the lateral rectus weakness. Aphasia is an inability to comprehend or produce speech or writing, or to read, that is usually related to lesions of the cerebral cortex. Dysarthria (difficulty producing speech), dysphagia (difficulty swallowing), and hoarseness (speech that has a gravelly sound) are usually seen in lesions that involve the roots of the IXth and Xth nerves or the lateral medulla where the nucleus ambiguus resides.

16 **The answer is C (Trigeminal motor nucleus, root of the trigeminal nerve).** This man has a loss of masticatory muscle function on the right and pain and thermal sense loss on the same side of his face; this combination of deficits clearly indicates a lesion involving the motor trigeminal nucleus and the root of the trigeminal nerve, both on the right (choice C). This represents a lesion of lower motor neurons and of primary sensory fibers of the trigeminal nerve. While this lesion also includes the principal sensory nucleus, the overwhelming loss of sensation on the right side of the face and oral cavity would be, by far, the most noticeable. Lesions A and E involve the medial lemniscus (plus the anterolateral system in A) and would result is corresponding sensory losses on the opposite side of the body. The lesion at B is on the left side; the deficits in the patient are on the right. The lesion at D includes the trigeminal motor nucleus and the anterolateral system but avoids the root of the trigeminal nerve; while on the right side, this area of damage does not match the man's deficits.

17 **The answer is C: Loss of facial movements.** The facial muscles arise from the 2nd pharyngeal arch and are innervated by the facial motor nucleus. Failure of these muscles to develop, or for their innervation to be properly established, would represent a serious condition, since the facial muscles participate in sucking and in the initial stages of swallowing. The muscles of mastication arise from the 1st pharyngeal arch and are innervated by the trigeminal motor nucleus; the stylopharyngeus muscle arises from the 3rd pharyngeal arch and is innervated by the glossopharyngeal nerve via cells located in the nucleus ambiguus. The nucleus ambiguus also contains cells that innervate the laryngeal and pharyngeal muscles that arise from the 4th pharyngeal arch. Proper development of muscles innervated by the 3rd and 4th arches is absolutely essential to swallowing. Muscles involved in breathing arise from mesoderm of the body; muscles that move the eyes and tongue arise from head mesoderm, not from the pharyngeal arches.

18 **The answer is B: Basilar artery.** This T2-weighted MRI clearly shows the basilar artery on the basilar pons, twisting slightly to the right as it passes rostrally, giving rise to the superior cerebellar arteries, and then branching into the posterior cerebral arteries (PCA); the PCA is most noticeable on the right. The vein of Galen is not located within the posterior fossa (as designated in the stem) and has no relationship to the basilar pons (which is obvious in the image). The vertebral arteries are bilateral structures and are located at medullary levels. There is no evidence in the history or examination, or in the image, to support the structure as being a clivus meningioma.

19 **The answer is C: Paramedian basilar branches at caudal levels.** This patient has a long tract deficit (hemiplegia) and a cranial nerve deficit; of these two, the cranial nerve deficit is the best localizing sign. Alternating, or crossed, deficits are cardinal signs of a brainstem lesion. The fact that the VIth nerve is involved places the lesion in the caudal and medial area of the pons; recall that paramedian branches of the basilar artery serve these exiting roots as well as the immediately adjacent corticospinal fibers within the basilar pons; the correct answer is C. This specific combination of deficits may also be called a middle alternating hemiplegia. The anterior spinal artery serves the pyramid, exiting fibers of the hypoglossal nerve, and the medial lemniscus. Long circumferential basilar branches serve lateral regions of the pons, completely avoiding structures involved in this patient. Paramedian branches at a more rostral pontine level and at the level of the basilar bifurcation will most likely involve corticospinal fibers but will not involve the abducens nerve. In fact, damage in the territory of paramedian branches from the basilar bifurcation may involve corticospinal fibers and the oculomotor root.

20 **The answer is C: Mesencephalic nucleus.** The temporalis muscle is one of the muscles of mastication. Stretching of the muscle spindles in the temporalis sends an impulse centrally on an A-alpha fiber; the cell body is located in the mesencephalic nucleus; collaterals of the afferent fiber are given off to the trigeminal motor nucleus bilaterally, and the temporalis contracts. This is a monosynaptic myotactic reflex. The facial motor nucleus innervates the muscles of facial expression and is involved in some cranial nerve reflexes, but not the muscle stretch reflex. The Edinger-Westphal nucleus contains preganglionic parasympathetic neurons that participate in the pupillary light reflex. The principal sensory nucleus receives discriminative touch information, and pain and thermal sensations are transmitted via the spinal trigeminal nucleus. These nuclei also participate in cranial nerve reflexes.

21 **The answer is E (Medial longitudinal fasciculus).** The abducens nucleus contains lower motor neurons, which innervate the lateral rectus on the same side, and interneurons whose axons cross the midline to enter the opposite medial longitudinal fasciculus (MLF). Some of these interneurons may arise from the paramedial pontine reticular formation immediately adjacent to the motor abducens nucleus. After these axons of these interneurons enter the MLF, they pass rostrally to terminate on oculomotor neurons that innervate the medial rectus muscle. The medial lemniscus (choice A) conveys discriminative touch and proprioception from the opposite side of the body; the mesencephalic tract (B) is involved in reflex functions mediated by the trigeminal nerve. The superior cerebellar peduncle (C) is a major pathway involved in motor functions; the central tegmental tract (D) contains many ascending and descending connections within the brainstem, such as rubroolivary fibers.

22 **The answer is D: Lateral lemniscus.** The fact that this lesion is in the pontine tegmentum is a general localizing sign. The additional information of a vague somatosensory deficit on the lower extremity strongly suggests that the damage is in the anterolateral area of the tegmentum. This is the area of the tegmentum generally containing the lateral part of the medial lemniscus (containing sensory input from the contralateral lower extremity) and the immediately adjacent superior olivary nuclei and the lateral lemniscus; the latter two are a tract and relay nucleus conveying auditory information. Damage to the cochlear nuclei would result in deafness in one ear; these nuclei are not in the pontine tegmentum but are at the pons-medulla junction. The central tegmental tract conveys information from the contralateral cerebellar nuclei, ipsilateral red nucleus, and other centers.

23 **The answer is A: Basilar pons.** The position at which the trigeminal nerve attaches to the lateral aspect of the pons, clearly seen in this image, marks the point at which the basilar pons is continuous with the middle cerebellar peduncle. The basilar pons is inferior (clinicians frequently refer to this as medial) to the trigeminal nerve, and the middle cerebellar peduncle is superior (clinicians frequently refer to this as lateral) to the trigeminal nerve; this lesion is in the basilar pons immediately inferior to the trigeminal root. There is no synaptic activity at this specific point, only a transition from one designated structure to another. This is an important concept when describing the location of lesions in this general area. The superior cerebellar peduncle, middle cerebellar peduncle, and pontine tegmentum are all clearly evident in this image but contain no portion of the lesion. The inferior cerebellar peduncle and the combination of the restiform body and the juxtarestiform body (= inferior cerebellar peduncle) are caudal to the plane of section represented by this image.

24 **The answer is A: Pharyngeal arch 1.** The muscles of mastication (temporalis, masseter, medial and lateral pterygoids), muscles that participate in swallowing (mylohyoid, anterior digastric), and the tensor tympani and tensor veli palatine arise from the 1st pharyngeal arch. The 2nd arch gives rise to the muscles of facial expression (innervated by the facial nerve), the stylopharyngeus muscle originates from the 3rd arch (innervated by the glossopharyngeal nerve), and the 4th arch gives rise to muscles of the larynx and pharynx (innervated by the vagus nerve), with some contributions from the 6th arch. In the human, there is either no 5th arch or it appears early, is quite rudimentary, and quickly disappears.

25 **The answer is B (Area containing the facial motor nucleus and the anterolateral system).** The areas outlined at B, D, and E all include the facial motor nucleus; these are the lower motor neurons that innervate the facial muscles on the same side of the face. Only the area outlined at B contains the facial motor nucleus and the anterolateral system (ALS). The outline at area D contains the facial nucleus and the abducens nucleus. The area outlined at E contains the facial nucleus and the ALS but also contains the portions of the medial lemniscus and corticospinal tract that are concerned primarily with the lower extremity. The area at A contains the entire corticospinal tract and much of the medial lemniscus, while the area at C encircles the vestibular nuclei, solitary tract and nucleus, and the abducens nucleus.

26 **The answer is D: Foville syndrome.** The Foville syndrome designates a lesion in the medial aspect of the pons at caudal levels that causes damage to corticospinal fibers in the basilar pons and to the exiting roots of the abducens. Both of these structures are in the territory of the paramedian branches of the basilar artery. This is a crossed deficit, also called a middle alternating hemiplegia. The Claude syndrome is resultant to damage in the midbrain tegmentum involving the red nucleus, roots of the oculomotor nerve, and crossed cerebellothalamic fibers. The Benedikt syndrome is the Claude syndrome plus damage to the crus cerebri; basically, Benedikt syndrome is Claude syndrome plus Weber syndrome. Collet-Sicard syndrome is one of the so-called syndromes of the jugular foramen (involving cranial nerve roots of IX, X, and XI) that also causes damage to the root of the XIIth nerve. The Gubler syndrome is also a lesion at the caudal pons, but it extends laterally to encompass the nucleus, or root, of the facial nerve rather than the root of VI.

27 **The answer is D (The superior cerebellar peduncle).** The superior cerebellar peduncle (brachium conjunctivum) is the major efferent pathway out of the cerebellum (choice D). As these fibers leave the cerebellum, they characteristically collect to form the superior cerebellar peduncle which, in turn, forms the lateral wall of the more rostral portions of the fourth ventricle (white in this image). This bundle of cerebellar efferent fibers continues rostrally and crosses the midline to the thalamus, and from the thalamus, information conveyed by this pathway goes to the motor cortex. The dark line, identified at label A, is the location of the medial lemniscus; the pontine tegmentum is indicated at label B. The position of the superior medullary velum (the rostral roof of the fourth ventricle that extends across the midline between the superior cerebellar peduncles) is indicated by C. Choice E not only identifies the general area of the basilar pons but also is specifically indicating the position of the corticospinal fibers located therein.

28 **The answer is B: Damage to the facial colliculus.** The elevation of the facial colliculus is formed by the abducens nucleus (innervates the ipsilateral lateral rectus muscle) and the internal genu of the facial nerve (innervates the ipsilateral facial muscles). Impingement by the tumor may damage these structures within the facial colliculus. It is true that a clivus meningioma may involve the abducens and facial roots, but this lesion is not within the fourth ventricle. Damage to the hypoglossal trigone will result in deviation of the tongue to the side of the weak genioglossus muscle on protrusion, and injury to the vagal trigone may result in visceral dysfunction or loss of vagal reflexes. The vestibular area overlies the vestibular nuclei, and damage to these nuclei results in nystagmus, equilibrium difficulties, and gait problems. Headache is commonly seen with brain tumors and, in this case, may reflect a partial interruption of egress of cerebrospinal fluid.

29 **The answer is A: Central tegmental tract.** Rubroolivary fibers arise from the parvocellular portion of the red nucleus, descend through the pons as part of the central tegmental tract, and end by fanning out over the lateral aspect of the principal olivary nucleus as the amiculum of the olive and terminating in this nucleus. The central tegmental tract also contains other important descending projections, including crossed cerebelloolivary, cerebelloreticular, and crossed and uncrossed cerebellovestibular fibers. The dorsal longitudinal fasciculus contains fibers that arise from the hypothalamus and descend to various brainstem centers, particularly the central gray and certain visceral nuclei. The lateral lemniscus is concerned with auditory information; the medial lemniscus is conveying discriminative touch, vibratory sense, and proprioception from the opposite side of the body. The medial longitudinal fasciculus within the pons contains a predominate population of ascending fibers, including those arising from the vestibular nuclei and influencing the motor nuclei of cranial nerves III, IV, and VI and the axons of abducens internuclear neurons.

Chapter 9

The Midbrain

Recall the Cautionary Tale: The images in this chapter are presented in a Clinical Orientation; this approach emphasizes how structure and function correlate with deficits in a clinical setting. MRI and CT have a universally recognized orientation and laterality. Line drawings, stained sections, and gross brain images in the Clinical Orientation are viewed here in the identical manner that one views an MRI: your right, as the physician-observer, is the patient's left, and the observer's left is the patient's right. The laterality of deficits and reference to connections follow accordingly.

QUESTIONS

Select the single best answer.

1 A 59-year-old woman presents with a sudden onset of diplopia, and with ataxia and tremor, most noticeable on her right side. CT reveals a small hemorrhage in the area indicated in the image below. This woman's signs and symptoms are most indicative of which of the following?

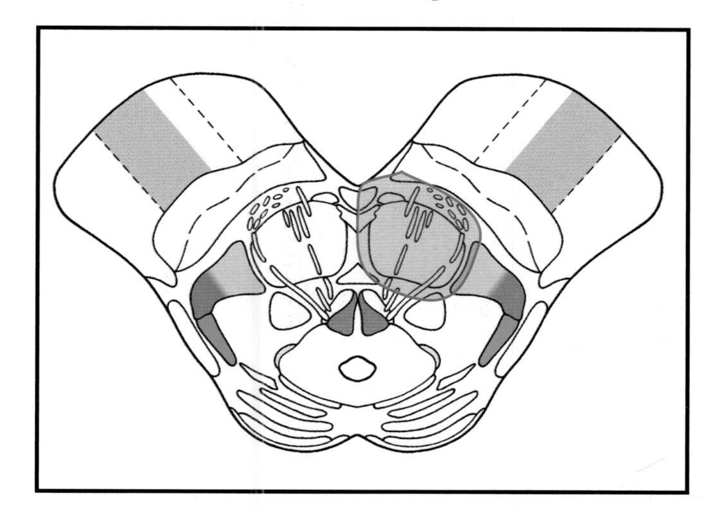

(A) Avellis syndrome
(B) Benedikt syndrome
(C) Claude syndrome
(D) Foville syndrome
(E) Weber Syndrome

2 A 41-year-old man presents with the complaint of double vision and persistent headache. The history reveals that his symptoms have waxed and waned over the past several months, but over the last 2 weeks have become noticeably worse. The examination shows that his right eye is deviated slightly outward and the pupil responds slowly to light. An MRI reveals an aneurysm impinging on one of the structures indicated by the white outlines in the image below. Damage to which of the following structures would most likely explain this patient's deficits?

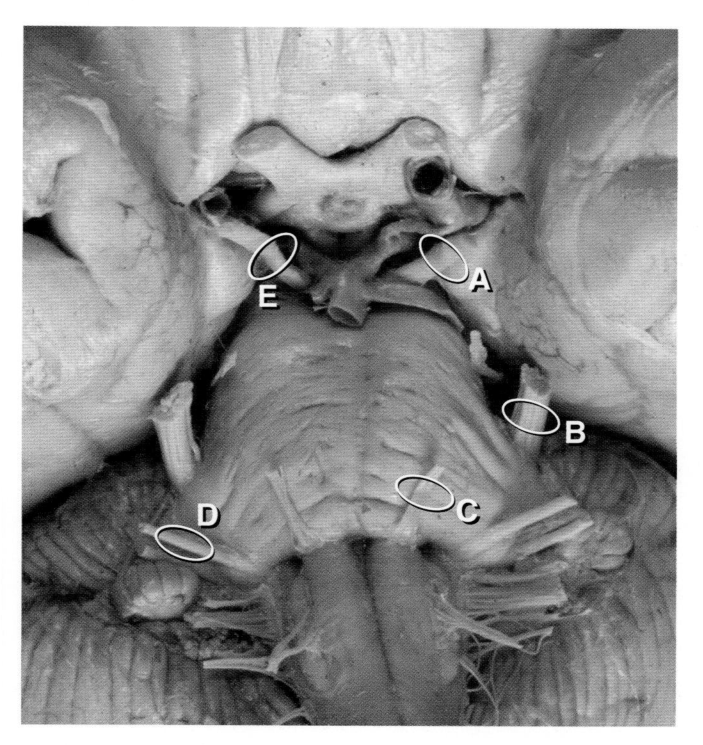

(A) A
(B) B
(C) C
(D) D
(E) E

3 A 71-year-old woman presents to the Emergency Department with the complaint of persistent headache and the recent onset of severe vomiting. The examination reveals that the woman is lethargic and has trouble remembering her address. MRI reveals a large tumor in the quadrigeminal cistern compressing the tectum and obstructing the cerebral aqueduct. A careful neurologic examination may also reveal deficits reflecting damage to which of the following?

(A) Abducens nerve
(B) Chorda tympani
(C) Maxillary nerve
(D) Oculomotor nerve
(E) Trochlear nerve

4 Which of the following structures contains axons that are an essential part of the pupillary light reflex pathway, but whose cells of origin are located in the retina?

(A) Brachium conjunctivum
(B) Brachium of the inferior colliculus
(C) Brachium of the superior colliculus
(D) Posterior commissure
(E) Ventral white commissure

5 A 73-year-old man is followed by his neurologist for progressive motor deficits that include dysmetria and a festinating gait. All of this man's signs and symptoms suggest that he is losing dopamine-containing cells in his brain. Which of the following contains the greatest concentration of these cells?

(A) Locus (nucleus) coeruleus
(B) Nucleus raphe magnus
(C) Substantia nigra, pars compacta
(D) Substantia nigra, pars lateralis
(E) Substantia nigra, pars reticulata

6 The MRI of a 52-year-old man reveals a dark area in the midbrain shown at the white arrow in the image below. The man does not have any frank symptoms. Which of the following would most likely explain this dark area?

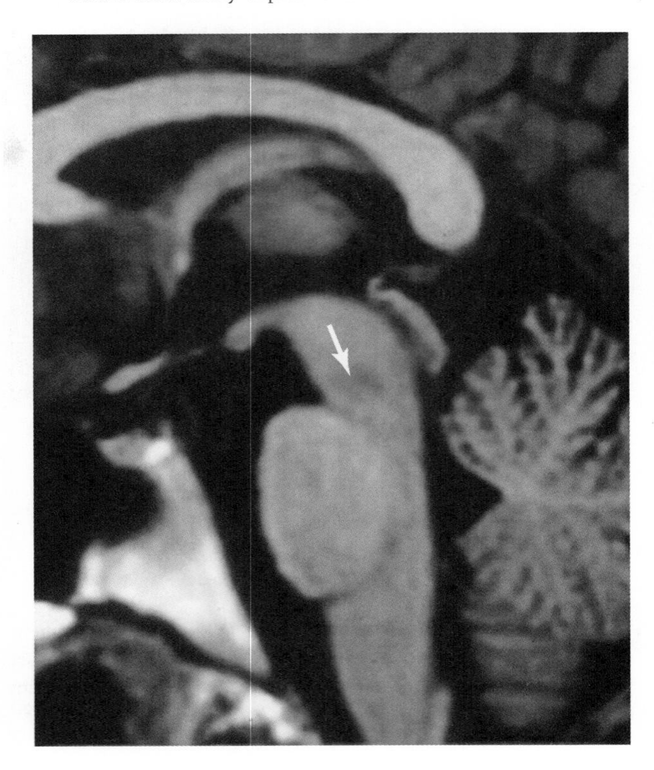

(A) Brainstem glioblastoma multiforme
(B) Crus cerebri
(C) Decussation of brachium conjunctivum
(D) Small hemorrhagic lesion
(E) Substantia nigra, pars compacta

7 Which of the following nuclei or areas contains numerous opiate receptors and cells that project to brainstem centers that, in turn, inhibit the transmission of pain information?

(A) Central nucleus
(B) Interpeduncular nucleus
(C) Nucleus raphe obscurus
(D) Nucleus raphe pallidus
(E) Periaqueductal gray

8 A 67-year-old woman is playing cards with her family when she suddenly moans and falls out of her chair onto the floor. She is unconscious upon arrival at the Emergency Department. The history (provided by her daughter) reveals that she is hypertensive, diabetic, and does not like to take her medications. The examination reveals an obese woman with dilated pupils who responds only to deep pain stimulation. After several days in the Neuroscience Intensive Care Unit, she is still unresponsive, has dilated pupils, regular slow respiration, and an EEG characteristic of a sleep state. Damage to which of the following midbrain areas would most likely explain this woman's condition?

(A) Central gray
(B) Red nucleus
(C) Reticular formation
(D) Substantia nigra
(E) Tectum

9 A 69-year-old man is brought to his physician's office by his wife with the complaint of sudden numbness on his left hand. The examination reveals that the man has a loss of pain and thermal sense on his left side, loss of discriminative and vibratory sense only on his left lower extremity, and all cranial nerves are normal. A lesion in which of the shaded areas would most likely result in these deficits?

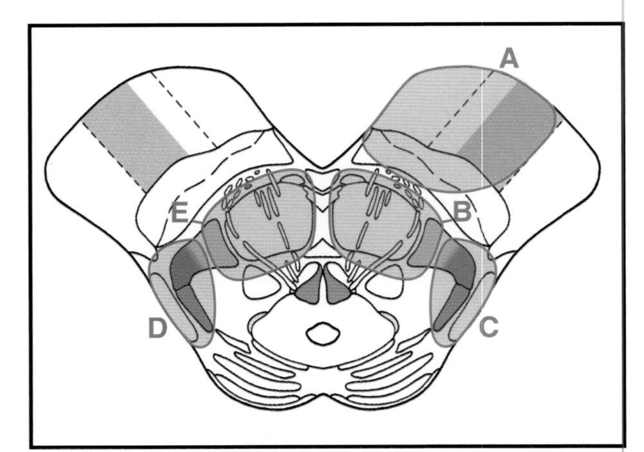

(A) A
(B) B
(C) C
(D) D
(E) E

10 A 57-year-old woman visits her family physician. She explains that her husband noticed that the "dark part" of her eye was larger on the left. The history is unremarkable. The examination reveals a right pupil diameter at about 2.5 mm and the left pupil at 5.0 mm. MRI shows a small aneurysm impinging on the left oculomotor root. Which of the following would explain this condition in this woman?

(A) Damage to large-diameter oculomotor root fibers
(B) Damage to preganglionic sympathetic fibers
(C) Damage to preganglionic parasympathetic fibers
(D) Damage to postganglionic sympathetic fibers
(E) Damage to postganglionic parasympathetic fibers

11 A 43-year-old woman complains to her primary care physician that she "can't hear as well as I used to." After a preliminary examination, she is referred to an audiologist who determines that the woman is not deaf in either ear, even though her perception of sounds and her ability to localize sound in space is altered. Which of the following midbrain structures might be affected to result in these changes?

(A) Brachium conjunctivum
(B) Crus cerebri
(C) Inferior colliculus
(D) Mesencephalic nucleus
(E) Superior colliculus

12 Testing the integrity of the papillary light reflex pathway is an integral part of a neurological examination. Which of the following structures, outlined in white on the image below, represents the location of essential synaptic contacts in this pathway?

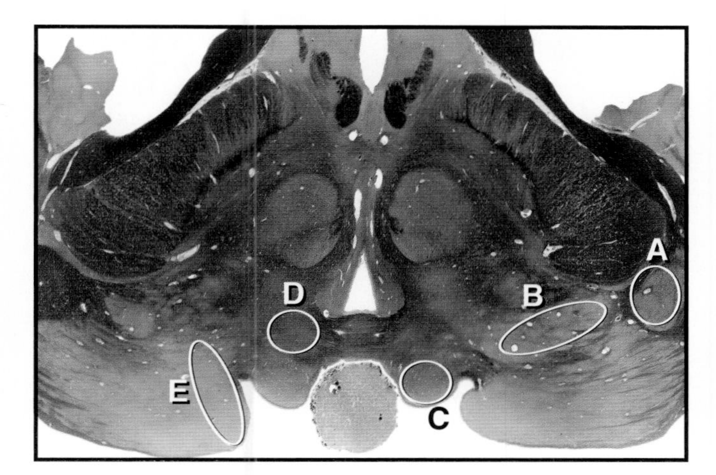

(A) A
(B) B
(C) C
(D) D
(E) E

13 A 73-year-old man presents with hearing and visual complaints. The examination reveals that the man has hearing problems and eye movement difficulties, but he is neither deaf nor blind. An MRI reveals an infarct of the superior and inferior colliculi on the right side. The artery that serves this infracted area arises from which of the following?

(A) A_1
(B) M_1
(C) M_2
(D) P_1
(E) P_2

14 During a routine neurological examination on a 49-year-old man, the physician notices that a light shined in the left eye does not elicit a direct pupillary reflex but does elicit a consensual pupillary reflex. Which of the following would most likely explain this deficit in this man?

(A) Damage to the left optic tract
(B) Damage to the left oculomotor nerve
(C) Damage to the left pretectal nucleus
(D) Damage to the right optic nerve
(E) Damage to the right oculomotor nerve

15 A 71-year-old woman is brought to the Emergency Department by her daughter. The history of the current event, given by the daughter, reveals that the woman suddenly became generally weak while working in her garden. During the examination, the physician notices that the woman's tongue deviates to the left on protrusion and that her uvula deviates to the right when she says "ah." MRI shows a lesion in the midbrain. Damage to which of the areas outlined on the image below would most likely explain this woman's deficits?

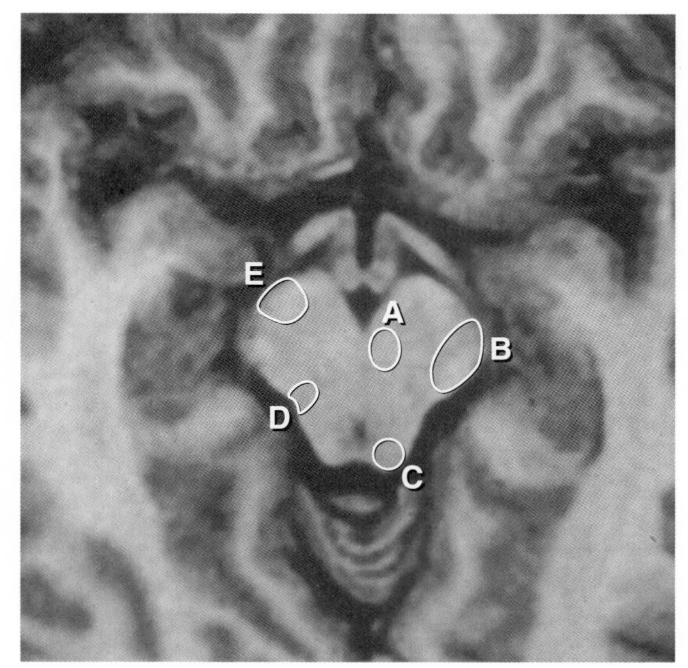

(A) A
(B) B
(C) C
(D) D
(E) E

16 A 57-year-old man visits his family physician with the complaint of clumsiness. The examination reveals that the man has a tremor in his left hand when he is sitting still in the chair, a noticeable decrease in spontaneous movements when he stands up, and he walks with short shuffling steps. Damage to, or loss of, which of the following midbrain structures would most likely explain this man's deficits?

(A) Crus cerebri of the basis pedunculi
(B) Red nucleus, pars magnocellularis
(C) Red nucleus, pars parvocellularis
(D) Substantia nigra, pars compacta
(E) Substantia nigra, pars reticulata

17 Which of the following structures contains rubrospinal fibers as they cross to the opposite side of the brainstem?

(A) Central tegmental tract
(B) Decussation of internal arcuate fibers
(C) Dorsal/posterior tegmental decussation
(D) Motor decussation
(E) Ventral/anterior tegmental decussation

18 A 25-year-old man is brought to the Emergency Department by EMS personnel who indicate that he was found lying on the sidewalk. The man is somnolent, but can be roused, and he complains of numbness. Observations of the man's arms reveal that he uses illegal drugs. The examination reveals that he has a loss of pin-prick sensation on the left side of his body, and that he has difficulty localizing sound, but is not deaf in either ear. MRI shows that he has a midbrain lesion. Damage to which of the areas indicated on the image below would most likely explain this man's deficits?

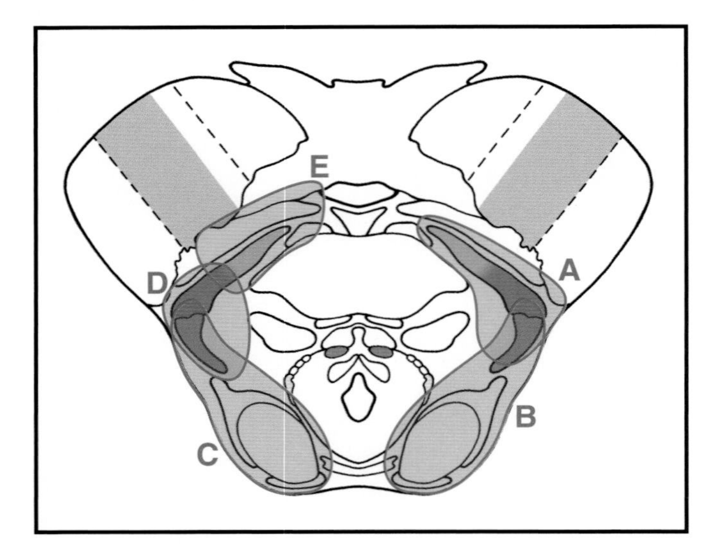

(A) A
(B) B
(C) C
(D) D
(E) E

19 The MRI of a 76-year-old woman reveals a small vascular defect in the middle third of the crus cerebri. Which of the following would be most directly affected by this lesion?

(A) Corticospinal fibers
(B) Frontopontine fibers
(C) Occipitopontine fibers
(D) Parietopontine fibers
(E) Temporopontine fibers

20 Descending projections from the midbrain participate in pathways that are directly involved in the inhibition of pain in the spinal trigeminal nucleus, in the pars caudalis, and in the posterior horn of the spinal cord. This region of the midbrain contains many cells with opiate receptors. Which of the following is the most likely location of these cells?

(A) Midbrain reticular formation
(B) Periaqueductal gray
(C) Red nucleus, pars magnocellularis
(D) Red nucleus, pars parvocellularis
(E) Substania nigra, pars reticulata

21 Which of the following structures is indicated by the white arrow in the image below?

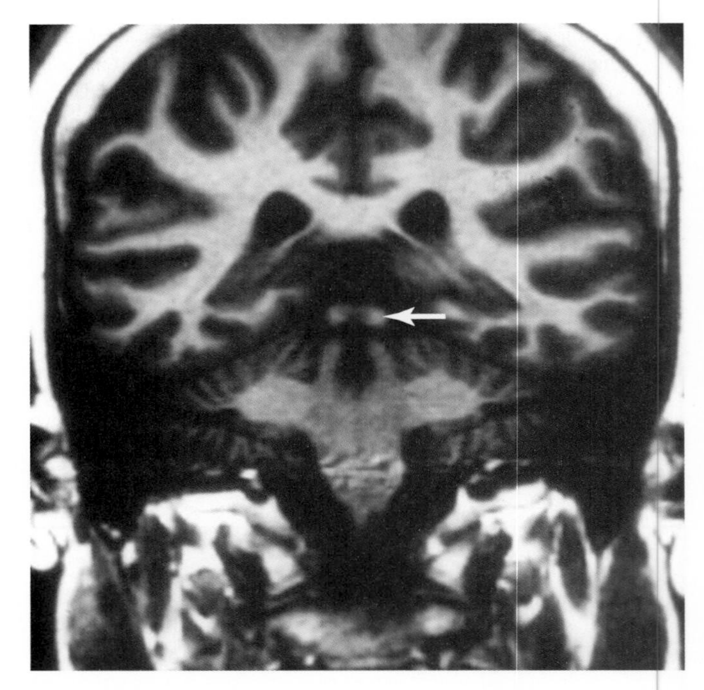

(A) Brachium conjunctivum
(B) Mammillary body
(C) Inferior colliculus
(D) Pineal body
(E) Superior colliculus

22 A 68-year-old man has a large hemisphere stroke. He is somnolent, and over the next several days becomes decorticate. Which of the following midbrain structures is responsible for the posturing of the upper extremities in this patient?

(A) Central tegmental tract
(B) Crus cerebri
(C) Red nucleus
(D) Reticular formation
(E) Substantia nigra

23 A 79-year-old woman is brought to the Emergency Department. The history of the current event, provided by her family, was of a sudden onset of weakness while watching TV. The examination reveals a profound weakness of the upper and lower extremities on the left side, and loss of most voluntary movements of the right eye, although she is able to abduct her right eye. Which of the following would most likely also be seen in this patient, subsequent to the careful neurological examination?

(A) Constriction, pinhole, of the left pupil
(B) Deviation of the tongue to the right on protrusion
(C) Deviation of the uvula to the left on vocalization
(D) Dilation of the left pupil
(E) Weakness/drooping of the left lower face

24 Which of the following structures is indicated by the white arrows in the image below?

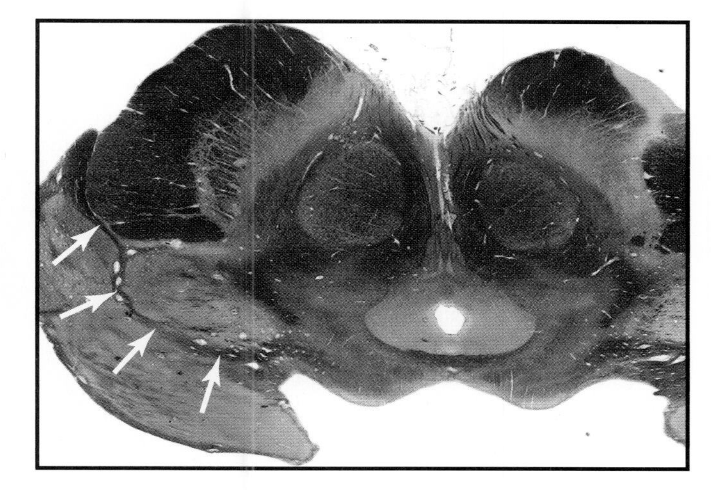

(A) Brachium of the inferior colliculus
(B) Brachium of the superior colliculus
(C) Decussation of the superior colliculi
(D) Lateral lemniscus
(E) Optic radiations

25 A 63-year-old man is brought to the Emergency Department after collapsing at his job. The examination reveals a frank right-sided weakness and diplopia. A CT reveals a hemorrhagic lesion involving the medial two-thirds of the crus cerebri and much of the substantia nigra. These two structures collectively comprise which of the following?

(A) Atrium
(B) Basis pedunculi
(C) Cerebral peduncle
(D) Tectum
(E) Tegmentum

ANSWERS

1 **The answer is C: Claude syndrome.** This woman's deficits and the position of her lesion indicate damage to the fibers of the oculomotor nerve as they are passing through the midbrain and to the cerebellothalamic fibers and the red nucleus. Recall that rubrospinal fibers cross (in this case to the right side) and that the cerebellothalamic fibers, through the thalamus, influence the left motor cortex, which, in turn, sends corticospinal fibers to the right side of the body. This is the Claude syndrome. The Weber syndrome involves much of the crus cerebri, portions of the substantia nigra, and the exiting oculomotor roots; this results in an alternating (or crossed) deficit, body on one side, cranial nerve on the other. The Benedikt syndrome is a combination of the Weber and Claude syndromes, while the Foville syndrome (sometimes called a syndrome of the caudal pontine base) involves the abducens root and the corticospinal fibers in the basal pons; this is also an alternating deficit, but with a different cranial nerve. The Avellis syndrome is one of several syndromes related to the jugular foramen with deficits that especially reflect damage to the cranial nerves passing through this foramen.

2 **The answer is E (Right oculomotor nerve).** Double vision (diplopia), right eye deviated outward (the lateral rectus is intact and the medial rectus is weak), and slowly responsive pupil (parasympathetic axons to sphincter pupillae muscle compromised) all point to the aneurysm affecting the right oculomotor root. Diplopia will result from damage to the left oculomotor nerve (choice A, deficits are on the wrong side) and from damage to the abducens nerve (diplopia, but no effect on pupil). In the case of abducens damage (C), the affected eye would also deviate slightly inward (right lateral rectus weak and the right medial rectus intact). Impingement on the trigeminal nerve (B) would result in loss of pain and thermal sense on the left side of the face and oral cavity and possible weakness of masticatory muscles on that side. The overwhelming sign resultant to damage of the facial nerve (D) would be a weakness of the muscles of facial expression on the right side.

3 **The answer is E: Trochlear nerve.** The trochlear nerve exits the brainstem immediately caudal to the inferior colliculus to enter the superior (quadrigeminal) cistern. It then passes around the lateral aspect of the midbrain as one of the important structures within the ambient cistern. The abducens nerve exits ventrally and passes through the prepontine cistern; the oculomotor nerve is related to the interpeduncular cistern. The chorda tympani and maxillary nerve are both peripheral nerves, the former conveying taste fibers from the lingual nerve to the facial nerve and the latter conveying general sensation from the mandible, mandibular teeth, and portions of the oral cavity into the brainstem via the trigeminal nerve.

4 **The answer is C: Brachium of the superior colliculus.** A light shined in the right eye initiates impulses in the optic nerve, which travel through the fibers of the optic chiasm and tract, pass directly into the brachium of the superior colliculus, and enter (and terminate) in the pretectal nucleus. The pretectal nucleus projects bilaterally to the parasympathetic preganglionic neurons of the oculomotor nerve which, after terminating in the ciliary ganglion, innervate the sphincter pupillae muscle. The brachium of the inferior colliculus transmits fibers from the inferior colliculus to the medial geniculate nucleus; the brachium conjunctivum contains cerebellar efferent fibers that are en route to the thalamus. The posterior commissure contains a variety of decussating fibers, including some that originate from the pretectal nucleus; the ventral white commissure contains crossing fibers in the spinal cord conveying pain and thermal information.

5 **The answer is C: Substantia nigra, pars compacta.** The dopamine-containing cells of the substantia nigra, pars compacta, are characterized by their black appearance (they contain melanin) in a fresh brain slice. These cells project to the neostriatum. A patient with Parkinson disease, especially if advanced, will have lost many of these dark cells and the dopamine contained therein. The other parts of the substantia, the reticulata (which is similar in size to the compacta) contain GABAergic cells that project mainly to the thalamus, tectum, and tegmental nuclei, and the lateralis projections are not well defined. The locus coeruleus (sometimes called the nucleus pigmentosus pontis because of its location and dark color) contains a small number of pigmented cells that send a noradrenergic projection to wide areas of the central nervous system; the nucleus raphe magnus sends a serotonergic projection to the nuclei of the brainstem and spinal cord.

6 **The answer is C: Decussation of the brachium conjunctivum.** The decussation brachium conjunctivum, also called the decussation of the superior cerebellar peduncle, is a large fiber bundle located on the midline at a cross-sectional level through the inferior colliculus; this plane will also frequently include the trochlear nucleus. The fact that the cerebral aqueduct is clearly seen in this sagittal image indicates that the plane of the image is on the midline. The fibers in this decussation originate from cells of the cerebellar nuclei and, after decussating, will terminate in the thalamus and brainstem. The crus cerebri and the substantia nigra are also seen in sagittal MRI, but the plane of the image will be several millimeters from the midline, will not show the cerebral aqueduct, and will show an obviously smaller basilar pons and a sharp appearance of the inferior colliculus. Lesions within the brainstem, especially one that would be this big, would be symptomatic. In addition, hemorrhagic lesions, or brain tumors, would have textures of light and dark and would most likely show evidence of edema.

7 **The answer is E: Periaqueductal gray.** This important region of the midbrain receives spinomesencephalic fibers and contains cells with many opiate receptors that project bilaterally to the nucleus raphe magnus (NRM). In turn, the NRM gives rise to raphespinal fibers that presynaptically inhibit the transmission of pain and thermal information in the posterior horn, entering on C- and A-delta fibers. The central nucleus (of the inferior colliculus) is concerned with auditory information; the interpeduncular nucleus receives the habenulointerpeduncular tract and has diffuse connections with the limbic system. The nuclei raphe pallidus and raphe obscurus are located in the medulla and provide serotonergic projections to the brainstem and spinal cord.

8 **The answer is C: Reticular formation.** The differential diagnosis of a profoundly unresponsive state (coma) includes structural, metabolic, or psychogenic causes. Anatomically, a sudden lesion resulting in coma is usually localizable to one of three areas: (1) large bilateral cortical damage; (2) large diencephalic lesions; or (3) interruption of the ascending reticular activating system. The midbrain reticular formation contains the cuneiform and subcuneiform nuclei and traversing fibers ascending from the reticular nuclei of lower brainstem levels. This area of the midbrain receives input from lower levels of the neuraxis and contains passing fibers that project to diencephalic centers that are involved in initiating and maintaining a wakeful state. The central gray influences the descending inhibition of pain, the red nucleus and substantia nigra are essential to motor function, and the tectum relates to auditory and vestibular systems.

9 **The answer is D (Lesion of the lateral part of the medial lemniscus and anterolateral system on the right).** The location of the lesion on the patient's right side correlates with the deficits on the left side of his body. This lesion includes the lateral portion of the medial lemniscus (the lower extremity is topographically represented in this area), entire anterolateral system, and the brachium of the inferior colliculus. The loss of discriminative touch and vibratory sense on the lower left extremity relates to the medial lemniscus damage, and the total left-sided loss of pain and thermal sense correlates with the destruction of the anterolateral system. Inclusion of the brachium of the inferior colliculus produces insignificant deficits. Lesion C is on the wrong side, and lesions A, B, and E all involve fibers of the oculomotor nerve; this man has no cranial nerve deficits. In addition, lesion A involves corticonuclear fibers, and there is no clinical evidence of damage to these fibers.

10 **The answer is C: Damage to preganglionic parasympathetic fibers.** The oculomotor root contains preganglionic parasympathetic fibers that arise from cells in the area of the Edinger-Westphal nucleus and terminate in the ciliary ganglion. Postganglionic fibers arise in the ciliary ganglion and innervate the sphincter pupillae muscle of the iris. These small diameter fibers may be affected first by pressure on the oculomotor root, not only due to their small size, but also due to their superficial location within the nerve. The larger-diameter motor axons are more centrally located in the nerve and are affected by pressure secondary to the visceromotor fibers. There are no sympathetic preganglionic nuclei associated with cranial nerves. The postganglionic sympathetic fibers that serve the head, and the dilator pupillae muscle of the iris, originate from the superior cervical ganglion.

11 **The answer is C: Inferior colliculus.** The inferior colliculus receives ascending auditory input via the lateral lemniscus and relays this information to the medial geniculate nucleus via the brachium of the inferior colliculus. At this point in the auditory pathway, the information is bilateral, and lesions may alter the perception, or localization, of sound, but not result in deafness in either ear. The brachium conjunctivum is conveying cerebellar efferents to the thalamus, and the crus cerebri contains corticospinal and corticonuclear fibers among others. The mesencephalic nucleus is concerned with proprioceptive information transmitted on the trigeminal nerve; the superior colliculus is related to visual motor function.

12 **The answer is D (pretectal, or pretectal olivary, nucleus).** The pretectal nucleus receives afferents via axons of retinal ganglion cells of both eyes (partial crossing of optic fibers in the optic chiasm) and projects bilaterally to the Edinger-Westphal nucleus and immediately adjacent cells. Preganglionic cells receiving input from the pretectal nucleus project to the ciliary ganglion via the oculomotor nerve, and from this ganglion to the sphincter pupillae muscles of the iris. The lateral geniculate nucleus (choice A) and the medial geniculate nucleus (B) are synaptic centers for visual and auditory information, respectively. The superior colliculus (the rostral part of which is indicated at C) and the medial pulvinar (E) are concerned with visual motor function and have complex connections with extrastriate visual cortices.

13 **The answer is D: P_1: The first segment of the posterior cerebral artery.** These designations for the main segments of the cerebral arteries (A = anterior cerebral; M = middle cerebral; P = posterior cerebral) are commonly used in clinical medicine. The colliculi are served by the quadrigeminal artery, which is a branch of P_1. This vessel arises from P_1, encircles the midbrain by passing through the ambient cistern (en route, it provides branches to the lateral midbrain), and enters the superior cistern where it branches to serve the colliculi. A_1 gives rise to many small penetrating vessels that serve the anterior hypothalamus and closely located optic structures; the lenticulostriate arteries are especially important branches of M_1. Those middle cerebral branches that are located on the insular cortex are M_2. In addition to the quadrigeminal artery, the thalamoperforating artery and small perforating vessels arise from P_1. Important P_2 branches include the medial and lateral posterior choroidal arteries and the thalamogeniculate artery.

14 **The answer is B: Damage to the left oculomotor nerve.** The afferent limb of the pupillary light reflex is via the optic nerve, and the efferent limb is via the oculomotor nerve. In this case, the afferent limb from the left eye is functional because there is a consensual response; if the patient was blind in his left eye, neither response, direct or consensual, would be present. In this man, the left efferent limb, the left oculomotor nerve, is affected; there is a loss of the direct response. On the other hand, the right oculomotor nerve is intact because there is a consensual response. Each pretectal nucleus projects bilaterally to both Edinger-Westphal nuclei. A patient with a lesion involving the left pretectal nucleus will still have a direct response; which this man has. A light shined in the left eye will activate retinal ganglion cells projecting to the left and right pretectal nuclei (recall the optic chiasm). The right pretectal nucleus influences the right and left Edinger-Westphal nuclei; there is a direct response in the left eye.

15 **The answer is E (Middle third of the crus cerebri containing corticospinal and corticonuclear fibers).** Corticospinal and corticonuclear fibers are located in the middle part of the crus cerebri. Damage to corticospinal fibers accounts for weakness (on the contralateral side); damage to the immediately adjacent corticonuclear fibers explains the cranial nerve deficits. These cranial nerve deficits are especially evident during attempted voluntary movements. The red nucleus (choice A) is involved in the motor system but not in cranial nerve function; the lateral parts of the crus cerebri (B) contain parieto-, occipito-, and temporopontine fibers, damage to which may result in a general ataxia, but no cranial nerve deficits. The superior colliculus (choice C) is functionally related to visual motor function and visual orientation. The anterolateral system and portions of the medial lemniscus are in the area represented in D; damage in this area would result in corresponding somatosensory deficits.

16 **The answer is D: Substantia nigra, pars compacta.** This man's deficits clearly point to the fact that he has early stages of Parkinson disease. The principal cause of this disease is the loss of dopamine-containing cells in the substantia nigra, pars compacta; these are also the melanin-containing cells of the substantia nigra. Parkinson disease is a neurodegenerative disease that may have multiple causes. The substanita nigra, pars reticulata does not contain dopamine cells; the crus cerebri contains corticopontine, corticonuclear, and corticospinal fibers; lesions of the latter two result in frank paralysis. The red nucleus, pars magnocellularis gives rise to rubrospinal fibers that primarily innervate flexor motor neurons of the cervical spinal cord, while the pars parvocellularis projects to other brainstem areas, such as the reticular formation and inferior olive, as one component of the central tegmental tract.

17 **The answer is E: Ventral/anterior tegmental decussation.** Rubrospinal fibers arise from the magnocellular portion (the caudal portion) of the red nucleus, cross the midline in the ventral or anterior tegmental decussation, and descend through lateral portions of the brainstem tegmentum and medulla. These fibers preferentially target flexor motor neurons at cervical cord levels. The central tegmental tract is located in the brainstem tegmentum between the levels of the red nucleus and principal olivary nucleus and contains fiber populations that interconnect various brainstem structures: sort of an intrabrainstem highway. This tract also contains projections from the cerebellum that leave the decussation of the brachium conjunctivum and descend within the brainstem. The decussation of the internal arcuate fibers is the crossing of fibers comprising the posterior column–medial lemniscus system; the motor decussation is the crossing of corticospinal tracts; both are in the caudal medulla. The dorsal/posterior tegmental decussation is the crossing of tectospinal fibers.

18 **The answer is C (Area of inferior colliculus and anterolateral system on the right).** The inferior colliculus processes auditory information that it receives from both ears; damage to this structure will not cause frank deafness in either ear but will alter the perception or localization of sounds. The anterolateral system (ALS) is damaged on the right side; this explains his left-sided loss of pain sensation. The area at A includes the ALS and the medial lemniscus (discriminative touch, vibratory sense, proprioception), but not auditory structures; the area at B has the correct structures, but they are on the wrong side in relation to the man's pain deficit. Area D contains the ALS on the correct side, but also contains the lower extremity portion of the medial lemniscus; the man has no deficits that correlate with damage to that structure. Area E encompasses the upper extremity portion of the medial lemniscus and much of the substantia nigra; this man has no deficits to correlate with such damage.

19 **The answer is A: Corticospinal fibers.** The middle one third of the crus cerebri of the midbrain contains corticospinal fibers; corticonuclear fibers are located in the medial part of this specific area. Damage to this particular area of the crus results in a profound weakness of the contralateral upper and lower extremities and may also result in certain important and characteristic cranial nerve deficits. The medial one-third of the crus contains frontopontine fibers; and parieto-, occipito-, and temporopontine fibers are located in the lateral one-third of the crus. Damage to larger parts of the corticopontine projection may result in ataxia, or in ataxia and other deficits reflecting damage to these projections plus adjacent structures.

20 **The answer is B: Periaqueductal gray.** Cells located in the periaqueductal gray (PAG) project to the nucleus raphe magnus (NRM) and, from there, the cells of the NRM project to the spinal trigeminal nucleus and the posterior horn for inhibition of pain transmission. This influence on pain transmission is mainly through the mechanism of presynaptic inhibition. The cells of these PAG-to-NRM projections have numerous opiate receptors. The reticular formation of the midbrain is small and participates primarily in ascending pathways. The magnocellular (large celled) and the parvocellular (small celled) portions of the red nucleus project to the contralateral cervical spinal cord and to the ipsilateral reticular formation and inferior olivary nucleus, respectively. The substantia nigra, pars reticulata, contains GABAergic cells that project to the striatum, thalamus, and superior colliculus.

21 **The answer is C: Inferior colliculus.** The plane of this T1-weighted MRI (CSF is black, it is hypointense) passes through the most caudal portions of the midbrain and includes both inferior colliculi; it also passes through the superior cerebellar peduncles (brachium conjuctivum), both of which are in the image below the two colliculi. The plane of the image also includes the middle cerebellar peduncle and portions of the pontine tegmentum. The mammillary bodies are paired structures located adjacent to each other on the inferior aspect of the hypothalamus. The pineal body is located in the superior cistern, as are the inferior colliculi, but is positioned on the midline at a more rostral plane. In fact, the pineal and superior colliculi would most likely appear, and together, in a more rostral coronal plane. An image through the superior colliculus would also include many brainstem structures including the medulla.

22 **The answer is C: Red nucleus.** In decorticate rigidity, the upper extremities are flexed due to the excessive drive through the rubrospinal projections that preferentially target flexor motor neurons in the cervical spinal cord. Rubrospinal projections arise from the large-celled part of the red nucleus and cross in the dorsal tegmental decussation. In this respect, rubrospinal fibers share, with corticospinal and tectospinal fibers, the characteristic of decussating to influence spinal motor neurons on the opposite side. The central tegmental tract conveys ipsilateral rubroolivary and contralateral cerebelloreticular and cerebelloolivary fibers. Clinically, the most important components of the crus cerebri are corticospinal and corticonuclear fibers; the reticular formation of the midbrain is not significantly involved in motor function. The substantia nigra projects to, among other targets, the neostriatum; damage to the nigra results in Parkinson disease from which this man is not suffering.

23 **The answer is E: Weakness/drooping of the left lower face.** The occurrence of a deficit of the corticospinal tract function coupled with oculomotor deficits signifies the following: (1) this is a midbrain lesion (oculomotor deficits are the best localizing sign); (2) it is called a superior alternating hemiplegia; (3) the eponym is the Weber syndrome; and (4) in addition to corticospinal fibers and the oculomotor nerve, the lesion also involves corticonuclear fibers, which influence the function of cranial nerve motor nuclei. The corticonuclear fibers within the crus cerebri are located medial to the corticospinal tract and are involved in the lesion when the Weber syndrome includes corticospinal and oculomotor fibers. Damage to corticonuclear fibers in the right crus (as in this woman) results in deviation of the tongue to the left (not the right, B), deviation of the uvula to the right (not the left, C), and dilation of the right pupil (not the left, D). An abnormally small left pupil (A) is not a feature of the Weber syndrome.

24 **The answer is B: Brachium of the superior colliculus.** The brachium of the superior colliculus is a small, but important, bundle of fibers that leaves the optic tract (seen in this image), courses in a shallow groove between the pulvinar and the medial geniculate body (also seen), and ends in the pretectal nucleus. These fibers represent a major part of the entire afferent limb of the pupillary light reflex pathway. The brachium of the inferior colliculus is located on the lateral aspect of the midbrain and extends between the inferior colliculus and the medial geniculate nucleus. The decussation of the superior colliculi is a connection across the midline between these two structures; the lateral lemniscus passes through the lateral area of the pontine tegmentum and caudal midbrain to end in the inferior colliculus. Optic radiations arise from cells of the lateral geniculate nucleus and course to the visual cortex as one fiber population of the retrolenticular limb of the internal capsule.

25 **The answer is B: Basis pedunculi.** The crus cerebri plus the substantia nigra together form the basis pedunculi. The tegmentum is internal to the substantia nigra and below a line drawn across the midbrain at the level of the cerebral aqueduct; everything between the nigra and this line is the tegmentum. Larger structures in the tegmentum include the decussation of the superior cerebellar peduncle, red nucleus, medial lemniscus, and anterolateral system. Smaller, but equally important, structures of the tegmentum include the oculomotor and trochlear nuclei, medial longitudinal fasciculus, and the central tegmental tract. The cerebral peduncle is the basis pedunculi plus the tegmentum; the tectum is that part of the midbrain above the line across the cerebral aqueduct, basically the nuclei forming the superior and inferior colliculi. An atrium is a chamber or cavity; there is an atrium of the lateral ventricle, atria of the heart, and atria in the nasal cavity.

Chapter 10

An Overview of Cranial Nerves of the Brainstem

The coverage in this chapter is related primarily to cranial nerves in the brainstem (nuclei and internal course of roots), foramina through which they exit the skull, and general principles concerning their peripheral distribution. Information on the cortical influences on cranial nerve function and the functional relationships of ascending and descending tracts to cranial nerves is covered in more depth in the chapters on sensory the motor pathways.

Recall the Cautionary Tale: The images in this chapter are presented in a Clinical Orientation; this approach emphasizes how structure and function correlate with deficits in a clinical setting. MRI and CT have a universally recognized orientation and laterality. Line drawings, stained sections, and gross brain images in the Clinical Orientation are viewed here in the identical manner that one views an MRI: your right, as the physician-observer, is the patient's left, and the observer's left is the patient's right. The laterality of deficits and reference to connections follow accordingly.

QUESTIONS

Select the single best answer.

1 A 21-year-old man is transported to the Emergency Department from the site of a motorcycle collision. He is unconscious upon arrival, and the initial neurological examination reveals a dilated left pupil. He has facial and scalp injuries, "road rash," and a fractured femur. Three hours later he is conscious, has had a CT, and further examination reveals a loss of all voluntary movement of the left eye, a drooping left eyelid, and loss of pinprick sensation on his left forehead, in addition to the still dilated left pupil. CT revealed a skull fracture. Significant damage to structures traversing which of the following would most likely explain this man's deficits?

(A) Infraorbital foramen
(B) Inferior orbital fissure
(C) Supraorbital foramen
(D) Superior orbital fissure
(E) Optic canal

2 Which of the following cranial nerves pass through the foramen identified by the circle in the image below?

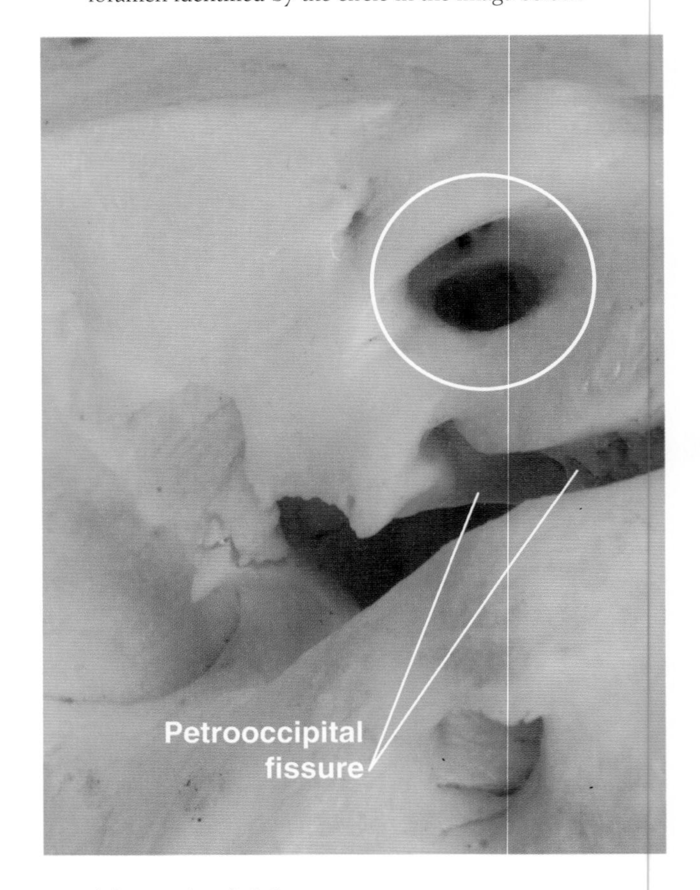

(A) Facial and abducens
(B) Hypoglossal and spinal accessory
(C) Glossopharyngeal and vagus
(D) Vagus and spinal accessory
(E) Vestibulocochlear and facial

3 A 41-year-old man is brought to the Emergency Department from the site of a motor vehicle collision. He is conscious, has facial abrasions, a broken nose, and an apparent shoulder dislocation. CT confirms the broken nose, and reveals a fracture of the proximal humerus, and a basal skull fracture that passes through the jugular foramen. Based on these observations, which of the following cranial nerves might be damaged?

(A) III, IV, VI
(B) V (motor + sensory)
(C) VII, VIII
(D) IX, X, XII
(E) IX, X, XI

4 A 23-year-old woman is brought to the Emergency Department from the site of a motorcycle collision. The initial examination reveals a compound fracture of the right humerus, significant "road rash" (extremities, right side of face), and a probable broken nose. CT also shows a fractured right clavicle and a basal skull fracture transecting the structure indicated by the arrow in the image below. Which of the following deficits would most likely be seen in this woman?

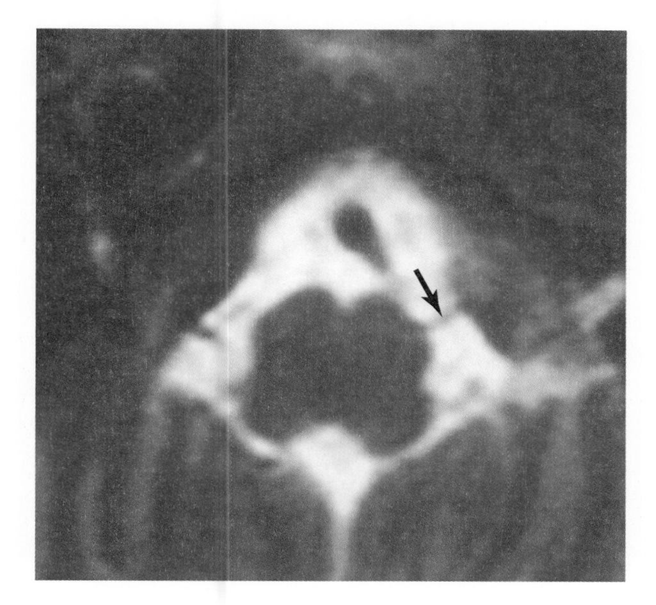

(A) Dysarthria and dysphagia
(B) Loss of facial movements on left side
(C) Tongue deviates to right on protrusion
(D) Tongue deviates to the left on protrusion
(E) Unable to elevate left shoulder

5 Which of the following combinations of cranial nerves may be collectively referred to as the cranial nerves (CNs) of the pons-medulla junction?

(A) CNs III, IV, VI
(B) CNs V, VI, IX
(C) CNs XI, X, XII
(D) CNs VI, VII, VIII
(E) CNs VIII, IX, X

6 You gently touch a wisp if cotton to the left cornea of a 23-year-old man during a routine neurological examination, and the left eye blinks (direct response) and the right eye blinks (consensual response). The cell bodies of the afferent limb of this reflex are located in which of the following?

(A) Corneal epithelium
(B) Mesencephalic nucleus
(C) Principal sensory nucleus
(D) Spinal trigeminal nucleus
(E) Trigeminal ganglion

7 A 62-year-old man visits his dentist with the complaint of gum and tooth pain. A thorough examination reveals no infection or lesion, and the man is referred to his physician. As part of the examination and evaluation, an MRI reveals an aberrant vascular loop impinging on the root of a cranial nerve as shown at the arrow in the image below. This man is most likely suffering which of the following?

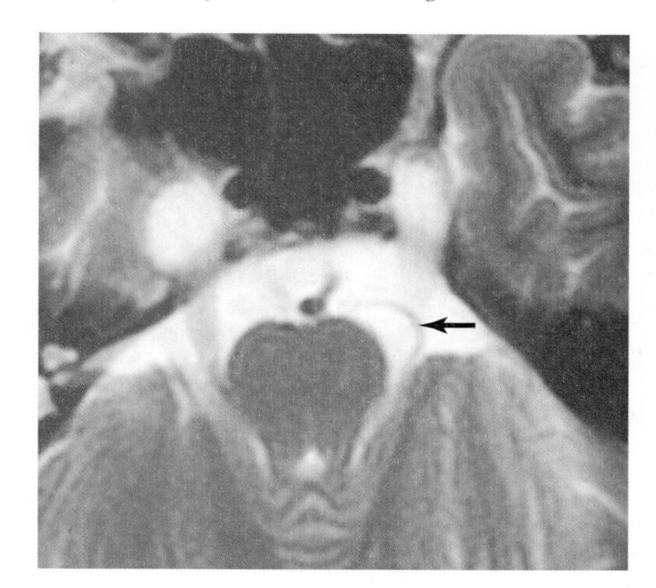

(A) Facial tic
(B) Geniculate neuralgia
(C) Glossopharyngeal neuralgia
(D) Sphenopalatine neuralgia
(E) Trigeminal neuralgia

8 A 43-year-old man presents with swelling and pain on the left side underneath the body of the mandible. The examination reveals a subcutaneous oval mass about 2 cm × 4 cm that is painful when palpated. In addition, when this man attempts to protrude his tongue, it deviates to the left. Which of the following is most likely paralyzed to produce this deficit?

(A) Genioglossus muscle
(B) Hyoglossus muscle
(C) Intrinsic tongue muscles
(D) Palatoglossus muscle
(E) Styloglossus muscle

9 A 62-year-old woman presents with deficits indicative of cranial nerve involvement. The results of a T2-weighted MRI reveal a lesion, represented by the circle in the given image, damaging the cranial nerve traversing the circle. Which of the following would you most likely expect to see if light was shined in the woman's right eye?

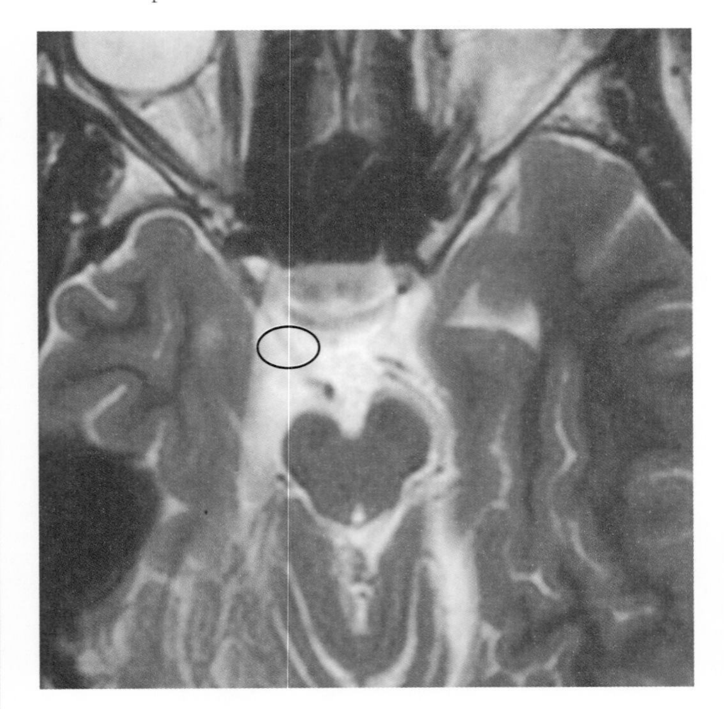

(A) Blindness in the right eye
(B) Right homonymous hemianopia
(C) Inability to abduct right eye
(D) Loss of the consensual pupillary reflex
(E) Loss of the direct pupillary reflex

10 A 60-year-old man presents to his family physician with what he calls a "problem seeing." The examination reveals that his vision is normal, but his right eye is deviated slightly inward, and does not abduct on attempted gaze to the right. There are no other findings. Which of the following is the most likely location of this lesion in this patient?
(A) Abducens nucleus on the left
(B) Abducens nucleus on the right
(C) Abducens root in the basilar pons on the right
(D) Abducnes root on the right
(E) Abducens root on the left

11 A 45-year-old woman presents to her otolaryngologist for throat pain. The examination reveals that the woman experiences lancinating pain in the posterior oral cavity (pharynx, base of tongue, ear) when swallowing or coughing. MRI reveals no overt pathology. The physician suspects that this patient is most likely suffering which of the following?
(A) Alternating hemianesthesia
(B) Glossopharyngeal neuralgia
(C) Polyps of the vocal folds
(D) Trigeminal neuralgia
(E) Wallenberg syndrome

12 A 23-year-old man is transported from the site of a motorcycle collision to the Emergency Department. He is conscious, has a broken femur, extensive skin damage on his upper extremity, and facial and scalp lacerations. The examination reveals that the man has difficulty swallowing, loss of sensation on the pharyngeal wall, hoarseness, and weakness of the trapezius muscle. CT reveals a basal skull fracture. This fracture went through which of the following foramina to produce these deficits?
(A) Foramen ovale
(B) Foramen rotundum
(C) Hypoglossal canal
(D) Jugular foramen
(E) Superior orbital fissure

13 A 32-year-old man is brought to the Emergency Department from the site of a motor vehicle collision. He is unconscious, has significant facial abrasions, possible broken nose, and an apparent dislocated hip. CT also shows a basal skull fracture that passes through the foramen indicated at the arrow in the image below. Assuming that the structure, or structures, traversing this opening are damaged, which of the following deficits would most likely be seen?

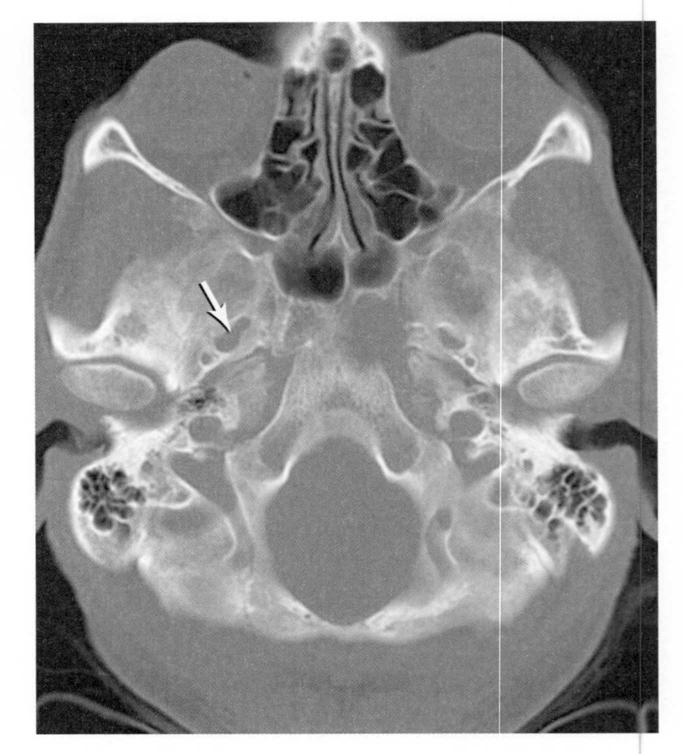

(A) Deafness
(B) Dysarthria, dysphagia
(C) Loss of sensation on the forehead
(D) Loss of sensation over the maxillary area
(E) Loss of sensation over the mandibular area

14 A 56-year-old woman complains of persistent headache that seems to be largely refractory to OTC medications. The examination reveals that these headaches are frontal and sometimes centered in the orbit. To test one aspect of visual system function, a light shined in both eyes results in a normal pupillary light reflex in both eyes. The efferent limb of this reflex is conveyed by which of the following?
(A) Abducens nerve
(B) Facial nerve
(C) Medial longitudinal fasciculus
(D) Oculomotor nerve
(E) Optic nerve

15 The cranial nerve indicated by the arrow in the image below contains sensory and motor fibers. Which of the following is the target of the motor fibers conveyed by this nerve?

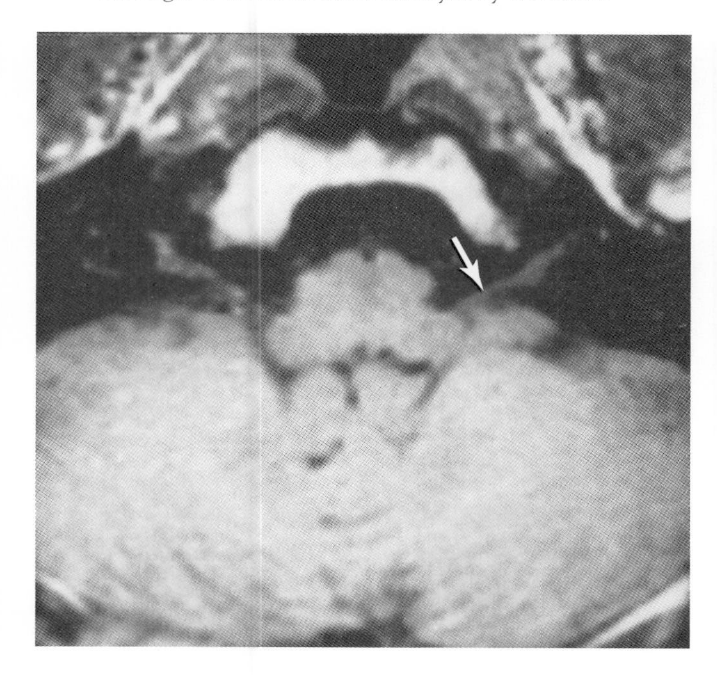

(A) Mylohyoid muscle
(B) Orbicularis oculi
(C) Pharyngeal musculature
(D) Stylopharyngeus muscle
(E) Vocalis muscle

16 A 39-year-old woman visits her family physician with the complaint that her meals "don't taste the same as they used to." Palpation suggests a tumor; MRI confirms that a tumor had encapsulated, and obviously damaged, a nerve. The physician indicates that this tumor is the likely cause of the woman's loss of sense of taste. Damage to which of the following would account for this woman's symptoms?
(A) Buccal nerve
(B) Lingual nerve
(C) Mylohyoid nerve
(D) Inferior alveolar nerve
(E) Infraorbital nerve

17 A 51-year-old man presents with a complaint of numbness on his face. The history reveals that this has been slowly progressive over several months, and the examination localizes this sensory deficit to the skin over the mandible, the mandibular teeth, floor of the oral cavity, and lower lip. In addition, the muscles of mastication are weak. MRI reveals a small tumor, probably a meningioma, invading one foramen in the base of the skull. Which of the following is most likely involved?
(A) Foramen rotundum
(B) Foramen ovale
(C) Foramen spinosum
(D) Infraorbital foramen
(E) Superior orbital fissure

18 Which of the following structures contains neurons whose axons end in terminal ganglia (sometimes, these may be called intramural ganglia)?

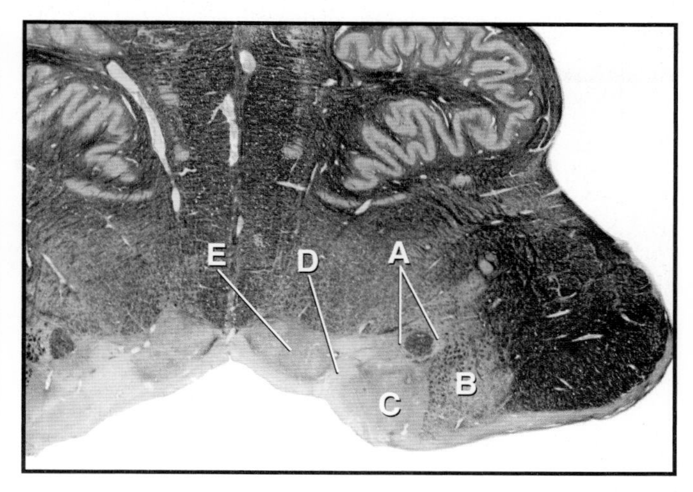

(A) A
(B) B
(C) C
(D) D
(E) E

19 A 45-year-old woman presents with recurring and persistent headache. The history is unremarkable other than the fact that the women noted that she thought the headaches started several weeks ago, but could not remember exactly when. MRI reveals an aneurysm arising from the posterior cerebral artery-posterior communicating artery junction and slightly impinging on the oculomotor nerve. Which of the following deficits would be initially seen in this patient?
(A) Constriction of the pupil only
(B) Constriction of the pupil + muscle paralysis
(C) Dilation of the pupil only
(D) Dilation of the pupil + muscle paralysis
(E) Paralysis of extraocular muscles only

20 The neonatal nurses inform the attending pediatrician that a 3-day-old male baby is having significant difficulty suckling and swallowing. The examination reveals that the baby has little to no facial and jaw movements and cannot swallow. Eye and tongue movements appear age-appropriate. Based on the physical examination, the physician concludes that this is a developmental defect within the brainstem. Which of the following would explain this combination of deficits?
(A) Absence of general somatic afferent cell columns
(B) Absence of general somatic efferent cell columns
(C) Absence of general visceral efferent cell columns
(D) Absence of special visceral afferent cell columns
(E) Absence of special visceral efferent cell columns

21 A 43-year-old welder is brought to the Emergency Department following an accident at a construction site. The EMS personnel indicate that the man fell about 30 feet from a scaffolding onto a concrete surface. The examination reveals broken limb bones and head injuries with bleeding from the mouth and nose; the man is unconscious. CT reveals a basal skull fracture passing through the foramen, indicated by the arrow in the image below, which damaged the structures traversing this opening. When he regains consciousness, which of the following deficits would be the most immediate clinical concern?

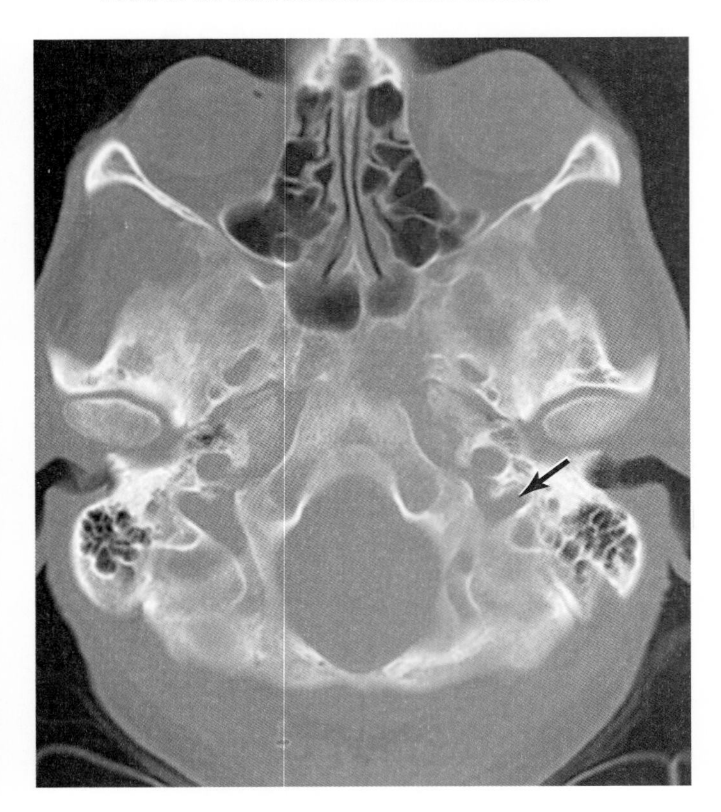

(A) Apnea
(B) Deafness in left ear
(C) Dysphagia
(D) Masticatory paralysis
(E) Tongue paralysis

22 A 38-year-old woman presents to her family physician with the complaint of difficulty seeing. The examination reveals diplopia and the following combination of motor deficits. On attempted lateral gaze to the right, neither the right nor the left eye will look toward the right. On attempted lateral gaze to the left, the left eye looks to the left, but the right eye does not look to the left. These combinations of deficits constitute which of the following?
(A) Claude syndrome
(B) Foville syndrome
(C) One-and-a-half syndrome
(D) Millard-Gubler syndrome
(E). Medial medullary syndrome

23 A 46-year-old man visits his family physician with the complaint of persistent dizziness. The history reveals that these symptoms have been getting slowly worse over several months. The examination shows, in addition to the dizziness, that the man has hearing loss and tinnitus in his left ear. MRI reveals a tumor on the left side, in the immediate vicinity of the internal acoustic meatus, that measures about 2.5 cm in diameter. Considering the location and size of this tumor, which of the following would this patient also most likely have?
(A) Loss of taste on the root of the tongue
(B) Numbness on the right side of the face
(C) Numbness on the left side of the face
(D) Weakness of facial muscles only on the lower left
(E) Weakness of the genioglossus muscle on the left

24 A 32-year-old woman visits her family physician with the complaint of a persistent low-grade discomfort in her eyes. The examination reveals that the conjunctiva of the woman's eyes is somewhat inflamed and that she has an apparent lack of tear production: a fact confirmed by the woman. This may be called the dry eye syndrome. Which of the following cranial nerve nuclei specifically participates in the secretory function that is compromised in this woman?
(A) Dorsal motor vagal nucleus
(B) Inferior salivatory nucleus
(C) Nucleus ambiguus
(D) Solitary nucleus
(E) Superior salivatory nucleus

25 Which of the following receives projections from the nucleus indicated at the arrow in the image below?

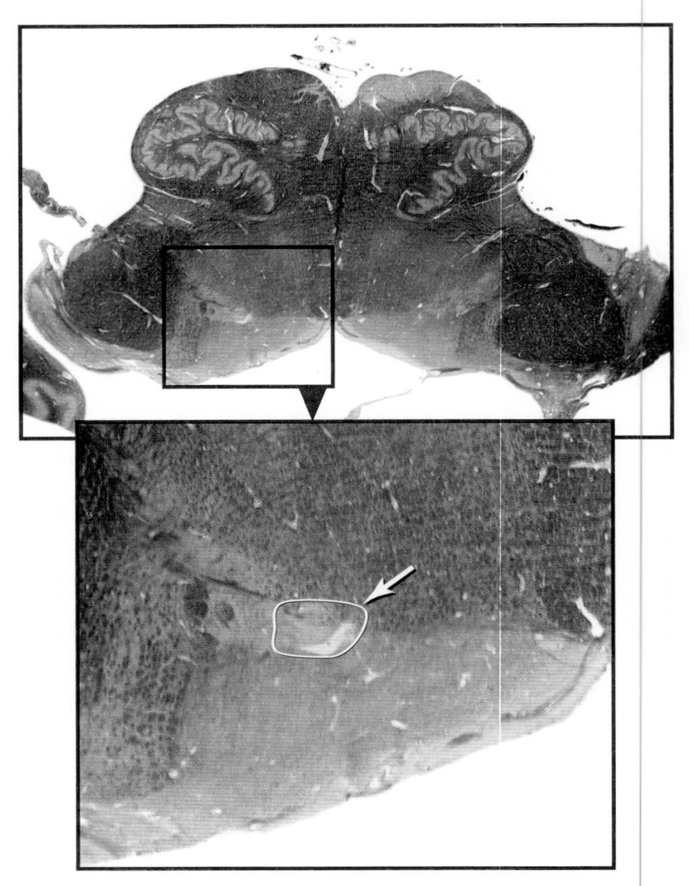

(A) Ciliary ganglion
(B) Otic ganglion
(C) Sphenopalatine ganglion
(D) Submandibular ganglion
(E) Terminal ganglia

26 A 31-year-old man is transported from the site of the motor vehicle collision to the Emergency Department. The examination reveals that he has facial injuries, a dislocated shoulder, and a fractured femur. While he has no immediate indicators of brain or spinal cord injury, the physician notices that the man's jaw deviates slightly to the right when he closes his jaw. CT shows a basal skull fracture traversing the right foramen ovale. Traumatic denervation of which of the following would explain this man's deficit?

(A) Anterior belly of the digastric
(B) Masseter muscle
(C) Mylohyoid muscle
(D) Pterygoid muscles
(E) Temporalis muscle

27 Which of the following cranial nerves, or major branches of cranial nerves, is located in the opening indicated by the arrow in the image below?

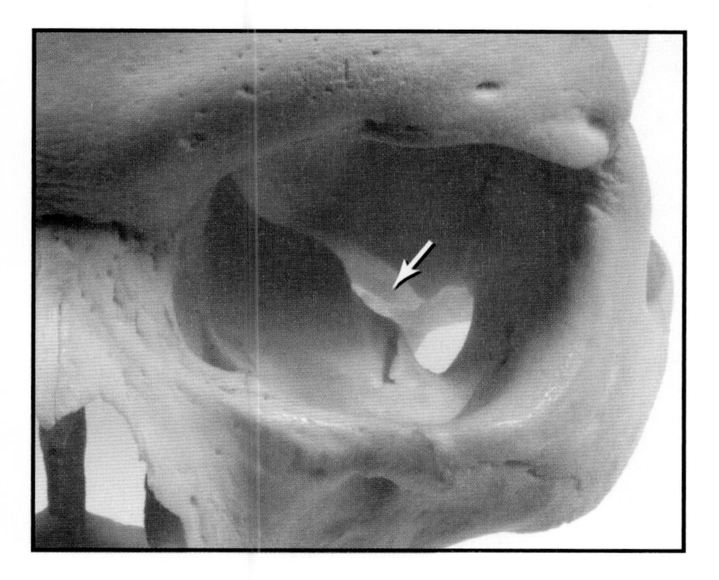

(A) Abducens nerve
(B) Mandibular nerve
(C) Maxillary nerve
(D) Ophthalmic nerve
(E) Trochlear nerve

28 During a routine neurological examination on a 39-year-old woman, the physician touches a wisp of cotton to the patient's cornea and her eyes blink. The afferent limb of this reflex is on the trigeminal nerve. Which of the following contains the efferent limb?

(A) Abducens nerve
(B) Facial nerve
(C) Oculomotor nerve
(D) Trigeminal nerve
(E) Trochlear nerve

ANSWERS

1 **The answer is D: Superior orbital fissure.** This patient has damage to the abducens, trochlear, and oculomotor nerves (loss of all eye movement), a dilated pupil (loss of parasympathetics in IIIrd nerve), and the ophthalmic nerve (sensory loss over the forehead). All of these structures, plus small vessels, traverse the superior orbital fissure. The inferior orbital fissure contains the maxillary nerve which is sensory to the maxillary area of the face, the nasal mucosa, and the maxillary sinus. The supraorbital and infraorbital foramina contain cutaneous branches of the ophthalmic and maxillary nerves; this injury does not involve the supraorbital fissure, as this would exclude all nerves innervating extraocular muscles. The optic canal contains the optic nerve; this man has no visual complaints.

2 **The answer is E: Vestibulocochlear and facial.** This is a view of the medial aspect of the petrous portion of the temporal bone and immediately adjacent parts of the occipital bone on the left side of the skull. The petrooccipital fissure is the location of the inferior petrosal sinus; this groove may also be called the sulcus for the inferior petrosal sinus. The opening immediately above this fissure (and in the circle) is the internal acoustic meatus. This particular foramina is characterized by the transverse crest (seen here) and the fact that it does not open directly to the external aspect of the skull. The vestibulocochlear nerve and the facial nerve enter this opening; the former originates from sensory receptors in the petrous portion of the temporal bone, while the latter portion passes through to eventually exit the stylomastoid foramen to innervate the muscles of facial expression. The hypoglossal nerve exits through the hypoglossal canal and the accessory, vagus, and glossopharyngeal nerves exit the skull via the jugular foramen. The abducens nerve, after traversing the cavernous sinus, passes through the superior orbital fissure.

3 **The answer is E: IX, X, XI.** Cranial nerves IX (glossopharyngeal), X (vagus), and XI (accessory) traverse the jugular foramen. Damage to these nerves will result in several deficits, the most noticeable of which will be dysarthria, dysphagia, the inability to elevate the shoulder against resistance (the shoulder may also droop), and an inability to rotate the head to the opposite side against resistance. Cranial nerves III (oculomotor), IV (trochlear), and VI (abducens) pass through the superior orbital fissure; VII (facial) and VIII (vestibulocochlear) enter the internal acoustic meatus; the facial nerve exits the skull via the stylomastoid foramen. The nerves arising from the trigeminal ganglion course through the superior orbital fissure (V_1—ophthalmic), the foramen rotundum (V_2—maxillary), and foramen ovale (V_3—mandibular + motor root); the XII nerve (hypoglossal) traverses the hypoglossal canal.

4 **The answer is D: Tongue deviates to the left on protrusion.** This T2-weighted axial MRI is through the midmedullary levels and shows the left root of the hypoglossal nerve exiting between the pyramid, which is anterior to the exit, and the inferior olivary eminence, which is posterior to the exit. This nerve passes through the hypoglossal canal as it exits the skull to serve the muscles of the tongue on the same side. Damage to the hypoglossal root results in the tongue deviating to the weak side on attempted protrusion, in this case to the left. Dysarthria, dysphagia, and the inability to elevate the shoulder, especially against resistance, would be seen in damage to the vagus, glossopharyngeal, and accessory nerves, all of which share the common feature of exiting the skull via the jugular foramen.

5 **The answer is D: CNs VI, VII, VIII.** The abducens (VI), facial (VII), and vestibulocochlear (VIII) nerves are lined up from medial to lateral at the caudal edge of the pons where it is continuous with the rostral medulla. The oculomotor (III) and trochlear nerves exit the inferior and superior aspects of the midbrain; the trigeminal (V) nerve exits the lateral surface of the pons. The glossopharyngeal (IX), vagus (X), and accessory (XI) nerves are associated with the postolivary sulcus of the medulla; the hypoglossal (XII) nerve exits the medulla via the preolivary sulcus.

6 **The answer is E: Trigeminal ganglion.** This is the corneal reflex. The afferent endings are naked nerve endings in the cornea; their cell bodies are located in the left trigeminal ganglion, and their central processes terminate in the left spinal trigeminal nucleus, pars caudalis. The spinal nucleus of V, in turn, sends crossed ascending fibers toward the thalamus (for recognition of pain) and collaterals to the motor facial nuclei on both sides. The efferent limb of this reflex originates in the facial nuclei; the left for the direct response, and the right for the consensual response. Cell bodies in the mesencephalic nucleus are unipolar and convey proprioceptive information; these fibers do not have cell bodies in the trigeminal ganglion and are involved in the jaw-jerk reflex. The principal sensory nucleus contains the second order neurons that convey discriminative touch on the trigeminal nerve. The spinal trigeminal nucleus contains the second-order neurons for the corneal reflex pathway; the first-order cells are in the trigeminal ganglion.

7 **The answer is E: Trigeminal neuralgia.** Neuralgia, also called neurodynia, is a severe lancinating and/or searing pain that may have trigger zones (areas where stroking or touching will initiate pain); the painful sensations in neuralgia are related to the distribution of sensory branches of a particular cranial nerve and are paroxysmal. In this image, the superior cerebellar artery loops over the root of the trigeminal nerve. This is a likely cause of trigeminal neuralgia, which is a sudden-onset pain that may arise from a trigger zone, frequently located at the angle of the mouth. The V_2 (about 20% of cases), V_3 (about 17%), and $V_2 + V_3$ (40+%) divisions of the trigeminal nerve are commonly involved. There are medical and surgical treatments for trigeminal neuralgia. Facial tic is the sudden involuntary contraction of facial muscles; geniculate neuralgia is pain from deep and superficial areas of the ear mediated by the facial nerve. Sudden severe pain from the throat or palate (swallowing may be a trigger) is glossopharyngeal neuralgia; similar pain from the area of the nose, maxillary teeth, ears, and sinuses signals sphenopalatine neuralgia.

8 **The answer is A: Genioglossus muscle.** The genioglossus muscle has its origin from the superior genial (mental) spine and fans out to enter the base of the tongue. Each genioglossus muscle, when it contracts, pulls that side of the tongue toward the midline; when they contract together, the tongue protrudes symmetrically (straight) out of the mouth. When one genioglossus muscle is weak or paralyzed, the opposite pulls toward the midline and the tongue will deviate to the weak side. The styloglossus and palatoglossus muscles draw the tongue upward (elevate) and slightly backward, and the hypoglossus muscle depresses (pulls downward) the tongue. The intrinsic tongue muscles consist of longitudinal, transverse, and vertical fascicles. These muscles change the shape of the tongue, and are very important for speech.

9 **The answer is E: Loss of the direct pupillary reflex.** The oculomotor nerve is clearly seen in this axial MRI almost from its site of origin at the oculomotor sulcus of the midbrain to the point where it enters the orbit. A lesion of the right oculomotor nerve interrupts the efferent limb of the pupillary light reflex pathway (the direct response is lost); the afferent limb is conveyed by the optic nerve. A light stimulus to the right eye is received at the pretectal nucleus which projects to both Edinger-Westphal nuclei bilaterally. However, the lesion of the right oculomotor nerve interrupts the parasympathetic fibers on that side, but the left efferent limb is not affected. The black spot located medially adjacent to the nerve, just anterior to the midbrain, is the P_1 segment of the posterior cerebral artery. Since this is not a lesion of the optic nerve, the woman is not blind in the right eye nor does she have a hemianopia. Only lesions of optic structures caudal to the optic chiasm will produce a hemianopia involving the opposite halves of both visual fields.

10 **The answer is D: Abducens root on the right.** The only deficit experienced by this patient is the diplopia (his "problem seeing") resulting from the inward deviation of his right eye; this would be exaggerated by attempted lateral gaze to the right. This indicates a right abducens root lesion: a right lateral rectus paralysis only. A lesion of the right abducens nucleus would result in right lateral rectus paralysis plus a paralysis of the left medial rectus. This is due to the fact that an abducens nucleus lesion will affect internuclear neurons, located within the right abducens nucleus, that project to medial rectus motor neurons on the left via the medial longitudinal fasciculus. In a similar manner, a lesion of the abducens root in the basilar pons on the right will result in a right lateral rectus paralysis plus a left hemiparesis due to involvement of the immediately adjacent corticospinal fibers. Lesions on the left are on the wrong side.

11 **The answer is B: Glossopharyngeal neuralgia.** Severe pain following stimulation of the posterior aspects of the oral cavity, such as when swallowing, is related to the distribution of sensory branches of cranial nerve IX to this area, and is called glossopharyngeal neuralgia. The causes are largely unknown, although multiple sclerosis or aberrant vascular loops pressing the IXth root are possibilities. Alternating hemianesthesia is a sensory loss on one side of the body and on the opposite side of the face; this, along with dysarthria, dysphagia, and ataxia, are components of the Wallenberg (posterior inferior cerebellar artery [PICA] or lateral medullary) syndrome. Trigeminal neuralgia is a severe excruciating pain originating from the face usually from the vicinity of the corner of the mouth (V_2, V_3) that is usually set off by a variety of actions (chewing, shaving, putting on lipstick, brushing teeth, even wind on the face). Essentially it is a pain that is identical to that of glossopharyngeal neuralgia, but of a different distribution. Polyps cause discomfort, coughing, and hoarseness, but not extreme, sudden, and unpredictable pain.

12 **The answer is D: Jugular foramen.** The collection of symptoms experienced by this man clearly indicate damage to the glossopharyngeal, vagus, and accessory nerves. All three of these nerves traverse the jugular foramen and can be injured by trauma to this foramen. In fact, there are several syndromes of the jugular foramen. The foramen ovale contains the mandibular nerve and the trigeminal motor root; the maxillary nerve traverses the foramen rotundum. The superior orbital fissure contains the ophthalmic, abducens, oculomotor, and trochlear nerves, while the hypoglossal canal contains the nerve after which it is named. It should be remembered that all of these openings in the skull base also transmit small blood vessels.

13 **The answer is E: Loss of sensation over the mandibular area.** The arrow is pointing to the foramen ovale which transmits the maxillary division of the trigeminal nerve, fibers comprising the trigeminal motor root, and the accessory meningeal artery. Damage to the contents of this foramen would result in a loss of pain and thermal sense over the mandibular region, a comparable sensory loss inside the oral cavity including the mandibular teeth, and weakness of the masticatory muscles; all of this would be on the right side. Deafness in one ear would require involvement of the internal acoustic meatus, and dysarthria (difficulty speaking) and dysphagia (difficulty swallowing) would be seen if a basal skull fracture involved the jugular foramen. The ophthalamic division of the trigeminal nerve traverses the superior orbital fissure, then passes through the supraorbital foramen (there may be more than one, in which case they are foramina) to fan out over the forehead. If the supraorbital foramen, or foramina, are not completely formed, this may appear as, and be called, a notch. The maxillary division of the trigeminal nerve traverses the foramen rotundum and exits on the maxillary region of the face via the infraorbital foramen; this nerve is also sensory to the maxillary teeth and palate.

14 **The answer is D: Oculomotor nerve.** The afferent limb of the papillary light reflex is via the optic nerve, to the pretectal nucleus, and bilaterally to the Edinger-Westphal (EW) nucleus. The efferent limb originates in the EW nucleus, travels via the oculomotor nerve to the ciliary ganglion, and from this structure to the sphincter muscle of the pupil. The abducens nerve innervates the ipsilateral lateral rectus muscle; the facial nerve innervates the muscles of facial expression and other targets. The medial longitudinal fasciculus contains, among other fibers, abducens interneurons involved in the internuclear ophthalmoplegia pathway.

15 **The answer is D: Stylopharyngeus muscle.** The plane of this image is immediately caudal to the medulla-pons junction (note the large size of the restiform body) and is at the level of the cerebellopontine angle (note the obvious appearance of the flocculus on the left side). Consequently, at this location the cranial nerve in the image is the most rostral nerve exiting the postolivary sulcus: cranial nerve IX. The glossopharyngeal nerve exits from the postolivary sulcus rostral to the root of the vagus nerve; consequently, of these two nerves, it is the one closest to the medulla-pons junction. The stylopharyngeus muscle is the only muscle innervated by the glossopharyngeal nerve. This muscle originates from the 2nd pharyngeal arch, has a Special Visceral Efferent functional component, and is part of the efferent limb of the gag reflex. The mylohyoid muscle is innervated by the trigeminal nerve, and the orbicularis oculi by the facial nerve. The pharyngeal musculature and the vocalis muscle are innervated by the vagus nerve.

16 **The answer is B: Lingual nerve.** This branch of the trigeminal nerve is sensory to the anterior surface of the tongue, floor of the oral cavity, and the mandibular gingivae; it conveys taste fibers originating from the anterior two-thirds of the tongue and connects with the facial nerve through the chorda tympani. In this respect, taste fibers travel on this branch of the trigeminal nerve prior to joining the facial nerve. Although traveling on the fifth nerve for a short distance, taste fibers are always considered part of the seventh nerve. The buccal nerve contains sensory branches to the inner aspects of part of the oral cavity and motor fibers to the lateral pterygoid muscle, and the mylohyoid nerve supplies the mylohyoid muscle and the anterior belly of the digastric. The inferior alveolar nerve is sensory for the mandibular teeth and gives rise to the mylohyoid nerve. The infraorbital nerve is sensory from the lower eyelid, skin of the maxilla, medial nose, and the medial canthus and structures related thereto.

17 **The answer is B: Foramen ovale.** The mandibular nerve (V_3) and the motor root of the trigeminal nerve pass through the foramen ovale. The deficits clearly implicate this specific foramen as the location of the lesion. The maxillary nerve (V_2) traverses the foramen rotundum and the foramen spinosum contains the middle meningeal artery. The infraorbital foramen is not in the skull base; it transmits the infraorbital nerve which is sensory to the maxillary region of the face. Cranial nerves III, IV, VI and V_1 are located in the superior orbital fissure.

18 **The answer is D (Dorsal motor nucleus of the vagus nerve).** This plane of section is through mid-to-rostral portions of the hypoglossal nucleus (choice E); and, in the section, the laterally adjacent dorsal motor nucleus (D) and the solitary nucleus (A) are clearly seen. The dorsal motor vagal nucleus (and other parasympathetic nuclei) appears very light in myelin-stained sections due to the fact that these preganglionic fibers are very lightly myelinated. The preganglionic parasympathetic fibers arising from this nucleus distribute, via branches of the vagus nerve, to terminal ganglia located in thoracic and abdominal viscera. From the terminal ganglia, short postganglionic fibers distribute to smooth muscle and glandular epithelium in the target organs. The inferior (B) and medial (C) vestibular nuclei function in balance, equilibrium, and regulation of eye movement, while the hypoglossal nucleus (E) serves the motor function to the ipsilateral side of the tongue. The solitary nucleus (A) is the visceral center (general and special visceral sensation) of the brainstem.

19 **The answer is C: Dilation of the pupil only.** The oculomotor nerve contains parasympathetic preganglionic fibers (these are general visceral efferent [GVE] fibers and general somatic efferent [GSE] fibers) that innervate the extraocular muscles served by this nerve. The GVE preganglionic fibers are small diameter, lightly myelinated fibers that are distributed in the periphery of the oculomotor nerve; the GSE motor fibers are larger diameter, heavily myelinated fibers that are located in the more central regions of the nerve. Consequently, a lesion impinging on the oculomotor root will compromise the superficially located small-diameter GVE fibers first and the pupil dilates. As the lesion impinges further on the nerve root, the more centrally located larger diameter GSE fibers are recruited in the area of damage; eye movement deficits are seen in addition to the pupil dilation. When the GVE parasympathetic fibers are damaged, the sympathetic input takes over and the pupil dilates.

20 **The answer is E: Absence of special visceral efferent cell columns.** This baby has motor problems related to masticatory function (muscles innervated by the trigeminal nerve), facial muscles (innervated by the facial nucleus), and swallowing (muscles innervated by the nucleus ambiguus). These nuclei form an interrupted cell column in the pons and medulla consisting of special visceral efferent cells that innervate muscles that arise within pharyngeal arch mesoderm in the following sequence; arch 1 = trigeminal; arch 2 = facial; arches 3, 4 = ambiguus (glossopharyngeal, vagus, respectively). General somatic afferent cell columns in the brainstem are sensory (pain and thermal sense) and general somatic efferent cell columns are motor to muscles not derived from pharyngeal arches (cranial nerves III, IV, VI, XII). The solitary nucleus is the special visceral afferent cell column of the medulla receiving taste input via cranial nerves VII, IX, and X; the general visceral efferent cell column is made up of the parasympathetic nuclei of cranial nerves VII, IX and X.

21 **The answer is C: Dysphagia.** The left jugular foramen is identified by the arrow. This foramen transmits the glossopharyngeal, vagus, and accessory cranial nerves. In addition, it contains two small arteries and the continuations of the inferior petrosal sinus and the sigmoid sinus with the internal jugular vein. Injury to these cranial nerves at this location will result in hoarseness, dysphagia, dysarthria, increased heart rate, sensory losses (minor) representing the peripheral branches of IX and X, and weakness of the trapezius and sternocleidomastoid muscles. Of these, the most immediate concern will be dysphagia or difficulty swallowing. With the laryngeal and pharyngeal muscles on the left side paralyzed, including the vocalis on the left, there is the real danger of swallowed solids or fluids entering the trachea (held open by cartilaginous rings and with the vocal folds not closing completely) and resulting in aspiration pneumonia. Apnea is not a feature of syndromes of the jugular foramen, and this patient will not be deaf in his left ear since the auditory meatus, internal or external, is not involved. Masticatory weakness would be present in damage to the foramen ovale, and tongue paralysis would result from damage to the hypoglossal canal; neither is damaged in this case.

22 **The answer is C: One-and-a-half syndrome.** In this case, this woman has lost (on attempted voluntary movement) the functions of the right lateral and medial recti muscles and the left medial rectus muscle; only the left lateral rectus functions during attempted horizontal gaze in either direction. The loss

of function of the right lateral and medial recti muscles and of the left medial rectus muscles (three muscles paralyzed) coupled with the surviving left lateral rectus muscle (the one surviving muscle) is the basis for the name of this syndrome. The lesion is on the right side in the pons and involves the abducens nucleus (motor to right LR, interneurons to left MR) and axons originating from interneurons located in the left abducens nucleus that enter the right medial longitudinal fasciculus to serve the right MR. The Claude syndrome is a midbrain lesion involving the oculomotor fibers, red nucleus, and adjacent cerebellothalamic fibers; the Foville syndrome is a pontine lesion involving the abducens root and corticospinal fibers. Millard-Gubler (or just Gubler syndrome) is a lateral pontine lesion that involves the corticospinal fibers and root of the facial nerve, while the medial medullary syndrome is a lesion in the medulla involving the hypoglossal root, corticospinal fibers, and medial lemniscus.

23 **The answer is C: Numbness on the left side of the face.** All factors (dizziness, tinnitus, loss of hearing) and the location of the mass clearly suggest that this man is suffering from a vestibular schwannoma (frequently and incorrectly called an acoustic neuroma). These lesions present with the deficits experienced by this man (and include vertigo), and when they are larger than 2 cm, they frequently impinge on the trigeminal root with an ipsilateral sensory loss and absent corneal reflex. Although the facial nerve shares the internal acoustic meatus, facial weakness in seen in only about 10% to 12% of cases. Taste from the root of the tongue is carried on the glossopharyngeal nerve; the hypoglossal nerve innervates the genioglossus; neither is involved in vestibular schwannomas. Weaknesses of facial muscles only on the lower left of the face are the result of a central lesion; numbness of the right side of the face is on the wrong side.

24 **The answer is E: Superior salivatory nucleus.** The superior salivatory nucleus is located adjacent to the fibers of the facial nerve distal to the internal genu of the facial nerve. The preganglionic parasympathetic fibers that arise from this nucleus exit the brainstem on the facial nerve, travel on the greater petrosal nerve, and proceed to the pterygopalatine ganglion as the nerve of the pterygoid canal. Postganglionic fibers from the pterygopalatine ganglion travel on the zygomaticotemporal nerve (a branch of the maxillary nerve), then for a short distance on the lacrimal nerve (a branch of the ophthalmic nerve), before ending in the lacrimal gland. The solitary nucleus is the visceral afferent nucleus of the brainstem for both general and special (taste) visceral input. The dorsal motor vagal nucleus is the source of preganglionic fibers for thoracic and abdominal viscera; the inferior salivatory nucleus is the source of preganglionic fibers that end in the otic ganglion for the eventual innervation of the parotid gland. Cells in the nucleus ambiguus innervate a variety of skeletal muscles of the head that are served by the glossopharyngeal and vagus nerves.

25 **The answer is B: Otic ganglion.** The inferior salivatory nucleus, as indicated in this image, is located immediately medial to the rostral portions of the solitary tract and nucleus. The parasympathetic preganglionic fibers that originate from this nucleus travel on the glossopharyngeal nerve and, via the tympanic nerve and lesser petrosal nerve, end in the otic ganglion. Postganglionic fibers from the otic ganglion innervate the parotid gland. The ciliary ganglion receives preganglionic fibers, via the oculomotor nerve, from the Edinger-Westphal nucleus which, in turn, projects to the sphincter pupillae muscle of the eye. The sphenopalatine and submandibular ganglia receive preganglionic parasympathetic input that originates from the superior salivatory nucleus and travels on the facial nerve. These ganglia serve the lacrimal gland and the submandibular and sublingual salivary glands, respectively. Terminal ganglia receive parasympathetic preganglionic input from the dorsal motor vagal nucleus (obviously via the vagus nerve) and send their postganglionic fibers to the thoracic and abdominal viscera.

26 **The answer is D: Pterygoid muscles.** The medial and lateral pterygoid muscles pull the mandible toward the midline when they contract; when these muscle groups contract together, the jaw closes symmetrically. When these muscles are weak on one side, the healthy muscles will pull the jaw slightly toward the weak side when they contract, and the jaw deviates toward the weak side. The masseter and temporalis muscles are jaw-closing muscles, participate in protraction/retraction of the mandible, but have little to no effect on medial-lateral movements. The mylohyoid elevates the floor of the oral cavity on swallowing, and the anterior belly of the digastric participates in depressing the mandible. All of these muscles are innervated by the trigeminal nerve.

27 **The answer is C: Maxillary nerve.** The overall opening indicated by the arrow is the inferior orbital fissure. The maxillary nerve leaves the trigeminal ganglion between the ophthalmic and mandibular branches of the ganglion, traverses the inferior orbital fissure and infraorbital groove, to exit onto the face from the infraorbital foramen (as the infraorbital nerve) to serve the maxillary part of the face and lateral aspect of the nose. The infraorbital groove and foramen are clearly seen in this image. The maxillary nerve is sensory to the maxillary sinus, portions of the palate, maxillary teeth and gums, and the maxillary area of the face. The abducens and trochlear nerves, and the ophthalmic nerve, a major branch of the trigeminal nerve, all pass through the cavernous sinus, then exit the cranial cavity by passing through the superior orbital fissure. The mandibular nerve arises from the lateral part of the trigeminal ganglion, exits the skull via the foramen ovale, and is sensory to the mandibular teeth and gums, tongue, and the mandibular area of the face. The motor fibers of the trigeminal nerve also travel with the mandibular nerve.

28 **The answer is B: Facial nerve.** This is the corneal reflex. The afferent limb is conveyed by the ophthalmic division of the trigeminal nerve, the afferent cell body is in the trigeminal ganglion, and the central process ends in the spinal trigeminal nucleus, pars caudalis. Fibers arising from the spinal trigeminal nucleus cross the midline, ascend toward the thalamus and, en route, send collaterals to the motor facial nuclei. The efferent limb arises from the facial nucleus to innervate the orbicularis oculi muscle (ipsilateral preponderance). The abducens, oculomotor, and trochlear nerves innervate extraocular muscles; the trigeminal nerve innervates the muscles of mastication. None of these is directly involved in the corneal reflex.

Chapter 11

The Diencephalon (Dorsal Thalamus)

Recall the Cautionary Tale: The images in this chapter are presented in a Clinical Orientation*: this approach emphasizes how structure and function correlate with deficits in a clinical setting. MRI and CT have a universally recognized orientation and laterality. Line drawings, stained sections, or brain slices in the* Clinical Orientation *are viewed in the identical manner that one views an MRI: your right, as the physician-observer, is the patient's left, and the observer's left is the patient's right. The laterality of deficits and reference to connections follow accordingly.*

QUESTIONS

Select the single best answer.

1 Which of the following thalamic nuclei projects to the posterior paracentral gyrus?

(A) Dorsomedial
(B) Lateral geniculate
(C) Ventral lateral
(D) Ventral posterolateral
(E) Ventral posteromedial

2 A 70-year-old man presents with a sudden onset of motor symptoms. The examination reveals dysmetria and dysdiadochokinesia, most pronounced on his right side. MRI shows a lesion in the area outlined in the image below. Which of the following thalamic nuclei is the main target of the axons arising from the neurons damaged by this lesion?

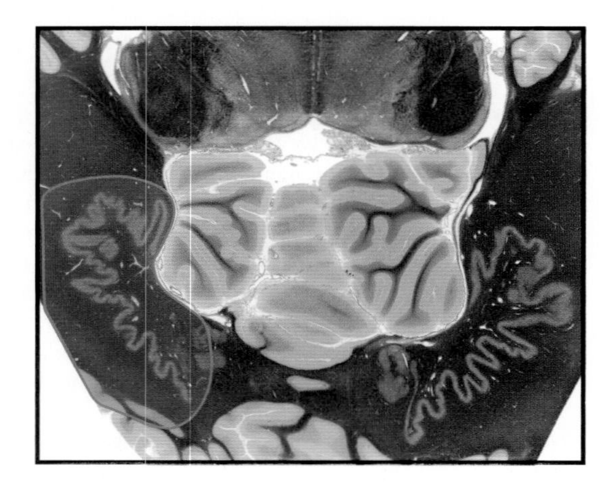

(A) Anterior
(B) Centromedian
(C) Dorsomedial
(D) Ventral lateral
(E) Ventral posterolateral

3 Which of the following vessels is the principal blood supply to the pulvinar nucleus of the dorsal thalamus?

(A) Lenticulostriate
(B) Medial striate
(C) Quadrigeminal
(D) Thalamogeniculate
(E) Thalamoperforating

4 A 77-year-old man is brought to the Emergency Department (ED) following a precipitous event at his home. His wife explains that he had a sudden severe headache, became nauseated, vomited, and passed out. He regained consciousness by the time he arrived at the ED. CT of this man's brain reveals the image below. Which of the following is indicated at the tip of the arrow in this image?

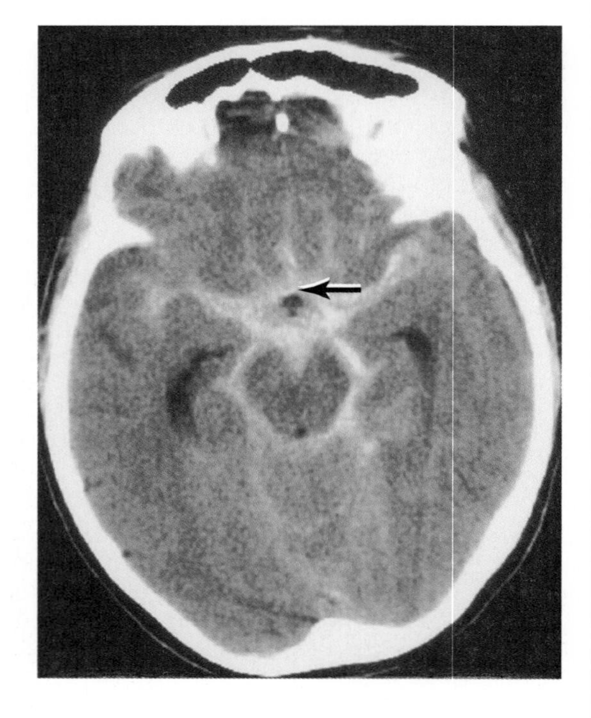

(A) Cerebral aqueduct
(B) Foramen of Magendie
(C) Interventricular foramen
(D) Supraoptic recess of third ventricle
(E) Suprapineal recess of third ventricle

5 A 34-year-old man presents with confusion and persistent headache. MRI reveals an enlarged lateral ventricle on the left side, presumably resulting from a dime-sized tumor located in the left interventricular foramen. Based on the location of this tumor, which of the following diencephalic structures is most directly impinged upon?
(A) Anterior nucleus
(B) Centromedian nucleus
(C) Dorsomedial nucleus
(D) Ventral anterior nucleus
(E) Ventral lateral nucleus

6 A 58-year-old man presents with the complaint of difficulty seeing. The history reveals that the man has hypertension and that he is largely noncompliant regarding his medications. The examination reveals that he has a right homonymous hemianopia, and an MRI shows a lesion of the primary visual cortex. Which of the following relays vital input to this damaged area of cortex?
(A) Lateral dorsal nucleus
(B) Lateral geniculate nucleus
(C) Medial geniculate nucleus
(D) Pulvinar nucleus
(E) Ventral lateral nucleus

7 A 67-year-old woman presents with the sudden onset of deficits affecting the right side of her body. The results of the examination clearly suggest a stroke; this is confirmed by the results of an MRI, which is shown below. Based on the location of this lesion, which of the following thalamic nuclei is most likely the major focus of this hemorrhagic event?

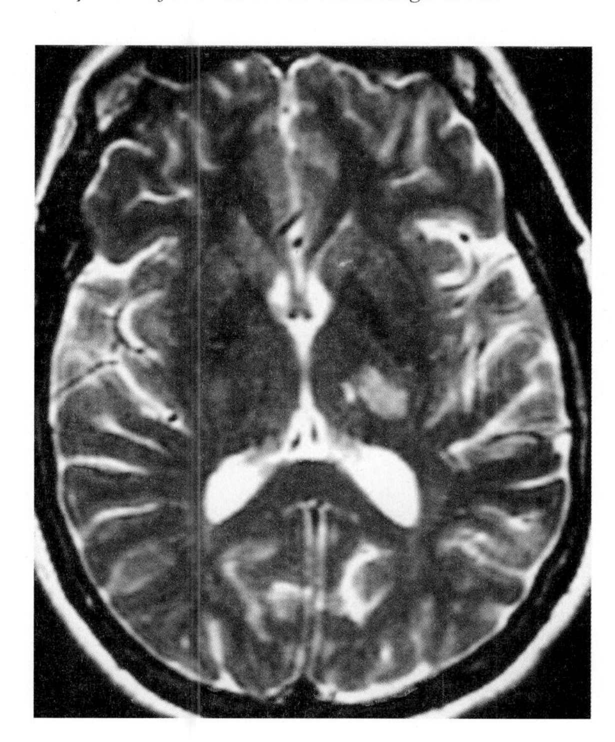

(A) Anterior
(B) Dorsomedial
(C) Pulvinar
(D) Ventral anterior
(E) Ventral posterolateral

8 A 59-year-old morbidly obese man is brought to the Emergency Department following an episode at his home. The history reveals that the man has uncontrolled diabetes and hypertension. The precipitating factor for this visit was the sudden onset of involuntary flailing movements of his right upper extremity, which occur intermittently without warning. MRI reveals a well-localized infarct in the diencephalon. This lesion is most likely located in which of the following nuclei of the diencephalon?
(A) Anterior
(B) Centromedian
(C) Subthalamic
(D) Ventral lateral
(E) Ventral posteromedial

9 Which of the following thalamic nuclei is located within the internal medullary lamina and, therefore, is commonly referred to as one of the intralaminar nuclei?
(A) Anterior
(B) Dorsomedial
(C) Centromedian
(D) Lateral dorsal
(E) Thalamic reticular

10 The MRI of a 53-year-old man reveals a lacunar infarct in the area of the thalamus that selectively projects to the cortical region indicated by the arrows in the image below. Which of the following thalamic nuclei represents the most likely location of this man's lesion?

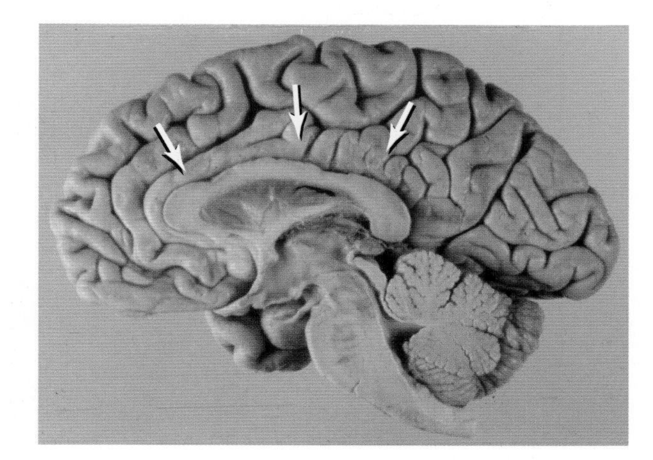

(A) Anterior
(B) Centromedian
(C) Dorsomedial
(D) Ventral posteromedial
(E) Ventral lateral

11 A 71-year-old man is brought to the Emergency Department after his wife discovered that she had significant difficulty awakening him in the morning. The history reveals that the man has suffered from hypertension for several years. MRI reveals an infarct involving the following thalamic nuclei: anterior, rostral part of the dorsomedial, and ventral anterior. Which of the following vascular territories is most likely represented by this lesion?

(A) Anterior choroidal artery
(B) Medial striate artery
(C) Quadrigeminal artery
(D) Thalamogeniculate artery
(E) Thalamoperforating artery

12 A 77-year-old woman is brought to the Emergency Department by her son. The history, provided by the son, revealed that she complained of a sudden loss of sensation. The examination reveals that she is oriented as to time and place, has normal cranial nerve function, but has a loss of pain sensation, discriminative touch, and vibratory sense on the right side of her body. MRI reveals a well-localized thalamic lesion. Which of the areas indicated in the image below represents the likely location of this woman's lesion?

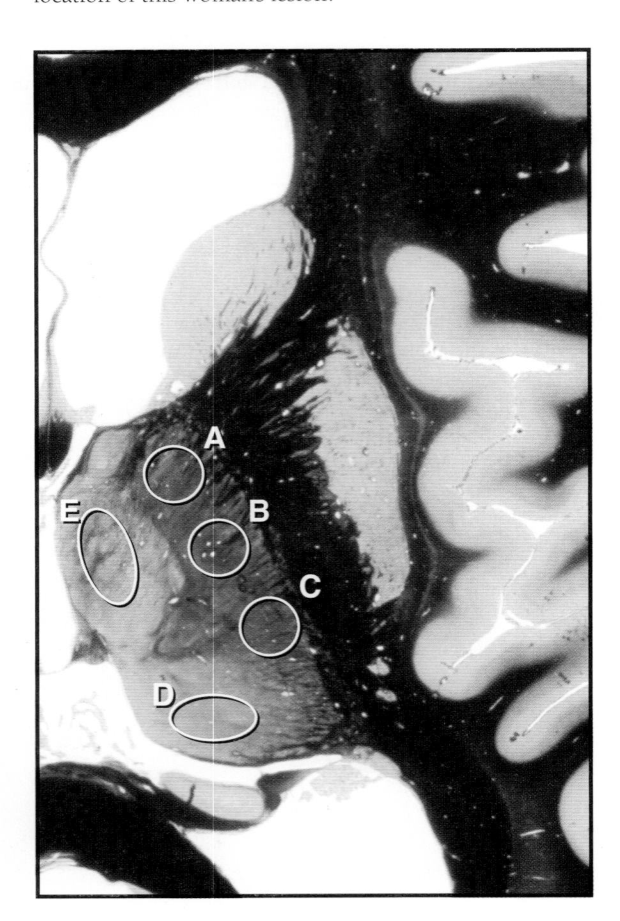

(A) A
(B) B
(C) C
(D) D
(E) E

13 Which of the following thalamic nuclei is primarily responsible for conveying pain, thermal, and discriminative touch sensations for the face to the appropriate part of the primary somatosensory cortex?

(A) Centromedian
(B) Pulvinar
(C) Ventral lateral
(D) Ventral posterolateral
(E) Ventral posteromedial

14 A 57-year-old woman presents to her ophthalmologist with the complaint of "not being able to see things to my left, I think I'm going blind." The examination revealed the following. First, the physician noticed that the woman, when in a resting attitude, tended to keep her head rotated to her right. Second, the woman ignored objects that entered her left visual fields: when the physician's nurse stepped into the woman's left visual space, the woman did not "see" her until the physician told her to look to her left. Third, when tested carefully, the woman was not blind in either her right or her left visual fields. MRI revealed a localized hemorrhagic lesion in the right thalamus. An infarct in which of the following would most likely account for this woman's deficit?

(A) Centromedian
(B) Dorsomedial
(C) Lateral geniculate
(D) Pulvinar
(E) Ventral lateral

15 A 58-year-old man presents with signs and symptoms indicative of an intracranial aneurysm. The physical examination and diagnostic studies confirm an aneurysm of 7 mm diameter, and it is surgically treated with a clip. The postoperative CT reveals bilateral hypodense regions at the arrows in the image below, suggesting that blood flow to these areas was inadvertently compromised. Based on the location of these hypodense areas, which of the following would be the most likely outcome in this case?

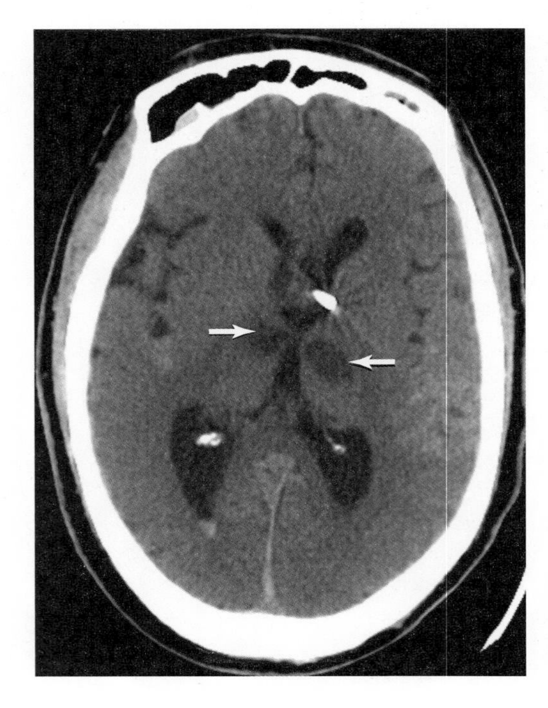

(A) Bilateral blindness in both eyes
(B) Bilateral loss of somatic sensation
(C) Difficulty regaining consciousness in the first 12 hours
(D) Persistent somnolence after regaining consciousness
(E) Unable to regain consciousness

16 An association nucleus of the thalamus is one that receives input from several different sources and projects to widespread cortical regions that are not specifically motor or sensory in function. Which of the following nuclei fits these criteria?
(A) Anterior
(B) Dorsomedial
(C) Lateral geniculate
(D) Ventral lateral
(E) Ventral posterolateral

17 Which of the following thalamic nuclei is an essential synaptic station in those circuits relayed through the dorsal thalamus (such as the Papez circuit) that are related to functions of the limbic system?
(A) Anterior
(B) Centromedian
(C) Dorsomedial
(D) Ventral lateral
(E) Ventral posteromedial

18 Which of the following thalamic, or hypothalamic, nuclei is the primary target of the fiber bundle indicated at the arrow in the image below?

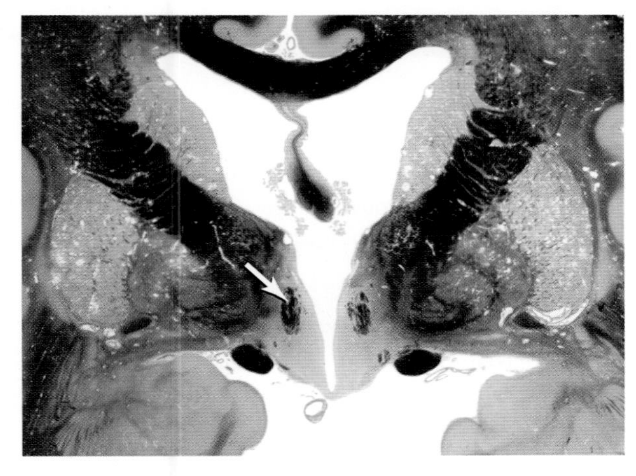

(A) Anterior
(B) Centromedian
(C) Habenular
(D) Mammillary
(E) Supraoptic

19 A relay nucleus of the thalamus is one that receives input from a specific source and projects to a specific cortical region that may have a motor or sensory function as broadly defined. Which of the following nuclei fits these criteria?
(A) Centromedian
(B) Dorsomedial
(C) Lateral dorsal
(D) Lateral posterior
(E) Ventral lateral

20 A 59-year-old man complains to his family of facial numbness. The history taken at the Emergency Department reveals that the man is diabetic. The examination confirms facial numbness over much of the left side of his face and reveals that the man is not compliant with his medication regimen. MRI reveals a focal lesion in the thalamus. Recognizing the deficits experienced by this man, this lesion is most likely located mainly in which of the following nuclei?
(A) Dorsomedial
(B) Ventral anterior
(C) Ventral lateral
(D) Ventral posterolateral
(E) Ventral posteromedial

21 Which of the following structures is identified by the arrow in the detail from the sagittal image below?

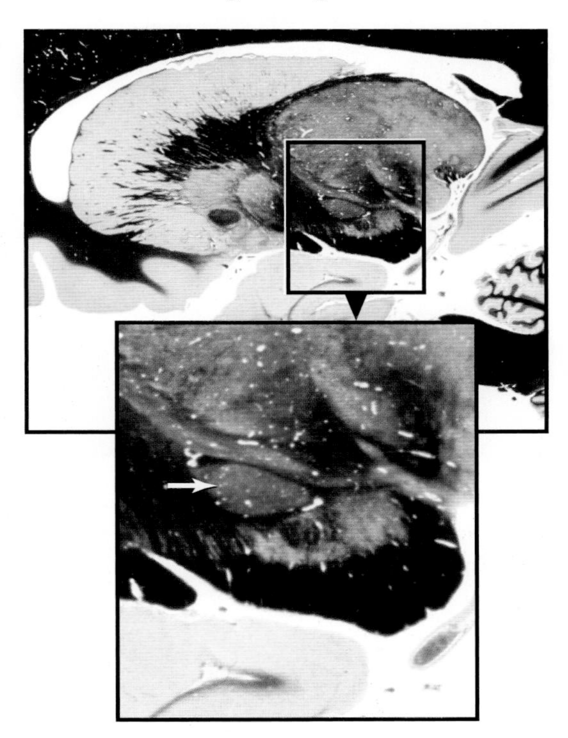

(A) Lenticular fasciculus
(B) Substantia nigra
(C) Subthalamic nucleus
(D) Thalamic fasciculus
(E) Zona incerta

22 A 27-year-old woman visits her family physician. The history and examination reveal the following: the woman had been quite athletic all of her life; over the last year, she has eaten excessively and gained a significant amount of weight, and she is no longer able to participate in sports. Enhanced MRI reveals well-localized bilateral hypothalamic lesions. Bilateral damage to which of the following would most likely be the cause of this woman's clinical course?
(A) Dorsomedial hypothalamic nucleus
(B) Lateral hypothalamic nucleus/area
(C) Mammillary nucleus
(D) Suprachiasmatic nucleus
(E) Ventromedial hypothalamic nucleus

23 Which of the following thalamic nuclei receives important projections from the internal segment of the globus pallidus and from the contralateral cerebellar nuclei?

(A) Anterior nucleus
(B) Lateral dorsal nucleus
(C) Ventral lateral nucleus
(D) Ventral posterolateral nucleus
(E) Ventral posteromedial nucleus

24 Which of the following functions is mediated through the structure indicated at the arrow in the image below?

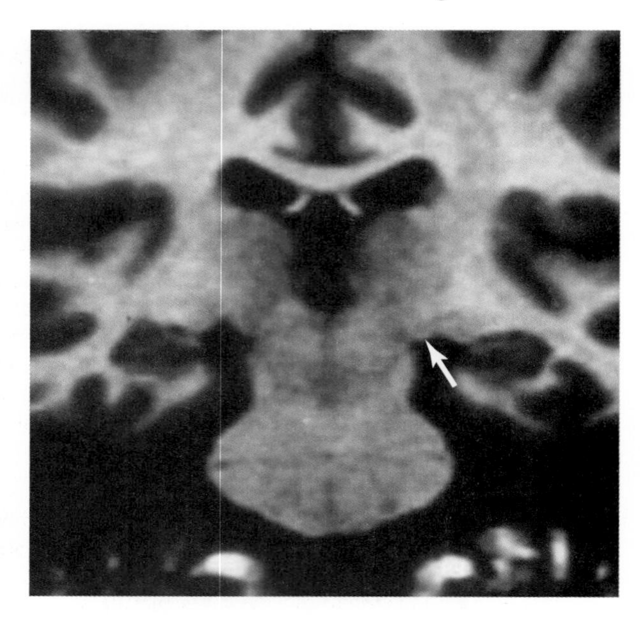

(A) Hearing
(B) Olfaction
(C) Somatosensory
(D) Taste
(E) Vision

25 A 74-year-old man is brought to the Emergency Department by his wife who explained that he had a sudden "spell." The examination reveals that the man has a left homonymous hemianopia and has some difficulty localizing sound in space, but he is not deaf. The man has no somatomotor or somatosensory deficits, and he has no trouble walking. MRI reveals a small infarct in his thalamus. Based on his deficits, this lesion is most likely in which of the following vascular territories?

(A) Anterior choroidal artery
(B) Lateral posterior choroidal artery
(C) Lenticulostriate artery
(D) Thalamogeniculate artery
(E) Thalamoperforating artery

26 The MRI of a 69-year-old woman reveals a small lesion in the region of the thalamus indicated by the outline in the image below. Occlusion of branches of which of the following vessels would most likely result in this lesion?

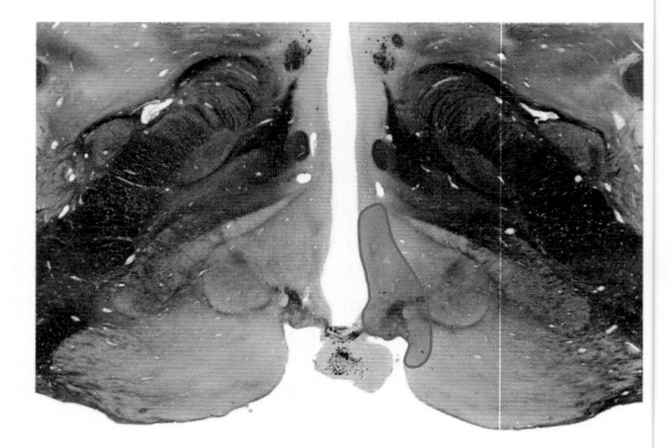

(A) Anterior choroidal artery
(B) Anterior inferior cerebellar artery
(C) Lateral posterior choroidal artery
(D) Medial posterior choroidal artery
(E) Thalamoperforating artery

ANSWERS

1 The answer is D: Ventral posterolateral. The posterior paracentral gyrus is the lower extremity portion of the primary somatosensory cortex; it receives thalamocortical projections from the ventral posterolateral (VPL) of the thalamus. Specifically, the more lateral part of the VPL nucleus selectively projects to the posterior paracentral gyrus. The other nuclei project as follows: dorsomedial to the cingulate gyrus and frontal lobe rostral to the precentral sulcus; lateral geniculate to the medial occipital cortex; ventral lateral to the somatomotor cortex; ventral posteromedial to the face area of the postcentral gyrus.

2 The answer is D: Ventral lateral. Axons arising from cells of the dentate, emboliform, and globose nuclei exit the cerebellum in the superior cerebellar peduncle, course through the decussation of this peduncle, and enter the ventral lateral (VL) nucleus of the dorsal thalamus. Lesions of these nuclei result in very characteristic motor deficits. While the VL also receives fibers from the basal nuclei, it projects to several regions of the cerebral cortex; its most important target is the somatomotor cortex (area 4). The anterior nucleus receives input from the limbic system and projects to the cingulate gyrus. Several brain regions, such as the amygdale, basal regions of the forebrain, and olfactory tubercle, project to the dorsomedial nucleus, which, in turn, projects to large areas of the orbital and frontal cortex. The centromedian nucleus receives input from the spinal cord, basal nuclei, cerebellum, and reticular formation and projects to the basal nuclei and wide areas of the cerebral cortex. The posterior column–medial lemniscus and anterolateral systems project to the ventral posterolateral nucleus, which, in turn, projects primarily to the somatosensory cortex (areas 3, 1, and 2).

3 The answer is D: Thalamogeniculate. As indicated by its name, the thalamogeniculate artery serves the medial and lateral geniculate bodies and the immediately adjacent structures: pulvinar, centromedian, ventral posterolateral and posteromedial nuclei, and others. The thalamogeniculate artery is a branch of the P_2 segment of the posterior cerebral artery. The lenticulostriate arteries, branches of the M_1 segment of the middle cerebral artery, serve the lenticular nucleus and adjacent posterior limb of the internal capsule. The medial striate (artery of Heubner) usually arises from the A_2 segment of the anterior cerebral artery and serves the head of the caudate nucleus and parts of the anterior limb of the internal capsule. The quadrigeminal and thalamoperforating arteries originate from P_1 and serve the tectum of the midbrain and anterior portions of the dorsal thalamus, respectively.

4 The answer is D: Supraoptic recess of third ventricle. The cavity of the diencephalon is the third ventricle; in this axial CT, it appears somewhat hourglass shaped with the rostral part being larger and the caudal part being somewhat smaller. This ventricle communicates caudally with the fourth ventricle via the cerebral aqueduct, which is seen in this image as a small black spot in the midbrain, and with the lateral ventricles through the interventricular foramina (there are two, one from each lateral ventricle into the midline third ventricle). The inferior portion of the third ventricle is located between the hypothalami (one hypothalamus on either side) and has two recesses. In this axial CT, the lowermost portion of the third ventricle is seen as two continuous areas: the more rostral and larger part of the black area is the supraoptic recess (at the arrow) and the more caudal and smaller part is the infundibular recess. This latter is the part of the ventricle that extends into the stalk of the pituitary. The foramen of Magendie is the opening of the fourth ventricle into the cisterna magna.

5 The answer is A: Anterior nucleus. The interventricular foramen, in which this tumor is located, is the space between the column of the fornix (which is rostromedial to the foramen) and the anterior nucleus of the thalamus (which is caudolateral to the foramen). Enlargement of the left lateral ventricle results from blockage of CSF flow from the left lateral ventricle into the third ventricle. The confusion experienced by this man may partially reflect the interruption of messages from this nucleus to the cingulate cortex as well as the increased intracranial pressure. The centromedian is located in the internal medullary lamina within the thalamus, and the dorsomedial nucleus is in the medial area of the thalamus bordering on the third ventricle; both project to broad areas of the cerebral cortex. The ventral anterior (VA) and ventral lateral (VL) nuclei are located in the lateral area of the thalamus adjacent to the internal capsule; the VA is a thalamic nucleus involved with the limbic system, and the VL is a relay nucleus intimately involved in motor function.

6 The answer is B: Lateral geniculate nucleus. The primary visual cortex receives input vital to vision from the lateral geniculate body on the same side via the optic radiations. Both eyes send information to each lateral geniculate body; this is why lesions caudal to the optic chiasm result in visual loss in both visual fields. The medial geniculate nucleus relays auditory information to the temporal lobe via the auditory radiations; the ventral lateral nucleus relays information from the cerebellum and basal nuclei to the somatomotor cortex. All of these thalamocortical systems pass through various limbs of the internal capsule. The pulvinar has connections with the visual cortex; these connections are not specifically concerned with vision, but with visual-motor function. The lateral dorsal nucleus is located in the upper portions of the thalamus and has no particular clinical importance.

7 The answer is E: Ventral posterolateral. This is a hemorrhage located within the territory of the thalamogeniculate artery, a branch of P_2. This vessel generally serves the geniculate nuclei (as its name implies), the pulvinar, the centromedian, and ventral posteromedial and posterolateral nuclei. This vascular territory may extend more rostrally to include caudal portions of the ventral lateral nucleus (VL pars caudalis). As is the case in this patient, and as is commonly seen in many cerebral strokes, only part of a vascular territory may be involved. This lesion is located primarily in the left ventral posterolateral nucleus; this correlates with the onset of major deficits on this woman's right side; this woman has major sensory losses. None of the other nuclei in this vascular region receives any significant input from the body. The anterior, ventral anterior, and dorsomedial (rostral portions) nuclei receive their blood supply predominately from the thalamogeniculate artery (a branch of P_1). Caudal portions of the dorsomedial nucleus are served by the medial posterior choroidal artery.

8 **The answer is C: Subthalamic nucleus.** Lesions of the subthalamic nucleus (these are usually vascular in nature) result in unpredictable forceful/flailing movements of the upper extremity, contralateral to the lesion. The subthalamic nucleus has no direct influence on lower motor neurons. Recall that the motor expression of damage to the subthalamic nucleus is expressed through the corticospinal tract. The anterior and centromedian nuclei do not have specific sensory or motor functions. The ventral lateral nucleus is a relay station for cerebellar and basal nuclei to influence the motor cortex; a lesion of this structure will result in deficits that reflect the fact that these major parts of the motor system send information to this nucleus. The ventral posteromedial nucleus is a sensory relay nucleus that receives pain, thermal sense, discriminative touch, and vibratory sense to the primary somatosensory cortex (areas 3, 1, and 2).

9 **The answer is C: Centromedian.** The intralaminar nuclei (these are also called the intralaminar nuclear group) consist of groups of cells insinuated within the internal medullary lamina (IML); the centromedian is the largest of these, is easily seen in brain slices and stained sections, and is sometimes discernable in MRI. The intralaminar nuclei are generally divided into a more rostral division consisting of the paracentral, central lateral, and central medial nuclei and a caudal division consisting of the centromedian, parafascicular, and subparafascicular nuclei. The anterior nuclear group actually consists of anteroventral, anterodorsal, and anteromedial nuclei but is commonly referred to as the anterior nucleus of the thalamus; this group of cells forms an important landmark: the anterior tubercle of the thalamus. The interventricular foramen (of Monro) is the space located between the anterior tubercle of the thalamus and the column of the fornix. The dorsomedial nucleus is located medial to the IML; the lateral dorsal nucleus is lateral to the IML and is part of the lateral nuclear group, which also includes the pulvinar. The thalamic reticular nucleus is a comparatively thin layer of cells that forms a shell on the lateral and inferior aspects of the thalamus; the neuron cell bodies forming this nuclear shell intermingle with the fibers forming the external medullary lamina.

10 **The answer is A: Anterior.** The anterior nucleus of the dorsal thalamus receives input from the mammillary nucleus and other areas and projects to the cingulate gyrus. These are parts of a major pathway that functions in emotions and behavior; it is a large part of the Papez circuit. The centromedian nucleus has projections to the basal nuclei, subthalamus, and substantia nigra, functioning in concert with the basal nuclei. The dorsomedial projects to wide areas of the frontal lobe including its orbital aspect. Somatosensory information from the face and oral cavity is relayed to the face area of the postcentral gyrus; the ventral lateral nucleus receives input from the cerebellum and basal nuclei and projects to the somatomotor cortex (precentral and anterior paracentral gyri).

11 **The answer is E: Thalamoperforating artery.** The thalamoperforating artery arises from the P_1 segment of the posterior cerebral artery, passes through the interpeduncular fossa, and penetrates the base of the midbrain. These branches of this artery basically pass straight upward to serve rostral portions of the thalamus. Some of the thalamic nuclei served by this vessel are involved in cortical arousal. The head of the caudate nucleus and parts of the anterior limb of the internal capsule are served by the medial striate artery; a number of important structures adjacent to the optic tract are served by the anterior choroidal artery. The quadrigeminal artery serves the tectum of the mesencephalon; the caudal parts of the thalamus, including the pulvinar and geniculate nuclei, are served by the thalamogeniculate artery.

12 **The answer is C (Ventral posterolateral nucleus).** The ventral posterolateral nucleus (choice C) receives pain and thermal sensations, discriminative touch, vibratory sense, and proprioception from the opposite side of the body and relays this onto the somatosensory cortex (postcentral and posterior paracentral gyri) on the same side. This information is conveyed by the anterolateral and the posterior column–medial lemniscus systems. The ventral anterior nucleus (A) has interconnections with the basal nuclei, and the ventral lateral nucleus (B) receives input from the cerebellum and basal nuclei and projects to the somatomotor cortex. The pulvinar (D) is involved in visual motor function and may also function in what may be called "blind sight": a patient with damage to the visual cortex may be able to localize a bright light in space under certain circumstances. The dorsomedial nucleus (E) has connections to the convexity of the frontal lobe (rostral to the precentral gyrus) and to its orbital surface.

13 **The answer is E: Ventral posteromedial nucleus.** The ventral posteromedial nucleus receives input from the contralateral spinal trigeminal nucleus, and bilaterally from the principal sensory nucleus, and relays this information onto the face region of the postcentral gyrus, the largest part of the primary somatosensory cortex on the same side. The centromedian nucleus projects to wide areas of the cerebral cortex; the pulvinar projects to visual association cortices in the occipital and temporal lobes. The primary somatomotor cortex receives thalamocortical projections from the ventral lateral nucleus, and the ventral posterolateral nucleus relays pain, thermal sense, and discriminative touch information from the contralateral side of the body to the appropriate areas of the somatosensory cortex.

14 **The answer is D: Pulvinar nucleus.** The pulvinar nucleus has extensive connections with visual association cortices, connections with the frontal and parietal eye fields, and connections with the temporal lobe, particularly the superior temporal gyrus. Damage to the pulvinar that avoids the optic radiations may, in some patients, contribute to the phenomenon of a hemispatial neglect to the contralateral side; right-sided lesion may produce, neglect of the left visual space. The pulvinar may also participate in the general area of visual-motor function and in orientation in visual space. It is through these diverse connections that the pulvinar may participate in visual function, as broadly defined, yet not specifically in the "seeing" of objects. In addition, there is evidence that the pulvinar nucleus may function in what is called "blind sight" in some, but not all, patients. In this situation, for example, the patient may have a lesion of the right visual cortex and acknowledge blindness in the left hemifields; the patient has no conscious vision in the blind hemifields. When a light is placed in the blind hemifield, the patient does not "see" the light, but when asked if there were a light where would it be, the patient points to the location of the light. The centromedian nucleus projects diffusely to the cortex, including the frontal lobe; the dorsomedial also targets much of the frontal lobe; the ventral lateral nucleus relays cerebellar and basal nuclei information to the somatomotor cortex. The lateral geniculate nucleus receives visual information from the retina and conveys it onto the visual cortex; lesions of this structure cause partial blindness in both eyes.

15 **The answer is E: Unable to regain consciousness.** The location of these bilateral lesions is in the territory of the thalamoperforating artery. This artery originates from the P_1 segment and is immediately adjacent to the basilar bifurcation, which is the primary site of aneurysms in the vertebrobasilar circulation. Inadvertent inclusion of the origin of the thalamoperforating artery within the aneurysm clip will result in the lesions seen here. The thalamoperforating artery serves the rostral portions of the dorsal thalamus including those areas of nuclei and traversing fibers that are involved in cortical arousal. Damage to these areas, especially bilaterally, as in this patient, results in persistent coma; the cerebral cortex cannot be aroused. Bilateral lesions may result if both arterial trunks are trapped within the clip or if the branches to both sides originate from a single common trunk (about 8% of cases). These lesions do not involve thalamic nuclei that relay visual or somatosensory information. Choices C and D are not applicable since the patient will not wake up.

16 **The answer is B: Dorsomedial nucleus.** The dorsomedial nucleus of the thalamus receives input from diverse areas such as the amygdaloid complex, olfactory tubercle and bulb, basal forebrain, and temporal, orbital, and piriform cortices. In turn, this nucleus projects to the frontal cortex rostral to the precentral gyrus, to the medial aspect of the frontal lobe, and to its orbital surface; it receives from diverse areas and projects to diverse areas. The other choices are not association nuclei but are relay nuclei; they receive information from specific sources and project to specific functional areas of the cerebral cortex.

17 **The answer is A: Anterior nucleus.** This nucleus is an essential station between the hippocampus and the cortex of the cingulate gyrus. The mammillary nucleus receives input from the hippocampus via the fornix (specifically its postcommissural portion) and projects to the anterior thalamic nucleus via the mammillothalamic tract. In turn, the anterior nucleus projects to the cingulate gyrus; this total pathway is the Papez circuit. The centromedian and dorsomedial nuclei are not principals in limbic system circuits; the ventral lateral and ventral posteromedial nuclei are related to motor (for the body) and sensory (for the face) functions, respectively.

18 **The answer is D: Mammillary nucleus of the hypothalamus.** The arrow is indicating the postcommissural portion of the column of the fornix as it courses through the hypothalamus (separating medial from lateral hypothalamic zones) toward its termination in the mammillary nucleus. The fornix arises from the hippocampus. The mammillary nucleus, in turn, projects to the anterior nucleus of the thalamus via the mammillothalamic tract. The centromedian nucleus is an intralaminar nucleus (that is also an association nucleus) that receives from diverse sources and projects to diverse targets. The habenular nucleus receives input from limbic structures in the temporal lobe and from preoptic areas and projects to the interpeduncular region of the midbrain. The supraoptic nucleus receives visual input and projects to other hypothalamic areas for the maintenance of circadian rhythms.

19 **The answer is E: Ventral lateral.** The ventral lateral (VL) nucleus of the thalamus receives a pallidothalamic projection from the medial segment on the same side and a cerebellothalamic projection from the cerebellar nuclei on the opposite side. In turn, the VL projects primarily to the somatomotor cortex (precentral and anterior paracentral gyri: area 4). The VL is a relay nucleus that is an essential synaptic station in the pathways that constitute the motor system. The other choices, centromedian, dorsomedial, lateral dorsal, and lateral posterior, are all association nuclei. These nuclei receive input from diverse sources and, in turn, project to diverse targets.

20 **The answer is E: Ventral posteromedial nucleus.** The spinal trigeminal nucleus (mainly the pars caudalis) on the right projects to the left ventral posteromedial (VPM) nucleus; the left VPM projects to the somatosensory nucleus on the same side. This pathway conveys somatosensory information (pain and thermal sense) from the face to the cerebral cortex. The dorsomedial and ventral anterior nuclei project to widespread areas of the frontal lobe, excluding the precentral gyrus. The ventral lateral nucleus relays information from the cerebellum and basal nuclei to the primary somatomotor cortex. The ventral posterolateral nucleus receives input from the contralateral side of the body via the anterolateral system and the posterior column–medial lemniscus system and, in turn, projects to the primary somatosensory cortex.

21 **The answer is C: Subthalamic nucleus.** The subthalamic nucleus is a part of the thalamus that influences motor activity through its interconnections with the basal nuclei. The lateral segment of the globus pallidus projects (via GABAergic fibers) to the subthalamic nucleus, which, in turn, projects (via glutaminergic fibers) to the medial segment of the globus pallidus. The subthalamic nucleus also receives a glutaminergic corticosubthalamic connection. The subthalamic nucleus has the following characteristics: (1) in axial, coronal, and sagittal planes, it is lens-shaped; (2) it is immediately adjacent to the substantia nigra; and (3) it is located superior to the nigra but inferior to the lenticular fasciculus; all of these characteristics are seen in this image. The dark line immediately superior to the subthalamic nucleus is the lenticular fasciculus. Above this line is a light area, the zona incerta, and superior to this structure, the dark line is the thalamic fasciculus.

22 **The answer is E: Ventromedial hypothalamic nucleus.** The ventromedial nucleus of the hypothalamus is a satiety center; experimental stimulation will result in a decrease in feeding behavior, while lesions will result in an increase in feeding behavior (hyperphagia). The dorsomedial hypothalamic nucleus is a behavioral center, stimulation of which causes sham rage, while lesions produce decreased aggression. The lateral hypothalamus is also a feeding center but with effects opposite those of the ventromedial nucleus: stimulation causes increased feeding; lesions result in decreased feeding. The suprachiasmatic nucleus regulates circadian rhythms; lesions of this nucleus will abolish these rhythms. The mammillary nucleus is an important synaptic center in limbic system circuits that participates in processing short-term memory into long-term memory.

23 **The answer is C: Ventral lateral nucleus.** The ventral lateral (VL) nucleus, pars caudalis, receives a projection from the cerebellar nuclei (cerebellothalamic fibers) on the contralateral side, while the VL nucleus, pars oralis, receives input from the ipsilateral internal segment of the globus pallidus (pallidothalamic fibers). The VL nucleus, in turn, projects to the primary somatomotor cortex. The anterior nucleus receives input from the mammillary nuclei and projects to the cingulate gyrus; the lateral dorsal nucleus receives input from nuclei of the limbic system and also projects to the cingulate gyrus. Ventral posterolateral (VPL) and ventral posteromedial (VPM) nuclei receive sensory input from the body (to VPL) and from the head through the trigeminal nerve (to VPM) and project to their respective areas of the primary somatosensory cortex.

24 **The answer is A: Hearing.** The elevation indicated by the arrow is the medial geniculate body; the neurons immediately internal to this elevation collectively form the medial geniculate nucleus. This nucleus is a major relay station in the thalamus for the transmission of auditory information onto the cortex (transverse temporal gyrus). At this point in the auditory pathway, the information is essentially bilateral. Thus, lesions at this point in the pathway may result in an alteration in sound perception, or localization, but not likely in deafness. Olfactory input is processed through structures in the basal forebrain (olfactory bulb, nuclei, and tubercle); somatosensory information is relayed by the ventral posterolateral and posteromedial thalamic nuclei. The ventral posteromedial nucleus is also the synaptic center that relays taste information onto the sensory cortex. Vision is relayed through the thalamus by the lateral geniculate nucleus; the elevation immediately lateral to the medial geniculate body in this image is the lateral geniculate body that overlies the corresponding nucleus.

25 **The answer is D: Thalamogeniculate artery.** The thalamogeniculate artery arises from the P_2 segment of the posterior cerebral artery, courses around the brainstem at its junction with the thalamus, and serves the pulvinar, medial (auditory) and lateral (visual) geniculate nuclei, and adjacent nuclei, such as the centromedian and ventral posterolateral nuclei. This man's symptoms correlate with damage to the geniculate nuclei. The anterior choroidal artery serves the optic tract, corticospinal fibers in the internal capsule at its junction with the crus cerebri, and inferior portions of the basal nuclei and some temporal lobe structures. The lateral posterior choroidal artery serves portions of the choroid plexus of the lateral ventricle; the lenticulostriate arteries are important branches of the M_1 segment of the middle cerebral artery that serve much of the basal nuclei and posterior limb of the internal capsule. The thalamoperforating artery, a branch of P_1, serves the anterior, ventral anterior, and the rostral half of the dorsomedial thalamic nuclei.

26 **The answer is D: Medial posterior choroidal artery.** The medial posterior choroidal artery originates from the P_2 segment of the posterior cerebral artery within the lateral part of the interpeduncular cistern, wraps around the brain at the area of the midbrain-thalamus junction, and enters the caudal aspect of the third ventricle through the cistern of the velum interpositum. Branches of the medial posterior choroidal artery enter the roof of the third ventricle to serve the choroid plexus. En route, this vessel also provides branches to thalamic structures seen in this image: the habenula, caudal and medial portions of the dorsomedial nucleus, and small areas of the medial portions of the pulvinar. The anterior choroidal artery serves the choroid plexus in the temporal horn and portions of the adjacent amygdale, hippocampus, ventral basal nuclei, and inferior parts of the posterior limb of the internal capsule. The anterior inferior cerebellar artery serves the inner ear (via the labyrinthine branch), portions of the brainstem, and much of the inferior surface of the cerebellum. The choroid plexus of the lateral ventricle is served by the lateral posterior choroidal artery; the anterior portions of the dorsal thalamus are served by the thalamoperforating artery.

Chapter 12

The Telencephalon

Recall the Cautionary Tale: The images in this chapter are presented in a Clinical Orientation: *this approach emphasizes how structure and function correlate with deficits in a clinical setting. MRI and CT have a universally recognized orientation and laterality. Line drawings, stained sections, and gross brain images in the Clinical Orientation are viewed here in the identical manner that one views an MRI: your right, as the physician-observer, is the patient's left, and the observer's left is the patient's right. The laterality of deficits and reference to connections follow accordingly.*

QUESTIONS

Select the single best answer.

1 A 71-year-old woman is brought to the Emergency Department by EMS. The examination reveals that cranial nerve function is within normal ranges, but she has a profound weakness of her left lower extremity (LE). MRI shows a well-localized cortical lesion in the right hemisphere. Based on her deficits, which of the following outlined areas would represent the most likely location of the cortical damage in this woman?

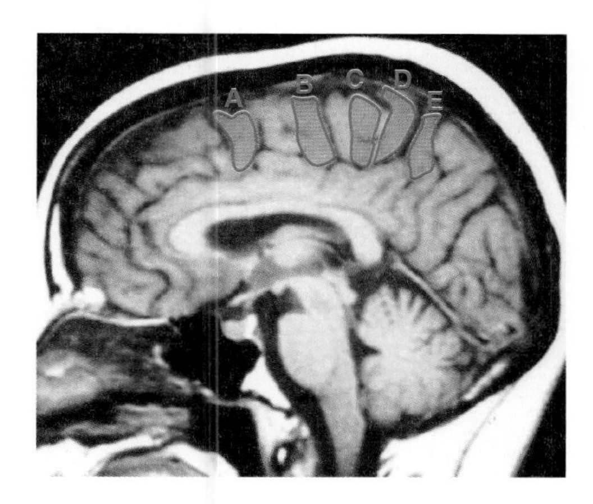

(A) A
(B) B
(C) C
(D) D
(E) E

2 A 57-year-old woman presents at the Emergency Department with signs and symptoms suggestive of a rapidly expanding lesion that will probably result in uncal herniation. A lesion that may have this particular outcome in this patient is most likely located in which of the following?

(A) Frontal lobe
(B) Insular lobe
(C) Occipital lobe
(D) Parietal lobe
(E) Temporal lobe

3 A 73-year-old man presents with signs and symptoms that suggest the possibility of a lesion in the cerebral cortex. His main deficit is an inability to express himself other than in single words. MRI reveals a small lesion in the pars opercularis which partially explains the man's deficits. Which of the areas indicated in the image below represents the most likely site of this man's lesion?

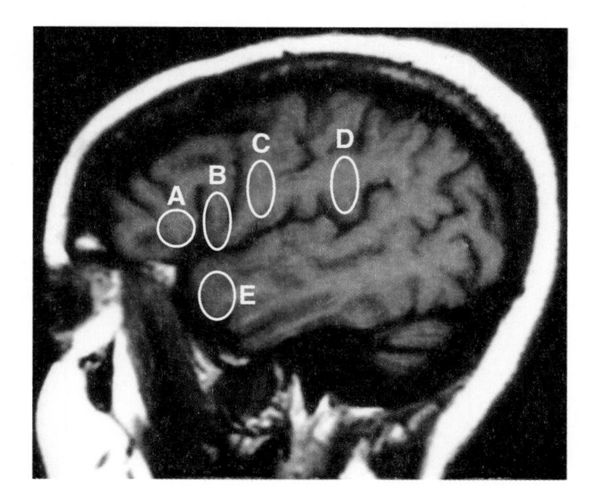

(A) A
(B) B
(C) C
(D) D
(E) E

4 The collateral sulcus is a consistent landmark that separates which of the two following structures?

(A) Cingulate gyrus from the corpus callosum
(B) Cuneus from precuneus
(C) Inferior parietal lobule from superior parietal lobule
(D) Parahippocampal gyrus from occipitotemporal gyrus
(E) Pars triangularis from pars orbitalis

5 A 69-year-old man is brought to the Emergency Department by the local EMS team. He is somewhat somnolent and complains of a headache. The examination reveals normal cranial nerve function but a profound right-sided weakness and loss of pain sensation and discriminative touch on his right upper and lower extremities. MRI reveals a hemorrhagic stroke in the internal capsule. Based on this man's deficits, this lesion is most likely located in which of the following?
(A) Anterior limb
(B) Genu
(C) Posterior limb
(D) Retrolenticular limb
(E) Sublenticular limb

6 Which of the following structures is identified by the two arrows in the detail of the image below?

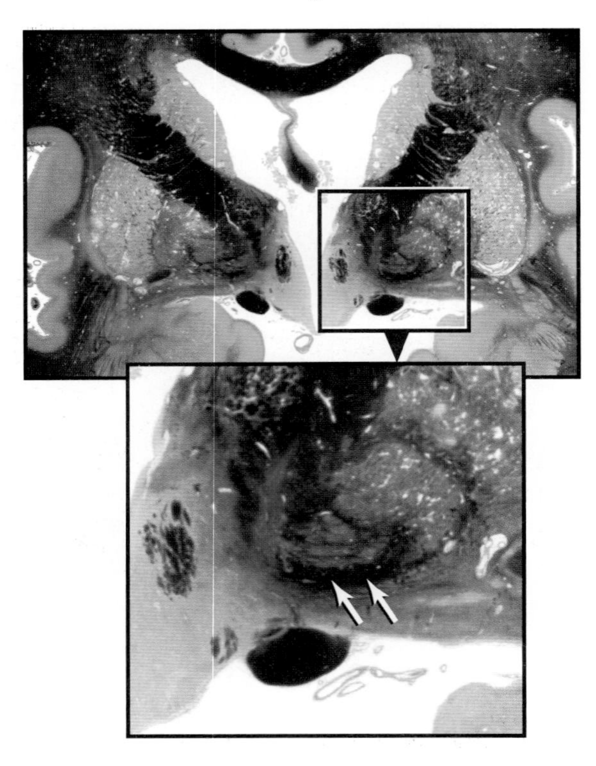

(A) Ansa lenticularis
(B) Anterior commissure
(C) Lenticular fasciculus
(D) Postcommissural fornix
(E) Thalamic fasciculus

7 A 77-year-old woman visits your office with the complaint of difficulty seeing. The history reveals that the woman believes that this came on rather suddenly. The examination reveals that she is not blind in either eye, but has some visual loss in both eyes. Suspecting a stroke, the physician orders a CT which shows a cortical lesion. This cortical defect is most likely found in which of the following?
(A) Angular gyrus
(B) Broca area
(C) Lingual gyrus
(D) Posterior paracentral gyrus
(E) Transverse temporal gyrus

8 A 71-year-old man is brought to the Emergency Department by a neighbor. The history reveals that the man was working in his garden, and suddenly "my arm went limp." The examination reveals a profoundly weak right upper extremity (UE), but no weakness of the left UE or of either lower extremities. Sensation is normal on all extremities and on his face. MRI reveals a localized cortical lesion in one of the outlined areas on the image below. Damage to which of these areas correlates with the deficits experienced by this man?

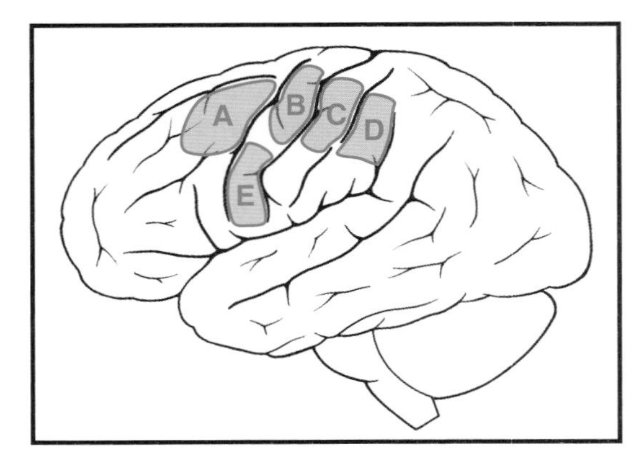

(A) A
(B) B
(C) C
(D) D
(E) E

9 Which of the following fiber bundles is found within the structure indicated at the arrow in the image below?

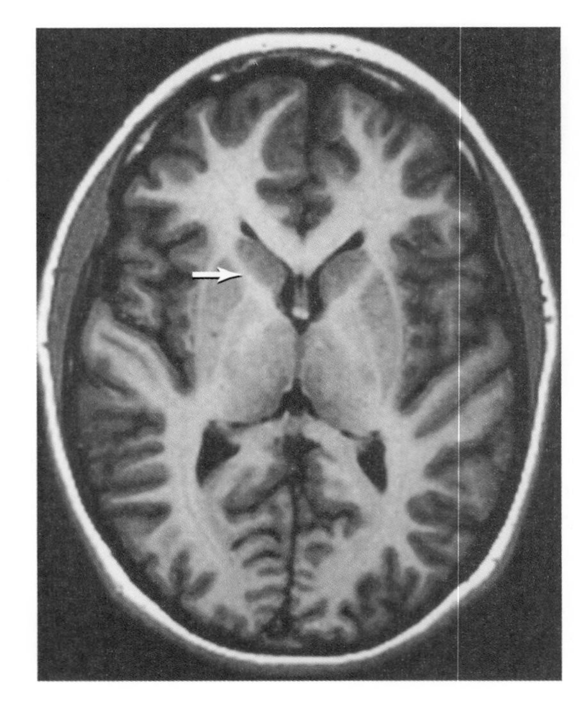

(A) Corticonuclear
(B) Corticospinal
(C) Frontopontine
(D) Optic radiations
(E) Parietopontine

10 The sagittal MRI of a 59-year-old man reveals a lesion in the inferior parts of the cuneus and in the adjacent area of the lingual gyrus. This patient is most likely suffering which of the following?
(A) Auditory deficits
(B) Motor (expressive) aphasia
(C) Sensory (receptive) aphasia
(D) Somatosensory deficits
(E) Visual deficits

11 The lenticular nucleus is composed of which of the following structures?
(A) Caudate nucleus + putamen
(B) Caudate nucleus + globus pallidus
(C) Putamen only
(D) Putamen + globus pallidus
(E) Putamen + claustrum

12 A 78-year-old woman has one small vascular lesion indicated by the arrows. This is an incidental finding. Which of the following vessels serves this particular area of the forebrain?

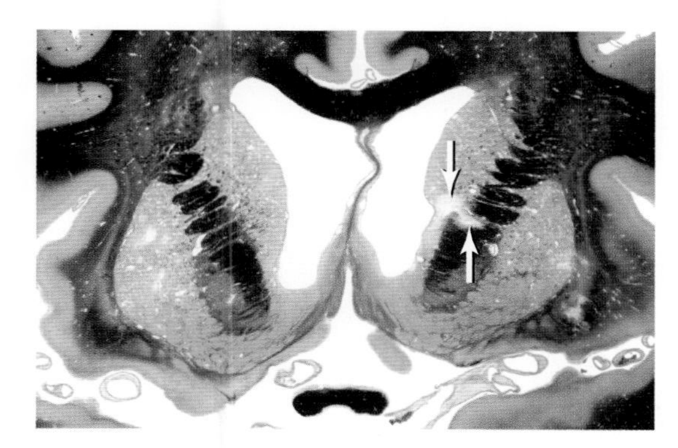

(A) Anterior choroidal artery
(B) Anterior communicating artery
(C) Callosomarginal artery
(D) Lenticulostriate artery(ies)
(E) Medial striate artery

13 A 77-year-old man collapsed at home and is transported to the Emergency Department. The history reveals that the man is being medicated for hypertension and diabetes, and that this event was of sudden onset. Results of the examination reveal that the man is hemiplegic and has a loss of all somatic sensation on the same side of the body as the weakness. The MRI reveals an infarct of the posterior limb of the internal capsule. This lesion is in the area of the telencephalon served by which of the following arteries?
(A) Anterior choroidal
(B) Calcarine
(C) Lateral posterior choroidal
(D) Lenticulostriate
(E) Medial striate

14 A 77-year-old woman has a sudden headache and loses consciousness. The next day, after she had regained consciousness, she was aphasic, a bit somnolent, was able to answer questions with "yes" or "no," but was unable to construct even a short sentence. The woman seems to be fully aware of her situation. MRI revealed a lesion in the cerebral cortex. This lesion is most likely located in which of the following?
(A) Angular gyrus + supramarginal gyrus
(B) Anterior + posterior paracentral gyri
(C) Cuneus + lingual gyrus
(D) Pars opercularis + pars triangularis
(E) Uncus + parahippocampal gyrus

15 A 39-year-old morbidly obese woman presents at the Emergency Department. The history reveals that she is diabetic and hypertensive, but refuses to take medication. The examination reveals deficits that are consistent with a focal lesion in the paleostriatum; this impression was confirmed in MRI. Which of the structures indicated in the image below is the most likely site of this lesion?

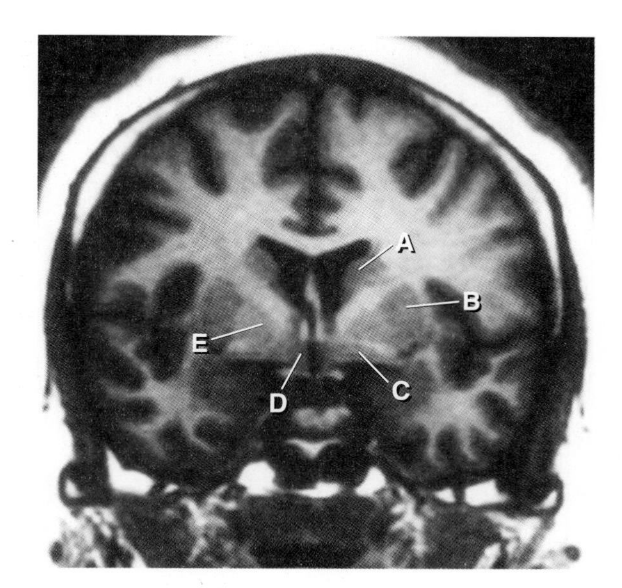

(A) A
(B) B
(C) C
(D) D
(E) E

16 A 51-year-old woman is brought to the Emergency Department by her husband, who indicates that she was working in the garden and suddenly became disoriented. Suspecting a stroke, the physician orders an MRI which reveals a large infarct in the retrolenticular limb of the internal capsule. Which of the following would be the most obvious deficit in this patient?
(A) Distortion of sound
(B) Paraesthesia on the extremities
(C) Paraesthesia on the face
(D) Paralysis of facial and tongue movements
(E) Visual deficits in the opposite hemifield

17 A 69-year-old man is brought to his family physician by his wife. The results of the examination reveal deficits that suggest that the man has had a precipitous vascular event. A CT is ordered, revealing a small stroke in the genu of the internal capsule. Which of the following fiber populations is most affected in this man's stroke?

(A) Corticonigral
(B) Corticonuclear
(C) Corticopontine
(D) Corticospinal
(E) Corticostriatal

18 Which of the structures indicated on the image below is the source of a direct GABAergic projection to the dorsal thalamus?

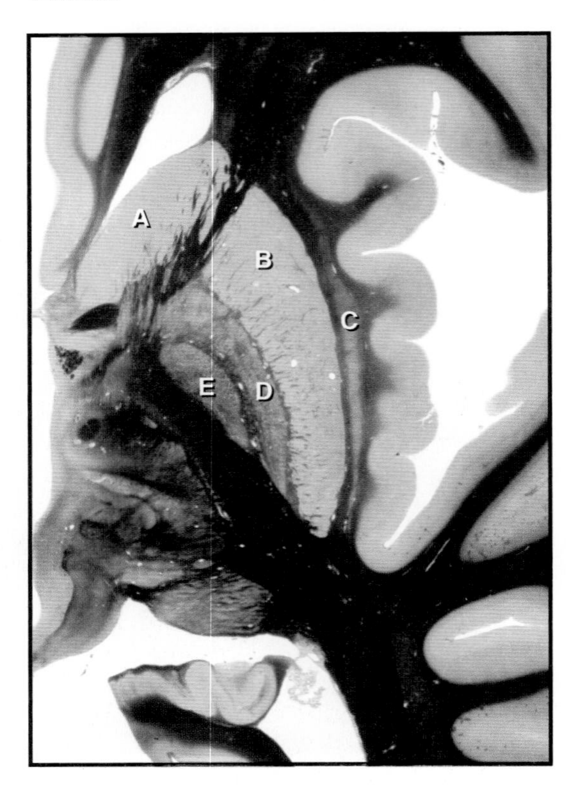

(A) A
(B) B
(C) C
(D) D
(E) E

19 The neostriatum, in whole or in part, is a target of several neurodegenerative diseases, such as Huntington and Parkinson diseases. Some of these are treatable, but are not curable. Which of the following constitutes this part of the forebrain?

(A) Caudate nucleus only
(B) Caudate nucleus + putamen
(C) Caudate nucleus + putamen + globus pallidus
(D) Putamen only
(E) Putamen + globus pallidus

20 A 34-year-old woman presents to her family physician with the complaint of increasingly persistent headaches that are largely refractory to OTC medications. MRI reveals a large fusiform aneurysm compromising the long (gyri longi) and short (gyri breves) gyri. This aneurysm is located on a vessel found on which of the following?

(A) Frontal lobe
(B) Insular lobe
(C) Occipital lobe
(D) Parietal lobe
(E) Temporal lobe

21 A 71-year-old man is brought to the Emergency Department from the site of a minor motor vehicle collision. The man indicted that he stopped at a rural intersection, did not see anyone coming, and proceeded, only to be hit on his right rear fender. During the examination, the physician discovered that this man had a right homonymous hemianopia of which he seemed unaware. MRI showed a small well-localized lesion in the white matter of the hemisphere. A lesion in which of the areas indicated on the image below would most likely explain this man's neurologic deficit?

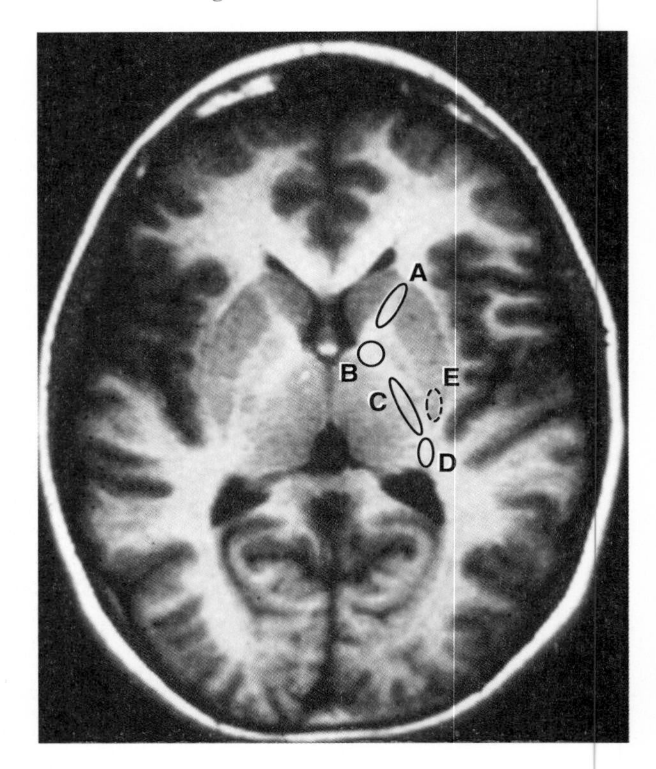

(A) A
(B) B
(C) C
(D) D
(E) E

22 Which of the following structures is indicated by the arrow on the given image?

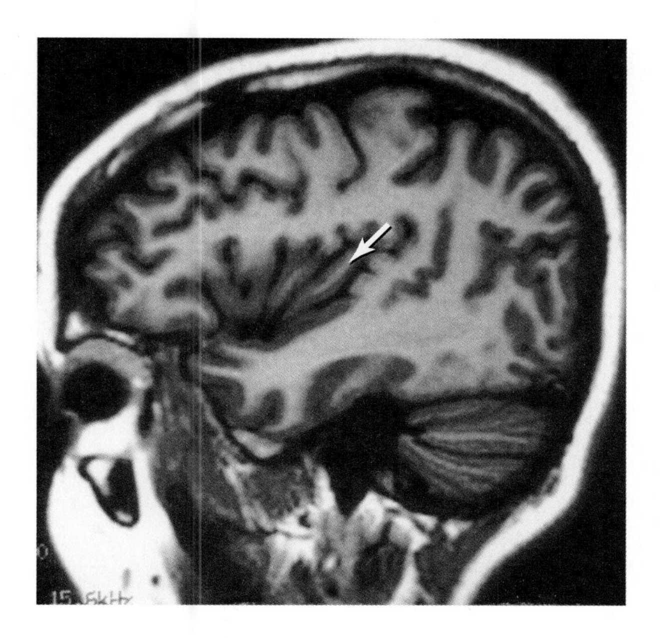

(A) Angular gyrus
(B) Gyrus brevis (short gyrus)
(C) Gyrus longi (long gyrus)
(D) Parietal operculum
(E) Transverse temporal gyrus

23 A 73-year-old man is transported to the Emergency Department by the local EMS team. The history, provided largely by the son, reveals that the father suddenly fell while raking leaves in the yard and was unable to get up. The man is on medications for mild diabetes and hypertension. The examination shows that the man has a profoundly weak left lower extremity, but he has age-appropriate strength and mobility in his upper extremities. MRI reveals a cortical stroke. Which of the following represents the most likely location of this lesion?
(A) Anterior paracentral gyrus on the left
(B) Anterior paracentral gyrus on the right
(C) Precentral gyrus, medial portion on the left
(D) Precentral gyrus, medial portion on the right
(E) Posterior paracentral gyrus on the right

24 A 69-year-old woman presents at the Emergency Department with the complaint of sudden onset of difficulty speaking. The examination reveals that the woman's tongue deviates to the right on attempted protrusion, and her uvula deviates to the left when she says "Ah." She has no other deficits, but her speech is slurred. MRI reveals a small brain lesion. Which of the following represents the most likely location of this woman's lesion?
(A) Anterior limb, internal capsule
(B) Genu, internal capsule
(C) Lateral medulla
(D) Medial medulla
(E) Posterior limb, internal capsule

25 A 19-year-old man is delivered to the Emergency Department from the site of a motor vehicle collision. The examination reveals facial and head injuries, a broken clavicle, and compound fractures of the radius and ulna. The EMS personnel indicate that his injuries are related to his being propelled into the steering wheel and dashboard at high speed. CT reveals facial fractures, bilateral damage to about the rostral 4 cm of the temporal lobes, and blood in the temporal horns. Damage to which of the following structures would have the most serious implications for this man's future?
(A) Amygdaloid nucleus
(B) Anterior thalamic nucleus
(C) Hippocampal formation
(D) Stria terminalis
(E) Tail of the caudate nucleus

26 Which of the following fiber bundles associated with the telencephalon contains fibers that directly exert an inhibitory influence on neurons of the dorsal thalamus?
(A) Amygdalofugal pathway
(B) Column of the fornix
(C) Lenticular fasciculus
(D) Subthalamic fasciculus
(E) Tapetum

27 A 78-year-old man is brought to the family physician by his wife. The history, provided largely by the wife, reveals that a couple of days ago he had a sudden headache for which he took an OTC medication and went to bed. The next morning, he seemed confused and his comments made little sense. The examination revealed that the man seems to not understand the questions being asked, and, when he responds, his answers are essentially unintelligible. He also has great difficulty reading and writing. MRI reveals a lesion in the cerebral cortex. Based on his symptoms, which of the following is the most likely location of this lesion?
(A) Anterior and posterior paracentral gyri
(B) Cuneus and lingual gyri
(C) Pars opercularis and pars triangularis
(D) Supramarginal and angular gyri
(E) Transverse temporal gyrus

ANSWERS

1 **The answer is C (The anterior paracentral gyrus).** The lower extremity (LE) is somatotopically represented in the anterior paracentral gyrus, which is the medial continuation of the precentral gyrus. Regions of the body from about the hip up are represented in the precentral gyrus. The small sulcus located between choices C and D is the medial terminus of the central sulcus; a line drawn from this point straight down to the cingulate sulcus makes the boundary between LE motor cortex (anterior paracentral gyrus, choice C), which is rostral, and LE sensory cortex (posterior paracentral gyrus, choice D), which is caudal. The marginal sulcus (or marginal ramus of the cingulate gyrus) is the caudal border of the posterior paracentral gyrus. Choices A and B are gyri of the medial surface of the superior frontal gyrus. Choice E is the rostral aspect of the precuneus, which is the largest portion of the parietal lobe on the medial surface of the hemisphere.

2 **The answer is E: Temporal lobe.** A rapidly expanding lesion in the temporal lobe will force the medial temporal lobe (uncus and possibly portions of the parahippocampal gyrus) over the edge of the tentorium cerebelli into the midbrain; this is uncal hernination. The most probable damage is to the crus cerebri and oculomotor root; this is a Weber syndrome which usually presents as an alternating hemiplegia. Masses in the frontal, parietal, and occipital lobes usually result in central, or transtentorial, herniation due to the fact that these tend to descend directly downward through the tentorial notch.

3 **The answer is B (Pars opercularis).** This man's lesion is in the inferior frontal gyrus which is comprised of the pars opercularis (B, identified in the stem), the pars triangularis (A), and the pars orbitalis (unlabeled). As the name implies, the orbital part is toward the orbital surface of the frontal lobe. Choice C is the lower part of the precentral gyrus; choice D is part of the parietal operculum overlying the most caudal portion of the insular cortex; choice E is a portion of the temporal pole. Lesions of the pars opercularis and the adjacent triangularis result in a Broca (expressive or motor) aphasia.

4 **The answer is D: Parahippocampal gyrus from occipitotemporal gyrus.** The inferior surface of the temporal lobe is largely formed by the medially located parahippocampal gryus and the more laterally located occipitotemporal gyrus or gyri (sometimes there are two) with the collateral sulcus in between these two major gyri. The sulcus of the corpus callosum separates the cingulate gyrus from the corpus callosum; the parietooccipital sulcus is insinuated between the cuneus and precuneus; superior and inferior parietal lobules are separated by the intraparietal sulcus. The anterior horizontal limb (or branch) of the lateral (Sylvian) sulcus is located between the pars triangularis and pars orbitalis; the anterior ascending ramus is located between the pars triangularis and the pars opercularis.

5 **The answer is C: Posterior limb (of internal capsule).** This man has motor and sensory losses on the same side of the body; this is generally a feature of forebrain lesions. This information traverses the posterior limb of the internal capsule as corticospinal fibers (motor) and thalamocortical radiations (sensory information traveling from the thalamus to the sensory cortex). Corticospinal fibers are located in about the caudal one-half of the posterior limb and arranged somatopically: upper extremity more rostral, trunk in the middle, and lower extremity more caudal. It is important to keep in mind that damage to the posterior limb may present clinically as a predominate motor deficit in many cases. Lesions of the anterior limb do not result in significant motor or sensory deficits and damage to the genu will produce noticeable and significant deficits related to cranial nerve motor function but not to the body. The sublenticular limb contains auditory radiations and the optic radiations traverse the retrolenticular limb.

6 **The answer is A: Ansa lenticularis.** The ansa lenticularis originates from the medial segment of the globus pallidus, arches medially around fibers of the internal capsule, and proceeds caudally to join fibers of the lenticular fasciculus in the prerubral field. This conjoined bundle (ansa lenticularis + lenticular fasciculus) enters the thalamus as the thalamic fasciculus. The ansa lenticularis contains GABAergic pallidothalamic fibers that terminate primarily in the ventral lateral nucleus of the thalamus. A small portion of the anterior commissure is seen just outside the box, and to its right, immediately inferior to the junction between the putamen and the lateral segment of the globus pallidus. The postcommissural fornix is seen in the detail and appears as many tightly packed small fascicles.

7 **The answer is C: Lingual gyrus.** The lingual gryus is the lower bank of the primary visual cortex, the gyrus on the upper bank being a portion of the cuneus. The calcarine sulcus separates these two gyri; consequently, the visual cortex is frequently called the calcarine cortex. Damage to the lingual gyrus will result in a quadrantanopia in the opposite visual field. For example, a lesion of the right lingual gyrus will result in a left superior quadrantanopia. The angular gyrus is at the caudal end of the superior temporal sulcus and, along with the supramarginal gyrus, forms the inferior parietal lobule. The Broca area is in the inferior frontal gyrus (sometimes called the Broca convolution) and may refer to mainly the partes opercularis and triangularis. The primary somatosensory cortex for the lower extremity is in the posterior paracentral gyrus; the auditory cortex is in the transverse temporal gyrus.

8 **The answer is B (About the middle one-third of the precentral gyrus).** In this view of the lateral aspect of the left hemisphere, the trajectory of the central sulcus is clearly seen passing in a slightly oblique direction to, and over, the superior and medial edge of the hemisphere. The precentral gyrus (motor cortex for the face, upper extremity (UE), and trunk to about the hip) is rostral to this sulcus, and the postcentral gyrus (sensory cortex for the face, UE, and trunk to about the hip) is caudal to this important landmark. Lesion B is in the UE region of the motor cortex, lesion C is in the UE area of the sensory cortex, and lesion E is in the face area of the motor cortex. Lesion D is in part of the inferior parietal lobule, caudal to the postcentral sulcus. Lesion A is in caudal parts of the middle frontal gyrus, but rostral to the precentral sulcus; this generally correlates with the location of the frontal eye field. Recall that the term "somatomotor cortex" is the precentral gyrus + the anterior paracentral gyrus; the term precentral gyrus is not synonymous with somatomotor cortex; it is only part of the overall somatomotor cortex.

9 **The answer is C: Frontopontine fibers.** The arrow indicates the anterior limb of the internal capsule, which contains the frontopontine fibers, anterior thalamic radiations, and some lesser fiber populations. The anterior limb of the internal capsule is located between the head of the caudate nucleus and the lenticular nucleus, the larger portion of which is the putamen; these relationships are clearly seen in this image. Corticonuclear fibers are located in the genu of the internal capsule; corticospinal fibers are found in about the caudal half of the posterior limb of the internal capsule. Both of these fiber populations are extremely important in clinical medicine; damage to these parts of the internal capsule results in characteristic deficits. Parietopontine fibers are associated with both the posterior and retrolenticular limbs of the internal capsule; the optic radiations, also called the geniculocalcarine radiations, traverse the retrolenticular limb of the internal capsule.

10 **The answer is E: Visual deficits.** The lower parts of the cuneus and upper portion of the lingual gyrus border on the calcarine sulcus; this is the primary visual cortex, and it receives the geniculocalcarine radiations. A lesion in this area results in a homonymous hemianopia on the opposite side. For example, a lesion of the right visual cortex results in a left homonymous hemianopia. The transverse temporal gyrus is the auditory cortex, and motor (expressive) aphasia is seen in lesions of the partes opercularis and triangularis of the inferior frontal gyrus; a lesion in this area is called a Broca aphasia, and the inferior frontal gyrus is sometimes called the Broca convolution. Somatosensory deficits may be seen in a variety of forebrain and brainstem lesions, but not in lesions of the visual cortex. Sensory (receptive) aphasia is seen in lesions of the inferior partietal lobule, which includes the supramarginal and angular gyri; this is also called a Wernicke aphasia.

11 **The answer is D: Putamen + globus pallidus.** The lenticular nucleus is composed of the putamen and globus pallidus. The combination of the caudate and the putamen is the neostriatum. The combination of the caudate and the lenticular nucleus (putamen + globus pallidus) is the corpus striatum (not a choice). The other combinations in choices B and E are not collectively known by any single name or designation.

12 **The answer is E: Medial striate artery.** The medial striate artery (artery of Heubner) originates at, or immediately proximal or distal to, the junction of the anterior communicating artery with the anterior cerebral artery. Its branches serve the head of the caudate nucleus and the adjacent portion of the anterior limb of the internal capsule. The lenticulostriate arteries (also called the lateral striate arteries) serve the lenticular nucleus, the lateral edge of the anterior limb, and much of the posterior limb. Penetrating branches of the anterior communicating artery largely serve anterior parts of the hypothalamus; the callosomarginal artery serves much of the medial aspect of the frontal lobe. The optic tract, inferior portions of the internal capsule, the choroid plexus in the temporal horn, and adjacent structures are served by the anterior choroidal artery.

13 **The answer is D: Lenticulostriate artery.** The lenticulostriate arteries (also called the lateral striate arteries) may arise as one or two major trunks that immediately branch into smaller perforating vessels that penetrate the brain, or they may arise as a series of smaller branches; in either case, they are branches of the M_1 segment of the middle cerebral artery. These vessels serve much of the lenticular nucleus and adjacent portions of the anterior and posterior limbs of the internal capsule. The anterior choroidal artery serves the transition from the internal capsule into the crus cerebri, the optic tract, inferior regions of the basal nuclei, and immediately adjacent structures. The calcarine artery is part of the P_4 segment of the posterior cerebral artery and serves the primary visual cortex. The lateral posterior choroidal artery arises from P_2 and serves part of the choroid plexus of the lateral ventricle; the medial striate originates from the anterior cerebral artery, at about its intersection with the anterior communicating artery, and serves the head of the caudate nucleus and part of the anterior limb of the internal capsule.

14 **The answer is D: Pars opercularis + pars triangularis.** This woman has had a stroke (sudden onset) that has given rise to an expressive (motor or Broca) aphasia as seen by her inability to formulate a coherent response. Patients with this type of lesion may clearly understand what they hear, but have great difficulty formulating a verbal response. This lesion is located in the inferior frontal gyrus which is comprised of the partes opercularis, triangularis, and orbitalis. Sometimes, the inferior frontal gyrus is called the Broca convolution. The angular and supramarginal gyri make up the inferior parietal lobule (the Wernicke area); the anterior and posterior paracentral gyri are the somatomotor and somatosensory areas for the lower extremity, respectively. The cuneus and lingual gyrus form the upper and lower banks, respectively, of the visual cortex (they are separated by the calcarine sulcus); the uncus and parahippocampal gyrus are located on the inferiomedial aspect of the temporal lobe. Lesions in all of these areas will result in significant deficits.

15 **The answer is E (The globus pallidus).** The terms paleostriatum, or pallidum, are widely used alternatives for globus pallidus. Paleostriatum is a term that also specifies that this part of the basal nuclei is phylogenetically older than the neostriatum (caudate + putamen). The globus pallidus is part of the lenticular nucleus (along with the putamen) and is the source of paleothalamic projections. The caudate nucleus (choice A) is located in the lateral wall of the lateral ventricle, and at this particular plane (level of the column of the fornix), it is transitioning from its head to its body. Choice B is the putamen, the lateral part of the lentiular nucleus; in combination with the caudate, it forms the neostriatum. The area comprising the ventral pallidum and the ventral striatum is indicated at C, and lesions that involve this area may correlate with cognitive decline. The hypothalamus is indicated at D.

16 **The answer is E: Visual deficits in the opposite hemifield.** The key to the correct answer is the fact that this lesion is located in the retrolenticular limb of the internal capsule. This large fiber bundle contains the optic (geniculocalcarine) radiations traveling from the lateral geniculate body to the visual cortex. All lesions of optic structures caudal to the optic chiasm (optic tract, lateral geniculate nucleus, optic radiations, visual cortex) give rise to deficits in the opposite hemifield or opposite quadrant of the visual field. For example, a lesion of the optic radiations on the right will result in a left homonymous hemianopia.

Auditory radiations travel through the sublenticular limb of the internal capsule; corticonuclear fibers that influence cranial nerve function traverse the genu of the internal capsule. Paraesthesias may arise from central or peripheral lesions at several different locations; thalamocortical radiations convey general sensory information that travels through the posterior limb of the internal capsule.

17 **The answer is B: Corticonuclear fibers.** The genu contains corticonuclear fibers that influence the activity of the nuclei of cranial nerves. While these fibers influence both motor and sensory nuclei, clinical deficits are only observable through the corticonucler-to-motor-nuclei pathway. The most noticeable deficits are weakness of the facial muscles on the lower face, deviation of the tongue on protrusion (both on the side opposite the lesion), and deviation of the uvula (toward the side of the lesion) on phonation. The patient may also have dysarthria and dysphagia. Corticopontine fibers are found in the anterior, posterior, and retrolenticular limbs; corticonigral and corticospinal fibers are located in the posterior limb. Some corticostriatal fibers traverse the anterior and posterior limbs, while others enter the neostriatum directly from the subcortical white matter.

18 **The answer is E (Medial segment of the globus pallidus).** The medial segment of the globus pallidus is the location of the cells of origin of a GABAergic pallidothalamic projection which passes through the ansa lenticularis and the lenticular fasciculus, then enters the thalamic fasciculus to end primarily in the ventral lateral nucleus of the thalamus. This projection represents a major, and important, efferent path from the basal nuclei. The neostriatum (caudate, A; putamen, B) gives rise to a GABAergic striatopallidal connection that is essential to the functional integrity of the basal nuclei, but it does not project directly to the dorsal thalamus. The lateral segment of the globus pallidus has connections with the subthalamc nucleus which, in turn, projects to the medial segment; the lateral segment does not project directly to the thalamus. The claustrum (C) has significant connections with various regions of the cerebral cortex, but no important connections with the basal nuclei or the dorsal thalamus.

19 **The answer is B: Caudate nucleus + putamen.** The neostriatum consists of the caudate nucleus and the putamen as a unit. In fact, these portions of the basal nuclei appear of the texture and gray color in MRI, are of the same color on myelin-stained preparations (see images in this book), and arise embryologically from the same anlage. The corpus striatum is the term used to describe the caudate, putamen, and globus pallidus as a unit. The putamen and the globus pallidus together form the lenticular nucleus. The caudate and putamen by themselves are just that, the caudate and putamen.

20 **The answer is B: Insular lobe.** The short and long gyri (gyri breves et longi) collectively constitute the surface of the insular lobe. While one could argue that the insula is not a lobe, it does, in fact, meet the criteria: it is made up of named structures and is separated from the surrounding named structures (the frontal, parietal, and temporal opercula) by a named sulcus or fissure (the circular sulcus of the insula). While all of the other choices/lobes are made up of named gyri and sulci, none of these is named the long and short gyri.

21 **The answer is E (Retrolenticular limb of the internal capsule).** Optic radiations (also called the geniculocalcarine radiations) arise in the lateral geniculate nucleus, traverse the retrolenicular limb of the internal capsule, and end in the visual cortex which is located on the upper and lower banks of the calcarine sulcus. The name of this part of the internal capsule indicates its location relative to the lenticular nucleus (retro-, behind). The sublenticular limb, its position indicated at D, originates in the medial geniculate nucleus, passes obliquely forward and laterally below the caudal portion of the lenticular nucleus (dotted line at D, and thus its name sub-) to end in the auditory cortex in the transverse temporal gryus. Among the most important fibers in the posterior limb of the internal (C) capsule are thalamocortical (from the ventral lateral and ventral posteromedial and ventral posterolateral nuclei) and corticospinal from the motor cortex. Corticonuclear fibers are located in the genu of the internal capsule. A range of important and characteristic deficits result following lesions of the genu and posterior limb. The anterior limb contains anterior thalamic radiations and frontopontine fibers.

22 **The answer is C: Gyrus longi (long gyrus).** In this sagittal MRI, the plane is deep enough in the hemisphere to expose the surface of the insular lobe: this is frequently referred to as the insular cortex. The caudal area of the insula is comprised of usually two obliquely oriented long gyri and the rostral portion of two, or three, vertically oriented short gyri (also seen here). These two areas are separated, one from the other, by the central sulcus of the insula. The angular gyrus is the cortex located around the upper and caudal end of the superior temporal gyrus and, along with the marginal gyrus, comprises the inferior parietal lobule. The parietal operculum lies over the insular cortex, much like the lips lie over the teeth. The transverse temporal gyrus is located on the upper aspect of the superior temporal gyrus and within the lateral sulcus; the primary auditory cortex is located in this gyrus. The insular cortex has been implicated in a variety of possible functions including addictive behavior.

23 **The answer is B: Anterior paracentral gyrus on the right.** The body is somatotopically represented in the motor cortex in the following pattern: lower extremity in the anterior paracentral gyrus, hip and trunk in the medial portion of the precentral gyrus, upper extremity, with emphasis on the hand, in the middle area of the precentral gyrus, and face in the lateral portion of the precentral gyrus. The posterior paracentral gyrus is the somatosensory cortex for the lower extremity. Recall that the precentral gyrus + the anterior paracentral gyrus = the somatomotor cortex; the precentral gyrus, on its own, is only part of the somatomotor cortex; the precentral gyrus does not have a somatotopical representation of the entire body.

24 **The answer is B: Genu, internal capsule.** The genu of the internal capsule contains corticonuclear fibers that originate from the face area of the motor cortex, pass through the genu of the internal capsule, and end in the motor nuclei of selected brainstem nuclei; for diagnostic purposes, the most important are the facial, ambiguous, hypoglossal, and accessory nuclei. The deficits in this woman indicate a lesion in the genu on the left; the tongue deviates to the right (weak genioglossus on that side) and the uvula deviates to the left (muscles are weak on the right, but the configuration of the uvular muscles is such that the uvula deviates away from the weak side). Lesions of the anterior limb of the internal capsule (choice A) do not cause focal cranial nerve deficits; lesions of the posterior limb (E) will result in extremity weakness and sensory loss, all on the same side of the body, but without cranial nerve deficits. Lesions of the lateral medulla (Wallenberg syndrome) will result in deviation of the uvula, but the patient will also have an alternating hemianesthesia; not seen in this woman. Lesions of the medial medulla (Déjèrine syndrome) will result in deviation of the tongue on protrusion, but will also present with contralateral motor and sensory losses; this combination is also not seen in this woman.

25 **The answer is A: Amygdaloid nucleus.** The only structure completely within the zone of damage, about the rostral 4 cm of the temporal lobes bilaterally, is the amygdaloid nucleus. Bilateral damage to this structure will give rise to the Klüver-Bucy syndrome, a complex of behavioral features that includes hyperactivity, hypersexuality, and unpredictable emotional responses. Taken collectively, the features of this syndrome make it extremely difficult to function in most social situations. The rostral end of the hippocampus may be damaged, but the deficits, if any, are not significant (memory problems) when compared to the significant consequences of bilateral amygdala damage. The anterior nucleus is far from the temporal horn; the caudate tail and stria terminalis are in the temporal horn, but damage to these structures causes no significant clinical deficits.

26 **The answer is C: Lenticular fasciculus.** The lenticular fasciculus contains GABAergic pallidothalamic fibers that originate from the medial segment of the globus pallidus, pass across the internal capsule, are positioned between the subthalamic nucleus and the zona incerta, pass through the prerubral field, then arch laterally to enter the thalamic fasciculus for distribution to the thalamus where they end primarily in the ventral lateral nucleus. These fibers are inhibitory to the thalamus. The amygdalofugal pathway contains fibers from the amygdaloid nucleus to the hypothalamus and brainstem; the column of the fornix contains fibers traveling from the hippocampus to the preoptic area, but most notably to the mammillary nuclei. The subthalamic fasciculus contains fibers passing between the subthalamic nucleus and the globus pallidus; the tapetum is made up of fibers that cross in the splenium of the corpus callosum, and arch downward in the lateral wall of the posterior horn of the lateral ventricle.

27 **The answer is D: Supramarginal and angular gyri.** These two gyri collectively form the inferior parietal lobule which is commonly called, in clinical parlance, the Wernicke area. Damage to these gyri results in a receptive or fluent aphasia (he can speak but it is nonsensical: this is paraphasic speech) that is usually accompanied by an inability to write (agraphia) and read (alexia). This collection of deficits is called Wernicke aphasia. The anterior and posterior paracentral gyri are the primary somatomotor and somatosensory cortices for the lower extremity; the cuneus and lingual gyri border on the calcarine sulcus and represent the primary visual cortex. The pars opercularis and the pars triangularis are portions of the inferior frontal gyrus, which is sometimes called the Broca convolution. Damage to these structures results in a nonfluent, or expressive, aphasia. The transverse temporal gyrus is the auditory cortex.

Chapter 13

General Sensory Pathways: Proprioception, Discriminative Touch, Pain, and Thermal Sensations

Recall the Cautionary Tale: The images in this chapter are presented in a Clinical Orientation: this approach emphasizes how structure and function correlate with deficits in a clinical setting. MRI and CT have a universally recognized orientation and laterality. Line drawings, stained sections, and gross brain images in the Clinical Orientation are viewed here in the identical manner that one views an MRI: your right, as the physician-observer, is the patient's left, and the observer's left is the patient's right. The laterality of deficits and reference to connections follow accordingly.

QUESTIONS

Select the single best answer.

1 During a routine neurological examination on a 27-year-old woman, the physician applies a tuning fork (128 Hz) to the right medial and lateral malleoli. The woman does not perceive vibration, but when the tuning fork is touched to her calf, she perceives that it is cold. Damage to which of the structures labeled in the image below explains this woman's deficit?

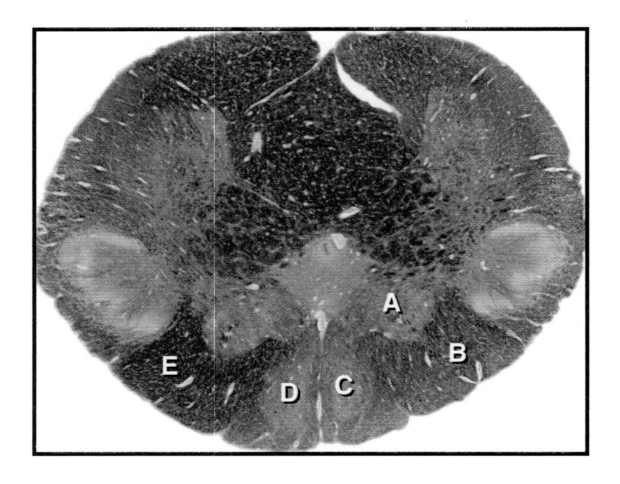

(A) A
(B) B
(C) C
(D) D
(E) E

2 Which of the following receptors is especially important to the visually impaired patient when reading Braille?

(A) Merkel cell complex
(B) Meissner corpuscle
(C) Naked nerve ending
(D) Pacinian corpuscle
(E) Ruffini ending

3 A 17-year-old boy is eating his favorite raspberry preserves, made by his grandmother, when he has a sudden sense of pressure in two of his lower teeth. Looking in a mirror, he sees a seed jammed in between two teeth. He extracts the seed, and the sensation goes away. Which of the following brainstem nuclei most likely received the sensory input in this situation?

(A) Gustatory nucleus
(B) Inferior salivatory nucleus
(C) Principal sensory nucleus
(D) Spinal trigeminal nucleus pars caudalis
(E) Spinal trigeminal nucleus pars oralis

4 A 77-year-old woman presents with the complaint of difficulty ambulating. The history indicates that this has become slowly more noticeable over about a 3-month time frame. After observing the woman's difficulty walking, the physician orders an MRI, and the results are shown in the given image. Based on the location and circumscribed extent of this lesion, which of the following deficits would you most likely expect to see?

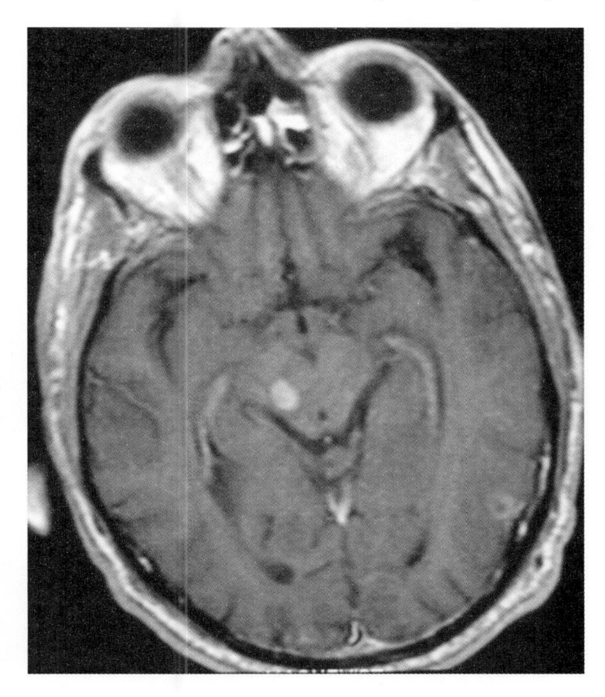

(A) Cerebellar ataxia of UE and LE on the right
(B) Hemiplegia of UE and LE on the left
(C) Hemiplegia of UE and LE on the right
(D) Loss of proprioceptive sense on UE and LE on the left
(E) Loss of proprioceptive sense on UE and LE on the right

5 One of the essential tests performed during a neurological examination is the application of a sharp object, such as a pin, to the skin to ascertain if pain sensation is intact. Which of the following represents the conduction velocity of the fibers conveying this information?
(A) About 80 to 120 m/s
(B) About 60 to 80 m/s
(C) About 35 to 60 m/s
(D) About 0.5 to 2.0 m/s

6 A 23-year-old woman is brought to the Emergency Department from the site of a motor vehicle collision. The patient is conscious, is in pain, and has a compound fracture of her femur. The neurological examination reveals a profound weakness of the right lower extremity and a loss of vibratory sense on the lower extremity and up to and including the iliac crest also on the right. She does not perceive pinprick on the left side beginning at the level of the umbilicus and down the left extremity. Which of the following represents the most likely level of the lesion in this patient?
(A) About T6 on the right
(B) About T8 on the right
(C) About T10 on the right
(D) About T8 on the left
(E) About T10 on the left

7 A 41-year-old man is transported to the Emergency Department from a home for challenged adults. The history of the current event, provided by one of his caretakers, indicated that the man suddenly began moaning and rubbing one side of his body. The man was unable to communicate exactly what was bothering him. Diagnostic imaging revealed the lesion at the arrow in the image below. Based on the location, size, and laterality of this lesion, this man is most likely suffering which of the following?

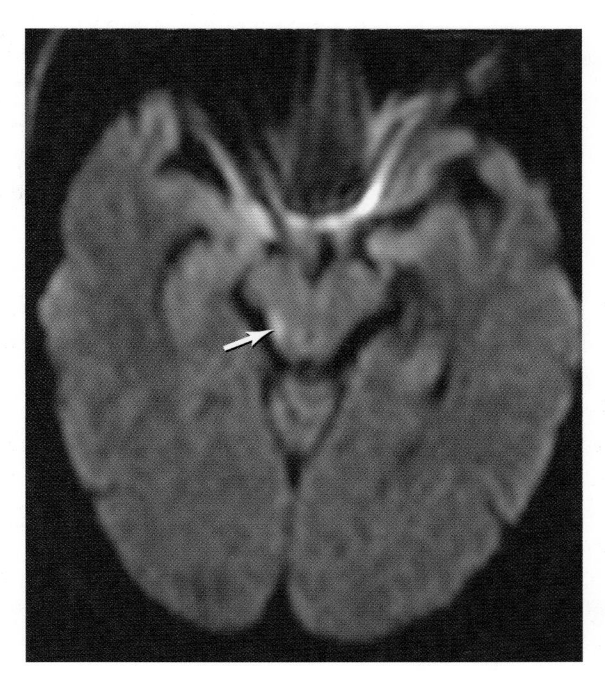

(A) Deafness in his right ear
(B) Loss of pain and thermal sense on his right side
(C) Loss of pain and thermal sense on his left side
(D) Loss of proprioception on his right upper extremity
(E) Loss of proprioception and pain on his left side

8 A 54-year-old woman presents with general sensory complaints that have developed over several weeks. The history reveals that her symptoms first appeared on her buttock and genitalia, followed by her foot, then on her lower extremity, and then on her hip, all on the left side. The examination reveals that the woman has a loss of pain and thermal sensations on her lower extremity but no cranial nerve or somatomotor deficits. MRI reveals a tumor within the vertebral canal. Which of the following would explain this pattern of deficits in this woman?
(A) Extramedullary tumor on the left above (posterior to) the denticulate ligament expanding medially
(B) Extramedullary tumor on the right above (posterior to) the denticulate ligament expanding medially
(C) Extramedullary tumor on the left below (anterior to) the denticulate ligament expanding medially
(D) Extramedullary tumor on the right below (anterior to) the denticulate ligament expanding medially
(E) Extramedullary tumor on the anterior midline expanding posteriorly

Questions 9 and 10 are based on this patient:

A 71-year-old man is brought to the Emergency Department by his son. The history of the current event, as provided by the son, is that the man complained of a sudden loss of sensation on his hand and face. The examination reveals normal motor functions of cranial nerves and of the extremities. The man does have a loss of pain and thermal sense on the left side of his body and, of the same modalities, on the right side of his face. MRI shows a lesion in the brainstem in one of the areas indicated in the image below.

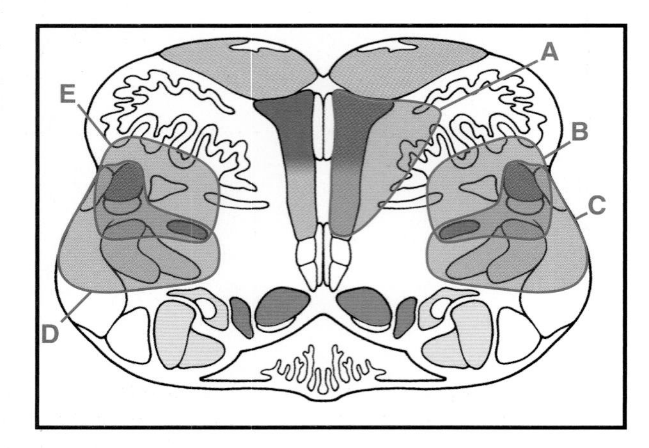

9 A lesion in which of the areas in this image would explain this man's deficits?
(A) A
(B) B
(C) C
(D) D
(E) E

10 Which of the following deficits would the physician also easily discover, assuming a thorough neurological examination?
(A) Apnea
(B) Aphasia
(C) Dysmetria
(D) Dysphagia
(E) Mutism

11 A 45-year-old man presents with a loss of only pain and thermal sensations beginning at T6 and extending caudally on his left side. All other sensations are normal. Which of the following would specify this deficit?
(A) Alternating hemianesthesia
(B) Anesthesia
(C) Conduction anesthesia
(D) Dissociated sensory loss
(E) Paraesthesia

12 A 39-year-old man is transported to the Emergency Department. The history reveals that this man is a heavy user of several illegal drugs, particularly cocaine. The examination and diagnostic studies, which include CT, reveal that this man has a hemorrhagic lesion in his cerebral cortex in the shaded area outlined in the image below. Which of the following would be the most obvious deficit in this man?

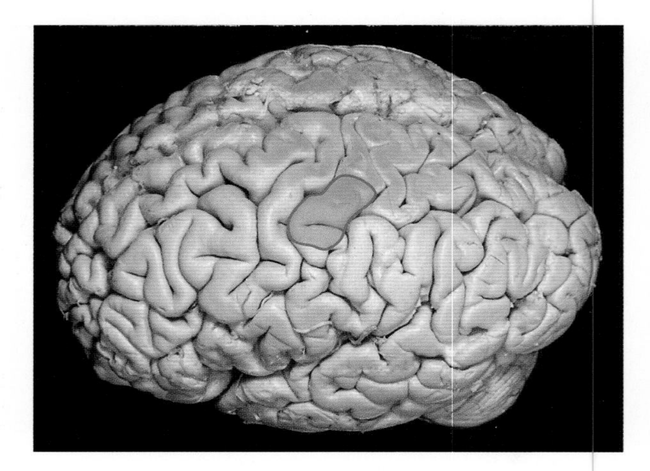

(A) Loss of sensation on the right side of the face
(B) Loss of sensation on the right trunk
(C) Loss of sensation on the right trunk and upper extremity
(D) Loss of sensation on the right upper extremity
(E) Loss of sensation on the right lower extremity

13 The receptive fields on the fingertips and around the oral cavity, especially on the lips, are small. Which of the following is characteristic of these receptive fields?
(A) A high receptor density and corresponding small area of representation in the somatosensory cortex
(B) A high receptor density and corresponding small area of representation in the posterior paracentral cortex
(C) A high receptor density and corresponding large area of representation in the somatosensory cortex
(D) A low receptor density and corresponding small area of representation in the somatosensory cortex
(E) A low receptor density and corresponding large area of representation in the somatosensory cortex

14 A 78-year-old man presents with hoarseness, ataxia, and an alternating sensory loss (left side of his body, right side of his face). The history revealed that these symptoms appeared suddenly. This combination of deficits is characteristic of which of the following syndromes?
(A) Claude
(B) Collet-Sicard
(C) Dejerine
(D) Vernet
(E) Wallenberg

15 A 65-year-old woman presents with the complaint of numbness on her face. The examination reveals that motor function of the cranial nerves and the body is intact, but the woman has a loss of pinprick sensation on the right side of her face. MRI shows a small lesion in the area indicated by the outline in the image below. Which of the following nuclei correlates with the deficit and location of the lesion in this woman?

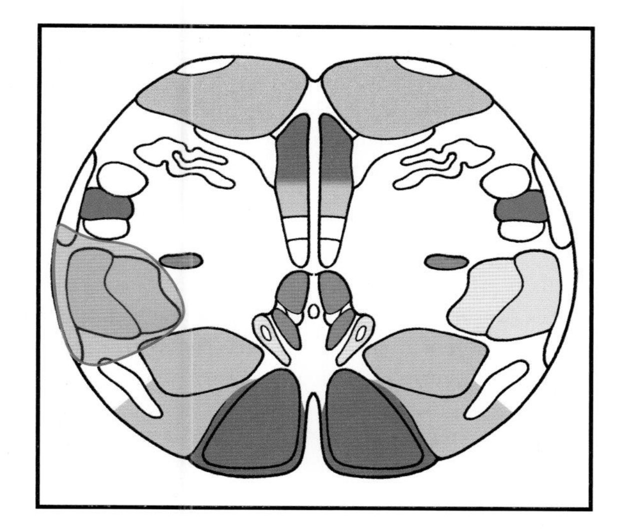

(A) Nucleus ambiguus
(B) Principal sensory nucleus
(C) Spinal trigeminal nucleus, pars caudalis
(D) Spinal trigeminal nucleus, pars interpolaris
(E) Spinal trigeminal nucleus, pars oralis

16 A 4-year-old girl is brought to the Emergency Department (ED) by her mother. The history is well known to the Chief of the ED. The current examination reveals that the girl has burns on both hands from touching a hot burner. She has an unusual neurological disorder (probably inherited) that greatly reduces her sensitivity to pain and temperature. Which of the following fiber types is not functioning properly in this girl?
(A) $A\alpha$
(B) $A\beta$
(C) $A\delta$ and C
(D) $A\delta$ only
(E) C only

17 A 70-year-old man is brought to the Emergency Department by his wife. The history of the current event revealed that the man experienced sudden numbness on his hand and face. The examination revealed a loss of pain and thermal sense on the right side of his face and on the left side of his body and weakness of facial muscles on the right side of his face. CT shows a defect in the brainstem that clearly suggests a small hemorrhagic lesion in one of the areas indicated in the given image. A lesion in which of these areas would explain this man's deficits?

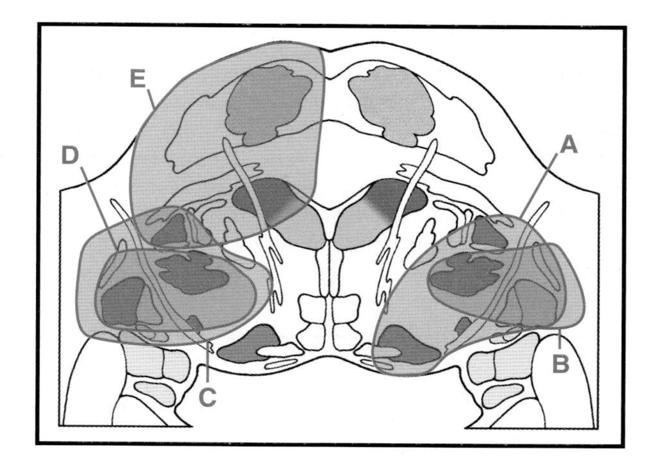

(A) A
(B) B
(C) C
(D) D
(E) E

18 During a surgical procedure on a 43-year-old woman to remove a schwannoma from the posterior root of the sixth cervical nerve, the fibers forming the medial division of the posterior root were damaged. Which of the following characterize these specific fibers?
(A) Finely myelinated, rapidly conducting, subserve pain and thermal sense
(B) Finely myelinated, slowly conducting, subserve vibratory sense and discriminative touch
(C) Heavily myelinated, slowly conducting, subserve vibratory sense and discriminative touch
(D) Heavily myelinated, rapidly conducting, subserve pain and thermal sense
(E) Heavily myelinated, rapidly conducting, subserve vibratory sense and discriminative touch

19 A 71-year-old man complains to his family physician of difficulty walking. The examination reveals no cranial nerve deficits, a distinct sensory loss, but no muscle weakness. The man tends to put his right foot down awkwardly, but forcibly, to the floor when he walks; the left foot and both upper extremities appear to be not affected. MRI reveals a small infarcted area in the man's brainstem interrupting information being relayed from the spinal cord to the cortex. Damage to which of the following would correlate with this man's particular deficit?
(A) Left accessory cuneate nucleus
(B) Left spinal trigeminal nucleus
(C) Right cuneate nucleus
(D) Right gracile nucleus
(E) Right spinal laminae I to IV

20 A 23-year-old man is brought to the Emergency Department from the site of a motorcycle collision. The examination reveals that the man is intoxicated, has intact cranial nerves (motor and sensory), and can move his lower extremities. Beginning about 8 to 10 weeks after this incident, the man is unable to detect pinprick beginning just below the umbilicus and extending down bilaterally over both lower extremities but sparing much of the buttocks and the anal region. MRI reveals a traumatic spinal cord lesion represented by one of the shaded regions in the image below. A lesion in which of these areas would most likely explain this man's deficits?

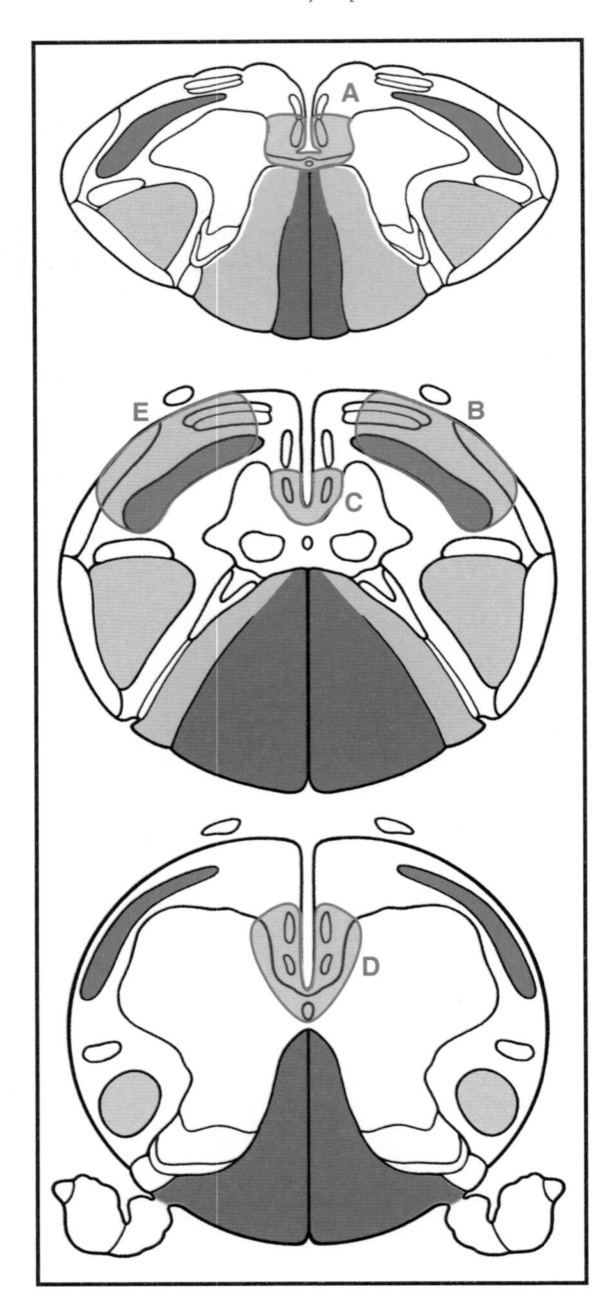

(A) A
(B) B
(C) C
(D) D
(E) E

21 Which of the following receptors is both rapidly adapting and slowly adapting?

(A) Free (naked) nerve ending
(B) Hair follicle receptor
(C) Meissner corpuscle
(D) Merkel cell receptor
(E) Pacinian corpuscle

22 The MRI of a 75-year-old woman reveals a lesion in the medulla, on the left side, that resulted from an occlusion of penetrating branches of the anterior spinal artery. The results of the entire evaluation suggest that embolus shed from a site of probable bacterial endocarditis was the likely precipitating event in this woman's case. This woman would most likely experience which of the following?

(A) Deviation of the tongue to the right
(B) Loss of discriminative touch of the left extremities
(C) Loss of discriminative touch of the right extremities
(D) Loss of pain and thermal sense on the left extremities
(E) Loss of pain and thermal sense on the right extremities

23 Which of the areas indicated in the image below contains neuronal cell bodies whose axons are involved in the descending inhibition of pain in the brainstem and in the spinal cord?

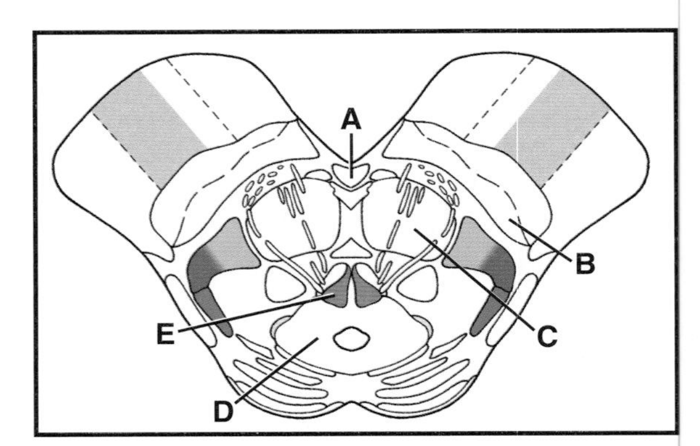

(A) A
(B) B
(C) C
(D) D
(E) E

24 A 21-year-old man is brought to the Emergency Department (ED) from the site of a motor vehicle collision. Upon arrival at the ED, he is conscious, has facial lacerations, is bleeding from his nose, and has a compound fracture of his humerus. The examination reveals that he is profoundly weak on his left side (lower extremity), somewhat weak on his left side (upper extremity), perceives vibration when a tuning fork is applied to his left clavicle, but not to his thumb or below that level, and perceives pinprick on his right thumb, index finger, and middle finger, but not on the little finger or below that level. This lesion is located at which of the following levels?

(A) About C4 on the right
(B) About C4 on the left
(C) About C6 on the right
(D) About C6 on the left
(E) About C8 on the left

25 A 12-year-old boy was a passenger in a vehicle involved in a collision. He was restrained in a seatbelt but came out of this restraint when the vehicle rolled over. He is conscious at the Emergency Department. The examination reveals a fractured ankle, broken nose, and numerous cuts and bruises, and the boy is weak in all extremities. CT and MRI reveal no spinal fractures or acute blood in the CNS. Over the next several weeks, the boy's motor function gradually improves, but he begins to experience pain and numbness over the C8–T3 dermatomes bilaterally. Which of the following would explain this clinical picture?

(A) Avulsion of posterior roots
(B) Cauda equina syndrome
(C) Congenital syringomyelia
(D) Syringobulbia
(E) Traumatic syringomyelia

Questions 26 and 27 are based on the following patient:

A 49-year-old man visits a neurologist at the suggestion of his family physician. The history reveals that the man has long-standing sensory deficits that have gotten progressively worse over the last 8 months. The examination reveals (1) normal cranial nerve function, (2) mild bilateral motor deficits, and (3) bilateral loss of all sensory modalities on the lower extremities. The man notes that these "funny sensory feelings" have progressed from nonexistent to severe over time, but he is not sure how long.

26 MRI reveals a single lesion, elliptical in shape, about 2 × 3 cm in size. This tumor is most likely which of the following?

(A) Anaplastic astrocytoma
(B) Glioblastoma multiforme
(C) Oligodendroglioma
(D) Meningioma
(E) Schwannoma

27 Impingement on which of the following structures would most likely explain the sensory deficits in this man?

(A) Anterior paracentral gyrus
(B) Internal capsule, posterior limb
(C) Medial medulla, bilaterally
(D) Postcentral gyrus
(E) Posterior paracentral gyrus

28 A 54-year-old woman presents at the Emergency Department with the complaint of a sudden onset of neurological symptoms. The examination confirms the woman's complaint, and MRI reveals the lesion shown in the given image. Based on the location of this lesion, which of the following is most likely this woman's primary deficit?

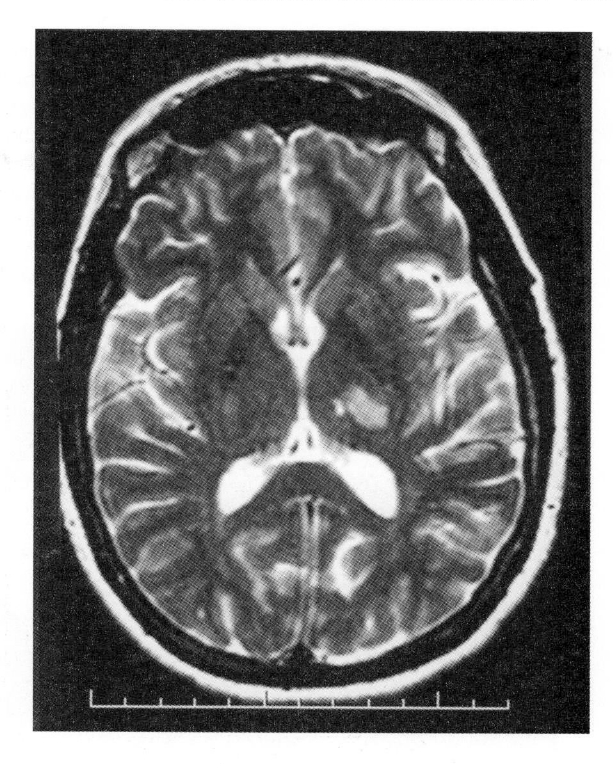

(A) Loss of all sensations on the left side of the body
(B) Loss of all sensations on the right side of the body
(C) Loss of discriminative touch and vibratory sense on the left side of the body
(D) Loss of discriminative touch and vibratory sense on the right side of the body
(E) Loss of pain and thermal sense on the left side of the body

29 Which of the following describes the characteristics of action potentials generated by rapidly adapting receptors such as Pacinian corpuscles?

(A) At the initiation of the stimulus, but not at its removal
(B) At the initiation of the stimulus, during the time it is present (but at a lower frequency), and at the removal of the stimulus
(C) At the initiation of the stimulus, during the time it is present (but at a higher frequency), and at the removal of the stimulus
(D) At the initiation of the stimulus and at its removal, but no action potentials are generated during sustained stimulation
(E) At the removal of the stimulus, but not at its initiation

30 A 49-year-old man presents at the Emergency Department (ED) with a sudden onset of difficulty speaking. The history of the current event revealed that the man was engaged in illegal drug use at the time and had been seen at this ED on numerous occasions for problems related to this habit. The examination substantiated that the man had clear-cut deficits, and CT revealed a hemorrhagic lesion in the area outlined in the image below. Based on the location of this lesion, which of the following deficits would this patient most likely have?

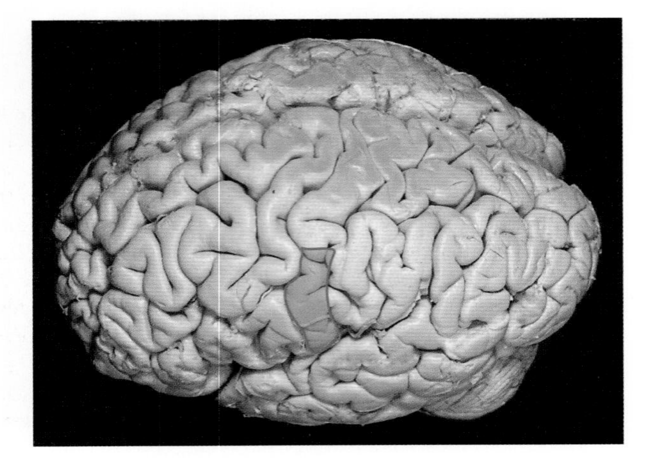

(A) Loss of sensation bilaterally on lower face and tongue
(B) Loss of sensation on left side of face and oral cavity
(C) Loss of sensation on right side of face and oral cavity
(D) Weakness of left side of tongue and lower face
(E) Weakness of right side of tongue and lower face

ANSWERS

1 **The answer is D (The right nucleus gracilis).** The right nucleus gracilis (D) receives input from the fasciculus gracilis; both are conveying vibratory sense, discriminative touch, and proprioception from the right lower extremity. The fasciculus (B) and nucleus (A) cuneatus convey vibratory sense and discriminative touch from the upper extremity, on the left side in this image. The left nucleus gracilis (C) receives information from the left lower extremity (the wrong side for this patient); the right cuneate fasciculus (E) conveys information from the right upper extremity. Recall that the fasciculi cuneatus and gracilis are made up of the central processes of primary sensory fibers (these are the 1st order neurons in this pathway) and that this sensory information does not cross the midline in the spinal cord. The nuclei cuneatus and gracilis contain the cell bodies of the 2nd order neurons in this pathway; these fibers cross in the sensory decussation to enter the medial lemniscus.

2 **The answer is A: Merkel cell complex.** These sensory complexes are very sensitive to small spatial/temporal skin indentations, are numerous in the fingertips and lips, have small receptive fields, and are slowly adapting receptors. The Merkel cell complex consists of the Merkel cell, a modified epithelial cell attached to adjacent epithelial cells by desmosomes, and an afferent nerve ending located at the inferior aspect of the Merkel cell. The Meissner and Pacinian corpuscles are endings that encode tap/flutter and vibration/pressure, respectively. While very important to touch, they do not provide the exquisitely detailed input essential to reading Braille. The naked nerve ending is sensitive to thermal and nociceptive input; the function of the Ruffini ending is not well known. While these other receptors do not encode the particularly sensitive information that is a feature of the Merkel cell complex, they obviously contribute to the overall ability of the blind patient to read Braille.

3 **The answer is C: Principal sensory nucleus of the trigeminal nerve.** The principal sensory nucleus of the trigeminal nerve receives discriminative touch from the face and oral cavity and input on position from the periodontal ligament. This input is basically the head equivalent of the posterior column–medial lemniscus system of the body. In this situation, the slight displacement of the teeth is perceived as a sensation of pressure or a change in position (position sense), but not as pain. This sensation is relayed to the thalamus and then onto the cortex, where the displacement is recognized and localized. The gustatory nucleus is the rostral portion of the solitary nucleus that is concerned with taste; the inferior salivatory nucleus supplies preganglionic parasympathetic fibers that travel with the glossopharyngeal nerve. The spinal trigeminal nucleus partes oralis et caudalis are parts of the trigeminal complex in the pons and medulla, respectively. The caudalis is specifically concerned with the transmission of pain and thermal sensations from the face and oral cavity, including the teeth.

4 **The answer is D: Loss of proprioceptive sense on upper and lower extremities (UE and LE) on the left.** This lesion is in the lateral midbrain on the right and in the exact territory of the medial lemniscus, which conveys vibratory, position, and proprioceptive senses from the left side of the body (UE and LE). This is the midbrain part of the overall posterior column–medial lemniscus system. The apparent motor difficulties are actually the result of an interruption of this important sensory input; this is frequently called a sensory ataxia. The woman's LE is not weak, but she lacks the sensory input that indicates where her LE is in space and on a substrate. This woman not only is ataxic but also may put her foot forcibly to the floor in an effort to create the missing input. Hemiplegia is not a viable choice because the lesion is clearly not in the crus cerebri on either side; ataxia of the cerebellar type results from lesions located more medially in the midbrain. Furthermore, the deficit in this choice (A) is also on the wrong side.

5 **The answer is D: About 0.5 to 2.0 m/s.** Pain and thermal sensations are conveyed by lightly myelinated (about 0.2- to 1.5-μm fiber diameter), slowly conducting fibers with comparatively short internodal distances; these are C fibers. While C fibers are called unmyelinated, they do lie in troughs on Schwann cells and are separated from the interstitial fluid. For C fibers, these Schwann cells serve the same function as a myelin sheath. These modalities are also carried by Aδ fibers (about 1 to 5 μm in diameter, about 5 to 30 m/s conduction velocity), but this choice is not available. The other choices are larger diameter, rapidly conducting fibers with greater internodal distances that convey proprioceptive input as broadly defined.

6 **The answer is B: About T8 on the right.** The loss of pinprick sensation is key in the localization of this patient's lesion for two reasons. First, fibers conveying this information ascend about two spinal segments as they cross the midline and then join the anterolateral system (ALS). Second, the dermatome at the umbilicus is T10. Consequently, a loss of pain sensation on the left side beginning at about T10 would indicate damage to the ALS on the opposite side (the right side) about two spinal segments higher (lesion at T8 on the right). The lesions at T10 are too low and/or on the wrong side and at T6 are too high. The former would result in a pain and thermal loss beginning at about T12, and the latter at about T8, both on the opposite side.

7 **The answer is C: Loss of pain and thermal sense on his left side.** In this interesting case, the patient has a sudden loss of his ability to perceive pain and thermal sensations of the left side of his body, indicating involvement of the anterolateral system (ALS). Recall that fibers in this tract cross the midline (in the anterior white commissure) within about two levels of their synapse in the posterior horn. Throughout its course in the brainstem, the ALS conveys information originating from the opposite side of the body. At this location within the midbrain (about the level of the inferior colliculus), the ALS is very close to the surface of the brain, and the lateral lemniscus is also close to the surface, as it enters the inferior colliculus. This lesion undoubtedly involved parts of the lateral lemniscus and perhaps portions of the brachium of the inferior colliculus; however, the profound somatosensory loss is overwhelmingly more noticeable to the patient. In fact, any auditory deficit would not be deafness but would most likely be in the ability to localize sound in space or as a change in the quality of the sound. Choice B is on the wrong side; there is no involvement of medial lemniscus (ML) fibers (D and E). The

ML is located medial to this relatively superficial lesion; at this level of the midbrain, the ML is immediately internal to, and parallel with, the orientation of the substantia nigra. Fibers conveying information from the opposite lower extremity are located more laterally in the ML and those conveying input from the opposite upper extremity are medially located.

8 **The answer is D: Extramedullary tumor on the right below (anterior to) the denticulate ligament expanding medially.** This extramedullary tumor (extramedullary because it is external to the spinal medulla) is located on the right, anterior to (below) the denticulate ligament, and, as it expands medially, it impinges on the spinal cord and on the anterolateral system within the cord. The anterolateral system conveys pain and thermal sensations from the contralateral side of the body; deficits on the left side of the body specify a lesion on the right side of the spinal cord. Within the anterolateral system, the body is somatotopically represented: the buttock and foot most laterally and the other portions of the lower extremity and hip at successively more medial locations. As the tumor enlarges, it results in a contralateral ascending sensory loss. A tumor that impinges on the spinal cord above (posterior to) the denticulate ligament impinges on the corticospinal tract and results in motor deficits. A ventrally located tumor expanding posteriorly would result in bilateral damage to the anterior horn with the corresponding motor deficits.

9 **The answer is D (The lesion involving the spinal trigeminal tract and the anterolateral system, both on the right).** This man has what is commonly called an alternating (alternate or crossed) hemianesthesia; this is a loss of sensation on one side of the body and on the opposite side of the face. In this case, it is a medullary lesion on the right that damaged fibers of the anterolateral system (ALS; pain/thermal sense from the left side of the body) and the spinal trigeminal tract (pain/thermal sense from the right side of the face). The lesion at A involves fibers conveying discriminative touch and vibratory sense from the right side of the body; trigeminothalamic fibers convey pain and thermal sense from the right side of the face; this combination is not present in this patient. The lesion at B results in pain/thermal loss of the right side of the body and the left side of the face, but only in the distribution of the ophthalmic nerve. The lesion at C is the mirror of that at D: ALS deficits on the right side of the body, trigeminal deficits on the left side of the face. The lesion at E is the mirror of that at B: pain/thermal losses on the left side of the body and on the right side of the face in the distribution of the ophthalmic nerve.

10 **The answer is D: Dysphagia.** The lesion in this man also encompasses the nucleus ambiguus that innervates pharyngeal and laryngeal musculature via the glossopharyngeal and vagus nerves; this also affects the vocalis muscle of the vocal folds. This patient is in danger of food or fluids entering the trachea when swallowing; the trachea is held open by cartilaginous rings, the vocal folds do not close properly, and the swallowed food or fluids seek the path of least resistance. Rather than entering the constricted esophagus, the food or fluids enter the trachea. The patient would be in danger of aspiration pneumonia. Apnea is the absence of breathing, and aphasia is an inability to comprehend or to produce speech; this is usually seen in patients with lesions of the cerebral hemisphere. Dysmetria is an inability to judge the speed or distance of a movement and may be divided into hypometria and hypermetria; this is usually a sign of cerebellar lesions. Mutism is the lack of speech that may result from organic disease or may be psychological in nature.

11 **The answer is D: Dissociated sensory loss.** This is a spinal cord lesion on the patient's right side at about T4; a right-sided T4 lesion results in a left-sided deficit beginning at about T6. Dissociated sensory losses refer to a loss of one sensation without the loss of another sensation: in this case, pain and thermal sensations, but no loss of discriminative touch and vibratory sense. Dissociated sensory losses are largely characteristic of damage to the spinal cord. Anesthesia simply refers to a loss of all sensations. One example, alternating hemianesthesia is seen in a lateral medullary syndrome; this is an alternating loss of a sensation, on one side of the face and the other side of the body. Conduction anesthesia is the administration of an anesthetic agent to block transmission (nerve block or block anesthesia); paraesthesia (also called paresthesia) is the spontaneous occurrence of an abnormal sensation that is perceived as prickling, burning, or scratching. Paraesthesia may occur spontaneously, even when there is no stimulus.

12 **The answer is D: Loss of sensation on the right upper extremity.** The central sulcus is clearly seen on this brain beginning on the lateral surface just above the lateral (Sylvian) sulcus. From this point, it passes over the surface of the brain in an oblique and slightly caudal direction and continues over the dorsomedial edge of the hemisphere. The precentral gyrus (motor cortex, area 4) is rostral to this sulcus, and the postcentral gyrus (sensory, areas 3, 1, and 2) is caudal to this consistent landmark. For convenience, these gyri can be generally divided into thirds: the face is represented in the lateral one-third, the upper extremity (particular emphasis on the hand) in the middle one-third, and the trunk in the medial one third. The lower extremity is represented in the posterior paracentral gyrus, which is on the medial aspect of the hemisphere. This man's lesion is in about the middle one-third of the postcentral gyrus of the left hemisphere; the patient perceives a loss of sensation (pain, thermal sense, discriminative touch, vibratory sense) on his right side. Recall that chronic drug abuse may compromise cerebral vessels.

13 **The answer is C: A high receptor density and corresponding large area of representation in the somatosensory cortex.** Small receptive fields, such as those on the fingertips and lips, have many receptors that are closely packed, and this part of the body has a large representation in its corresponding area of the sensory cortex. For example, the thumb has small receptive fields but has a proportionately very large cortical representation. Large receptive fields, such as those on the back, have fewer, more widely spaced receptors, and this body part has a very small representation in the somatosensory cortex. For example, the receptive fields on the back are very large, but the cortex dedicated to this body area is very small, much smaller than that for the thumb. The posterior paracentral cortex (gyrus) is the somatosensory cortex for the lower extremity.

14 **The answer is E: The Wallenberg syndrome.** This syndrome, also called the lateral medullary syndrome or the posterior inferior cerebellar artery (PICA) syndrome, is a result of damage, usually vascular, to the dorsolateral medulla. This medullary region is served by branches of PICA, and the deficits reflect damage to structures within this vascular territory as follows: hoarseness = nucleus ambiguus; ataxia = restiform body; sensory on the body = anterolateral system; sensory on the face = spinal tract of the trigeminal nerve. Dejerine is a syndrome of the medial medulla that involves the corticospinal fibers, medial lemniscus, and the hypoglossal nerve. Claude is a syndrome of the more central midbrain that affects the red nucleus, cerebellothalamic fibers, and the fibers of the oculomotor nerve. Vernet and Collet-Sicard are syndromes that are related to the jugular foramen. The former relates to the glossopharyngeal, vagus, and accessory nerves and presents with deficits reflecting damage to these three cranial nerves. The latter also involves the IXth, Xth, and XIth cranial nerves, plus the hypoglossal nerve (cranial nerve XII); this results in deficits of the IXth, Xth, and XIth with the added deficit of weakness of the tongue on the side of the lesion.

15 **The answer is C: Spinal trigeminal nucleus, pars caudalis.** The spinal trigeminal nucleus, pars caudalis, extends from the spinal cord–medulla junction rostrally to the level of the obex. In fact, this nucleus and the laterally adjacent spinal trigeminal tract interdigitate with the spinal laminae I to IV and with the posterolateral tract (of Lissauer) in the upper 1 to 2 levels of the cervical spinal cord, respectively. This nucleus is the primary station for the relay of nociceptive input entering on the trigeminal nerve; after relay in the pars caudalis, this information travels to the thalamus as anterior/ventral trigeminothalamic fibers. In addition to the trigeminal nerve, cranial nerves VII, IX, and X also make small nociceptive contributions to the spinal trigeminal tract and to the pars caudalis. The pars interpolaris of the spinal nucleus extends from the obex rostrally to the level of the rostral aspect of the hypoglossal nucleus, and the pars oralis extends from this point to the caudal aspect of the principal sensory nucleus. The principal sensory nucleus is concerned with discriminative touch and vibratory sense; it functions as the posterior column–medial lemniscus system of the brainstem. The nucleus ambiguus is motor to the muscles served by the glossopharyngeal and vagus nerves.

16 **The answer is C: Aδ and C fibers.** A δ fibers (conduction velocity of about 5 to 30 m/s) and C fibers (conduction velocity of about 0.5 to 2 m/s) respond to tissue damage, or to hot or cold, by steadily increasing their firing frequency. The peripheral endings of these fibers are called naked nerve endings or free endings. The pain resulting from tissue damage is indicated by a sustained firing rate of these fibers. C fibers are activated by tissue damage and are also activated by chemicals, such as bradykinins or insect venom, released at sites of injury. Aα fibers (conduction velocity of about 80 to 120 m/s) provide input from primary muscle spindles (spindles are the receptor for the muscle stretch reflex) and from Golgi tendon organs. Aβ fibers (conduction velocity of about 35 to 75 m/s) provide input from secondary muscle spindles and from skin mechanoreceptors.

17 **The answer is C (The lesion encompassing the anterolateral system, spinal trigeminal tract, and motor facial nucleus on the left).** The deficits experienced by this man are consistent with a lesion at C. Located within this lesion are the following structures and their functional correlates: right anterolateral system (ALS)—left-sided loss of pain and thermal sense; right spinal trigeminal tract—right-sided loss of pain and thermal sense on face and in the oral cavity; facial nucleus and root—weakness of facial muscles on the right side. Recall that a lesion of a cranial nerve motor nucleus or of the nerve root arising from that nucleus will usually result in the same deficits. The lesion at A encompasses the ALS, the spinal trigeminal tract, and the facial nucleus but also includes the abducens nucleus; this man does not complain of diplopia or of any horizontal gaze problems. The lesion at B involves the spinal tract and the facial nucleus but excludes the ALS; this would result in a pain and thermal sense loss on the left side of the face and oral cavity and facial paralysis on the left side. The lesion at D is the mirror of that at B with a corresponding reversal of deficits. The large lesion at E would, at the minimum, result in a left hemiplegia (corticospinal deficit), a left hemianesthesia (pain and thermal sense), a loss of proprioception of the left lower extremity, and paralysis of the right lateral rectus muscle.

18 **The answer is E: Heavily myelinated, rapidly conducting, subserve vibratory sense and discriminative touch.** The medial division fibers of the posterior root are the central processes of heavily myelinated (many are in the 12 to 20 μm range) fibers that have conduction velocities in the 80 to 120 m/s range; these are Aα fibers. Aβ fibers also contribute to the medial division of the posterior root (6 to 12 μm diameter, 35 to 75 m/s conduction velocity). In addition, the myelin coverings of these axons are also characterized by long intermodal distances. While these medial division fibers give off collaterals for reflexes, their main central process enters the ipsilateral posterior columns. The lateral division of the posterior root conveys lightly myelinated and unmyelinated Aδ and C fibers that are largely concerned with the transmission of pain and thermal sensations. These central processes enter the posterolateral tract (of Lissauer) and then distribute to the posterior horn. This anatomical arrangement explains the small ipsilateral sensory loss seen in some patients with spinal cord lesions. Generally speaking, heavily myelinated fibers are characterized by the following: (1) large diameter, (2) comparatively long intermodal distances, and (3) rapid conduction velocity. In similar manner, lightly myelinated fibers are characterized by the following: (1) small diameter, (2) comparatively short intermodal distances, and (3) slow conduction velocity.

19 **The answer is D: Right gracile nucleus.** This man has a sensory ataxia; the pathway conveying information on discriminative touch, vibratory sense, and proprioception from the body is the posterior column–medial lemniscus system. This man has sustained damage in this system at some point between the periphery and the cortex. In this man's case, the deficit is pronounced on the right lower extremity (LE) largely sparing the left LE and both upper extremities. The primary afferent cells conveying proprioceptive information from the right LE have their cells of origin in the posterior root ganglion at lower thoracic and lumbosacral levels; their central processes ascend as part of the gracile fasciculus to end in the

gracile nucleus, all on the same side. In this case, the right gracile nucleus is damaged, resulting in a right-sided deficit; the damage is before the sensory decussation, thus the deficits are on the same side. Damage to this system above the sensory decussation would result in deficits on the opposite side of the body. The accessory cuneate nucleus and the cuneate nucleus receive input from the upper extremity and project to the ipsilateral cerebellum and the contralateral thalamus, respectively. The left spinal trigeminal nucleus and right spinal laminae I to IV are concerned with the relay of pain and thermal information from the face and body, respectively.

20 **The answer is C (Posttraumatic syringomyelia at low thoracic levels).** The bilateral loss of pain sensation over the body with sparing of regions below the anesthetic level(s) is characteristic of lesions (syringomyelia) that involve the central area of the spinal cord with damage to the anterior (ventral) white commissure; pain and thermal sense fibers cross the midline in this small bundle. The signs and symptoms of posttraumatic syringomyelia may begin to appear weeks, or even months, after the causative event. The three cord levels are, from upper to lower, cervical (oval shape, large anterior and posterior horns), thoracic (round shape, small anterior and posterior horns), and lumbar (round shape, large anterior and posterior horns with a small amount of white matter). The sensory deficit in this man begins just below the umbilicus (which is the T10 dermatome) and extends to about the L4-5 dermatome level (the buttocks and anus are spared, these being the sacral and coccygeal dermatomes). These pain fibers ascend by about one spinal level as they cross the midline in the anterior white commissure and by about two spinal levels by the time they enter the anterolateral system. Recognizing this important feature of the trajectory of these fibers, the lesion in this man begins in the low thoracic level, at about T9, and extends to about the L3-4 level. The lesion at A is in the cervical level (lesion too high); the lesion at D is in the lumbar enlargement, would not produce a low thoracic deficit, and would probably result in a loss of pinprick sense over the buttock (lesion too low). The lesions at B and E result in sensory loss only on the side opposite the lesion, and with no caudal sparing.

21 **The answer is B: Hair follicle receptor.** The receptor associated with the hair follicle has characteristics of both rapidly and slowly adapting receptors. This receptor responds to a static stimulus (simply touching a few hairs on the arm with no movement) as well as to direction (passing the hand over the arm brushing only the hairs, one senses direction). Both Meissner corpuscles and Pacinian corpuscles are only rapidly adapting receptors. The Merkel cell receptor is a slowly adapting receptor only. Free or naked nerve endings are not classified as slowly or rapidly adapting; they simply respond to an adequate stimulus, and an action potential is produced.

22 **The answer is C: Loss of discriminative touch of the right extremities.** Bacterial endocarditis involves the heart valves; the leaflets may be damaged and may shed clusters of bacteria or small pieces of valve tissue into the circulation. Such material may enter the common carotid artery or the vertebral artery and lodge in the more distal branches of these main trunks. The territory served by the penetrating branches of the anterior spinal artery serves the pyramid (containing corticospinal fibers), exiting roots of the hypoglossal nerve, and the medial lemniscus. The fibers of the posterior column–medial lemniscus convey discriminative touch, vibratory sense, and proprioception from the contralateral side of the body. These fibers ascend as the central processes of primary sensory fibers in the posterior column, synapse in the posterior column nuclei, arch toward the midline as internal arcuate fibers, and cross the midline in the sensory decussation. The left-sided lesion results in a right-sided deficit. Pain and temperature are not conveyed via the medial lemniscus. A deviation of the tongue to the right would require a lesion on the right side.

23 **The answer is D (The periaqueductal gray (PAG)).** The periaqueductal gray (PAG), also called the central gray, is characterized by the following: (1) the anterolateral system contains ascending fibers that terminate in the PAG, (2) many cells in the PAG have opiate receptors, (3) some PAG cells give rise to descending projections to the nucleus raphe magnus (NRM), and (4) the NRM projects to the spinal trigeminal nucleus and to the posterior horn of the spinal cord for the inhibition of pain transmission at these sites. This is accomplished by NRM serotonergic fibers that activate enkephalinergic interneurons that act both presynaptically and postsynaptically to suppress the activity of Aδ and C fibers conveying pain information. The interpeduncular nucleus (choice A) receives input from the habenular nuclei and is considered one of the remote parts of the limbic system. The substantia nigra pars compacta (B) has well-known connections with the basal nuclei; the red nucleus (C) gives rise to a crossed rubrospinal projection (from its magnocellular part) and an uncrossed rubroolivary projection (from its parvocellular part). The oculomotor nucleus innervates four of the six main extraocular muscles.

24 **The answer is D: About C6 on the left.** The diagnosis in this case is based on the observation of left-sided motor deficits and left-sided vibratory sense loss beginning at C6 (vibratory sense is lost on the thumb, which is C6; vibratory sense is intact on the clavicle, which is C5) and below that level. The loss of pinprick sense begins at C8 (the little finger) and continues below that level on the right. The lesion is at C6 on the left; posterior column sense is lost at the level of the lesion on that side, and pain and thermal senses are lost about two levels below the lesion, on the opposite side. This is due to the fact that the 2nd order neurons conveying pain and thermal information ascend about two spinal levels as they cross the midline in the anterior white commissure. The posterior column deficits are on the left side because these fibers ascend uncrossed to the medulla; the modalities conveyed by these fibers cross the midline after synapsing in the medulla. This is a Brown-Séquard syndrome, commonly called a functional hemisection. Knowledge of key dermatomes is essential to diagnosis of many spinal cord lesions.

25 **The answer is E: Traumatic syringomyelia.** Traumatic syringomyelia, also called posttraumatic syringomyelia, may be seen in individuals who have experienced significant trauma that did not necessarily result in penetrating injury to the CNS. This may be seen in patients who have trauma that involves axial regions of the body. The patient has various motor deficits (paraplegia, quadriplegia) that may partially resolve, remain the same, or become worse over time. The pain and numbness first begin weeks or months after the precipitating injury. Avulsion of the posterior roots results in pain but may not have a motor component, especially a bilateral one. The cauda equina syndrome results in motor and sensory deficits of the buttocks and lower extremity; a congenital syringomyelia would appear much earlier than in this case. Syringobulbia is a brainstem cavitation and usually includes cranial nerve deficits.

26 **The answer is D: Meningioma.** This lesion is slow growing such that the patient is not even sure how long he has experienced symptoms. Also, the symptoms indicate some degree of localization in that only certain modalities are involved. The patient basically had no other deficits. Anaplastic astrocytomas and glioblastoma multiforme are aggressive tumors, especially the latter, that may present with signs of increased intracranial pressure (ICP), seizure, and headache and may result in death within time frames of 2 years or 1 year, respectively. Also, they are malignant tumors. Oligodendrogliomas frequently (50% to 80%) initially present with seizure and signs of ICP and have a predilection for white matter of the frontal lobe. Schwannoma is a lesion of peripheral nerves or of the roots of cranial nerves.

27 **The answer is E: Posterior paracentral gyrus.** The bilateral loss of all sensory modalities (in the face of no other appreciable deficits) from a single lesion greatly restricts the possible locations of the lesion. Pain, temperature, vibratory sensations, and discriminative touch all terminate in the posterior paracentral gyrus for the lower extremity. The two posterior paracentral gyri are separated from each other by a space of only a few millimeters, which contains the falx cerebri. A falcine meningioma is slow growing, impinges on the posterior paracentral gyri bilaterally, and involves only sensory input from the lower extremity. The anterior paracentral gyri are the somatomotor cortex for the lower extremity; a lesion here results in motor deficits of the lower extremity. Damage to the posterior limb of the internal capsule results in motor and sensory deficits on the opposite side of the body; bilateral damage to the medial medulla would result in quadriplegia (pyramids), loss of vibratory sense and discriminative touch on both upper and lower extremities (medial lemnisci), and paralysis of the tongue (hypoglossal nerves). The postcentral gyri receive sensory input for all parts of the body excepting the lower extremity; also, a single lesion cannot damage both postcentral gyri.

28 **The answer is B: Loss of all sensations on the right side of the body.** This lesion is on the left side and is largely coextensive with the position of the ventral posterolateral (VPL) nucleus of the thalamus. The VPL nucleus receives pain and thermal sensations via the anterolateral system (ALS) and discriminative touch, vibratory sense, and proprioception through the posterior column–medial lemniscus (PCML) system. Both of these systems convey their respective information from the opposite, in this case the right, side of the body. The VPL nucleus, when intact, relays both of these sensations to the primary somatosensory cortex on the same side. Consequently, with a left-sided thalamic lesion, the loss of all these sensations is on the right side of the body. ALS fibers cross the midline in the spinal cord, and fibers of the PCML system cross the midline in the caudal medulla at the sensory decussation. In the case of a VPL lesion, all sensations are lost, not just one or the other (choices C and E). Also, a left-sided lesion does not result in a left-sided deficit (A).

29 **The answer is D: At the initiation of the stimulus and at its removal, but no action potentials are generated during sustained stimulation.** Rapidly adapting receptors, such as Pacinian and Meissner corpuscles, respond to the initial application of a stimulus (such as pressure) with a burst of action potentials, are silent during sustained stimulation/pressure, and respond with a second burst of action potentials when the stimulus is removed. These receptors have little, or no, background activity before the stimulus is applied, during sustained stimulation, or after the stimulus is removed. Recall that hair follicle receptors may function as rapidly and slowly adapting receptors. Slowly adapting receptors, such as Merkel cell receptors and Ruffini endings, have a small amount of background activity before stimulation, a burst of activity at stimulation, continued action potentials during sustained stimulation (but at a higher level than background), and a second burst at removal of the stimulus, followed by a decrease to background levels.

30 **The answer is C: Loss of sensation on right side of face and oral cavity.** Two points are essential in the interpretation of this case. First, recognize that this lesion is in the left hemisphere; for orientation, the frontal pole, occipital pole, and cerebellum are easily seen. Second, recognize that this lesion is specifically located in the lateral third of the postcentral gyrus; for orientation, the lateral sulcus, central sulcus, and gyri bordering on anterior and posterior aspects of the central sulcus are clearly seen. The approximately lateral one-third of the postcentral gyrus is the primary somatosensory cortex for the face. In actuality, the face area represents about 40% of the postcentral gyrus. This cortical area receives input from the contralateral (in this case right) spinal trigeminal and principal sensory nuclei after a synaptic relay in the ventral posteromedial (VPM) nucleus. It is recognized that the principal sensory nucleus sends a small ipsilateral projection to the VPM nucleus, but this is not a major element in diagnosis. This deficit is not bilateral, and it is not on the left side of the face and oral cavity (A and B). Since this lesion is in the sensory cortex, muscle weakness is not a predominate finding (D and E). The perceived difficulty speaking (on the man's part) is related to the profound loss of sensation: one may recall the last time that he or she went to the dentist and had his or her mouth numbed for a filling.

Chapter 14

Special Sense I: The Visual System

Recall the Cautionary Tale: The images in this chapter are presented in a Clinical Orientation: this approach emphasizes how structure and function correlate with deficits in a clinical setting. MRI and CT have a universally recognized orientation and laterality. Line drawings, stained sections, and gross brain images in the Clinical Orientation are viewed here in the identical manner that one views an MRI: your right, as the physician-observer, is the patient's left, and the observer's left is the patient's right. The laterality of deficits and reference to connections follow accordingly.

More Caution Regarding the Visual System: There is a dichotomy regarding how visual structures, such as the optic tract or radiations, are viewed in MRI and CT and how visual fields and deficits thereof are represented. As explained in detail in the Preface and in the Introduction, a Cautionary Tale of Right and Left, when a physician is looking at an MRI or CT, his/her right is the patient's left and his/her left is the patient's right. This is a universal convention. Equally universal is the fact that visual fields are represented as the patient looks at the environment: the patient's right visual field is on the right and the patient's left visual field is on the left. In this chapter, the visual field of the right eye (Oculus Dexter) is specified OD and that of the left eye (Oculus Sinister) is specified OS. It is essential to understand these realities of clinical medicine when viewing representative lesions of visual structures in MRI and correlating the laterality of the lesion in the MRI with the corresponding visual field deficit.

QUESTIONS

Select the single best answer.

Questions 1 and 2 are based on the following case:

A 43-year-old woman presents with the complaint of difficulty seeing. The history reveals that this patient is overweight, has high blood pressure that is poorly controlled, and an A1C that has varied between 9.7 and 12.3 over the last 2 years. An ophthalmological examination reveals a visual loss, and MRI shows a lesion in the area indicated on the image below.

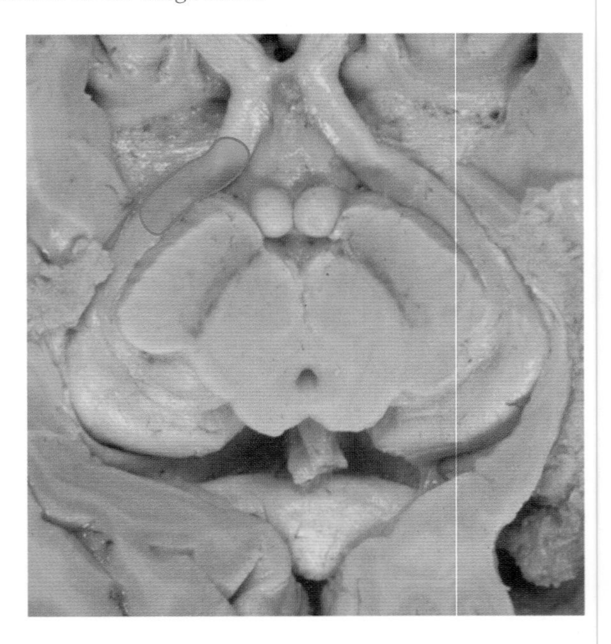

1 Which of the following would most specifically describe the symptoms experienced by this woman?

(A) Blindness in the left eye
(B) Left homonymous hemianopia
(C) Left superior quadrantanopia
(D) Right homonymous hemianopia
(E) Right superior quadrantanopia

2 Assuming this lesion to be the result of a vascular occlusion, which of the following vessels is most likely involved?

(A) Left anterior choroidal artery
(B) Left ophthalmic artery
(C) Right anterior choroidal artery
(D) Right medial posterior choroidal artery
(E) Right ophthalmic artery

3 During a neurological examination on a 31-year-old woman, the examining physician discovers that the pupillary light reflex is absent in the woman's left eye. The afferent limb of this reflex traverses which of the following?

(A) Brachium conjunctivum
(B) Brachium of the inferior colliculus
(C) Brachium of the superior colliculus
(D) Oculomotor nerve
(E) Posterior commissure

4 A 68-year old woman complains to her daughter of a sudden headache accompanied by a change in her vision. A visit to her physician confirms a partial visual loss, and MRI reveals a vascular lesion in the right hemisphere in the area outlined in the image below. This woman is most likely experiencing which of the following deficits?

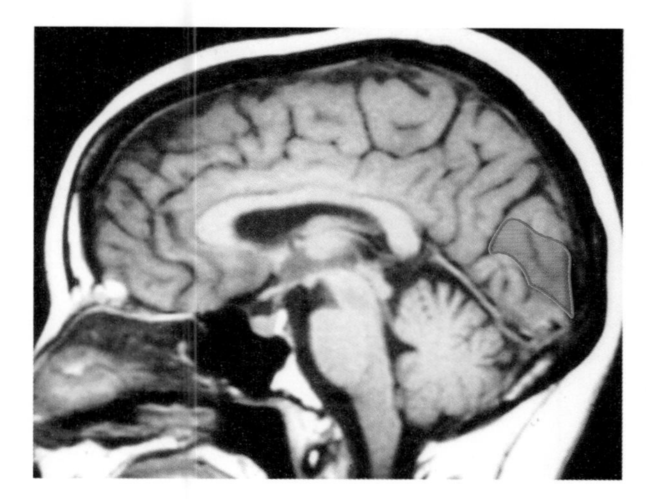

(A) Left homonymous hemianopia
(B) Left inferior quadrantanopia
(C) Left superior quadrantanopia
(D) Right inferior quadrantanopia
(E) Right superior quadrantanopia

5 A 45-year-old man visits his physician's office because he noticed changes in his vision over the last week. The history reveals that the man was elbowed in the side of his head while playing a pickup game of basketball. The next day the man noticed what he called "sparklers" in his right eye and, very shortly thereafter, he noticed a "dark spot" at the edge of the same eye that seemed to be getting larger. The ophthalmoscopic examination reveals a detached retina. This damage most likely took place at which of the following locations?

(A) Between the amacrine cells and the bipolar cells
(B) Between the ganglion cells and the bipolar cells
(C) Between the horizontal and amacrine cells and receptors
(D) Between the pigment epithelium and the choroid layer
(E) Between the receptor cells and pigment epithelium

6 A 31-year-old man presents to his family physician with the complaint of occasional problems with his eyes. The history reveals that the man sees what he calls "dark areas" as he looks around that will "go away then come back" and occasionally he sees "two of everything." He thinks that the cycles are becoming more frequent, but he is not sure. The examination reveals a visual field deficit, diplopia, and a decreased appreciation of vibratory sense on his right lower extremity. Suspecting multiple sclerosis, the physician orders an MRI which shows demyelination within the area of the brain outlined on the image below. Which of the following visual field deficits would most likely correlate with the location of this man's lesion?

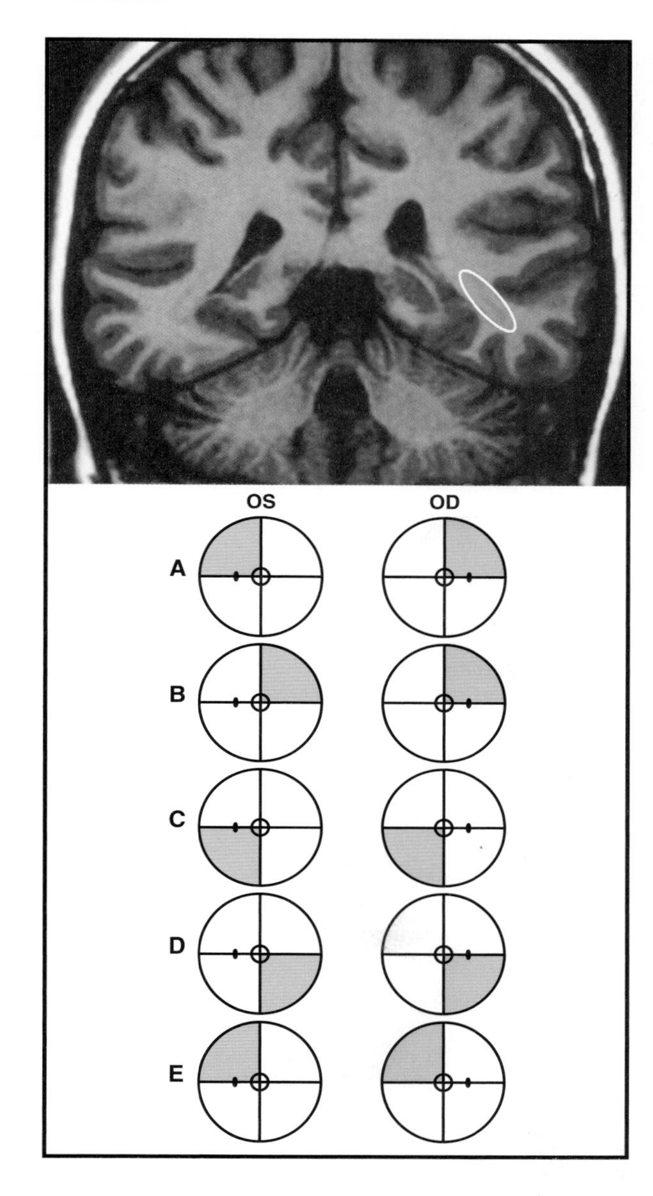

7 A 12-year-old boy undergoes a neurological examination as part of his application to participate on a sports team. Visual fields, vision, motion detection, stereoscopic vision, intraocular pressure, and the pupillary light reflex are all normal. However, when the Ishihara test plates are used, he is unable to recognize a rose-colored "6" made up of dots on a background of green dots. Which of the following is the most likely cause of this deficit?

(A) Loss of blue (S) cone opsin
(B) Loss of green (M) cone opsin
(C) Loss of red (L) cone opsin
(D) Loss of L and M cone opsin
(E) Loss of rhodopsin

8 A 59-year-old man is examined by his ophthalmologist. The examination reveals the following: (1) eye movements are normal; (2) the pupillary light and corneal reflexes are normal; and (3) and the optic nerves appear normal and symmetrical. The right eye shows a small hemorrhagic region and "cotton wool spots" near the optic disc. The visual field of this patient is seen as shown below. Which of the following is the most likely diagnosis?

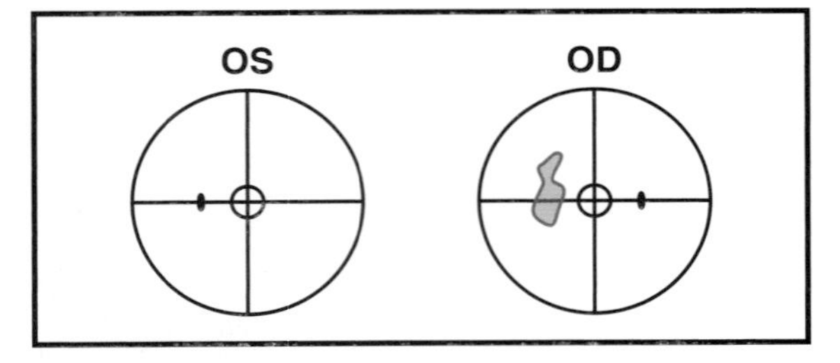

(A) Glaucoma
(B) Left homonymous hemianopia
(C) Pituitary tumor
(D) Retinal detachment
(E) Scotoma

9 A 2-month-old baby has congenital cataracts. The history reveals that the 39-year-old mother was never vaccinated against common childhood diseases. The mother is seeking genetic counseling concerning the probability of having another child with this same condition, since this condition is not present in her family or her husband's family. Which of the following would most likely explain this occurrence?
(A) Birth trauma
(B) Exposure to ultraviolet light
(C) Maternal exposure to rubella
(D) Maternal lack of folic acid
(E) Older primigravida

10 A 79-year-old woman complains to her physician that she is having trouble seeing; she has no history of visual problems. The visual field examination reveals a right superior quadrantanopia, and the ophthalmoscopic examination shows no frank abnormalities of the optic disc or retinal vasculature. Which of the following represents the most likely location of the lesion in this woman?
(A) Left lateral geniculate nucleus
(B) Left Meyer loop
(C) Left optic radiations
(D) Right Meyer loop
(E) Right optic tract

11 A 29-year-old woman visits her family physician with the complaints of diminished vision. The history and examination reveal that the woman also has persistent headaches that are not controlled by OTC medications, strikingly irregular menstrual periods, and her eating habits have recently changed. MRI shows a dumbbell-shaped pituitary tumor impinging on the areas indicted in the given image. Which of the following visual field deficits would most likely correlate with the location of this woman's lesion?

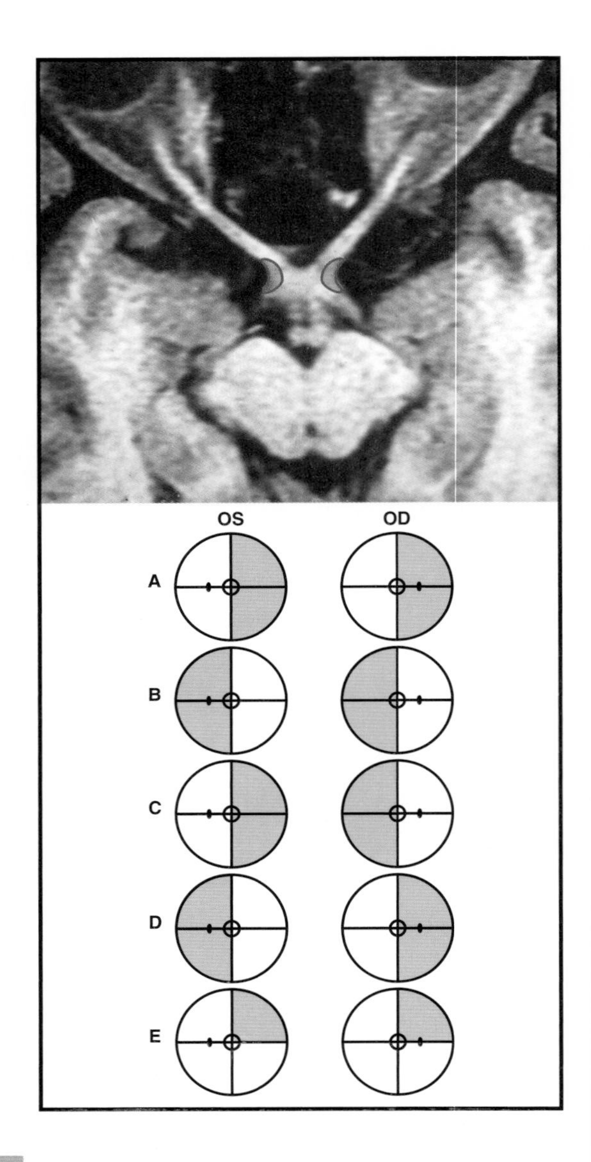

12 Which of the following is indicated by the arrows in the image below?

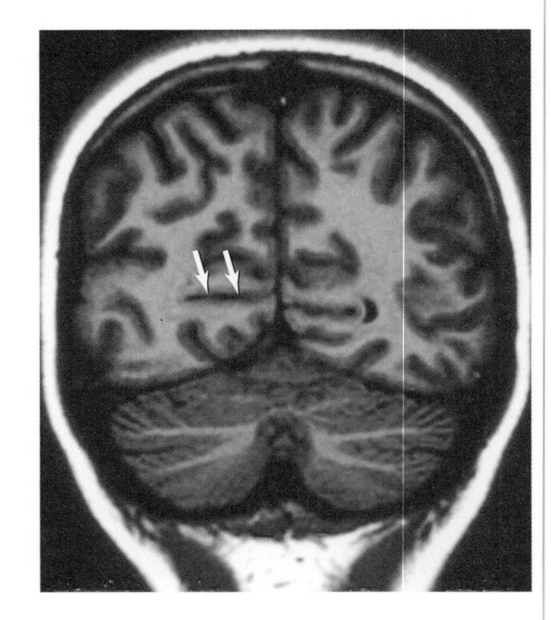

(A) Calcarine sulcus
(B) Collateral sulcus
(C) Marginal sulcus
(D) Occipital sulcus
(E) Parietooccipital sulcus

13 A 12-year-old boy is brought to the ophthalmologist by his mother. His intraocular pressure, visual fields, stereoscopic vision, and pupillary light reflexes are normal. When presented with Ishihara test plates (letters or numbers of one color in a dot pattern on a pattern of dots of another color), he is unable to detect a green-colored "N" in a dot pattern on a background of red dots. What is the likelihood that his 7-year-old sister would have the identical deficit?
(A) About 1:1
(B) About 1:2
(C) About 1:3
(D) About 1:4
(E) Almost zero

Questions 14 and 15 are based on the following patient.

A 24-year-old man visits his family physician with what he calls "embarrassing problems." The history and examination reveal that this man has been suffering persistent headaches that are refractory to OTC medications, has almost no interest in sex, and is concerned that he may be impotent. The general physical examination does not reveal any obvious cause for this man's symptoms. A neuro-ophthalmological examination reveals the visual field deficits in the image below.

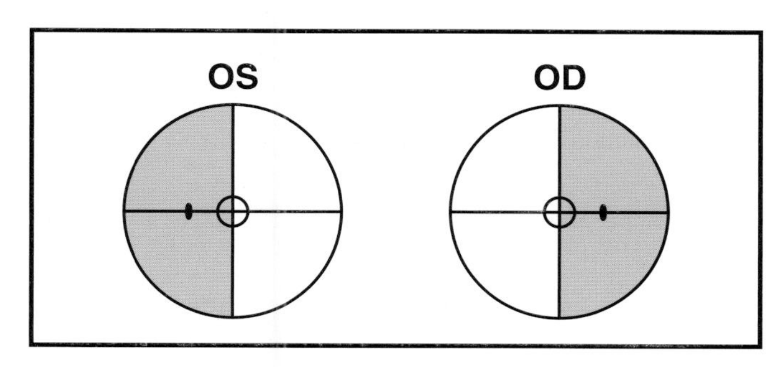

14 Which of the following describes the visual field losses in this man?
(A) Bilateral temporal scotoma
(B) Binasal hemianopia
(C) Bitemporal hemianopia
(D) Left homonymous hemianopia
(E) Right homonymous hemianopia

15 Which of the following is the most likely cause of the symptoms and deficits experienced by this man?
(A) Cushing syndrome tumor
(B) Gonadotrope tumor
(C) Malignant hypertension
(D) Prolactinoma
(E) Sellar meningioma

16 Which of the following describes the sequence of physiological events that take place in a rod when it is exposed to photons of light?
(A) Depolarization with a corresponding increase in glutamate release
(B) Depolarization with no change in glutamate release
(C) Hyperpolarization with no change in glutamate release
(D) Hyperpolarization with a corresponding decrease in glutamate release
(E) Hyperpolarization with a corresponding increase in glutamate release

17 Which of the following specifies the elevation identified by the line in the image below?

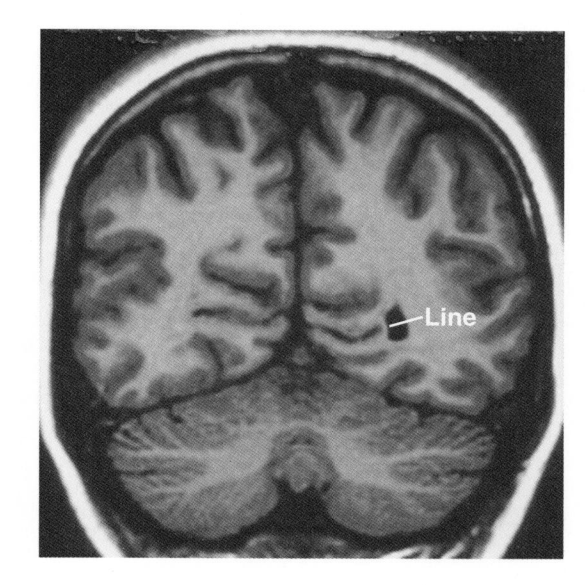

(A) Arcuate eminence
(B) Calcar avis
(C) Collateral trigone
(D) Medial eminence
(E) Müller trigone

18 A 61-year-old woman is diagnosed with glaucoma, a condition that, if left untreated, will most likely result in blindness. The ophthalmologist performs a small outpatient procedure that will increase the egress of aqueous humor from the anterior chamber into the venous circulation. This procedure increases flow through which of the following?
(A) Bichat canal
(B) Canal of Schlemm
(C) Dorello canal
(D) Optic canal
(E) Wirsung canal

19 A 62-year-old woman presents with signs and symptoms of increased intracranial pressure (headache, lethargy, emesis). The woman states that she has been experiencing these problems off-and-on for "weeks" and that they seem to be getting worse. The visual field examination is normal and the woman has no specific (or focal) neurological signs. Which of the following would you expect to see in this patient?

(A) Abnormal retinal pigmentation
(B) Macular degeneration
(C) Retinal detachment
(D) Papilledema
(E) Scotoma

20 A 31-year-old man visits his family physician with the complaint of difficulty walking. The history reveals that the man's walking problems have been ongoing for several weeks and seem worse sometimes than others; he thinks that it may be slightly worse in the evenings, but he is not sure. The examination reveals that muscle stretch reflexes are normal, and that muscle strength is also normal, in all extremities. When questioned further, the man indicates that his problem is especially noticeable when he walks down stairs, steps down over a curb, or engages in similar activity. A visual field examination by confrontation detects a partial loss of vision, and an MRI reveals frank demyelination in the areas outlined in the image below. Which of the following visual field deficits would most likely correlate with the location of this man's lesion?

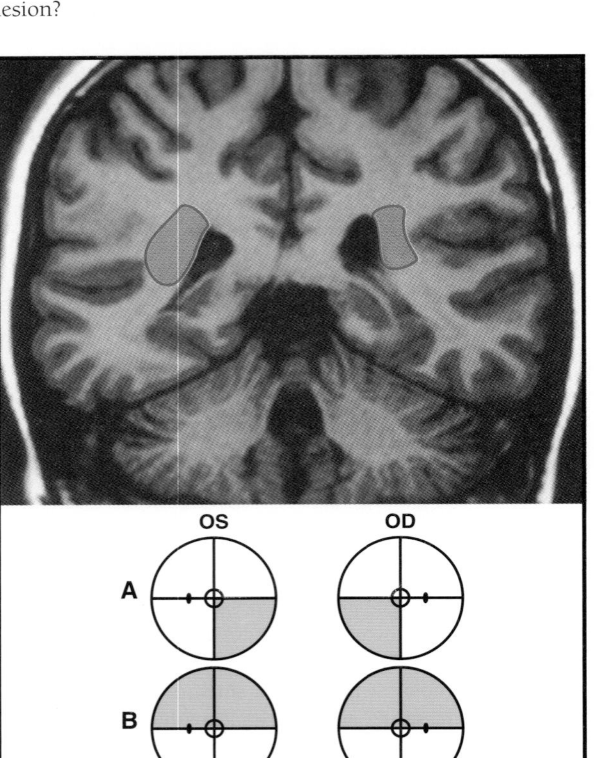

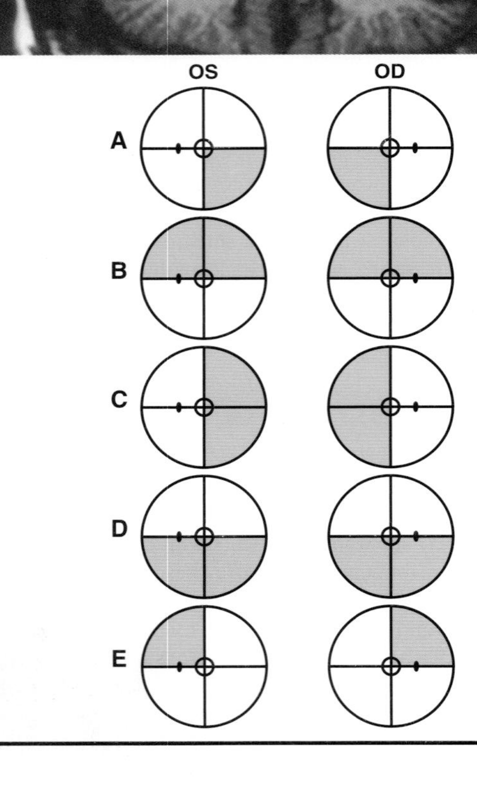

21 Which of the following is concerned with the circuits that allow the patient to discern fine detail, or form, as well as color?

(A) M (or Y) visual pathway
(B) Medial temporal cortex
(C) P (or X) visual pathway
(D) Posterior parietal cortex
(E) W visual pathway

22 A 24-year-old woman is brought to the Emergency Department from the site of a motor vehicle collision. The examination reveals that she has a compound fracture of her femur and tibia, facial injuries, and a probable crush injury to the orbit. CT reveals fractured orbital walls, blood in the orbit, and optic nerve damage as shown in the image below. When this patient is conscious and a full neurological examination is conducted, and a light is shined in the right eye, which of the following will be most noticeable?

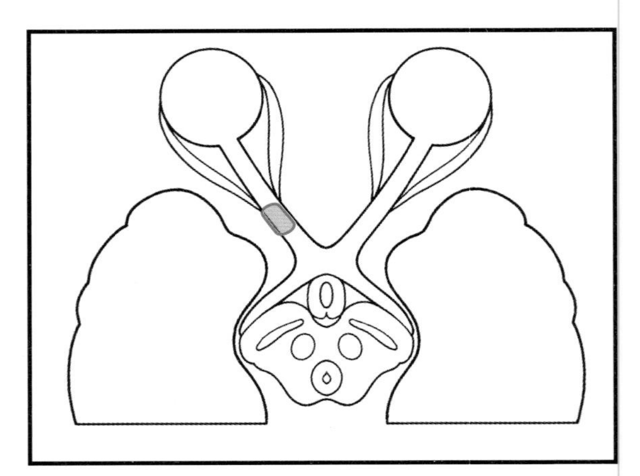

(A) Blindness in the left eye
(B) Blindness in the right eye
(C) Loss of corneal reflex in right eye
(D) Loss of pupillary light reflex in left eye only
(E) Loss of pupillary light reflex in right eye only

23 A 47-year-old man suddenly realizes the he must hold papers at arm's length in an attempt to get them into focus. Which of the following specifies this condition in this man?

(A) Amblyopia
(B) Mydriasis
(C) Myopia
(D) Presbyopia
(E) Visual agnosia

24 A-62-year-old man visits his family physician. His complaint is that a few days ago he "felt really bad" but rested, took OTC medications for headache and upset stomach, and felt better a couple of days later. However, he reports that, beginning right after this event, he began to run into things as he walked through his house; he ran into things he knew were there because he put them there. A visual field examination by confrontation revealed a deficit, and MRI showed evidence of a recent hemorrhagic event within the outlined area represented on the given image. Which of the following is the deficit that this man is most likely experiencing?

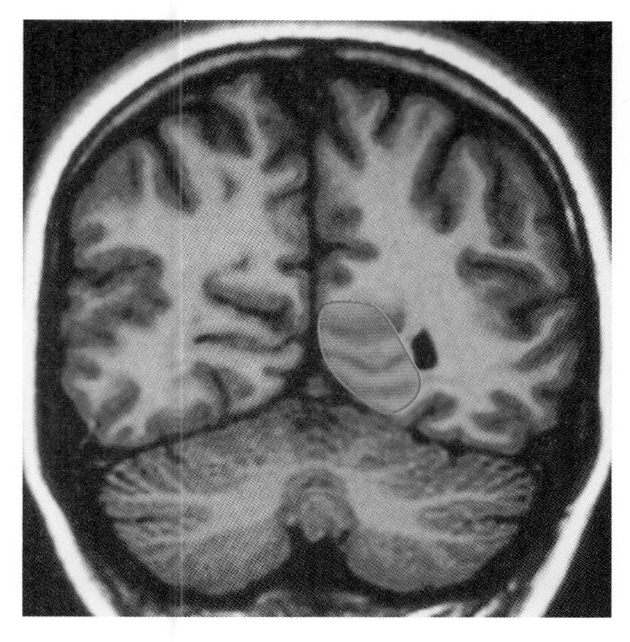

(A) Bitemporal hemianopia
(B) Left homonymous hemianopia
(C) Left superior quadrantanopia
(D) Right homonymous hemianopia
(E) Left inferior quadrantanopia

25 Which of the following cells of the neural retina have a center-surround type of receptive field organization?
(A) Amacrine cells
(B) Cone photoreceptors
(C) Ganglion cells
(D) Horizontal cells
(E) Rod photoreceptors

26 A 43-year-old man is brought to the Emergency Department from the site of a motor vehicle collision. He has facial injuries and fractures, and the CT clearly suggests that he has a probable transected left optic nerve. Which of the following, regarding the pupillary light reflex, would confirm this observation when a light is shined in the left eye?
(A) Direct and consensual response intact
(B) Direct response intact, consensual absent
(C) Direct response absent, consensual intact
(D) Direct and consensual responses intact but diminished
(E) Direct and consensual responses absent

27 A 29-year-old woman visits her family physician with the main complaint of irregular menses. The examination also reveals that the woman has had trouble sleeping (usually not an issue), and has generally not felt well for the last several weeks. A visual field examination by confrontation reveals a deficit, and subsequent MRI shows a small pituitary lesion that is impinging on the area indicated by the outline in the image below. Based on the position of this lesion, which of the following visual field deficits would you most likely expect to see in this woman?

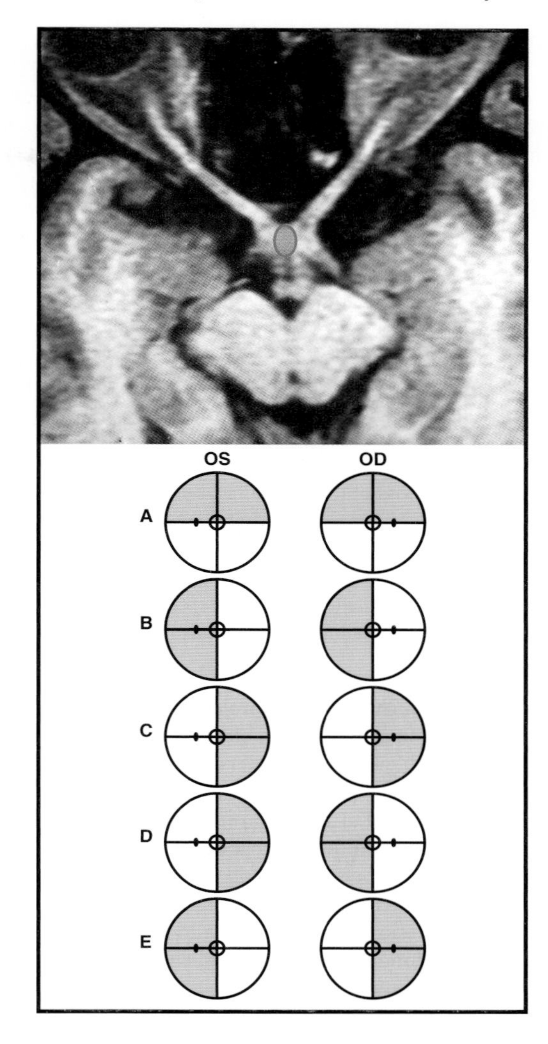

28 Which of the following is a correct synopsis of the characteristics of the fovea centralis of the macula lutea?
(A) At the posterior pole of the eye, contains rods and cones, has significant blood supply, point of least visual acuity, all retinal layers are present
(B) At the posterior pole of the eye, contains only rods, has few or no blood vessels, point of least visual acuity, all retinal layers present
(C) At the posterior pole of the eye, contains only cones, has few or no blood vessels, point of greatest visual acuity, all retinal layers present
(D) At the posterior pole of the eye, contains only cones, has few or no blood vessels, point of greatest visual acuity, only layer of cones present
(E) At the posterior pole of the eye, contains rods and cones, has few or no blood vessels, point of greatest visual acuity, only layer of rods present

29 The optic nerve is composed of axons arising from which of the following cells of the retina?
(A) Amacrine cells
(B) Bipolar cells
(C) Ganglion cells
(D) Horizontal cells
(E) Rod cells

ANSWERS

1 **The answer is B: Left homonymous hemianopia.** This lesion is in the right optic tract. Lesions of optic structures rostral to the chiasm result in visual deficits in the eye on that side. On the other hand, lesions caudal to the chiasm produce a partial loss of vision in both eyes. This is due to the fact that the axons of ganglion cells in the temporal retina (receiving input from the nasal visual field) do not cross in the chiasm, while ganglion cells in the nasal retina (receiving input from the temporal visual fields) cross in the chiasm. Consequently, damage to the right optic tract results in a loss of input from the nasal visual field in the right eye and from the temporal visual field in the left eye. Lesions in the optic tract result in a partial loss of vision in the opposite half (the left half in this case) of both visual fields. Hence, a lesion of the right optic tract results in a loss of the left half of both visual fields: a left homonymous hemianopia. Blindness in one eye is not correct, since the lesion is caudal to the chiasm; the right homonymous hemianopia is on the wrong side; quadrantanopias represent lesions of smaller portions (representing a quadrant) of the visual pathway, caudal to the chiasm.

2 **The answer is C: Right anterior choroidal artery.** The anterior choroidal artery, usually a branch of the internal carotid artery, serves the optic tract and a number of immediately adjacent, and important, structures. Since the lesion is on the right, the right anterior choroidal artery is involved. The ophthalmic arteries also arise from the internal carotid artery, but course rostrally to serve the optic nerve and eyeball on the same side. The ophthalmic artery gives origin to the central retinal artery. The medial posterior choroidal artery arises from the P_2 segment, arches around the midbrain, and enters the caudal portion of the third ventricle to serve the choroid plexus located therein. This vessel provides small penetrating branches to those midbrain areas that it circumvents.

3 **The answer is C: Brachium of the superior colliculus.** The afferent limb of the papillary light reflex originates from ganglion cells of the retina. The axons of these cells course through the optic nerve and tract, deviate from the tract to enter and traverse the brachium of the superior colliculus, and then enter the pretectal nucleus where they terminate. The pretectal nucleus projects to the ipsilateral nucleus of Edinger-Westphal (directly) and to this same nucleus on the contralateral side via the posterior commissure. Fibers from the Edinger-Westphal nucleus travel on the oculomotor nerve as the preganglionic component of the efferent limb of the pupillary light reflex. The brachium conjunctivum contains the efferent projections from the cerebellar nuclei; the brachium of the inferior colliculus contains fibers from the inferior colliculus to the medial geniculate nucleus. This structure may also contain some fibers that travel from brainstem auditory relay nuclei to the medial geniculate nucleus.

4 **The answer is B: Left inferior quadrantanopia.** This lesion is located in the superior bank of the calcarine sulcus (the cuneus) which is the location of the opposite (in this case the left) inferior quadrant of both visual fields; the deficit is a left inferior quadrantanopia. The lower bank of the calcarine cortex, the lingual gyrus, represents the opposite upper/superior quadrant of the visual fields; if the lesion is on the right side, it would be a left superior quadrantanopia. The combination of upper and lower quadrants on the same side of the visual fields constitutes a hemifield. A single lesion involving fibers from upper and lower quadrants results in a hemianopia: homonymous when it is the same sides of both visual fields (both right or both left) or heteronymous when it is not the same sides of both visual fields (one temporal, one nasal). The term heteronymous is infrequently used since such deficits are usually called bitemporal (both temporal visual fields) or binasal (both nasal visual fields) hemianopia.

5 **The answer is E: Between the receptor cells and pigment epithelium.** In the case of retinal detachment, the rod and cone layer (along with the other retinal layers) tears away at the interface of the outer segments of the receptors with the pigment epithelium. This interface in the adult is the remnant of the ventricular space that is present in the developing eye cup. Since the receptor cells depend on their contact with the pigment epithelium, this type of defect needs to be surgically corrected, and quickly, to avoid permanent loss of vision. In most cases, this repair is easily accomplished and results in no appreciable visual loss. Damage may take place at other locations in the retina; such lesions are usually not amenable to surgical repair and may lead to some permanent loss of vision.

6 **The answer is B (Right superior quadrantanopia).** Recall how MRI is viewed versus how visual fields are viewed. This lesion is in the inferior portion of the optic radiations on the left side; these fibers convey information that originates in the right superior quadrant of both visual fields: a right superior quadrantanopia. Also recall that lesions caudal to the optic chiasm result in deficits in the opposite visual field that may present as a hemifield defect or quadrant defect. The other quadrant deficits shown here may result from lesions in other locations, but not from this particular lesion.

7 **The answer is C: Loss of red (L) cone opsin.** This boy has an inherited condition, commonly called color blindness, in which the cone type for red (L-cone, red-absorbing) is absent and the boy does not see the red 6 on a field of green. This is called protanopia. Loss of other cone opsins, due to an absence of the corresponding cone that is capable of absorbing that color, would result in a corresponding inability to see that color. The genes for L- and M-cones (red-absorbing and green-absorbing) are on the X chromosome and, therefore, color-blindness is more common in males than females. Loss, or alteration, of the S-cone (blue-absorbing) is relatively rare.

8 **The answer is E: Scotoma.** A scotoma is an area of variable shape, frequently somewhat irregular, or size within the visual field; the patient may perceive an area of diminished vision (relative scotoma) or total visual loss (absolute scotoma). Scotoma may be seen in a variety of clinical situations such as migraine or retinitis pigmentosa. Glaucoma is a disease of the eye in which there is an increase in intraocular pressure, in addition to eventual visual loss; if left untreated, the optic nerve head is altered in size and shape (not seen in this case). Homonymous hemianopia is a loss of both left or both right portions of both visual fields, not a small loss in one field, and is usually indicative of lesions caudal to the optic chiasm. There is no indication in the examination of endocrine problems, or of bilateral visual loss, both characteristic of pituitary lesions. Retinal detachment frequently presents as sparkles in the peripheral region of a visual field followed by a progressively encroaching loss of vision.

9 **The answer is C: Maternal exposure to rubella.** Maternal exposure to the rubella virus, particularly during the first trimester, has a high incidence of birth defects, including congenital cataracts. Since this mother did not receive her childhood vaccinations, she was at risk of exposure; it is reasonable to assume that she came into contact with children of the appropriate age for rubella. While birth trauma may result in defects both great and small, cataracts is not one that is seen. Exposure to ultraviolet light does not result in cataracts; adequate amounts of folic acid in the maternal diet will guard against dysraphic defects that are seen during the first trimester, but will cause defects. Although this woman had her first child later in her reproductive life, her child's cataracts are not related thereto; defects related to later pregnancies are more global in nature.

10 **The answer is B: Left Meyer loop.** A general rule that is easy to remember is that lesions of optic structures (tract, lateral geniculate nucleus, all parts of the optic radiations) caudal to the chiasm result in a loss in some part of the opposite visual fields. For example, a lesion of the right optic tract (choice E) would produce a left homonymous hemianopia: a loss of the left portions of both visual fields. In this woman's case, her right superior quadrantanopia indicates that her lesion is on the left side. Her deficit results from damage to the left optic radiations that pass forward from the lateral geniculate nucleus to pass through the temporal lobe then loop abruptly caudally (the Meyer, or Meyer-Archambault, loop) to enter the inferior bank of the calcarine cortex. Damage to the left lateral geniculate nucleus, or to the left optic radiations, would result in a right homonymous hemianopia. Damage to the right Meyer loop would produce a left superior quadrantanopia.

11 **The answer is C (Binasal hemianopia).** These lesions are impinging on the lateral aspects of the optic chiasm. The fibers located in this part of the chiasm arise from ganglion cells in the temporal parts of the retina which, in turn, receive input from the nasal visual fields. Consequently, damage to the lateral portions of the chiasm results in a deficit of vision from both nasal visual fields: a binasal hemianopia (it is also a heteronymous hemianopia, but that term is rarely used because it does not include the term nasal). Choices A and B are right and left homonymous hemianopia which would be seen following certain lesions caudal to the chiasm and on the left and right sides, respectively. In like manner, the right superior quadrantanopia (E) represents a more restricted lesion on the left side, caudal to the chiasm. The bitemporal hemianopia (D) is considered later in this chapter.

12 **The answer is A: Calcarine sulcus.** The calcarine sulcus is located on the medial aspect of the occipital lobe, is comparatively straight and deep, and at its depth forms a named elevation in the medial wall of the posterior horn of the lateral ventricle. The primary visual cortex (area 17) is located on the upper (cuneus gyrus) and lower (lingual gyrus) banks of the calcarine sulcus. The marginal sulcus (also called the marginal ramus of the cingulate sulcus) separates the posterior paracentral gyrus from the precuneus; there are several occipital sulci, but they are all located on the lateral surface of the occipital lobe. The parietooccipital sulcus is also on the medial aspect of the hemisphere separating the precuneus from the cuneus. Occipitotemporal gyri (there are usually two) are separated from the parahippocampal gyrus by the collateral sulcus.

13 **The answer is E: Almost zero.** The abnormalities for color vision of red (L-cones, red-absorbing) and green (M-cones, green-absorbing) are encoded on the X chromosome. Patients who have normal color vision have one copy of the gene for the L pigment and may have several copies of the gene for the M pigment. The inability to recognize the green N on the red background is manifestly more common in males than in females. An inability to detect the green wavelengths is deuteranopia. Further investigation revealed that the boy's father, from whom the family is estranged, and his grandfather have the same visual deficit.

14 **The answer is C: Bitemporal hemianopia.** In this man, the visual loss involves the temporal visual fields of both eyes; both nasal visual fields are intact; this is a bitemporal hemianopia. This type of deficit may result from a lesion that involves the midline portion of the optic chiasm. The fibers that cross the midline in the optic chiasm arise from ganglion cells in the nasal regions of the retinae; this represents the temporal parts of the visual fields. While these are bilateral temporal visual field deficits, these are large field losses and are not characteristic of scotomata (choice A), which are frequently small irregular areas of vision loss. Left and right homonymous hemianopia are deficits seen resultant to lesions caudal to the optic chiasm and involving structures on the right and left sides, respectively.

15 **The answer is D: Prolactinoma.** This man is suffering a combination of deficits (visual and probably hormonal) that clearly indicates a lesion in the immediate area of the optic chiasm, or immediately adjacent nerves or tracts, and the pituitary. Recall that the optic chiasm is immediately superior to the anterior aspect of the pituitary, as the latter sits in the sella turcica. A prolactinoma is a pituitary tumor that produces excessive prolactin; in women, this results in galactorrhea and amenorrhea; in men, the symptoms are as experienced by this man. The tumor in the Cushing syndrome results in an overproduction of corticotrophin (truncal obesity, violaceous striae, cervical hump); a gonadotrope tumor may result in precocious puberty, or may remain silent until visual deficits or mass effect is seen. Malignant hypertension is of relatively sudden onset, progresses rapidly, and may cause death via kidney failure or cerebral hemorrhage. A sellar meningioma may, depending on its location and size, most certainly cause a bitemporal hemianopia. However, it would not cause the sexual dysfunction seen in this case.

16 **The answer is D: Hyperpolarization with a corresponding decrease in glutamate release.** Under normal conditions, there is a constant release of the neurotransmitter glutamate at the rod spherule. When the rod is stimulated by photons of light, the sodium (Na) current is blocked, the cell membrane is hyperpolarized (moving from about −40 to about −60 mV), and the amount of glutamate being released is decreased. Photoreceptors are the only sensory receptors that hyperpolarize in response to an appropriate, and adequate, stimulus.

17 **The answer is B: Calcar avis.** The calcar avis, also called the calcarine spur, is the elevation on the medial wall of the lateral ventricle that is the result of the deepest extent of the calcarine sulcus apposing the wall of the posterior horn. In this image, the full extent of the calcarine sulcus can be seen from the surface of the occipital lobe to its terminus at the ventricular wall. The arcuate eminence is the elevation on the petrous portion of the temporal bone signifying the position of the underlying superior semicircular canal; the medial eminence is the elevation in the floor of the fourth ventricle along the midline and rostral to the facial colliculus. The collateral trigone is the elevation in the floor of the atrium of the lateral ventricle, at the transition of the temporal horn to the occipital horn, resulting from the deepest extent of the collateral sulcus. The floor of the supraoptic recess of the third ventricle is the Müller trigone.

18 **The answer is B: Canal of Schlemm.** The canal of Schlemm is located at the circumference of the anterior chamber at the angle formed by the inner surface of the cornea and the iris. The procedure to enlarge these channels will increase the flow of aqueous humor from the anterior (and posterior) chamber(s) into the venous circulation. This is a common procedure that saves the vision of many individuals. The Bichat canal and the Wirsung canal are the eponyms for the quadrigeminal cistern and the pancreatic duct, respectively. The Dorello canal and the optic canal transmit, respectively, the abducens nerve and a part of the inferior petrosal sinus and the optic nerve.

19 **The answer is D: Papilledema.** Although this woman has no focal neurological signs at this time, she clearly has evidence of increased intracranial pressure. The increase in intracranial pressure is transmitted through the cerebrospinal fluid located within the sleeve of meninges surrounding the optic nerve. This increase in pressure causes the optic disc to rise up toward the examiner (looking at the disc through an ophthalmoscope), a sign commonly called a choked disc. The result is a stasis of axoplasmic flow at the head of the optic nerve with resultant swelling (papilledema). Macular degeneration, retinal detachment, and scotoma, will all cause various degrees of visual loss, but no signs or symptoms of increased intracranial pressure. Retinal pigmentation may have no associated visual losses and will also not be correlated with increased intracranial pressure.

20 **The answer is D (Bilateral, right + left, inferior quadrantanopias).** This case is an interesting dichotomy: the patient perceives one probable deficit (difficulty walking), when in actuality he has another deficit of which he is essentially unaware. This man has bilateral lesions in the upper portions of the optic radiations, also called geniculocalcarine radiations, which represent the lower portions of the visual field. Recall that, caudal to the optic chiasm, the relationship between lesion and deficit is reversed: right-sided lesions cause left-sided deficits, and superior lesions (e.g., in the optic radiations in this case) cause inferior deficits in the visual fields. In this case, the man's apparent difficulties walking, particularly under the conditions of stepping downward, are related to the fact that he does not "see" objects in the lower parts of his visual fields. Even though it is basically an unconscious action, when individuals walk down stairs, they glance downward to get quick information that will result in a coordinated and productive descent. When the lower half of each visual field is gone, the patient does not have that useful information; the descent is not especially coordinated leaving the patient with the perception that he/she is having difficulty walking. The patient does not realize that the root of the problem is missing visual input. Bilateral inferior or superior quadrantanopias are frequently referred to as altitudinal defects due to the fact that they lie below or above, and may respect, the horizontal meridian.

21 **The answer is C: P (or X) visual pathway.** The P, or X, visual pathway encodes fine details and information concerning color. This pathway originated from medium-sized retinal ganglion cells, also called beta cells, that project to the small-celled laminae (parvocellular laminae; laminae 3 to 6) of the lateral geniculate nucleus. This information is processed through areas 17 and 18 that extend into the medial temporal visual cortex.

22 **The answer is B: Blindness in the right eye.** This is a very straightforward situation. This woman has a damaged right optic nerve and, when tested, she will be blind in her right eye; this will be the most noticeable deficit to both the patient and physician. The corneal reflex in her right eye will be intact due to the fact that the afferent limb is conveyed via the ophthalmic branch of the trigeminal nerve, and the efferent limb is part of the peripheral branches of the facial nerve. When a light is shined in the right eye, the direct (right eye) and consensual (left eye) pupillary light reflexes are absent because the afferent limb of this reflex, the right optic nerve, is damaged. A light shined in the left eye in this patient will result in a direct response (left eye) and a consensual response (right eye) due to the fact that the afferent limb, in this case (the left optic nerve), is intact and the efferent limb on the right (ophthalmic nerve) is also intact. The pupillary light reflex will not be selectively lost in either the left or right eye in this case.

23 **The answer is D: Presbyopia.** Presbyopia, the inability to focus on near targets, is related to the loss of elasticity of the lens (and accommodation) with age. There is no loss of vision. This condition may appear earlier or later in life. Presbyopia is generally recognized to occur when the near point of focus has reached a distance of 22 to 24 cm (about 9 to 10 in) or greater. This visual problem is easily treated with corrective lenses. Myopia is nearsightedness: this is when the object is in focus only when excessively close to the eyes; myopia is also correctable with lenses. Mydriasis is dilation of the pupil; the patient with visual agnosia is able to see the object, but cannot recognize the object; the latter is the result of bilateral damage in the parietooccipital area. Amblyopia is poor vision in one eye resultant to abnormal visual input to that eye during development that caused abnormal (and altered) formation of cells and synaptic relationships with the visual cortex representing the affected eye.

24 **The answer is D: Right homonymous hemianopia.** The lesion in this case is located in the left visual cortex and involves both the upper and lower banks of the calcarine cortex; the calcarine sulcus is clearly seen within the outline representing the area of damage. This lesion produces blindness in the right halves of both visual fields, and these patients frequently complain that they bump into things, especially objects in their right visual fields. A bitemporal hemianopia is seen in patients with damage to the crossing within the optic chiasm; a left homonymous hemianopia would be seen in cases of lesions of the optic tract, geniculate nucleus, optic radiations, or visual cortex, all on the right side. A quadrantanopia may be superior (damage to the lower bank of the visual cortex), inferior (damage to the upper bank of the visual cortex), and in the visual field opposite the side of the lesion.

25 **The answer is C: Ganglion cells.** Of the choices available in this question, only the ganglion cells have both a center and a surround component to its receptive field. Ganglion cells are the efferent cells of the human (and mammalian retina); their axons collect to form the optic nerve of each eye.

26 **The answer is E: Direct and consensual responses absent.** In this patient, damage to the left optic nerve would transect the afferent limb of the pupillary light reflex; the optic nerve is the afferent limb. If the afferent limb is gone, both efferent limbs (to both eyes) are nonefunctional, and both the direct (left eye) and consensual (right eye) responses are absent. This would confirm that the man is blind in his left eye. In addition, the man would not "see" the light shined into his left eye. On the other hand, a light shined in the man's right eye would result in both an intact direct (right eye) and consensual (left eye, the man's blind eye) response. In this example, the right afferent limb (the right optic nerve is intact) and the efferent limb to both eyes are intact; the efferent limb travels on the oculomotor nerve. The oculomotor nerve travels a different route into the orbit than does the optic nerve. The pupillary light reflex is a constriction of the pupil in response to light.

27 **The answer is E (A bitemporal hemianopia).** This lesion is on the midline of the optic chiasm and impinges on the crossing fibers in the chiasm. These fibers arise from ganglion cells in the nasal retina that represents visual input from the temporal visual fields; this lesion is producing a bitemporal hemianopia. A bilateral superior quadrantanopia (choice A) would be seen in cases of bilateral lesions of the inferior bank of the visual cortex or of the inferior portion of the optic radiations at some place in their course through the hemisphere. Left homonymous hemianopia (B) and right homonymous hemianopia (C) are likely outcomes of lesions of the optic tract, lateral geniculate, optic radiations, and both banks of the visual cortex on, respectively, the right and left sides. A binasal hemianopia (D) is seen in the case of a lesion (or lesions) that impinges on the lateral aspect of the optic chiasm. At these lateral locations, the damaged fibers originate from ganglion cells in the temporal retina that receives input from the nasal visual fields.

28 **The answer is D: At the posterior pole of the eye, contains only cones, has few or no blood vessels, point of greatest visual acuity, only layer of cones present.** The macula lutea (the eponym is the Soemmerring spot) is a small, slightly oval, area at the posterior pole of the eye that is located just lateral to the optic disc. The fovea centralis is the depression in the center of the macula; the fovea centralis is the point of greatest/sharpest visual acuity, consists of only a layer of cones (all other retinal layers are absent), and is considered to have few or usually no blood vessels. While cones are concentrated in the fovea, rods are most populous in the peripheral parts of the retina, but in contrast to the fovea, where only the layer of cones is present; in the peripheral retina, all retinal layers are present. In similar fashion to the somatosensory system, where small receptive fields have a large cortical representation, visual input from about the central 20° of the retina represents about 80% of the terminations in the primary visual cortex.

29 **The answer is C: Ganglion cells.** The ganglion cells are the output cells of the retina. Their axons collect to form the nerve fiber layer located just external to the internal limiting membrane, and converge toward the optic disc where they collect to pass through the lamina cribrosa of the sclera to form the optic nerve. Amacrine cells have small cell bodies, no obvious axons, and dendrites with a fairly restricted distribution. These cells sense changes in an ongoing motion. Bipolar cells participate in complicated retinal processing that basically serves to sharpen images and textures (including edges) at different locations in the visual field. Horizontal cells have dendrites that distribute close to the cell body and axons that travel to varying distances parallel to the flat orientation of the layers of the retina; these cells sense change within the receptive field. Rods are the receptors of the retina that are activated in dim-lighting (or black-white) conditions.

Chapter 15

Special Sense II: The Auditory and Vestibular Systems

Recall the Cautionary Tale: The images in this chapter are presented in a Clinical Orientation: this approach emphasizes how structure and function correlate with deficits in a clinical setting. MRI and CT have a universally recognized orientation and laterality. Line drawings, stained sections, and gross brain images in the Clinical Orientation are viewed here in the identical manner that one views an MRI: your right, as the physician-observer, is the patient's left, and the observer's left is the patient's right. The laterality of deficits and reference to connections follow accordingly.

QUESTIONS

Select the single best answer.

1 A 41-year-old woman presents with vertigo, positional nystagmus, and episodic tinnitus. The woman states that these attacks can be severe and accompanied by nausea and vomiting. An audiogram reveals a low-frequency hearing deficit that the woman says "comes and goes." Which of the following represents the most likely diagnosis and corresponding treatment?

(A) Hypertensive cardiovascular disease, reserpine
(B) Meniere disease, diuretics, and salt restriction
(C) Psychosomatic disorder, psychiatric consult
(D) Temporal bone fracture, rest
(E) Vestibular schwannoma, surgery

2 A 57-year-old man presents to his family physician with the complaint of a sudden onset of deafness. The examination reveals total deafness in his left ear and some gait disturbances. Magnetic resonance angiography (MRA) reveals a vessel that stops abruptly, suggesting that it is obstructed and that territories distal to this point are deprived of blood. Occlusion of which of the following vessels would correlate with this man's primary deficit?

(A) Left anterior inferior cerebellar artery
(B) Left long circumferential branches of the basilar artery
(C) Left posterior cerebral artery
(D) Left posterior inferior cerebellar artery
(E) Left posterior spinal artery

3 A 16-year-old boy is transported to the Emergency Department from the site of a motorcycle collision. The examination reveals compound fractures of the left thigh and leg bones and head injuries; he was not wearing a helmet. The boy complains that he is not able to detect movements of his head forward and backward (flexion, extension), but can tell when he is moving his head to the right or left. Electronystagmography (ENG) shows that compensatory eye movements are normal with right and/or left head turns, but are greatly reduced with flexion and extension of the head. A CT is ordered. Which of the following would most likely explain the deficits seen in his boy?

(A) An occipital fracture involving the cerebellum bilaterally
(B) Temporal bone fractures that damage only the horizontal semicircular canals
(C) Temporal bone fractures that damage only the vertical semicircular canals
(D) Temporal bone fractures that sever the vestibulocochlear nerves
(E) Temporal subdural hematoma in the posterior fossa

4 Which of the following structures is indicated by the arrows in the image below?

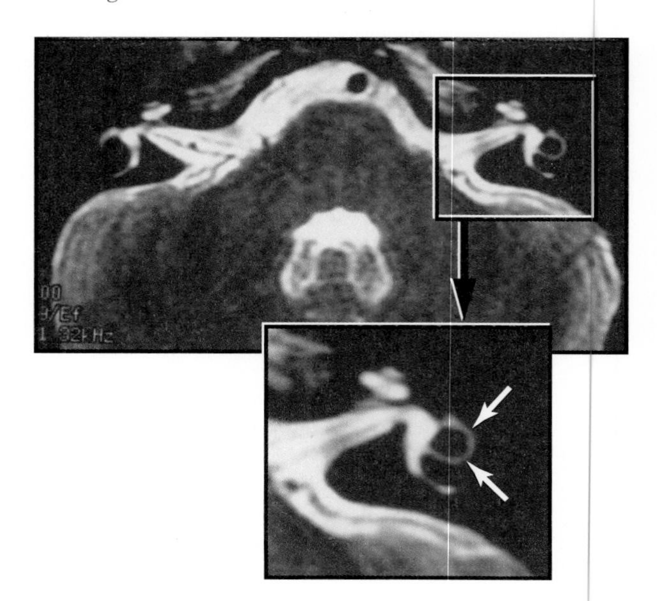

(A) Cochlea
(B) Lateral semicircular canal
(C) Posterior semicircular canal
(D) Superior semicircular canal
(E) Spiral ganglion

5 A 37-year-old man visits his family physician after returning from a scuba-diving Caribbean vacation. The history revealed that the man noticed his eyes were "twitching" and that he was "dizzy"; this was followed, after about 3 days, by a sudden hearing loss in his right ear. He is referred to an otologist who discovers that perilymph leaked from the cochlea into the middle ear resulting in a compromised auditory function. Which of the following is the most likely source of this leak?
(A) Basilar membrane
(B) Round window
(C) Spiral ligament
(D) Tectorial membrane
(E) Tympanic membrane

6 The head of a 19-year-old man is tilted toward the right; the otolith organs of the vestibular system produce a compensatory oculomotor response. This response is observed by watching the man's eye movement during the head-tilt procedure. Which of the following most likely represents the compensatory response in this man?
(A) A left beating horizontal nystagmus
(B) A rightward and upward eye deflection
(C) An upward eye deflection
(D) Counterroll (torsion) of the eyes to the right
(E) Counterroll (torsion) of the eyes to the left

7 A 42-year-old actor notices that his smile is not symmetrical. An examination at his physician's office reveals weakness of facial muscles at the angle of the mouth on the right, and lateralization of the tone from a vibrating tuning fork applied to the vertex of his skull to the left (Weber test). MRI showed a vestibular schwannoma on the left side. It was removed; the facial weakness slowly resolved. An audiogram reveals normal hearing in the right ear, but a high-frequency hearing loss in the left ear. Which of the following would most likely be more difficult for this man?
(A) Detection of high-frequency sounds
(B) Detection of low-frequency sounds
(C) Localization of high-frequency sounds
(D) Speech perception in a quiet room
(E) Vocalization in a quiet room

8 A 61-year old man presents with symptoms that suggest disease or damage to the eighth cranial nerve. The examination and diagnostic tests, which include CT, reveal damage to the structure indicated at the arrow in the given image. Which of the following deficits would be most likely experienced by this man?

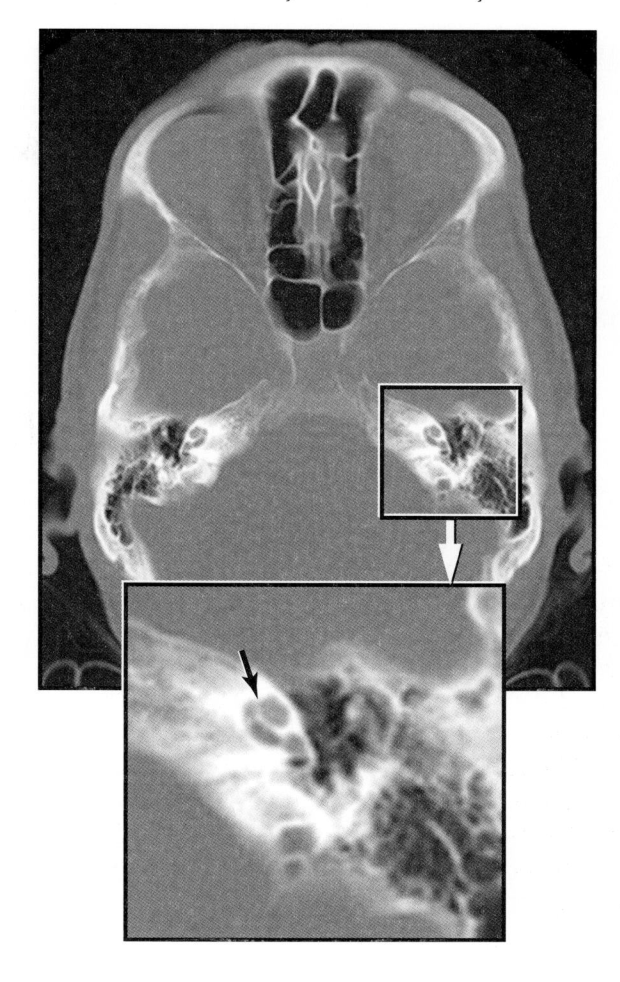

(A) Loss of compensatory eye movement upon head tilt to the left
(B) Loss of compensatory eye movement upon head tilt to the right
(C) Loss of hearing in the left ear
(D) Loss of hearing in the right ear
(E) Loss of low-frequency hearing in the left ear

9 Feedback to outer hair cells that influence the relative sensitivity of the cochlea is mediated by circuits that arise primarily from which of the following?
(A) Anterior cochlear nucleus
(B) Facial nucleus
(C) Nuclei of the lateral lemniscus
(D) Periolivary nuclei
(E) Posterior cochlear nucleus

10 A change in peak III of the auditory brainstem response would most likely suggest interruption of action potentials in which of the following portions of the central auditory pathway?
(A) Auditory cortex
(B) Auditory nerve
(C) Cochlear nuclei
(D) Inferior colliculus
(E) Superior olivary nuclei

11 Vestibular function is tested in a 19-year-old man using the caloric test. Which of the following is most likely to occur when warm water is introduced into this man's right ear?

(A) Horizontal nystagmus, fast phase directed left
(B) Horizontal nystagmus, fast phase directed right
(C) Ocular counterroll to the right
(D) Vertical nystagmus, fast phase directed down
(E) Vertical nystagmus, alternating fast phases

12 Which of the following structures in the image below is the primary source of fibers that form the medial vestibulospinal tract?

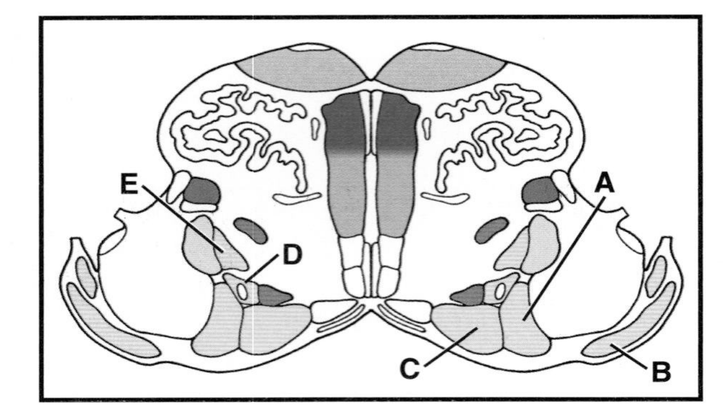

(A) A
(B) B
(C) C
(D) D
(E) E

13 A 39-year-old man presents to the otolaryngology clinic with the complaint of dizziness and a slight "buzzing" in his ears. Caloric irrigation of the right ear produces normal responses, while the same test in the left ear produces no responses. The man also has significant instability when attempting to stand on his left leg; MRI reveals the causative lesion. Which of the following would most likely explain the lesion causing this man's deficits?

(A) Cerebellar glioma on the left
(B) Fourth ventricle medulloblastoma
(C) Lateral medullary syndrome on the left
(D) Vestibular schwannoma on the left
(E) Vestibular schwannoma on the right

14 Hair cells are the receptor cells of the vestibular system and characteristically have a number of stereocilia and one kinocilium on their apical surfaces. Movements of these structures result in depolarization (and excitation) or hyperpolarization (and inhibition). For excitatory signal transduction in hair cells, which of the following events is most likely to occur?

(A) The stereocilia are compressed inward toward the cell surface, K^+ channels in the stereocilia close, Ca^{2+} channels in the basolateral cell membrane close, and excitatory neurotransmitter is released
(B) The stereocilia are bent away from the kinocilium, K^+ channels in the stereocilia close, Ca^{2+} channels in the basolateral cell membrane close, and excitatory neurotransmitter is released
(C) The stereocilia are rigid and do not move, endolymph motion forces K^+ channels in the stereocilia to open, Ca^{2+} channels in the basolateral cell membrane close, and excitatory neurotransmitter is released
(D) The stereocilia are bent toward the kinocilium, K^+ channels in the stereocilia open, Ca^{2+} channels in the basolateral cell membrane open, and excitatory neurotransmitter is released
(E) The stereocilia are bent perpendicular to the kinocilium, K^+ channels in the stereocilia close, Ca^{2+} channels in the basolateral cell membrane open, and excitatory neurotransmitter is released

15 In T2-weighted MRI, blood vessels appear black against a near-white background; the black representing blood flow in the vessel, not the actual blood cells. Which of the following vessels is indicated by the arrow in the image below?

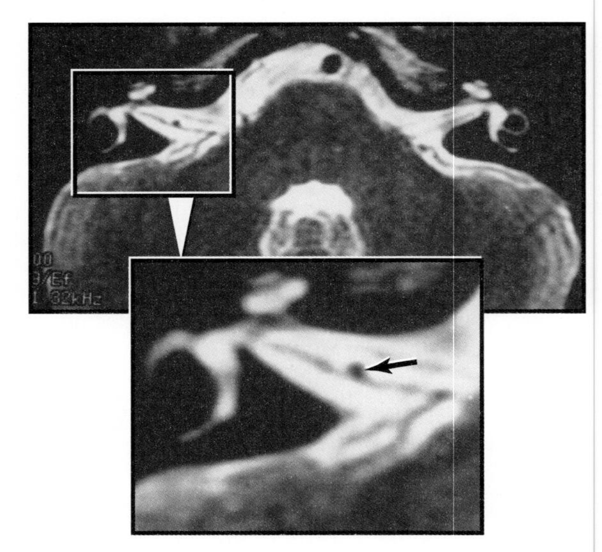

(A) Anterior inferior cerebellar artery
(B) Labyrinthine artery
(C) Long circumferential branch of basilar artery
(D) Posterior inferior cerebellar artery
(E) Posterior spinal artery

16 Inhibition is important in the binaural circuits processed through the superior olivary nuclei that are important for sound localization in auditory space. Which of the following neurotransmitters is used by interneurons that participate in these specific circuits?

(A) Aspartate
(B) Enkephalin
(C) Gamma-aminobutyric acid
(D) Glycine
(E) Taurine

17 Which of the following represents the location of the primary auditory cortex?

(A) Angular gyrus
(B) Long gyrus of the insula
(C) Superior temporal gyrus
(D) Supramarginal gyrus
(E) Transverse temporal gyrus

18 A 5-year-old girl is brought to her pediatrician by her mother. The history reveals that the girl has complained that her "ears hurt" for the last 2 days, and the examination reveals bilateral otitis media. An audiology report indicates a moderate diminution in hearing in both ears. Which of the following is most directly involved in this child?

(A) External auditory canal
(B) Horizontal semicircular canal
(C) Incus and stapes
(D) The cochlea
(E) Utricle

19 A 68-year-old man presents at the Emergency Department with the complaint of sudden sensory loss and difficulty walking. The examination reveals an alternating hemianesthesia and a loss of discriminative touch and vibratory sense on the left lower extremity, but normal strength in all extremities. In addition, the man has some weakness of his masticatory muscles on the right (jaw deviates slightly to the right on closing), but no frank paralysis. MRI shows a lesion, certainly vascular in nature, in the area indicated in the image below. In addition to his more obvious deficits, which of the following deficits may he also have based on the location and extent of the lesion?

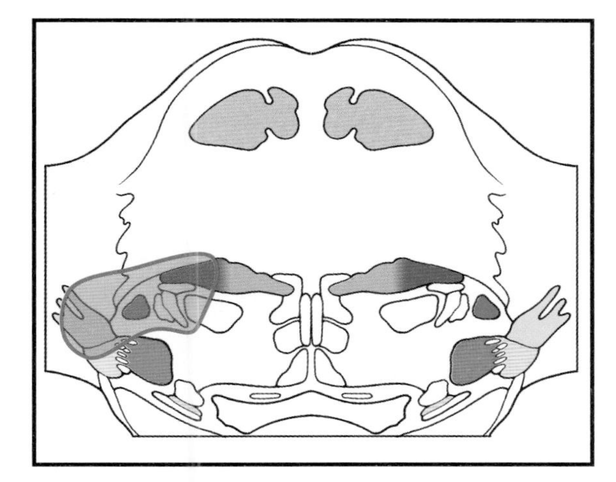

(A) Deafness in his left ear
(B) Deafness in his right ear
(C) Difficulty localizing sounds
(D) Dysequilibrium to the left
(E) Dysequilibrium to the right

20 Damage to medial vestibulospinal fibers originating in the left vestibular nuclei would most likely produce which of the following deficits?

(A) Bilateral postural ataxia in lower extremites
(B) Difficulty moving left upper extremity
(C) Postural instability of the head only
(D) Postural instability of the body on the right
(E) Tremor of left upper and lower extremities

21 Which of the following is indicated by the arrow in the image below?

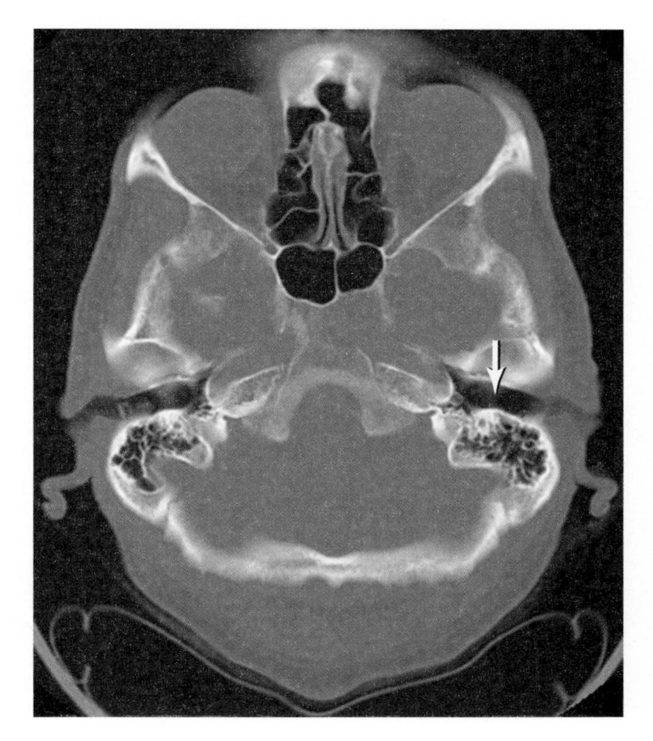

(A) External auditory canal
(B) External ear
(C) Inner ear
(D) Mastoid air cells
(E) Middle ear

ANSWERS

1 **The answer is B: Meniere disease, diuretics, and salt restriction.** These symptoms are characteristic of Meniere disease, which is caused by endolymphatic hydrops in the vestibular and cochlear membranous labyrinth. Attacks may be a few times a year to several times a week, may cycle between obvious symptoms and remissions, and may lead to permanent hearing loss. Standard treatment is with diuretics, taken regularly until symptoms abate, and a reduction of caffeine and salt intake. Vestibular schwannoma (these are sometimes called an acoustic neuroma, which is incorrect on two counts: it is not acoustic and not a neuroma) is a Schwann cell tumor of the vestibular root. This tumor is characterized by hearing loss (about 98% of cases), tinnitus (70%), and significant equilibrium difficulties (65%), and is commonly called dysequilibrium. If these tumors exceed 2.0 cm in size, the root of the trigeminal nerve may be recruited into the area of damage with the expected deficits. Hypertensive cardiovascular disease causes characteristic clinical problems, but not of a focal neurological nature. Psychosomatic disorders may have auditory components, but it is highly unlikely that these would be confirmed by audiogram. Temporal bone fracture could certainly cause auditory and vestibular dysfunction, or losses, but there is no indication of trauma in this patient.

2 **The answer is A: Left anterior inferior cerebellar artery.** The labyrinthine artery is a branch of the anterior inferior cerebellar artery, commonly called AICA in the clinical setting; this vessel enters the internal acoustic meatus and supplies the vestibular and cochlear structures. In 85% to 100% of cases, the labyrinthine artery arises from AICA; in those relatively rare cases when it does not originate from AICA, it is a branch of the basilar. Occlusion of AICA will, in the vast majority of cases, result in a total interruption of blood supply to the inner ear, producing deafness (reflecting cochlear damage) and dysequilibrium (reflecting vestibular damage). In addition, occlusion of AICA will result in cerebellar signs that are usually transient in nature. The long circumferential branches of the basilar artery serve the posterior and lateral regions of the pons; the posterior cerebral artery supplies blood supply to portions of the diencephalon, to inferomedial areas of the temporal lobe, and to medial portions of the occipital lobe. The posterior inferior cerebellar artery (PICA) serves the lateral medulla and medial portions of the inferior cerebellar surface. The posterior spinal artery, a branch of PICA in about 75% of cases, supplies blood to the posterior column nuclei and, after descending through the foramen magum, to the posterior columns.

3 **The answer is C: Temporal bone fractures that damage only the vertical semicircular canals.** Damage to the vertical semicircular canals would significantly compromise the patient's ability to detect and interpret movements of the head forward and backward (flexion and extension). Head movements to the right and left are detected through the horizontal semicircular canals and are intact in this patient. Fracture of the occipital bone involving the cerebellum bilaterally may give rise to cerebellar deficits; assuming they involve only the cortex, the characteristic cerebellar deficits would be transient in nature. Subdural hematoma within the posterior fossa, if large, would result in signs of increased intracranial pressure and possible compression of the brainstem. Fracture of the temporal bone that results in transection of the vestibulocochlear nerve would produce not only vestibular deficits but also total deafness in that ear.

4 **The answer is B: Lateral semicircular canal.** In this T2-weighted MRI, all fluid compartments appear white, or near-white, and the brain appears shades of gray-black. Consequently, the cerebrospinal fluid, fluid within the cochlea, and the fluid within the semicircular canals are of various shades of white. The arrows identify the lateral semicircular canal; a large part of the posterior semicircular canal is seen immediately caudal to the lateral canal. The cochlea, appearing somewhat like a flying saucer, is quite obvious at a position opposite that of the semicircular canals. The superior semicircular canal is not seen in the plane of this image. The spiral ganglion (of the cochlea) is made up of bipolar cell bodies that are located in the spiral canal of the modiolus; their central processes enter the brainstem at the pons-medulla junction as the cochlear part of the vestibulocochlear nerve; their peripheral processes arise from the organ of Corti.

5 **The answer is B: Round window.** The scala vestibule, which is located above the Reissner membrane, and the scala tympani, which is located below the basilar membrane, contain perilymph fluid. The scala tympani ends at the round window; this is an opening between the middle ear and the cochlea that is normally closed by a membrane. Damage to the round window may result in leakage of perilymph into the middle ear; auditory function is compromised from two aspects: fluid leakage out of the cochlea and fluid buildup in the middle ear. The basilar membrane is located between the osseous spiral lamina and the lateral wall of the cochlea, where it attaches at the spiral ligament; this membrane supports the organ of Corti. The tectorial membrane extends from the limbus outwardly over the sensory epithelium of the organ of Corti, where the taller stereocilia of the hair cells contact, or are embedded in, this membrane. The tympanic membrane, commonly called the ear drum, is a structurally complex, but thin, membrane that separates the tympanic cavity (middle ear) from the external acoustic meatus.

6 **The answer is E: Counterroll (torsion) of the eyes to the left.** For tilts of the head to either the right or left, there is an ocular counterroll as a compensatory eye movement; this is not a result of damage, but is a normal compensatory movement. This counterroll consists of a torsional eye movement toward the side of the head opposite the tilt. For example, if the head is tilted to the right, the compensatory counterroll of the eyes is to the left; if the head is tilted to the left, the compensatory eye movement is to the right. A variety of eye movement disorders may occur with lesions at various regions in the nervous system.

7 **The answer is C: Localization of high-frequency sounds.** Localization of sound depends largely on binaural cues; sounds arriving at the ears at slightly different times provide information that is essential to localizing the position of the sound in auditory space. The time differential in the arrival of sounds at the two ears is a factor that contributes to sound localization as it is processed within the brain. A loss of high-frequency perception in one ear will result in difficulty localizing the position of such sounds in space. This man has a high-frequency hearing loss in one ear but can perceive such frequencies in the other ear; he is unable to localize, but is not totally deaf to high-frequencies. Human speech is made up of relatively low frequencies, and also requires only monaural input, if not masked by background noise. Vocalization is not directly affected by auditory deficits, since the apparatus of vocalization receives no input from the eighth cranial nerve. However, patients who feel that they may be speaking too quietly or too loudly, because of auditory pathology, may voluntarily speak more loudly or more quietly in an effort to compensate.

8 **The answer is C: Loss of hearing in the left ear.** This CT clearly shows the coils of the cochlea within the petrous portion of the temporal bone of the left side. The contents of the cochlea are extremely delicate and function within very precise tolerances; damage to the cochlea results in deafness in the ear on the side of the damage, the left in this example. Low-frequency hearing loss results from damage to the apex of the cochlea, while high-frequency tones are encoded at the base of the cochlea. Loss of compensatory eye movements in response to head tilt is controlled by the semicircular canals, not the cochlea. The hardness of the petrous portion of the temporal bone protects the cochlea as well as the semicircular canals and related structures.

9 **The answer is D: Periolivary nuclei.** The vestibulocochlear nerve is classically considered, and rightfully so, a sensory nerve. However, there are efferent fibers that originate from cells of the periolivary nuclei and send their axons bilaterally to terminate on inner and outer hair cells; this is the olivocochlear tract or the bundle of Rasmussen. This projection suppresses, or dampens, the receptivity of the hair cells in the organ of Corti and may serve a protective function under conditions of excessive noise. Specific clusters of the periolivary nuclei give rise to predominately crossed and others to predominately uncrossed olivocochlear fibers; all forming the bundle. The anterior and posterior cochlear nuclei receive auditory input from the bipolar cells of the spiral ganglion and project to various relay nuclei of the overall auditory system. The facial nucleus innervates the muscles of facial expression and the posterior belly of the digastric. The nuclei of the lateral lemniscus are synaptic way-stations in the ascending auditory pathway.

10 **The answer is E: Superior olivary nuclei.** The auditory brain response (ABR) is a neurophysiological measure of electrical activity in the various portions of the central auditory pathway, beginning with the auditory nerve (peak I) and ending in the auditory cortex (peak VII). Peak III measures activity at the level of the superior olivary nuclei, commonly called the superior olive; these cells are found in the anterolateral portion of the pontine tegmentum between the levels of the facial colliculus and the motor trigeminal nucleus. Damage to this part of the auditory pathway may alter peak III of the ABR. Peak II measures activity in the cochlear nuclei; peak V measures responses in the inferior colliculus. The ABR is usually a computer-averaged series of responses to repetitive stimuli.

11 **The answer is B: Horizontal nystagmus, fast phase directed right.** The caloric test is the irrigation of the external auditory canal with either warm water or cold water, and observing the resultant eye movement. Warm water introduced into the external ear canal produces nystagmus with a fast phase directed toward the same side. Cold water introduced into the external ear canal produces nystagmus with a fast phase directed to the opposite side. This relationship between water temperature and direction of the fast phase is easily remembered by the mnemonic COWS: Cold Opposite, Warm Same. If this man had a vestibular disease, the response of the eyes would be noticeably reduced or possibly completely absent. This test is also called the Bárány caloric test. Ocular counterroll is seen in situations of damage to the semicircular canals, and vertical nystagmus may be seen in a variety of diseases of the labyrinth, cerebellum, and/or the brainstem.

12 **The answer is C (Medial vestibular nucleus).** The medial vestibular nucleus contains cells that give rise to axons that descend bilaterally, with an ipsilateral preponderance, to end mainly in spinal laminae VII to IX of cervical levels of the spinal cord. These descending fibers travel via the medial longitudinal fasciculus and exert excitatory and inhibitory influences on flexor and extensor muscles of the neck. For example, if a person falls backward, the labyrinth signals the position of the head in space to the medial vestibular nucleus; descending fibers from this cell group excite neck flexors and inhibit neck extensors in an effort to right the head and to protect it, should the fall continue. Although the inferior (or spinal) vestibular nucleus (choice A) makes a small contribution to the medial vestibulospinal tract, by far the most significant influence is from the medial nucleus. The cochlear nuclei (B) convey auditory input into the brain; the solitary nucleus (D) receives both general and special visceral afferent information; at this level, it is largely special visceral afferent, taste information. The structure at E is the spinal trigeminal nucleus, pars oralis.

13 **The answer is D: Vestibular schwannoma on the left.** This man has no caloric response to irrigation of his left ear (warm or cold) and, basically, could not stand on his left foot when he lifted his right foot. He has a loss of most, if not all, vestibular input on the left side. While the man did not specifically complain of any auditory deficits, testing revealed a significant diminution of hearing in his left ear, particularly to higher frequencies. The most common deficits seen with vestibular schwannoma are hearing deficits, tinnitus, and vertigo (sometimes called dysequilibrium). Vertigo may be objective, if the patient perceives that objects are moving around him/her, or subjective, if the patient perceives that he/she is moving and objects in the environment are not. Patients who complain of dizziness are usually describing vertigo of one type or another. Cerebellar glioma would result in characteristic cerebellar signs on the same side but have no effect on the caloric test. Medulloblastoma is a tumor of children, and, even if seen in an adult, would present with signs and symptoms of increased intracranial pressure (not a complaint in this patient). The lateral medullary syndrome on the left would present with equilibrium problems, but also with several signs indicative of cranial nerve involvement (spinal trigeminal tract, nucleus ambiguus). The schwannoma on the right is not consistent with the deficits.

14 **The answer is D: The stereocilia are bent toward the kinocilium, K^+ channels in the stereocilia open, Ca^{2+} channels in the basolateral cell membrane open, and excitatory neurotransmitter is released.** Stereocilia number 40–80 on any given hair cell, progress from being shortest, further from the kinocilium to being longest close to the kinocilium; the internal core of the stereocilium contains actin filaments. In addition, the apices of the cilia are connected one-to-the-other by delicate filaments so that when they bend, they do so as a unit. The single kinocilium is larger than the tallest stereocilium and its internal structure is characterized by the 9 + 2 pattern of microtubules seen in cilia at other locations in the body; the kinocilium is not very motile. The stereocilia are structurally and functionally polarized: when bent toward the kinocilium, the cell is depolarized and its firing rate increases, while bent away from the kinocilium, the cell is hyperpolarized and the firing rate decreases. At the same time, excitation requires that K^+ channels in the stereocilia open (with resultant K^+ flow into the cell from the surrounding endolymph). This K^+ inflow, and consequent depolarization, causes Ca^{2+} channels in the basolateral regions of the cell to open, Ca^{2+} flow into the cell, and causes synaptic vesicles to empty their neurotransmitter (glutamate or aspartate) into the synaptic cleft.

15 **The answer is B: Labyrinthine artery.** The labyrinthine artery (also called the internal auditory artery) is usually a branch of the anterior inferior cerebellar artery (AICA); 85% in some studies, up to 100% in other studies. The labyrinthine artery, after arising from the AICA, enters the internal acoustic meatus, which is fairly large in this individual, and branches to serve the labyrinth (cochlea + vestibule + semicircular canals). The AICA is seen on both sides of this image, especially on the left, as it passes laterally on the inferior surface of the cerebellar hemisphere. Long circumferential branches of the basilar artery serve posterior and lateral aspects of the pons, but are not visible in this image and, more importantly, do not follow the course of those vessels that are visible in this image. The posterior inferior cerebellar artery, also commonly abbreviated PICA in the clinical environment, is a branch of the basilar artery, serves the lateral medulla and the inferomedial portions of the cerebellum plus the choroid plexus of the fourth ventricle, and usually gives rise to the posterior spinal artery. The PICA is inferior to this axial plane. This latter vessel serves the posterior column nuclei and exits the skull to serve the posterior columns of the spinal cord.

16 **The answer is D: Glycine.** Glycine is a simple amino acid that is particularly prominent as an inhibitory neurotransmitter in the spinal cord and brainstem where it is mainly found in interneurons that form local circuits. This neurotransmitter is found in neurons associated with the superior olive that function to clarify the localization of sound in space. Taurine is an inhibitory neurotransmitter found in cells of the cerebral cortex and cerebellum, but is in very low concentrations in the brainstem and spinal cord. The localization and interactions of this neuroactive substance are not fully understood. Enkephalin, an inhibitory neurotransmitter, is found throughout most areas of the central nervous system, and its role in pain inhibition is well understood. Gamma-aminobutyric acid is probably the most important inhibitory neurotransmitter in the central nervous system, and is widely distributed. Its role in many pathways, particularly those of the basal nuclei, is well understood. Aspartate is an excitatory neurotransmitter.

17 **The answer is E: Transverse temporal gyrus.** The primary auditory cortex (also called AI or Brodmann area 41) is located in the transverse temporal gyrus, also called the gyrus of Heschl. The designation of primary auditory cortex as AI is consistent with the designation of the primary somatosensory cortex as SI and primary somatomotor cortex as MI. The secondary auditory cortex is Brodmann area 42. The transverse temporal gyrus is within the lateral sulcus; spreading the sulcus open reveals the gyrus. The superior temporal gyrus is on the surface immediately below the lateral sulcus, the angular gyrus surrounds the upper and caudal end of the superior temporal sulcus, and the supramarginal gyrus has the same relationship to the lateral sulcus. Collectively, the angular and supramarginal gyri form the inferior parietal lobule. The long gyrus of the insula (there are usually two, long gyri) is the caudal portion of the insular cortex and is separated from the more rostrally located short gyri by the central sulcus of the insula.

18 **The answer is C: Incus and stapes.** Otitis media is an inflammation of the middle ear; the contents of the middle ear are the malleus, incus, and stapes. Pain, due to pressure within the middle ear cavity resultant to the infection, is a very common symptom; young patients may rub or pull at their ears. The malleus attaches to the ear drum and to the incus, the incus attaches to the stapes, and the footpiece of the stapes inserts into the oval window. A decrease, or loss, in hearing may result from otitis media due to a compromise of the flexibility of the tiny joints between these ossicles. In fact, serious or untreated middle ear infections at an early age may result in permanent hearing loss in adulthood. Blockage of the external auditory canal may result in temporary hearing deficits, and injury to the cochlea may also result in hearing loss, but not in the case of otitis media as is the issue in this girl. The utricle and the horizontal semicircular canal are concerned with balance and equilibrium.

19 **The answer is C: Difficulty localizing sounds.** This lesion involves the trigeminal root and the anterolateral system that explains the alternating (or crossed) hemianesthesia: a right-sided loss of pain and thermal sense on the face and a left-sided loss of the same sensations on the left side of the body (excluding the head). The apparent difficulty walking is not related to actual weakness, but is related to the loss of discriminative touch from the left lower extremity; this man has a sensory ataxia. Masticatory weakness may be seen due to the fact that the fibers arising from the motor trigeminal nucleus exit immediately adjacent to the sensory root. In addition, this lesion includes the lateral lemniscus and portions of the superior olivary nuclei on the right. Central lesions of auditory pathways usually do not result in deafness in either ear, but lead to a difficulty localizing sound in auditory space. Deafness in either ear results from a lesion involving the dorsal and ventral cochlear nuclei, the cochlear nerve, or total damage to the cochlea itself. This lesion does not lead to dysequilibrium toward either side; sensory ataxia should not be confused with dysequilibrium.

20 **The answer is C: Postural instability of the head only.** Fibers that comprise the medial vestibulospinal tract arise mainly

from the medial vestibular nucleus (some arise in the inferior vestibular nucleus). They descend bilaterally, with an ipsilateral preponderance, within the medial longitudinal fasciculus to end primarily, and bilaterally, in cervical levels of the spinal cord; sparse numbers of these fibers may extend into upper thoracic levels. These descending fibers influence the activity of neck flexor and extensor motor neurons, as well as those spinal neurons that contribute to the propriospinal system. Lateral vestibulospinal fibers originate from the lateral vestibular nucleus, descend uncrossed, and project to all levels of the spinal cord. Bilateral postural difficulties of the lower extremities, or of the right side of the body, may occur under specific conditions, but they are not mediated by medial vestibulospinal fibers. Difficulty moving an extremity or tumor are characteristic of lesions of specific parts of the motor system; corticospinal for the former, basal nuclei or cerebellum for the latter.

21 **The answer is A: External auditory canal.** This CT clearly shows the full extent of the external auditory canal: it extends from the tympanic membrane, through the tympanic part of the temporal bone, to the point where it opens on the lateral aspect of the skull. As is evident in this image, this canal consists of an inner bony part and an outer cartilaginous part, both of which are lined by skin that continues over the auricle. The canal (sometimes also called the external acoustic meatus) plus the pinna (also called the auricle), collectively constitute the external ear; only the canal is labeled, not the pinna even though it is clearly evident in this image. The middle ear contains the ear ossicles, and the inner ear consists of the cochlea, semicircular canals, and related structures. The air cells of the mastoid bone are seen immediately caudal to the external auditory canal.

Chapter 16

Special Sense III: Olfaction and Taste

Recall the Cautionary Tale: The images in this chapter are presented in a Clinical Orientation: this approach emphasizes how structure and function correlate with deficits in a clinical setting. MRI and CT have a universally recognized orientation and laterality. Line drawings, stained sections, and gross brain images in the Clinical Orientation are viewed here in the identical manner that one views an MRI: your right, as the physician-observer, is the patient's left, and the observer's left is the patient's right. The laterality of deficits and reference to connections follow accordingly.

QUESTIONS

Select the single best answer.

1 While recovering from a motorcycle collision, a 27-year-old man complains that everything he eats "tastes funny." Which of the following specifies the perverted or distorted sense of taste experienced by this man?

(A) Ageusia
(B) Alexia
(C) Anosmia
(D) Dysgeusia
(E) Hyposmia

2 Olfactory receptor cells have a life span of about 30 to 60 days and, consequently, are replaced throughout life. Aging may affect receptor cell replacement as well as the ability to detect odors. Which of the following is the stem cell from which receptor cells arise?

(A) Basal cell
(B) Mitral cell
(C) Periglomerular cell
(D) Sustentacular cell
(E) Tufted cell

3 Taste buds are located in specialized structures called papillae. Which of the following represents the most widely distributed papillae in the oral cavity that have taste buds?

(A) Bowman gland
(B) Filiform
(C) Foliate
(D) Fungiform
(E) Vallate

4 Which of the following structures receives input that originates from taste buds and relays this information to the thalamus?

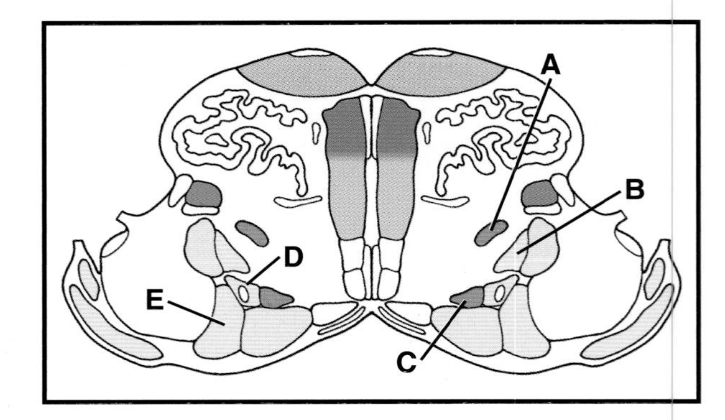

(A) A
(B) B
(C) C
(D) D
(E) E

5 A 25-year-old man is brought to the Emergency Department from the site of a motor vehicle collision. The examination reveals facial injuries, fractured clavicle and humerus, and lacerations. CT shows bilateral temporal lobe contusions that appear to involve the temporal pole and the periamygdaloid, piriform, and entorhinal cortices. Which of the following sensory deficits would correlate with this injury?

(A) Altered sense of smell only
(B) Altered sense of smell and taste
(C) Bilateral inferior quadrantanopia
(D) Right homonymous hemianopia
(E) Loss of balance and equilibrium

6 A 71-year-old woman is transported to the Emergency Department by EMS personnel from her residence. The history and examination reveal that she had a sudden loss of pain and thermal sensations and of discriminative touch and vibratory sense on the left side of her body and face, and some weakness of her left upper and lower extremities, but no frank paralysis. Cranial nerves are normal. In addition to these more obvious deficits, a careful neurological examination would also most likely reveal which of the following?

(A) Deviation of the tongue to the left on protrusion
(B) Deviation of the tongue to the right on protrusion
(C) Loss of most olfactory sense
(D) Loss of taste on the left side of the tongue
(E) Loss of taste on the right side of the tongue

7 Which of the following is the location of the cell bodies for the primary afferent neurons conveying taste from the anterior two-thirds of the tongue?
(A) Geniculate ganglion
(B) Inferior ganglion of IX
(C) Inferior ganglion of X
(D) Otic ganglion
(E) Pterygopalatine ganglion

8 Which of the following functions is associated with the outlined area in the image below?

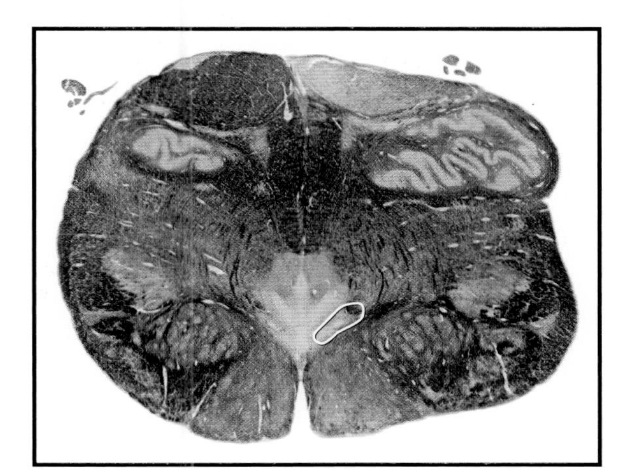

(A) Incoming general somatic sensation from the oral cavity
(B) Incoming general visceral sensation via cranial nerves VII, IX, and X
(C) Incoming proprioceptive information from the masticatory muscles
(D) Incoming taste information conveyed on cranial nerves VII, IX, and X
(E) Preganglionic parasympathetic input to terminal ganglia

9 Which of the following neurotransmitters is associated with the synapse between the olfactory receptor neurons and the neurons, such as tufted cells or mitral cells, which form the olfactory tract?
(A) Acetylcholine
(B) Dopamine
(C) GABA
(D) Glutamate
(E) Histamine

10 A 21-year-old man is transported to the Emergency Department (ED) from the site of a motorcycle collision. The EMS personnel explain to the ED physician that the man was wearing an open-face helmet, which explains some of his injuries. The examination reveals fractures of the mandible and maxilla (both bilateral), broken teeth, fractured femur, and probable nerve damage. CT and X-ray confirm these observations. During his recovery, the man complains that he has no sense of taste, but no other difficulties. Damage to which of the following would most likely explain this man's deficit?
(A) Alveolar nerves
(B) Facial nerves
(C) Lingual nerves
(D) Maxillary nerves
(E) Mylohyoid nerves

11 The parents of a 4-year-old girl, who are recent immigrants from Poland, bring their daughter to a pediatrician for an initial visit. The history reveals that the family is of the Jewish faith and that the girl was small at birth and had many medical problems during her first 4 years. The examination reveals a small child with diminished muscle stretch reflexes, no corneal reflexes, tachycardia, and of small stature for her age. The mother notes that the girl does not eat well and does not seem to like food. Suspecting a genetic cause, the physician orders an appropriate test revealing a defect in chromosome 9 (specifically 9q31 to q33). In this syndrome, there is a lack of important structures related to taste. Which of the following may be absent in this girl?
(A) Absence of all taste buds
(B) Absence of the geniculate ganglion
(C) Absence of filiform papillae
(D) Absence of foliate papillae
(E) Absence of fungiform papillae

12 Which of the following structures is identified by the arrow in the image below?

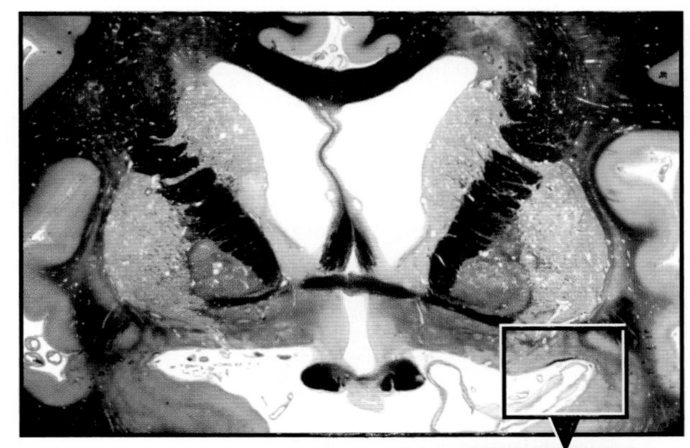

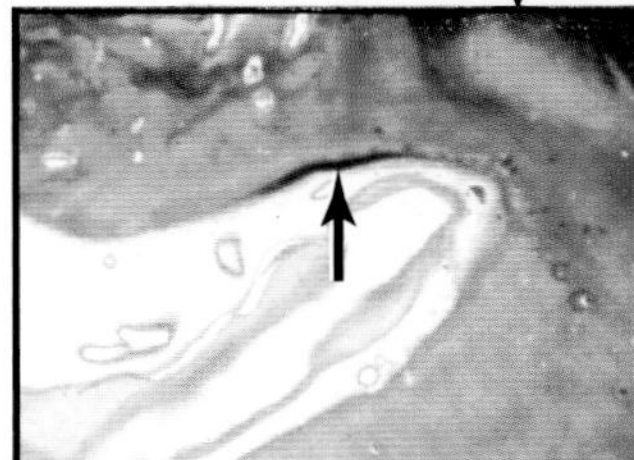

(A) Amygdalofugal pathway
(B) Lateral olfactory stria
(C) Medial olfactory stria
(D) Olfactory tract
(E) Stria terminalis

13 A 27-year-old man is transported to the Emergency Department from the site of a motor vehicle collision. Information provided by the EMS personnel, and as indicated by the man's injuries, reveals that his face/head hit either the steering wheel or the windshield or both. It is likely that this man has sustained damage to the olfactory nerve as a consequence to this collision. As this man recovers, which of the following is most likely his primary complaint?

(A) Dysosmia
(B) Hyposmia
(C) Loss of smell only
(D) Loss of taste and smell
(E) Olfactory hallucination

14 Olfactory receptor neurons have cilia which are essential to the receipt and transduction of odorant cues. Which of the following steps summarize the events that lead to the production of an action potential in these receptor cells?

(A) Odorant binds to receptor on mitral cell, activates second messenger pathway, which activates adenyl cyclase to block cAMP, anion channels are opened and K^- flows into the cell, generator potential then action potential
(B) Odorant binds to receptor on mitral cell, inactivates second messenger pathway, which inactivates adenyl cyclase to block cAMP, anion channels are opened and K^- flows into the cell, generator potential then action potential
(C) Odorant binds to G protein–coupled receptor, inactivates second messenger pathway, which inactivates adenyl cyclase to produce cAMP, cation channels are opened and Ca^{2+} flows out of the cell, generator potential then becomes an action potential
(D) Odorant binds to G protein–coupled receptor, activates second messenger pathway, which activates adenyl cyclase to produce cAMP, cation channels are opened and Ca^{2+} flows into the cell, generator potential then becomes an action potential
(E) Odorant binds to receptors on mitral and tufted cells, activates second messenger pathway, which activates adenyl cyclase to produce cAMP, cation channels are opened and Ca^{2+} flows into the cell, generator potential then action potential

15 Which of the following structures is indicted at the arrow in the image below?

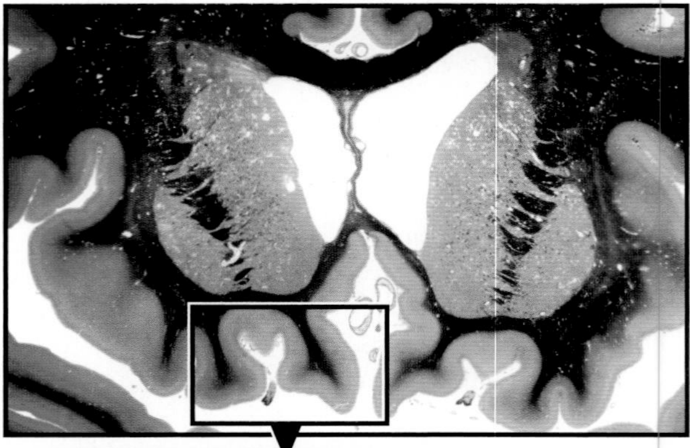

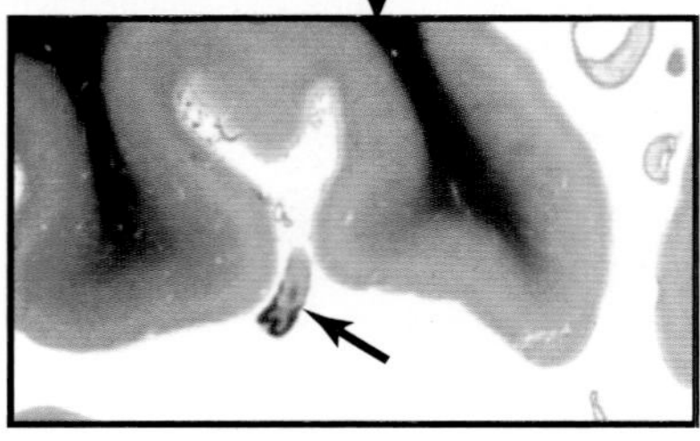

(A) Lateral olfactory stria
(B) Medial olfactory stria
(C) Olfactory bulb
(D) Olfactory tract
(E) Olfactory trigone

ANSWERS

1 **The answer is D: Dysgeusia.** Dysgeusia (also called parageusia) is an unpleasant taste perception when the substance in the mouth normally has a pleasant taste. There may be an unpleasant taste perception when no substance is in the mouth; this may be called a gustatory hallucination. Ageusia is the loss of taste; this may be to all substances or to only certain tastants, and may be due to difficulties at the taste bud or in peripheral or central pathways. A complete absence of taste may be a developmental or inherited defect. Anosmia is the complete absence of the ability to detect odors; this may be general (all odors), selective to only certain odors, and caused by damage to the olfactory apparatus or to central olfactory pathways. Hyposmia (also called olfactory hypesthesia or microsmia) is a decreased sense of smell and, like anosmia, it may be general or selective to only certain odors. Anosmia or hyposmia may result from developmental defects, may be acquired, or may result from trauma. Alexia is the inability to comprehend the meaning of words and is most frequently the result of lesions in the forebrain.

2 **The answer is A: Basal cell.** Basal cells are located at the inner aspect of the olfactory epithelium where they sit on a basal lamina. These basal cells undergo mitotic division and are, consequently, the stem cells for replacement of olfactory receptor cells. The other choices are cells within the olfactory epithelium. Mitral and tufted cells are found within the mitral cell layer and the external plexiform layer, respectively, gather input from the olfactory glomerulus, and give rise to axons that enter the olfactory tract. The periglomerular cells are comparatively small interneurons located within the glomerular layer and interconnecting the olfactory glomeruli. Sustentacular cells are the supporting cells of the olfactory epithelium.

3 **The answer is D: Fungiform papillae.** The surface of the tongue is generally divided into an oral (also called the presulcal) portion and a pharyngeal (also called the postsulcal) portion. The dividing line between these two portions of the tongue is the terminal sulcus, a V-shaped line beginning just in front of the foramen caecum and running in an oblique direction toward the lateral aspect of the tongue. Fungiform and foliate papillae are numerous on the presulcal portion of the tongue, but only fungiform papillae have taste buds. Foliate papillae are small, have a keratinized epithelium that participates in food processing, and do not have taste buds. Vallate (also called circumvallate) papillae are large (1 to 2 mm in diameter), are located immediately rostral to and parallel with the terminal sulcus, and form a row of five to seven elevations on either side of the tongue. The bowman gland is located in the olfactory epithelium.

4 **The answer is D (Rostral portions of the solitary nucleus).** The plane of this section is right at the medulla-pons junction; note the lack of a hypoglossal nucleus, large restiform body, position of the cochlear nuclei, and overall rectangular shape. All of these, particularly the presence of the cochlear nuclei, indicate that this level is through the rostral medulla, immediately caudal to the pons. This level is through the rostral portion of the solitary tract and nucleus; this nucleus is called the gustatory nucleus. At this level, the solitary nucleus is inferior to the vestibular nuclei, lateral to the inferior salivatory nucleus (choice C), and superior, and slightly medial, to the spinal trigeminal nucleus, pars oralis (B). Recall that cranial nerves VII, IX, and X convey special visceral afferent (SVA) taste sensation into the rostral portions of the solitary nucleus and general visceral afferent (GVA) sensations into the caudal portions of the solitary nucleus; this caudal part is called the cardiorespiratory nucleus. Cells of the gustatory nucleus project to the ventral posteromedial nucleus on the same side. Choice A is the nucleus ambiguus, a motor nucleus; E is the spinal, or inferior, vestibular nucleus.

5 **The answer is B: Altered sense of smell and taste.** Olfactory fibers project to the amygdaloid complex and to the adjacent periamygdaloid, piriform, and entorhinal cortices. Lesions in these areas of the temporal lobe may result in an altered or perverted perception of odors (parosmia). However, because smell and taste function in concert (and have the same functional component, special visceral afferent), this patient would also have an altered sense of taste. Recall how a bad cold, with its consequent diminution of a sense of smell, also radically affects the patient's sense of taste. Damage to the anterior and medial temporal lobes may also result in other deficits that are characteristic of damage to the amygdaloid complex or hippocampal formation. Injury to the temporal lobe may cause a bilateral quadrantanopia, due to damage to the Meyer loop, but this would be a bilateral superior quadrantanopia. A right homonymous hemianopia would result from a left-sided lesion caudal to the optic chiasm. The functions of balance and equilibrium are not centered in the temporal lobe.

6 **The answer is E: Loss of taste on the right side of the tongue.** Deficits on the same side of the body with no cranial nerve involvement indicate two things. First, the damage is in the forebrain; the pattern of sensory loss in this patient indicates involvement of the ventral posterolateral (VPL) and ventral posteromedial (VPM) thalamic nuclei. Second, left-sided sensory (and motor) deficits indicate a right-sided lesion; this lesion involves the caudal portions of the thalamus and impinges somewhat on the adjacent posterior limb of the internal capsule. Taste input to the solitary nucleus is ipsilateral, and the relay of this taste information to the VPM of the thalamus is also predominately ipsilateral. Taste input from the right side of the tongue goes through the right VPM and to the right sensory cortex; taste information from the left side of the tongue goes through the left VPM and to the left sensory cortex. A right-sided lesion, as in this case, produces left-sided sensory symptoms and a loss of taste on the right side of the tongue. Olfaction does not traverse any part of the thalamus en route to the cortex. A capsular lesion that results in tongue deviation would include the genu of the internal capsule; there are no indicators (tongue or other wise) of genu involvement.

7 **The answer is A: Geniculate ganglion.** The geniculate ganglion contains the cell bodies for all afferent fibers that travel on the facial nerve, both the somatic afferent and the visercal afferent; the latter includes Special Visceral Afferent (SVA; taste) fibers. These primary sensory fibers end in the rostral portion of the solitary nucleus (also called the gustatory nucleus), and from here are relayed to the ventral posteromedial nucleus of the same side. The inferior ganglia of IX and X receive SVA fibers from peripheral branches of the glossopharyngeal and vagus nerves, respectively, that arise from taste buds. These fibers convey input arising from taste buds on the posterior one-third and from the root of the tongue and epiglottis on IX and X. When these primary afferent fibers from IX and X that are conveying taste enter the brainstem, they also end in the gustatory nucleus. The otic ganglion contains General Visceral Efferent (GVE) cell bodies the fibers of which serve the parotid gland; the pterygopalatine ganglion contains GVE cell bodies that innervate the lacrimal gland.

8 **The answer is B: Incoming general visceral sensation via cranial nerves VII, IX, and X.** The features of this plane of section that characterize it as being through the caudal medulla are (1) it is caudal to the obex; (2) presence of the posterior column nuclei; (3) presence of the internal arcuate fibers; (4) very small restiform body; and (5) a comparatively small olivary nucleus. The outlined area encompasses the solitary tract and the adjacent solitary nucleus. Because this is a more caudal level of the medulla, the incoming information to the solitary nucleus at this level consists of general visceral afferent information from structures such as salivary glands and thoracic and abdominal viscera. General somatic afferent information from the oral cavity enters the brainstem through the trigeminal nerve and, at this level, is found with the tract of the Vth nerve and the spinal trigeminal nucleus, pars caudalis. Proprioceptive information from the masticatory muscles relates to the principal sensory and mesencephalic trigeminal nuclei. Taste information conveyed on cranial nerves VII, IX, and X enters the rostral portion of the solitary tract and nucleus; this part is commonly called the gustatory nucleus. Preganglionic parasympathetic input to terminal ganglion originates from the dorsal motor nucleus of the vagus that is immediately medial to the solitary tract and nucleus.

9 **The answer is D: Glutamate.** The synaptic contact between olfactory receptor cells and tufted cells, mitral cells, and periglomerular cells is excitatory. The neurotransmitter at these synaptic contacts is glutamate; these are glutaminergic contacts. Acetylcholine is an excitatory neurotramsmitter that is widely distributed in the nervous system (alpha and gamma motor neurons, preganglionic visceromotor neurons, reticular formation, basal forebrain, etc.) but is not found at this particular site. Dopamine-containing cells are found in the mesencephalon and in the forebrain; perhaps the best known are the dopaminergic nigrostriatal projections. This particular pathway may exert either an excitatory or inhibitory influence depending on the postsynaptic receptor. Glutamate is an excitatory neurotransmitter that is found at numerous sites in the central nervous system, most notably the cerebral cortex, the hippocampus and related cortex, some cerebellar nuclear efferents, and others. Histamine-containing cells have a limited distribution, being almost exclusively restricted to the hypothalamus, but distribute widely to the forebrain, cerebellum, and brainstem.

10 **The answer is C: Lingual nerves.** The fibers arising from taste buds in the anterior two-thirds of the tongue enter the lingual nerve, a branch of the mandibular nerve, then join the facial nerve through the chorda tympani, a branch of the facial nerve. Fractures of the mandible may injure the lingual nerve as it lies against the medial mandibular surface. These taste fibers enter the brainstem as components of the facial nerve; they are not considered parts of the trigeminal nerve. It is true that damage to the facial nerve proximal to its junction with the chorda tympani will result in loss of taste. However, such an injury would result in other significant symptoms such as paralysis of the muscles of facial expression. This is not the case in this man. The alveolar nerves (superior and inferior) convey pain and thermal sensations from the maxillary and mandibular teeth. The maxillary nerves, one of the three large branches from the trigeminal ganglion, provide significant innervation to the maxillary region of the face, both superficial and deep, but they do not convey taste fibers. The mylohyoid nerve supplies the muscle of the same name and the anterior belly of the digastric muscle.

11 **The answer is E: Absence of fungiform papillae.** This child has Riley-Day syndrome, also commonly known as familial dysautonomia. This is a relatively rare inherited disease that is almost unknown outside of the Eastern European Ashkenazi Jewish community. Other siblings within the same family may not have this disease. The diagnosis is based on the characteristic and variety of deficits and the results of genetic testing. One of the usual features of this disease is the absence of fungiform papillae and their related taste buds. Some vallate papillae may be affected, but the foliate papillae appear to be not affected. Consequently, not all taste buds (or all taste) are absent, only that transduced by the missing taste buds. The geniculate ganglion is unaffected; the filiform papillae do not have taste buds.

12 **The answer is B: Lateral olfactory stria.** The olfactory tract divides into a lateral olfactory stria and a medial olfactory stria immediately anterior to the anterior perforated substance; this triangular shaped area is sometimes called the olfactory trigone. Fibers of the lateral stria proceed laterally on the surface of the brain to eventually enter and terminate in the piriform cortex and in the corticomedial nuclei of the amygdala. Many of the axons comprising the lateral stria originate from mitral cells of the olfactory bulb. Fibers of the medial stria pass medially, cross in the anterior commissure, and pass rostrally in the contralateral olfactory tract to terminate in the anterior olfactory nucleus. Recall that olfactory sense is the only sensation that reaches the cerebral cortex without synapse in any portion of the thalamus; as a perceived sensation it is unique in this characteristic. The amygdalofugal pathway is a diffuse population of fibers that arise in the amygdala, course medially through the area containing the nucleus basalis of Meynert and the diagonal band of Broca, to terminate in the hypothalamus. It is not located on the surface of the basal forebrain. The stria terminalis arises from the amygdala, courses between the caudate nucleus and dorsal thalamus, and distributes to the hypothalamus and nuclei of the basal forebrain.

13 **The answer is D: Loss of taste and smell.** This injury is a type of acceleration-deceleration in which the olfactory nerves may be sheared off as they pass through the cribriform plate. Acceleration, as the head suddenly moves forward at the time of vehicular impact and the brain follows milliseconds later, and deceleration, as the head suddenly stops and the

brain continues to move forward for a few milliseconds. The olfactory nerves can be damaged during both the acceleration and the deceleration. In this type of injury, it is intuitive to think that loss of olfactory sense would be the major, or only, complaint. However, the most common complaint of these patients is a loss, or significant decrease, in the sense of smell and of taste; the sense of taste depends, to a significant degree, on substances that volatize from food when placed into the mouth or during chewing. When smell is compromised, taste is also compromised; both deficits are obvious to the patient. Recall that the functional component for both olfaction and taste is Special Visceral Afferent (SVA). Dysosmia (distorted sense of smell), hyposmia (smell may be normal, but diminished), and olfactory hallucination (perceiving an odor, either good or bad, when none is present) are not the main complaints by these individuals.

14 **The answer is D: Odorant binds to G protein–coupled receptor, activates second messenger pathway, which activates adenyl cyclase to produce cAMP, cation channels are opened and Ca^{2+} flows into the cell, generator potential then becomes an action potential.** This is the sequence of events that generally forms the chemical/molecular pathway from an odorant arriving at the mucus layer lining the olfactory epithelium and the ultimate production of an action potential in the receptor neuron. The receptor neuron, in turn, has synaptic contacts with mitral and tufted cells; axons of these cells form the olfactory tract. As is the case for taste, odorant molecules must be somewhat soluble to be effective at the binding site. Potassium is not the ion involved in the development of an action potential, and inactivation of either the second messenger of adenyl cyclase pathways would lead to a decrease in the probability of development of an action potential. It is likely that there is a receptor map within the olfactory epithelium, portions of which are preferentially related to certain odorants.

15 **The answer is D: Olfactory tract.** The olfactory tract runs from the olfactory bulb caudally to the location of the olfactory trigone. En route, it is found in the olfactory sulcus, which is between the medially located gryus rectus (straight gyrus) and the laterally located orbital gyri. The olfactory tracts convey all olfactory information from the receptor cells in the olfactory bulb to centers in the forebrain. Recall that olfaction, as a sensation, does not have a synaptic station in the thalamus; it travels directly to telencephalic targets. The olfactory tract proceeds caudally to the point where it attaches to the inferior aspect of the forebrain and forms the olfactory trigone. At the trigone, some fibers of the tract course laterally, as the lateral olfactory stria to targets in the temporal lobe, and other fibers course medially to enter the anterior commissure, where they cross the midline. The olfactory bulb is the oval enlargement at the rostral end of the olfactory tract.

Chapter 17

General Visceral Sensory and Motor Pathways

Recall the Cautionary Tale: The images in this chapter are presented in a Clinical Orientation*: this approach emphasizes how structure and function correlate with deficits in a clinical setting. MRI and CT have a universally recognized orientation and laterality. Line drawings, stained sections, and gross brain images in the* Clinical Orientation *are viewed here in the identical manner that one views an MRI: your right, as the physician-observer, is the patient's left, and the observer's left is the patient's right. The laterality of deficits and reference to connections follow accordingly.*

QUESTIONS

Select the single best answer.

1 Which of the following receives input from the cell group indicated by the arrow at the outlined area in the image below?

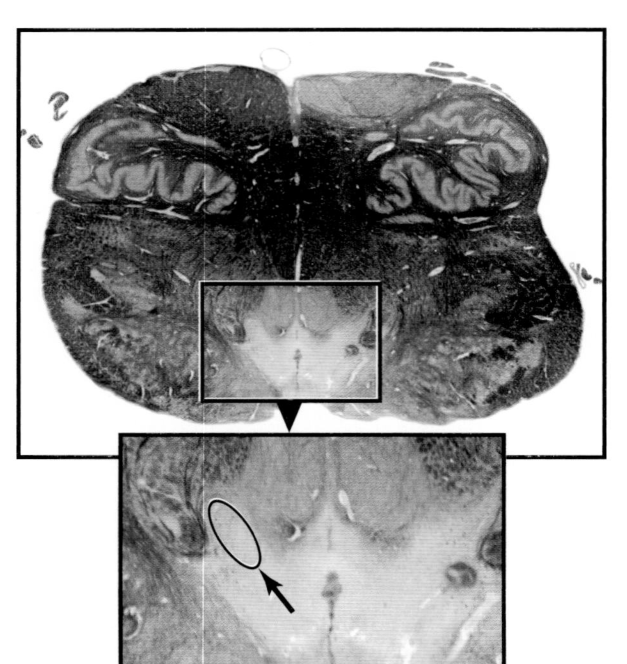

(A) Otic ganglion
(B) Pharyngeal musculature
(C) Pterygopalatine ganglion
(D) Smooth muscle
(E) Terminal ganglion

2 An 11-year-old boy is brought to the pediatrician by his mother with the complaint of "stomach" pain. The examination reveals a mild fever and tenderness in the lower right region of the abdomen when pressure is exerted in the immediate area of the McBurney point; these observations suggest appendicitis. The pain perception in this patient is most likely transmitted through which of the following?
(A) Greater splanchnic nerve
(B) Glossopharyngeal nerve
(C) Lesser splanchnic nerve
(D) Lumbar splanchnic nerve
(E) Vagus nerve

3 Which of the following represents the most likely location of the receptors that monitor internal body temperature and serum osmolarity?
(A) Anterolateral medulla
(B) Carotid body
(C) Hypothalamus
(D) Nucleus ambiguus
(E) Solitary nucleus

4 A 57-year-old man presents with the complaint of "dizziness" and being light-headed. The examination reveals that the man is overweight, does not exercise, and that his symptoms are most obvious when he arises from a supine to a standing position (as when getting out of bed). Laboratory tests reveal that he is also diabetic. Which of the following is an essential structure involved in the pathway mediating this man's primary complaint?
(A) Accessory nucleus
(B) Anterolateral medulla
(C) Inferior salivatory nucleus
(D) Solitary nucleus
(E) Spinal trigeminal nucleus

5 A 29-year-old woman suffered a severe spinal cord injury (not a transection) at C7 as a result of a skiing accident. After several weeks, she begins to experience autonomic dysreflexia: that is, noxious stimuli to the trunk and lower extremities evoke piloerection, sweating, and hypertension. Which of the following represents the most likely cause of these responses?

(A) Diminished parasympathetic reflexes
(B) Diminished sympathetic reflexes
(C) Exaggerated nociceptive reflexes
(D) Exaggerated parasympathetic reflexes
(E) Exaggerated sympathetic reflexes

6 Which of the following responses would be seen after activation of the cell group indicated by the arrow at the outlined area in the image below?

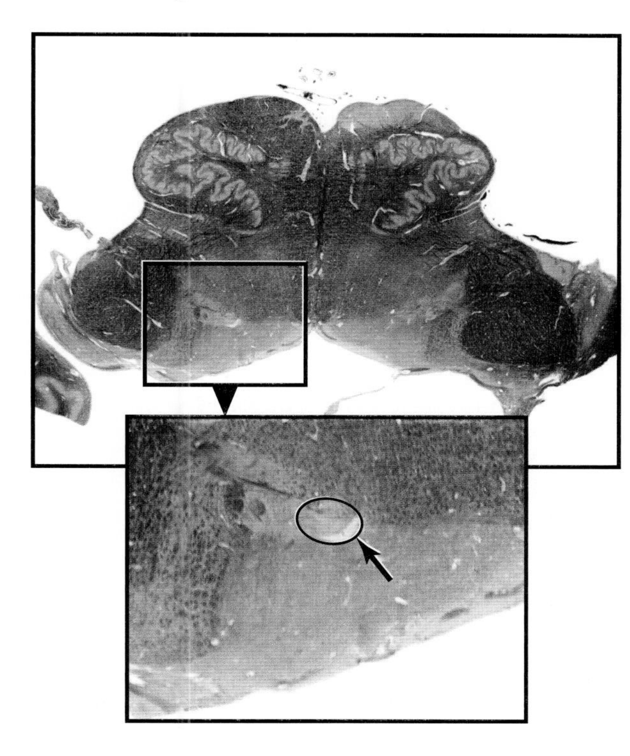

(A) Decreased secretion of lacrimal gland
(B) Decreased secretion of parotid gland
(C) Increased secretion of lacrimal gland
(D) Increased secretion of parotid gland
(E) Increased secretion of submandibular gland

7 A 67-year-old man undergoes a carotid endarterectomy; he had an 87% blockage at the bifurcation of the right common carotid into the internal and external carotid arteries. Following this surgery, the man experiences difficulty speaking, but the quality of his voice is normal. These symptoms suggest that an unintentional consequence of the surgery was injury to which of the following?

(A) Anterior roots of C5 and C6
(B) Facial nerve
(C) Glossopharyngeal nerve
(D) Hypoglossal nerve
(E) Vagus nerve

8 Which of the following is the location of the cell bodies whose axons provide direct inhibitory innervation to the detrusor muscle of the urinary bladder?

(A) Anterior horn at S2-4
(B) Dorsal motor vagal nucleus
(C) Inferior mesenteric ganglion
(D) Intermediolateral cell column at L1-2
(E) Superior mesenteric ganglion

9 A 43-year-old man visits his family physician with the complaint of pain. The history reveals that the man has significant job pressures in his position as an executive in a large company. The examination reveals no external evidence of trauma or disease, but localizes the man's pain to the epigastric region. The examining physician suspects that the man is suffering referred pain from an internal organ. Further tests confirm this fact. Which of the following is most likely the primary source of this man's symptoms?

(A) Diaphragm
(B) Esophagus
(C) Heart
(D) Kidney
(E) Stomach

Questions 10 and 11 are based on the following patient.

A 7-week-old boy child is brought to his pediatrician. The mother reports that he does not have regular bowel movements, seems unusually fussy at this age (she has two other children for comparison), and appears to have an enlarged abdomen. Diagnostic tests reveal a small constricted rectum and sigmoid colon and an enlarged descending and transverse colon.

10 This child is most likely suffering which of the following?

(A) Addison disease
(B) Hanson disease
(C) Hirschsprung disease
(D) Huntington disease
(E) Wilson disease

11 Which of the following most likely explains the morphology of the colon in this patient?

(A) The constricted segment has excess circular muscle layers
(B) The constricted segment has a lack of the myenteric ganglion cells
(C) The constricted segment has excess myenteric ganglion cells
(D) The enlarged segment has no circular muscle layers and is dilated
(E) The enlarged segment has a lack of myenteric ganglion cells

12 Which of the following is the most likely target of the cell bodies comprising the structure indicated by the arrows in the given image?

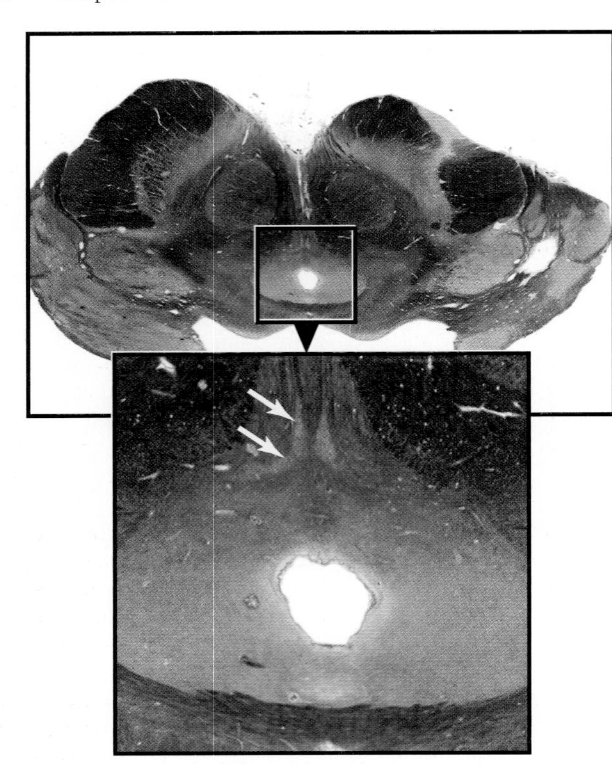

(A) Ciliary ganglion
(B) Dilator pupillae muscle
(C) Levator palpebrae muscle
(D) Sphenopalatine ganglion
(E) Sphincter pupillae muscle

13 A 21-year-old medical student experiences stage fright when she begins an oral presentation to her class and a few faculty who are also present. Which of the following is most likely to be a physiological manifestation of her anxiety?

(A) Decreased blood flow to the skin
(B) Decreased heart rate
(C) Decreased sweating
(D) Excessive salivation
(E) Increased blood flow to the gut

14 The long-term effects of vagotomy on the motility of the patient's intestinal tract are relatively mild. Which of the following would most likely explain this finding?

(A) The central and peripheral pathways of the sympathetic nervous system are intact and provide the major excitatory input that promotes peristalsis
(B) The central descending rubrospinal, reticulospinal, and vestibulospinal tracts within the nervous system can subserve this function in some cases
(C) The intrinsic neurons within the wall of the gut can function autonomously to generate waves of peristalsis in response to the presence of chyme in the gut
(D) The motility of the intestines is chiefly a function of striated muscles with their alpha motor neurons being located in the anterior horn of the spinal cord
(E) The parasympathetic innervation of the small intestines is conveyed by the pelvic nerves which originate from the sacral spinal cord

15 A 39-year-old woman is experiencing signs and symptoms of autonomic dysfunction. To determine whether or not this patient has an intact sympathetic innervation, a small skin sample is surgically removed and prepared for immunostaining. Antibodies to which of the following would represent the best choice for this evaluation?

(A) Aspartate
(B) Dopamine
(C) Epinephrine
(D) Neuropeptide Y
(E) Neurotensin

ANSWERS

1 **The answer is E: Terminal ganglion.** The outlined area is the dorsal motor nucleus of the vagus; these are parasympathetic preganglionic cell bodies whose axons travel, via the vagus nerve, to terminal ganglia located in viscera in the thorax and abdomen to the level of the splenic flexure. Terminal ganglia are sometimes called intramural ganglia because most are located on, or in, the wall of the visceral structure. Postganglionic fibers arising in terminal ganglia project to smooth muscle and glandular epithelium in the organ innervated. The otic ganglion receives parasympathetic preganglionic fibers from the inferior salivatory nucleus and projects to the parotid gland; the pterygopalatine ganglion receives the same type of fibers from the superior salivatory nucleus and projects to palatine and lacrimal glands. Pharyngeal musculature is innervated by motor neurons in the nucleus ambiguus; smooth muscles are innervated by postganglionic neurons located close to, on, or in the wall of organs.

2 **The answer is C: Lesser splanchnic nerve.** The sensory fibers associated with the splanchnic nerves convey mainly nociceptive information (pain). The distribution of the splanchnic nerves is generally as follows: greater splanchnic to the lower esophagus, stomach, liver, pancreas, and duodenum; the lesser splanchnic to the ileum, jejunum, appendix and ileocolic junction, ascending colon, and proximal half of the transverse colon; the lumbar splanchnic to the distal half of the transverse colon, descending and sigmoid colon, and rectum. Consequently, pain fibers conveying visceral pain from the appendix travel via the lesser splanchnic nerve. Visceral pain from other portions of the intestinal tract is conveyed on the appropriate splanchnic nerves. The glossopharyngeal nerve does not serve any of the viscera below the neck. The sensory fibers associated with the vagus nerve serve primarily physiological functions and not pain. In general, visceral pain is poorly localized and may be referred to superficial body areas; in some cases, this referred pain may signal the visceral structure involved (heart as one example).

3 **The answer is C: Hypothalamus.** The hypothalamus is the master visceral center of the brain. It contains neurons that directly monitor temperature and electrolyte concentrations in circulating blood. If there are changes from normal in these parameters, the modified signals are used by the hypothalamus to maintain homeostasis by alternating signals to both endocrine and visceromotor (autonomic) centers to reestablish normal levels. The anterolateral medulla receives input from the solitary nucleus and uses this input in the vasopressor response (increased heart rate and elevated blood pressure). The visceral input to the solitary nucleus, particularly its more caudal portions, is via the glossopharyngeal (carotid body) and vagus (visceral sensory input for thoracic and abdominal viscera) nerves. The carotid body is a specialized receptor located at the bifurcation of the common carotid artery that responds to decreased oxygen levels, increased carbon dioxide levels, and/or excessive concentrations of hydrogen ions. Changes in these levels will result in physiological responses to correct the imbalance. The nucleus ambiguus is motor to laryngeal and pharyngeal muscles.

4 **The answer is D: Solitary nucleus.** Orthostatic hypotension, an acute decrease in blood pressure when suddenly assuming an upright position (from which this man is suffering), indicates an interruption of the baroreceptor reflex. The glossopharyngeal and vagus nerves project to the solitary nucleus, specifically its cardiorespiratory (more caudal) part; the afferent limbs of this reflex are conveyed centrally via these two nerves. In turn, this part of the solitary nucleus influences two pathways which function to either increase or decrease heart rate and blood pressure. First, increases in blood pressure activate baroreceptors which feed into the solitary nucleus and, through vagal pathways, decrease heart rate and blood pressure. Second, decreases in blood pressure activate baroreceptors which feed into the solitary nucleus which, in turn, influences the anterolateral medulla and, ultimately, the intermediolateral cell column; the result is an increase in heart rate and blood pressure. The interplay between these pathways maintains blood pressure (in the absence of some overriding pathology) within normal physiological ranges. The anterolateral medulla is not the correct choice because it is specifically a part of the pathway that produces increased blood pressure (hypertension); this man is suffering hypotension. The accessory nucleus is motor to the trapezius and sternocleidomastoid muscles; the inferior salivatory nucleus contains preganglionic neurons that serve the otic ganglion. The spinal trigeminal nucleus receives somatosensory input from the face and auricle.

5 **The answer is E: Exaggerated sympathetic reflexes.** These exaggerated sympathetic responses are the result of a disconnect of spinal centers from brainstem areas (such as the solitary nucleus or reticular formation) and the hypothalamus that influence these centers. Severe spinal cord injuries at cervical levels disrupt the descending pathways that regulate (up or down) sympathetic preganglionic neurons, resulting in the loss of suprasegmental control of the corresponding visceral responses. As a consequence of this loss of significant descending control, and after a period of spinal shock, segmental sympathetic responses will be exaggerated. Visceromotor reflexes are not diminished, and nociceptive reflexes are not significantly altered. Spinal shock is the highly variable period of time beginning immediately after the injury, during which all function (somatic and visceral) distal to the level of the lesion is absent. An interesting tidbit is that in the clinical environment, autonomic dysreflexia may be seen with spinal cord injury above the T6 level.

6 **The answer is D: Increased secretion of parotid gland.** The outlined area encompasses the inferior salivatory nucleus; it is immediately inferior to the medial vestibular nucleus and medial to the solitary nucleus (this being, at this level, the gustatory nucleus). The inferior salivatory nucleus contains parasympathetic preganglionic neurons of the glossopharyngeal nerve whose axons travel, via the tympanic nerve (a branch of the glossopharyngeal nerve) and the lesser petrosal nerve (a branch from the tympanic plexus), to end in the otic ganglion. Postganglionic fibers arise from the otic ganglion to innervate smooth muscle and glandular epithelium of the parotid salivary gland. Decreased secretion of any salivary gland (choices A and B) would be effected by sympathetic innervation, not parasympathethic. Increased secretions of the lacrimal and submandibular glands are mediated by preganglionic parasympathetic fibers that arise from the superior salivatory nucleus, which is part of the facial nerve. The postganglionic fibers to these glands arise from the sphenopalatine and submandibular ganglia, respectively.

7 **The answer is D: Hypoglossal nerve.** Speaking involves the cerebral process of formulating speech (the concept and effort) and the mechanical movements of the oral cavity, face, and laryngeal structures that make speech happen. In addition, the muscles involved need to be coordinated; for example, cerebellar speech is an explosive slurred type of speech that reflects uncoordinated movements of the speech apparatus. The facial, hypoglossal, and vagus nerves are involved in speech. The facial and hypoglossal through movements of the facial muscles, especially those around the mouth, and the hypoglossal through movements of the tongue are all essential to articulate speech. The vagus also affects speech through its innervation of the muscles of the vocal folds. Surgery of this type more commonly affects the hypoglossal nerve because it is within the general area of the field; inadvertent pressure on the nerve by traction may be a contributing factor. The vagus nerve is also in the field, but when this man does speak, the quality of his voice is abnormal. Damage to the vagus results in a hoarse, gravely voice; the quality of the speech is significantly altered. The facial and glossopharyngeal nerves are not even close to the surgical field, and the anterior roots have no role in speech. It is appropriate to note that the carotid artery is close to the sympathetic trunk and has postganglionic sympathetic fibers on its surface. However, symptoms of damage to these structures (ptosis, myosis, facial anhydrosis) are almost never seen as a complication of this surgery.

8 **The answer is C: Inferior mesenteric ganglion.** The smooth muscle of the bladder wall, the detrusor muscle, receives a sympathetic and a parasympathetic innervation. Postganglionic sympathetic neurons, with their cell bodies in the inferior mesenteric ganglion, provide input to the bladder by way of the hypogastric plexus. During periods of urine storage, this sympathetic innervation inhibits the smooth muscle of the bladder wall directly and indirectly by also inhibiting local parasympathetic (postganglionic) neurons that are excitatory to the detrusor muscle. When the bladder fills, stretch receptors in the bladder wall are activated and eventually reach a threshold. The central processes of these afferents will (1) increase the activity through the parasympathetic (excitation of the bladder wall) pathway with contraction of the bladder and (2) inhibit the motor neurons innervating the external sphincter so that this sphincter relaxes, allowing the bladder to empty. The anterior horn at S2-4 is not the source of inhibitory input to the bladder wall; the intermediolateral cell column, while containing preganglionic sympathetic cell bodies, does not contain cell bodies that directly innervate the bladder wall. The dorsal motor nucleus of the vagus does not project to any portion of the pelvic viscera.

9 **The answer is E: Stomach.** Pain that is referred to the epigastric region originates, in most instances, from the stomach. Basically, referred pain is pain that originates from an internal organ but may be perceived by the patient as coming from the body wall. The circuits are as follows: a visceral afferent enters the posterior horn and gives rise either to a collateral, or to a main, branch that enters groups of tract cells that convey somatosensory information from the overlying body wall; activation of these tract cells gives rise to ascending information that eventually ends up in the appropriate area of the somatosensory cortex; the brain interprets this input as coming from the body wall because it is those tract cells that were activated. Referred pain from the diaphragm is perceived as coming from the area of the shoulder; referred pain from the esophagus is perceived as coming from the area over the sternum. Referred pain to the chest, shoulder, and proximal portions of the upper extremity is widely recognized as coming from the heart. Referred pain from the kidney is usually perceived as lower back pain that passes around the side and to the front of the lower region of the abdomen. Referred pain, if subtle but persistent, may represent important cues as to the origin of the disease process.

10 **The answer is C: Hirschsprung disease.** This patient has congenital megacolon also known as Hirschsprung disease. This is a congenital disease that more commonly affects males and may result in the death of the patient. The cause is a failure of proper innervation of the hindgut. Addison disease is also known as chronic adrenocortical insufficiency; this damage to the adrenal gland may be idiopathic, disease related, or autoimmune in nature. The patient presents with fatigue, skin pigmentation, weight loss, low blood pressure, and chronic nausea. Hanson disease (leprosy) is a chronic infection by *Mycobacterium leprae* that usually affects cooler regions of the body such as the toes, fingers, and skin. Huntington disease is an inherited condition of the basal nuclei that results in characteristic motor and cognitive deficits; it is not treatable or curable. Wilson disease is also a condition that involves the basal nuclei; this is an inherited disorder of copper metabolism that is treatable and curable.

11 **The answer is B: The constricted segment has a lack of the myenteric ganglion cells.** This child is suffering from congenital megacolon. The constricted segment of the colon, the sigmoid part, and the rectum have an absence (aganglionic) or a significantly reduced (hypoganglionic) number of myenteric ganglion cells. The aganglionic segment is in a chronically constricted state, sometimes referred to as a paralyzed condition. The normally innervated segment is proximal to the aganglionic segment, contains all the components of the myenteric plexus and ganglia, and becomes distended as a result of the failure of normal peristaltic activity through the aganglionic segment. Enterocolitis, an inflammation of the lining of the intestines, is a serious consequence of this disease, and it has a significant mortality rate. This disease is a result of the early failure of neural crest cells to migrate to their normal positions within the digestive tract during development. The muscle layers, circular and longitudinal, are basically normal. Although the clinical picture is very characteristic, a definitive diagnosis is made by examining a histological specimen and confirming the lack of ganglion cells.

12 **The answer is A: Ciliary ganglion.** The structure identified is the nucleus of Edinger-Westphal, the cell group classically identified as being the location of parasympathetic preganglionic

cells associated with the oculomotor nerve; the axons of these cells project to the ciliary ganglion. This is the first part of the pathway for innervation of the sphincter pupillae muscle and constriction of the pupil. It is now known that cells located in the inferior area of the periaqueductal gray, outside of the classically identified Edinger-Westphal nucleus, may also give rise to axons that end in the ciliary ganglion. Further research will certainly clarify this pattern in the human. The ciliary ganglion contains postganglionic parasympathetic cells whose axons project to the sphincter pupillae muscle; the dilator pupillae muscle receives postganglionic sympathetic input from the superior cervical ganglion. The levator palpebrae muscle is innervated by the oculomotor nucleus. The sphenopalatine ganglion is associated with the facial nerve, and contains postganglionic parasympathetic cells that project to the lacrimal gland and palatine glands. All of these projections are ipsilateral.

13 **The answer is A: Decreased blood flow to the skin.** Anxiety, or basically any kind of stress, evokes a cascade of reactions that reflect a global mobilization of the sympathetic part of the visceromotor system. These sympathetic responses may include cool skin due to a decreased blood flow, but an increase in sweating, increased heart rate and bronchodilation, a decrease in blood flow to the gut, and a decrease in peristalsis. It is also likely that the sympathetic response immediately increases blood flow to skeletal muscle; there is obviously a marked increase of blood to skeletal muscle if the individual actually needs to fight or flee. Collectively, these responses mobilize the body for immediate action and the ability to respond; the so-called "fight or flight" response. The other choices are either responses to increased parasympathetic outflow (choices B, D, and E) or not necessarily affected by parasympathetic fibers (C).

14 **The answer is C: The intrinsic neurons within the wall of the gut can function autonomously to generate waves of peristalsis in response to the presence of chyme in the gut.** A significant amount of the parasympathetic innervation of the gut would be interrupted by a bilateral vagotomy. However, a large population of neurons within the wall of the gastrointestinal tract, the enteric nervous system, is organized into complex circuits that are capable of mediating basic functions such as peristalsis and secretion, in response to ingested material in the lumen. This can occur independently of regulation by parasympathetic or sympathetic input. Although the sympathetic outflow to the gut would be intact after vagotomy, activation of this pathway results in decreased motility, secretion, and blood flow in the gut; sphincter tone is increased.

15 **The answer is D: Neuropeptide Y.** Antibodies to neuropeptide Y would be the best choice because it serves as a neurotransmitter, or neuromodulator, that is released by many sympathetic postganglionic fibers, including those that innervate blood vessels of the skin. Demonstrating a diminution or loss of neuropeptide Y would indicate a loss of postganglionic sympathetic fibers, or of their synaptic effectiveness, and correlate with visceromotor dysfunction. Generally, sympathetic ganglion cells use norepinephrine, rather that epinephrine, as a neurotransmitter. On the other hand, epinephrine is synthesized and released by most of the endocrine cells (modified sympathetic ganglion neurons) of the adrenal medulla. Aspartate is an excitatory neurotransmitter found at many CNS sites and often in association with glutamate. Dopamine is recognized as a neurotransmitter of the substantia nigra, pars compacta, and relates to functions of the basal nuclei. Neurotensin is a neurotransmitter found in the hypothalamus, amygdaloid nucleus, and the posterior horn of the spinal cord, and is involved in the perception of noxious stimuli.

Chapter 18

Motor Pathways: Corticospinal, Corticonuclear, and Other Influences on Motor Neurons

Recall the Cautionary Tale: The images in this chapter are presented in a Clinical Orientation: *this approach emphasizes how structure and function correlate with deficits in a clinical setting. MRI and CT have a universally recognized orientation and laterality. Line drawings, stained sections, and gross brain images in the* Clinical Orientation *are viewed here in the identical manner that one views an MRI: your right, as the physician-observer, is the patient's left, and the observer's left is the patient's right. The laterality of deficits and reference to connections follow accordingly.*

QUESTIONS

Select the single best answer.

1 A 79-year-old man is brought to the Emergency Department by his son after collapsing at a family picnic. The history reveals that he is being treated for high blood pressure and has been diabetic most of his life. The examination reveals a profound weakness of his left lower extremity, the function of both upper extremities is intact, and cranial nerves are not affected. MRI shows a localized forebrain lesion. Which of the following outlined areas represents the most likely location of this lesion?

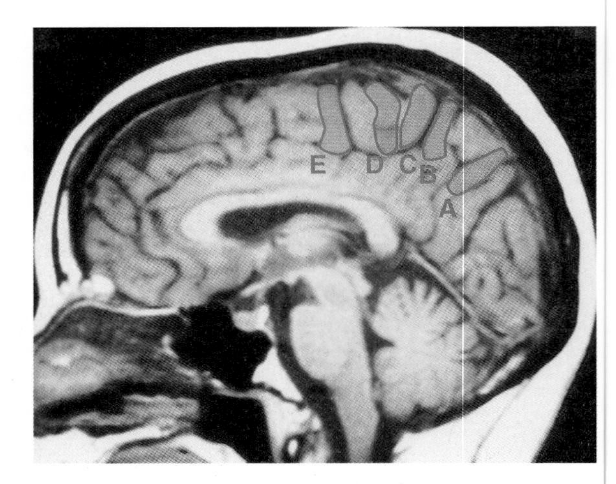

(A) A
(B) B
(C) C
(D) D
(E) E

2 A 47-year-old welder presents at the Emergency Department with sudden onset of difficulty speaking. The examination reveals slurred speech, deviation of the tongue to the right on attempted protrusion, and weakness of the lower facial muscles also on the right. CT shows a well-localized hemorrhagic lesion that is most likely located in which of the following?

(A) Anterior limb of the internal capsule on the left
(B) Genu of the internal capsule on the left
(C) Genu of the internal capsule on the right
(D) Posterior limb of the internal capsule on the left
(E) Posterior limb of the internal capsule on the right

3 Which of the following is the receptor for the afferent limb of the patellar tendon reflex or the jaw jerk reflex?
(A) Golgi tendon organ
(B) Merkel cell ending
(C) Muscle spindle
(D) Pacinian corpuscle
(E) Ruffini ending

4 Which of the following is a sign that would most likely be seen in a patient with a lower motor neuron lesion?
(A) Hyperreflexia
(B) Hypertonia
(C) Muscle fasciculations
(D) Muscle groups affected
(E) Muscle spasticity

5 The MRI of a 71-year-old man suggests a small vascular lesion affecting the structure indicated by the arrow in the image below. Which of the following would you most likely expect to see in this patient?

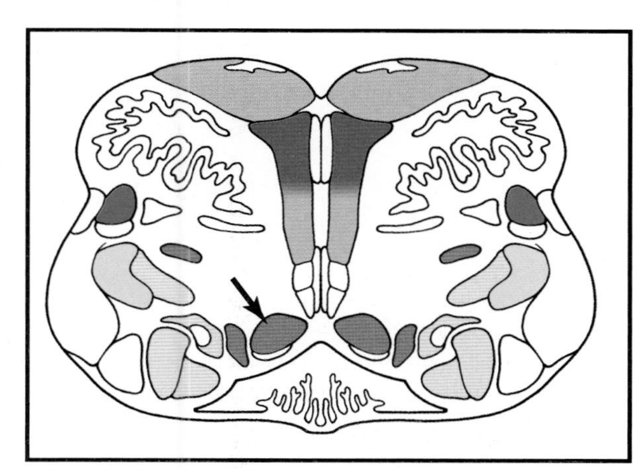

(A) Left internuclear ophthalmoplegia
(B) Paralysis of lateral rectus on gaze to the left
(C) Paralysis of lateral rectus on gaze to the right
(D) Tongue deviates to the left on protrusion
(E) Tongue deviates to the right on protrusion

6 The MRI of a 56-year-old man presents with a sudden loss of motor function of his right hand. The MRI reveals a circumscribed cortical lesion, most likely hemorrhagic in origin. Which of the following represents the most likely location of this lesion?
(A) Anterior paracentral gyrus
(B) Lateral third of the precentral gyrus
(C) Medial third of the precentral gyrus
(D) Middle third of the precentral gyrus
(E) Middle third of the postcentral gyrus

7 A Neurologist sees three patients in her clinic. These are a 23-year-old man with a Brown-Séquard syndrome resulting from trauma at C3-5, a 69-year-old woman with a Wallenberg syndrome, and a 79-year-old man with a hemorrhagic stroke in the lateral portions of the pontine tegmentum. In addition to the expected deficits, a careful examination would also reveal which of the following in all three patients?
(A) A loss of discriminative touch on the forehead
(B) A loss of pain and thermal sensation on one side of the face
(C) Constriction of the pupil (miosis) on the side of the lesion
(D) Deviation of the tongue and uvula on attempted phonation
(E) Weakness of the digits on the side opposite the lesion

8 Which of the following encodes the rate of change in muscle length during contraction of a skeletal muscle?
(A) Dynamic nuclear bag fibers
(B) Golgi tendon organ
(C) Merkel cell ending
(D) Nuclear chain fibers
(E) Static nuclear bag fiber

9 A 37-year-old woman presents to her family physician with intermittent physical problems. She is referred to a neurologist and, after a thorough examination, a tentative diagnosis of multiple sclerosis is made. MRI shows an area of demyelination in the region indicated by the arrow in the image below. Which of the following deficits would correlate most specifically with the location of this lesion?

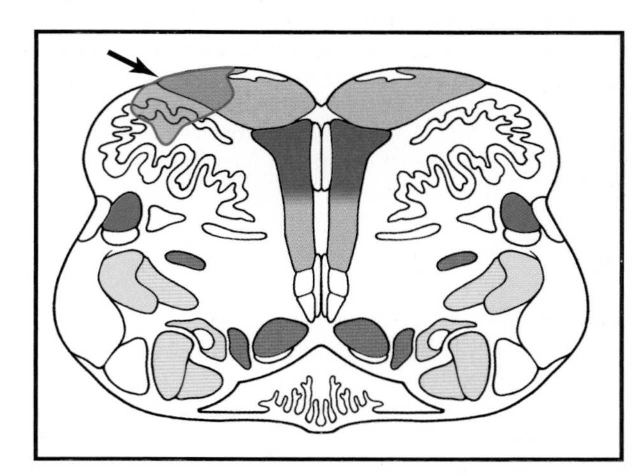

(A) Tongue deviates to the left, weak left upper and lower extremities
(B) Tongue deviates to the left, weak left lower extremity
(C) Tongue deviates to the right, weak left upper and lower extremities
(D) Tongue deviates to the right, weak left lower extremity
(E) Tongue deviates to the right, weak left upper extremity

10 A 67-year-old man presents at the Emergency Department with a medial medullary (Dejerine) syndrome. He has a right-sided weakness of the UE and LE, a right-sided loss of vibratory sense and discriminative touch, and a deviation of the tongue to the left on attempted protrusion. Weakness of which of the following muscles would explain the asymmetrical tongue movement in this man?

(A) Chondroglossus
(B) Genioglossus
(C) Hyoglossus
(D) Palatoglossus
(E) Styloglossus

11 A 47-year-old man is brought to the Emergency Department from the site of a motor vehicle collision. The examination reveals facial injuries with a probable broken nose, a compound fracture of the left humerus, and large bruises/contusions on his left thigh. CT confirms these observations and shows a skull fracture in the left frontal region, extensive intracranial hemorrhage, a fracture through the left orbit, and a fractured pelvis. The man is unconscious and within 6 hours exhibits decorticate posturing. Excessive action in which of the following tracts/systems would explain the flexion of the upper extremities in this man?

(A) Anterolateral
(B) Corticospinal
(C) Reticulospinal
(D) Rubrospinal
(E) Vestibulospinal

12 A 71-year-old woman presents to her family physician with difficulty swallowing and some numbness on her face. The history reveals that this was a sudden onset (2 days ago), and that the woman not only had difficulty swallowing but also complained that she got food "down the wrong tube." The examination revealed a loss of sensation on the right side of the face, some moderate sensory loss on the left side of her body, and a weak right vocal fold (endoscopic observation). MRI showed a lesion in the territory of the posterior inferior cerebellar artery (PICA). Damage to which of the following structures would explain this motor deficit in this patient?

(A) Hypoglossal nucleus
(B) Nucleus ambiguus
(C) Restiform body
(D) Solitary nucleus
(E) Spinal trigeminal nucleus

13 A 51-year-old woman awakens in the morning and has some difficulty talking. Her daughter takes her to the Emergency Department. The history reveals that the woman takes medication for diabetes and for elevated blood pressure, and is significantly overweight. The results of the examination suggest the probability of a small stroke; MRI shows a lesion localized to the area indicated by the arrow in the image below. Which of the following deficits would most likely be seen in this patient based on the location and extent of the damage?

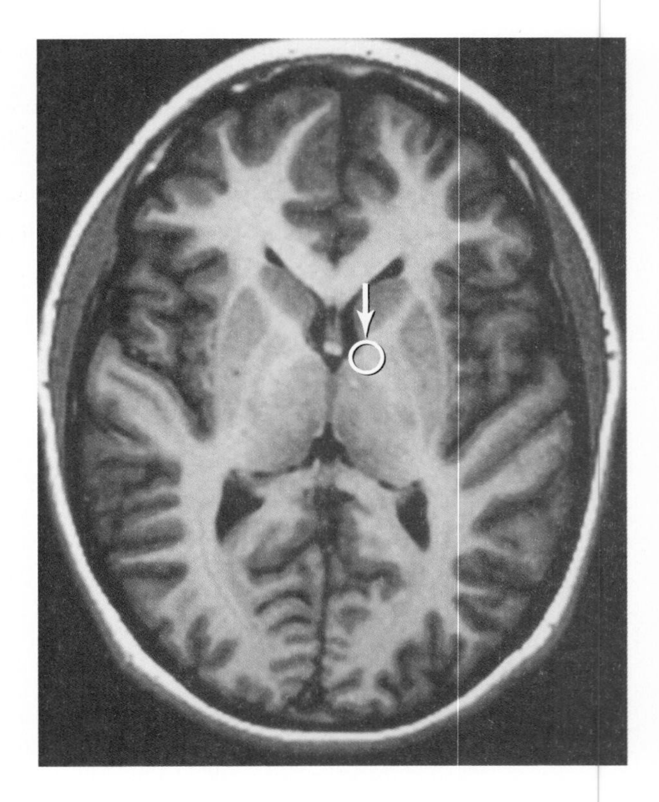

(A) Deviation of the tongue to the left on attempted protrusion
(B) Deviation of the uvula to the right on phonation
(C) Difficulty elevating the right shoulder against resistance
(D) Weakness of facial muscles on the lower right side of the face
(E) Weakness of masticatory muscles of the left

14 A 59-year-old man is transported to the Emergency Department (ED) after collapsing at his desk in a local office building. The history, provided by a colleague, revealed that the man was working, suddenly moaned loudly, and slumped out of his chair onto the floor. He is unconscious upon arrival at the ED. MRI reveals a massive cerebral hemorrhage. At about 7 hours after the initial event, he presents with features of decorticate posturing; about 10 hours later, he converts to a decerebrate state. Which of the following most likely takes place to explain this change in clinical status?

(A) Decreased excitatory outflow of cerebellothalamic axons
(B) Decreased corticospinal excitation of spinal flexor motor neurons
(C) Increased reticulospinal excitation of spinal flexor motor neurons
(D) Increased reticulospinal excitation of spinal extensor motor neurons
(E) Increased rubrospinal excitation of spinal extensor motor neurons

15 Stretch of a Golgi tendon organ will, through the appropriate synaptic circuits, ultimately result in which of the following?

(A) Excitation of alpha motor neurons innervating the muscle associated with the activated tendon organ
(B) Excitation of gamma motor neurons innervating muscle spindles in the muscle associated with the activated tendon organ
(C) Inhibition of alpha motor neurons innervating the muscle associated with the activated tendon organ
(D) Inhibition of gamma motor neurons innervating muscle spindles in the muscle associated with the activated tendon organ
(E) Monosynaptic inhibition of alpha motor neurons innervating the muscle associated with the activated tendon organ

16 A 63-year-old man falls from a scaffolding at his place of work and is transported to the Emergency Department. The man is semicomatose. The examination reveals that cranial nerve function is within normal ranges, the man has weakness of both lower extremities, and all somatosensory input (pain, thermal, two-point discrimination, vibratory sense) from the body is also within normal ranges. MRI shows a small single lesion. Which of the following represents the most likely location of this lesion?
(A) Bilateral caudal basilar pons
(B) Bilateral medial medulla
(C) Caudal part of motor decussation
(D) Complete motor decussation
(E) Rostral part of motor decussation

Questions 17 and 18 are based on the following patient.

A 71-year-old farmer is transported to the Emergency Department after collapsing while working in his barn. According to EMS personnel, the man's wife found him after he failed to come to the house for a noonday meal. The man was conscious and coherent, and indicated that his problem "came on real sudden." The examination revealed a profound weakness of his left upper and lower extremities, dilation of the right pupil, and loss of most voluntary movement of his right eye. MRI showed a localized lesion presumably of vascular origin.

17 A lesion in which of the outlined areas in the image below would most explain the deficits experienced by this man?

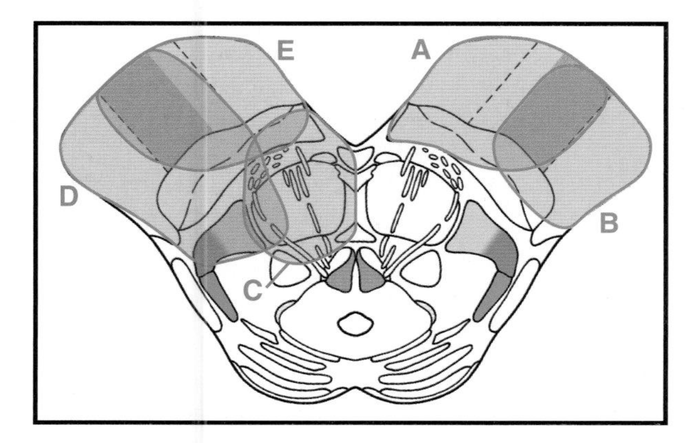

(A) A
(B) B
(C) C
(D) D
(E) E

18 A thorough neurological examination would also most likely reveal which of the following deficits in this man?
(A) Deviation of the tongue to the right on attempted protrusion
(B) Deviation of the uvula to the right on phonation
(C) Difficulty elevating the left shoulder against resistance
(D) Weakness of facial muscles on the lower right side of the face
(E) Weakness of masticatory muscles of the left

19 A 66-year-old man is transported to the Emergency Department by his son. The history of the current event, largely provided by the son, revealed that the man suddenly fell and had a loss of some motor control. The examination revealed normal cranial nerve function, normal sensation on the body and face, but extremity weakness. MRI reveals a lesion in the outlined area below that was localized to the rostrolateral portion of the motor decussation. Based on the location and extent of the lesion in this man, which of the following deficits would this patient most likely experience?

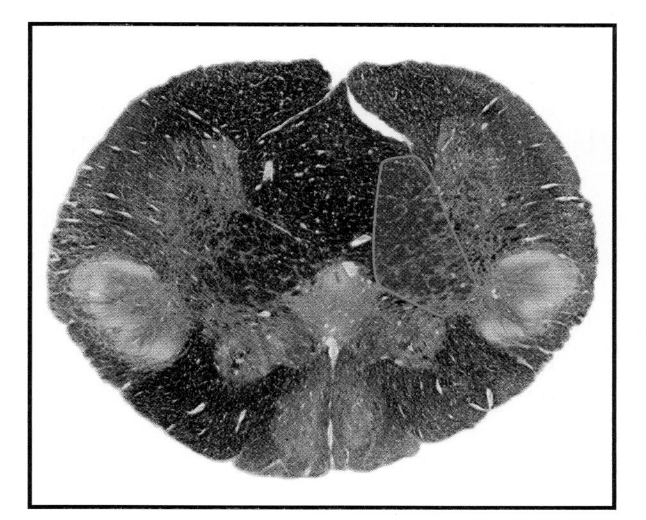

(A) Bilateral weakness of upper and lower extremities
(B) Weak left upper and lower extremities
(C) Weak left lower extremity and right upper extremity
(D) Weak right upper and lower extremities
(E) Weak right lower extremity and left upper extremity

20 A 71-year-old man is diagnosed with a middle alternating (or crossed) hemiplegia. Which of the following combination of deficits would most likely be seen in this patient?

(A) Corticospinal deficits and deviation of the tongue to the opposite side
(B) Corticospinal deficits and dilation of the pupil on the opposite side
(C) Corticospinal deficits and loss of somatosensory information on the same side
(D) Corticospinal deficits and weakness of lateral gaze to the opposite side
(E) Corticospinal deficits and weakness of most eye movement on the same side

Questions 21 and 22 are based on the following patient.

A 27-year-old woman visits her family physician with a generally vague complaint of "feeling real tired" intermittently. The examination reveals no obvious deficits, other than the general complaint, but the presence of several enlarged lymph nodes is noted by the physician. A hematological examination reveals the presence on immunoglobulin antibodies to acetylcholine receptors. This observation triggers a histological examination of a lymph node biopsy which reveals medullary hyperplasia within germinal centers present within the sample.

21 This woman is most likely suffering from which of the following?
(A) Amyotrophic lateral sclerosis
(B) Foster Kennedy syndrome
(C) Guillain-Barré syndrome
(D) Multiple sclerosis
(E) Myasthenia gravis

22 Assuming this woman's disease progresses, which of the following would most likely be the earliest and most obvious symptoms to appear and persist?
(A) Weakness of the eyelids and ocular muscles
(B) Weakness of the lower extremity muscles
(C) Weakness of the pharyngeal musculature
(D) Weakness of the tongue musculature
(E) Weakness of the upper extremity muscles

23 A 76-year-old man is brought to the Emergency Department by EMS personnel. The history, provided by his grandson who is a physician, indicates that the man became suddenly weak on his left side and also complained of numbness. The examination confirmed that the man was profoundly weak on his left side and had a left-sided loss of pain and discriminative sense; cranial nerve motor function was intact (when asked to smile, protrude his tongue), but he had a loss of sensation on the left side of his face. MRI showed a supratentorial lesion, presumably vascular in origin. Which of the regions outlined on the given image represents the most likely location of this man's lesion?

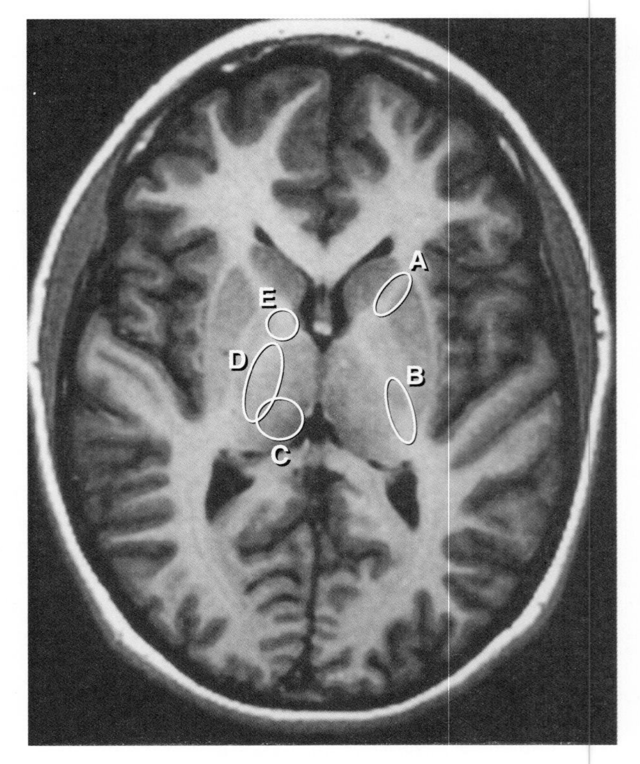

(A) A
(B) B
(C) C
(D) D
(E) E

24 An 82-year-old woman complains to her daughter that she suddenly started to have trouble "lifting things." An examination by her family physician revealed that this woman could not elevate her right shoulder against resistance, but that she could perform this movement with her left shoulder. MRI showed a small lesion in the caudal medulla on the right and extending into the cervical spinal cord at the junction of the anterior horn and adjacent white matter. Which of the following deficits would also most likely be experienced by this woman?
(A) Deviation of the tongue to the right on protrusion
(B) Difficulty swallowing and speaking
(C) Inability to rotate the head to the left against resistance
(D) Inability to rotate the head to the right against resistance
(E) Inability to smile on the right side of the face

25 Descending fibers from supraspinal centers excite motor neurons within the anterior horn that innervate intrafusal muscle fibers. Contraction of these muscle fibers results in increased activity in Aα (also called Ia) fibers which activate alpha motor neurons that excite extrafusal muscle fibers. What is this circuit called?

(A) Babinski reflex
(B) Crossed extensor reflex
(C) Flexor reflex
(D) Gamma loop
(E) Reciprocal inhibition

26 The largest fibers within the corticospinal tract have a diameter in the range of 11 to 20 μm. Which of the following represents the approximate range of the conduction velocity of these fibers?
(A) 30 to 35 m/s
(B) 40 to 45 m/s
(C) 50 to 55 m/s
(D) 65 to 70 m/s
(E) 80 to 100 m/s

27 A 74-year-old woman (retired dermatologist) experiences a sudden weakness of her upper extremity and has difficulty speaking; she contacts her daughter who immediately transports the mother to their family physician. The history of the current event (sudden onset, resolving within about 15 minutes) clearly suggests that this woman had a transient ischemic attack. A carotid angiogram is performed, and during this procedure, the woman experiences a sudden and profound weakness of her left upper extremity and slight slurring of her speech; the latter resolved. CT confirmed a cortical lesion most likely caused by thrombi knocked loose during the procedure. Which of the following outlined areas in the image below represents the most likely location of the lesion in this woman?

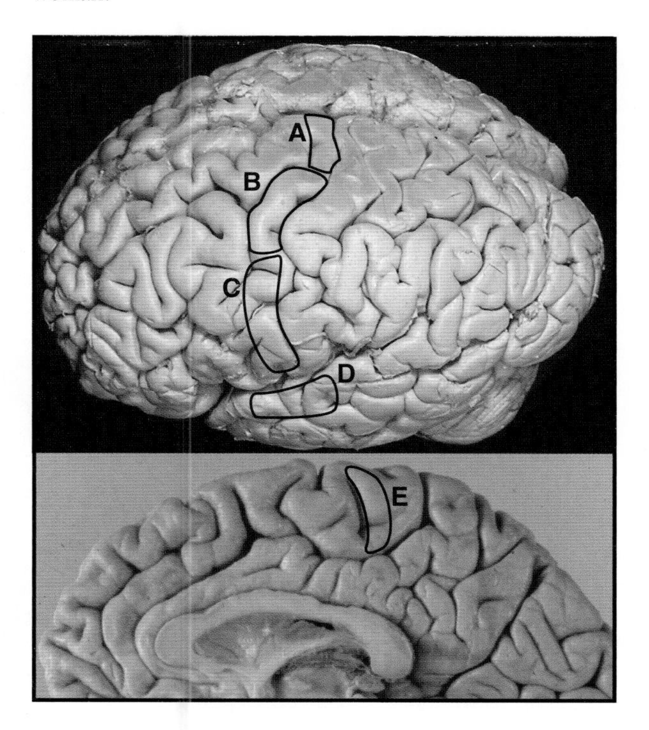

(A) A
(B) B
(C) C
(D) D
(E) E

28 A 45-year-old man is transported to the Emergency Department (ED) after collapsing at his home. This man is well known to the ED staff, having visited several times before for problems related to abuse of recreational drugs and alcohol. The man's current complaint is of a sudden onset of weakness of his upper and lower extremities (UE and LE) on his left side. The examination further reveals a loss of discriminative touch on the man's left side (UE + LE); when he protrudes his tongue, it deviates to the right. This man is most likely suffering which of the following?
(A) Benedikt syndrome
(B) Dejerine syndrome
(C) Dejerine-Roussy syndrome
(D) Foville syndrome
(E) Wallenberg syndrome

29 The circuitry that comprises the gamma loop is essential to maintenance of muscle tone and muscle stretch reflexes. Which of the following flow charts correctly specifies the connections that constitute this important circuit?
(A) Supraspinal activation of gamma motor neurons→ intrafusal muscle fibers do not contract→ spindle equator not stretched→ Ia fiber activity increases→ alpha motor neuron activated→ extrafusal muscle fibers contract
(B) Supraspinal activation of gamma motor neurons→ intrafusal muscle fibers contract→ spindle equator stretched→ Ia fiber activity decreases→ alpha motor neuron inactivated→ extrafusal muscle fibers do not contract
(C) Supraspinal activation of gamma motor neurons→ intrafusal muscle fibers contract→ spindle equator not stretched→ Ia fiber activity increases→ alpha motor neurons activated→ extrafusal muscle fibers contract
(D) Supraspinal activation of gamma motor neurons→ intrafusal muscle fibers contract→ spindle equator stretched→ Ia fiber activity increases→ alpha motor neurons activated→ extrafusal muscle fibers contract
(E) Supraspinal inactivation of gamma motor neurons→ intrafusal muscle fibers do not contract→ spindle equator not stretched→ Ia fiber activity decreases→ alpha motor neurons inactivated→ extrafusal muscle fibers do not contract

30 A 57-year-old man is brought to the Emergency Department by paramedics. They report that he was hunting, did not return home, and was found unconscious at his deer camp by his wife and son. It is not known how long he was unconscious. His CT is shown in the image below. Hyperdense areas in the temporal pole and occipital lobe on the right suggest acute blood, and an extensive hypodensity throughout the right temporal and occipital lobes indicate a probable hemisphere infarction. In addition, there is a herniation (H) of portions of the right medial temporal lobe into the midbrain and a consequent shift of the midbrain to the left and impingement of the crus cerebri onto the edge of the tentorium cerebelli on the left (arrow). Which of the following deficits would most likely be seen in this patient recognizing his particular clinical condition?

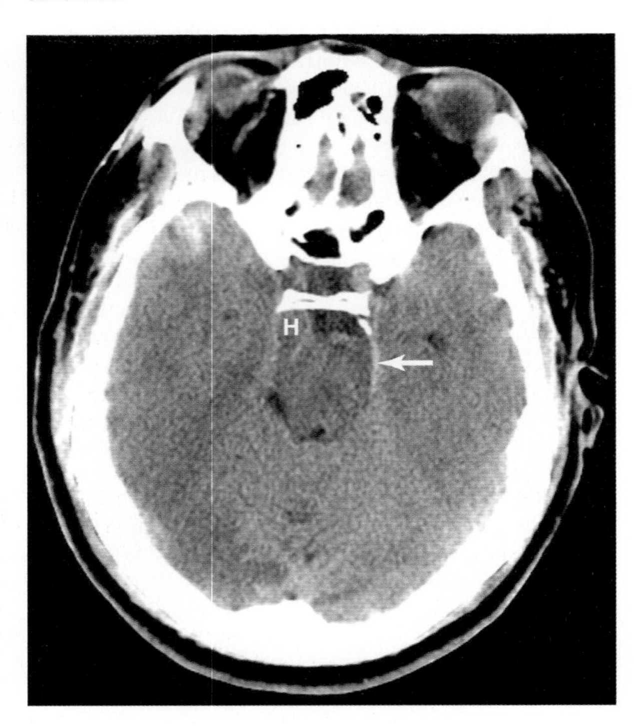

(A) Weakness of all extremities, bilateral oculomotor palsy
(B) Weakness of the extremities on the left, oculomotor palsy on the left
(C) Weakness of the extremities on the left, oculomotor palsy on the right
(D) Weakness of the extremities on the right, oculomotor palsy on the left
(E) Weakness of the extremities on the right, oculomotor palsy on the right

31 Which of the following represents the location of the cells of origin for the fibers located in the shaded area on the image below?

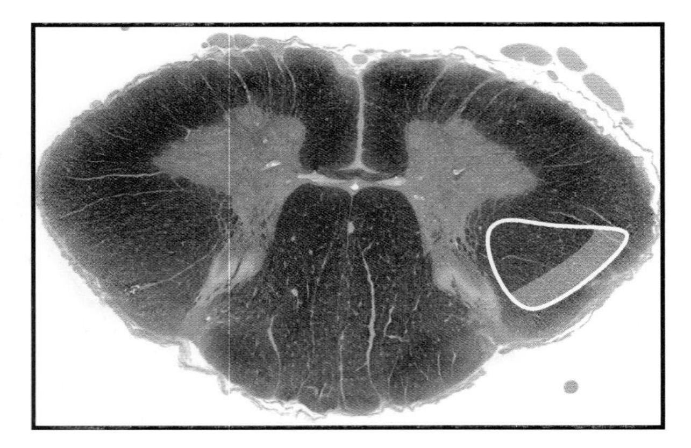

(A) Left anterior paracentral gyrus
(B) Left posterior paracentral gyrus
(C) Left precentral gyrus
(D) Right anterior paracentral gyrus
(E) Right precentral gyrus

32 A 31-year-old woman presents to her family physician with a complaint of "seeing two of everything." She is referred to a neurologist. The history reveals that this problem has been more noticeable at some times and less noticeable at other times; she has experienced this over several weeks. The examination reveals that the woman has weakness of ocular movement in both eyes, more pronounced on the left, drooping of both eyelids, also more pronounced on the left, and slight weakness of the left upper extremity when compared to the right. She also notes that she feels worse later in the day. The evidence clearly suggests a disease related to a failure of neurotransmitter function. Which of the following is most likely involved in this woman's disease?
(A) Acetylcholine
(B) Dopamine
(C) Histamine
(D) Glutamate
(E) Glycine

33 Lateral vestibulospinal tract fibers extend the length of the spinal cord, terminate at all levels based on a somatotopic pattern, and preferentially excite extensor motor neurons innervating paravertebral and proximal limb musculature. Which of the following is the primary source of these tract cells?
(A) Inferior vestibular nucleus
(B) Lateral vestibular nucleus
(C) Medial vestibular nucleus
(D) Medial and inferior vestibular nuclei
(E) Superior vestibular nucleus

34 A 63-year-old man is transported to the Emergency Department following a sudden onset of motor and sensory deficits. MRI reveals defects, vascular in origin, in the territories represented by the outlines in the image below. Which of the following deficits would most likely be experienced by this man?

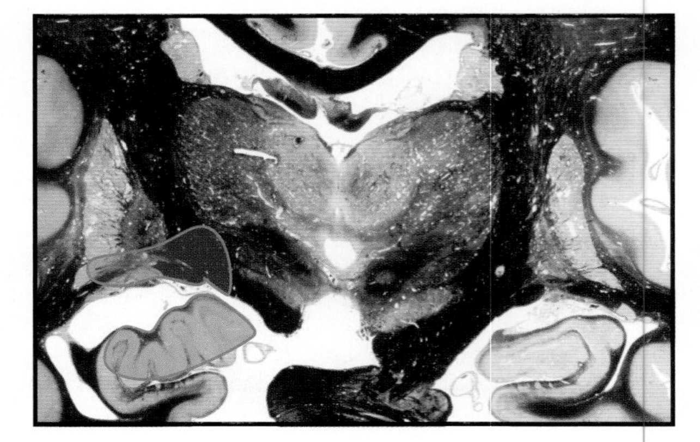

(A) Left hemianesthesia, bitemporal hemianopia
(B) Left hemiplegia, left homonymous hemianopia
(C) Left hemiplegia, right homonymous hemianopia
(D) Right hemiplegia, left homonymous hemianopia,
(E) Right hemiplegia, binasal hemianopia

ANSWERS

1 **The answer is D (The anterior paracentral gyrus).** The small dimple at the upper edge of the hemisphere between D and C is the small medial extension of the central sulcus; this sulcus is at the junction of the precentral gyrus with the postcentral gyrus. A line from this point down to the cingulate sulcus marks the junction of the anterior paracentral gyrus (D, lower extremity region of the somatomotor cortex) with the posterior paracentral gyrus (C, lower extremity region of the somatosensory cortex). Damage to cortical area D results in weakness of the lower extremity on the opposite side of the body. Lesion B is immediately caudal to the marginal sulcus in the medial aspect of the parietal lobe; lesion A is also in the medial aspect of the parietal lobe, but slightly rostral to the parietooccipital sulcus. E is in the medial aspect of the superior frontal gyrus, but significantly rostral to the lower extremity area of the motor cortex.

2 **The answer is B: Genu of the internal capsule on the left.** Corticonuclear fibers that influence cranial nerve motor nuclei (especially the facial, hypoglossal, and accessory nuclei and the nucleus ambiguus) travel through the genu of the internal capsule. After these fibers descend into the brainstem, they project to the part of the facial nucleus on the opposite side that innervates the lower facial muscles, to the opposite hypoglossal nucleus (genioglossus muscle), and to the opposite nucleus ambiguus (pharyngeal and laryngeal muscles). In other words, fibers originating from the left cerebral cortex, traverse the left genu, and influence lower motor neurons of cranial nerve nuclei on the right side. When movements are attempted (protruding the tongue, puckering the lips, or smiling), the motor deficit is seen on the right side, as is the case in this patient: lesion on the left, deficits on the right. In this case, the deficits of facial movements are also called a central seven. The main fiber bundles of the anterior limb of the internal capsule are anterior thalamic radiations and frontopontine fibers. The posterior limb of the internal capsule contains many important thalamocortical radiations and, in about its caudal half, corticospinal fibers that influence spinal motor neurons innervating the opposite upper and lower extremities. A lesion of the posterior limb results in a hemiplegia of the extremities on the opposite side.

3 **The answer is C: Muscle spindle.** The patellar tendon reflex, commonly called the knee jerk or knee jerk reflex, and the jaw jerk reflex are muscle stretch reflexes. Although they are sometimes called deep tendon reflexes, this is a misnomer: the receptors for these reflexes are in the muscle not in the tendon. The afferent fiber arising from the muscle spindle is an Aα (also called Ia) fiber. The Aα fiber is a heavily myelinated (13 to 20 μm), rapidly conducting (80 to 120 m/s) fiber that enters the posterior root and passes to lower motor neurons in the anterior horn that innervate the extrafusal fibers of the muscle from which the afferent volley originated; activation of the spindle results in a sudden contraction of the muscle containing the spindle. Such reflexes are also commonly called monosynaptic myotatic reflexes. It is important to remember that all reflexes, even those that are strictly motor in nature, have sensory endings on their afferent limbs. The Golgi tendon organ measures tension in muscle, and when activated, results in autogenic inhibition; the alpha motor neurons exciting the muscle under tension are inhibited and the muscle relaxes. The Merkel cell ending senses fine touch and pressure, and is an ending especially important to reading Braille. The Pacinian corpuscle senses vibration; the Ruffini ending senses joint position.

4 **The answer is C: Muscle fasciculations.** A lower motor neuron lesion separates the alpha motor neuron in, for example, the spinal cord from its muscle targets in the periphery. This may be a traumatic lesion or a disease process that may affect the cell body, axon, or neuromuscular junction. The deficits that are characteristic of a lower motor neuron lesion are (1) muscle fasciculations or fibrillations (involuntary contractions of muscle fibers innervated by individual motor units or groups of motor units); (2) flaccid paralysis; (3) eventual muscle atrophy; (4) hypotonia; and (5) hyporeflexia or areflexia. Muscle fasciculations represent the infrequent and irregular muscle contractions that are caused by equally irregular firing patterns of lower motor neurons as their cell bodies or axons die. Hyperreflexia, hypertonia, muscles that are initially flaccid but become spastic, and deficits of muscle groups are all features of an upper motor neuron lesion. Muscle spasticity, in this case, is seen as increased resistance to passive movement; the more rapid the movement, the greater the resistance.

5 **The answer is E: Tongue deviates to the right on protrusion.** The arrow is pointing to the right hypoglossal nucleus. Damage to this structure will result in weakness of the right genioglossus muscle and a deviation of the tongue to the right on attempted protrusion. Normally, each genioglossus muscle pulls its half of the tongue out of the mouth and slightly toward the midline; when both muscles work together, the tongue protrudes straight out of the mouth. In the case of this patient, the lower motor neurons innervating the right genioglossus muscle are damaged, and this muscle is weak; when an attempt is made to protrude the tongue, it deviates toward the weak (right) side. Internuclear ophthalmoplegia is seen consequent to a lesion in the medial longitudinal fasciculus between the abducens and oculomotor nucleus; weakness of the lateral rectus muscle (either right or left) results from damage to the abducens nucleus or nerve.

6 **The answer is D: Middle third of the precentral gyrus.** The motor and sensory cortex is somatotopically organized such that the precentral and postcentral gyri can generally be divided into about thirds; the face represented in about the lateral third, the upper extremity (UE) in about the middle third, and the trunk in about the medial third. The lower extremity (LE) is represented in the anterior paracentral gyrus (for motor) and the posterior paracentral gyrus (for sensory). In this case, the loss of motor function of the man's hand correlates with a lesion in the precentral gyrus in about the middle third. Remember, lesions of the UE and LE regions of the somatomotor cortex give rise to deficits on the side of the body opposite the lesion. The same general principle applies to lesions of the somatosensory cortex.

7 **The answer is C: Constriction of the pupil (miosis) on the side of the lesion.** The lesions in all three of these patients interrupt the descending hypothalamospinal fibers that are passing from the hypothalamus through the lateral area of the pontine tegmentum, lateral region of the medulla, and the lateral funiculus of the spinal cord to end in the intermediolateral cell column (IMLCC) in upper thoracic levels. Damage to these fibers will result in a Horner syndrome (including miosis) on the ipsilateral side. The complete pathway is ipsilateral and as follows: the hypothalamus to IMLCC, this cell group sends preganglionic fibers to the superior cervical ganglion (SCG), the SCG sends postganglionic fibers to the head where some distribute to the dilator pupillae muscle of the iris. Interruption of this pathway will remove the innervation of the dilator pupillae muscle, and the pupil will constrict due to the unopposed action of the sphincter pupillae muscle which receives innervation from the Edinger-Westphal nucleus associated with the oculomotor nerve. The complete Horner syndrome includes ptosis, anhidrosis, and miosis due to a lesion in the sympathetic pathway. None of the other choices is seen in all three patients.

8 **The answer is A: Dynamic nuclear bag fibers.** The muscle spindle contains two main types of fibers: nuclear bag fibers which are subdivided into dynamic nuclear bag fibers and static nuclear bag fibers, and nuclear chain fibers. The former are characterized by a central cluster of nuclei in the fiber, and the latter by a single central row of nuclei in the fiber. These constituent fibers within the muscle spindle may be activated either by attempts to elicit a reflex or during normal ambulatory activity. As the extrafusal muscle fibers are stretched, the dynamic bag fibers encode the rate at which the change in length is taking place. When the stretch is complete and a new muscle length established, the static bag fibers encode the new change in muscle length. Dynamic bag fibers provide information on the rate of change (but not just the change itself); static bag fibers provide information that a change has taken place (but not on the rate of that change). These complimentary bits of information are relayed into the CNS during a movement and contribute to the maintenance of posture. Nuclear chain fibers also encode change in length, but not the rate of that change. Golgi tendon organs convey information on tension within a muscle and serve a protective function in certain situations; the Merkel cell endings convey fine touch and pressure and are important for reading Braille.

9 **The answer is D: Tongue deviates to the right, weak left lower extremity.** The area of demyelination encompasses the lateral portion of the pyramid (containing corticospinal fibers), a portion of the ventral lamella of the principal olivary nucleus, and the exiting root fibers of the hypoglossal nerve, all on the right side. These latter fibers pass between the pyramid and the principal olive and exit the brainstem via the preolivary sulcus. The somatotopic arrangement of corticospinal fibers within the pyramid is such that fibers arising from the lower extremity area of the cerebral cortex descend predominately through its lateral area and corticospinal fibers from the upper extremity cortex are medially located. Consequently, the tongue deviates to the right on attempted protrusion, and the patient's left lower extremity is weak. Both extremities are not involved since the area of demyelination does not involve the entire pyramid. Also, the corticospinal deficit is on the opposite side due to the fact that the lesion is rostral to the motor decussation. Multiple sclerosis is an autoimmune disease of unknown etiology, more common in patients 20 to 50 years of age (50% to 60% of cases), more common in women than men (3:1), and may present as acute or chronic lesions.

10 **The answer is B: Genioglossus.** The genioglossal muscle arises from the superior mental spine on the inner aspect of the mandibular symphysis. The superior fibers of this muscle enter the basal aspect of the full length of the tongue. Each genioglossal muscle pulls obliquely toward the midline as the tongue is protruded. Weakness/paralysis of one genioglossus muscle will result in a deviation of the tongue toward the weak side on attempted protrusion. In this case, the pattern of motor deficits, cranial nerve on one side, corticospinal on the opposite, is called an alternate (or alternating) hemiplegia; this may also be called a crossed hemiplegia. The chondroglossus arises from the lesser cornu and part of the hyoid, enters the tongue along with the hyoglossus and genioglossus, and assists in depressing the tongue. The hyoglossus originates from the hyoid bone, enters the tongue close to the styloglossus, and its action is to depress the tongue. The palatoglossus arises from the soft palate, inserts into the side of the tongue, and its action is to elevate the back of the tongue. The styloglossus originates from the styloid process, enters the longitudinal tongue muscle adjacent to the hyoglossal, and its action is to elevate the tongue upward and backward.

11 **The answer is D: Rubrospinal.** In the situation of decorticate posturing, the cerebral cortex, and therefore corticospinal fibers, is effectively removed from any influence on lower motor neurons of the brainstem or spinal cord. This lack of cortical influence also extends to brainstem nuclei that, in turn, influence spinal lower motor neurons. The lesion is supratentorial and, due to the pressure dynamics of an expanding hemorrhagic lesion, frequently affects both sides of the forebrain. The red nucleus is infratentorial and, therefore, below the zone of destruction; this nucleus is spared, but is deprived of input from the cerebral cortex. Rubrospinal fibers arise from the red nucleus, cross the midline, and preferentially exert an excitatory influence on flexor motor neurons in cervical levels of the spinal cord. The level of activity of rubral neurons is increased due to the fact that other infratentorial influences on this nucleus are intact: for example, the excitatory cerebellorubral pathway. Reticulospinal and vestibulospinal fibers are also intact, but they have an excitatory influence on extensor motor neurons, and are the primary contributors to the extensor rigidity seen in these patients. The anterolateral system is the pathway for pain and thermal sensations from the body. In a decorticate patient, activation of this pathway may give rise to brainstem reflexes, but will not reach a level of conscious perception.

12 **The answer is B: Nucleus ambiguus.** The deficits seen in this patient are characteristic of a posterior inferior cerebellar artery (PICA) syndrome, also called a lateral medullary syndrome or Wallenberg syndrome. The PICA territory encompasses the anterolateral system, spinal trigeminal tract and nucleus, solitary tract and nucleus, inferior and medial vestibular nuclei, restiform body, and the nucleus ambiguus. The nucleus ambiguus, via the vagus nerve, innervates the muscles of the vocal folds, the pharyngeal (including the large constrictor) muscles, intrinsic laryngeal muscles, and the upper portion of the esophagus. Unilateral paralysis of these muscles will cause significant deficits (dysarthria, dysphagia), and bilateral damage may become a life-threatening medical emergency because the vocal folds collapse and the patient cannot breathe. Damage to the restiform body will contribute to the ataxia characteristically seen in these patients. Involvement of the spinal trigeminal tract will result in a sensory loss (pain and thermal sense) on the ipsilateral side of the face; damage to the anterolateral system will result in a loss of the same modalities on the opposite side of the body. Damage to the solitary nucleus, a visceral afferent center, will result in minimal deficits. It is true that hypoglossal nerve damage will affect swallowing due to difficulties manipulating the bolus of food in the mouth. However, the hypoglossal nucleus is not in the PICA vascular territory, and is unaffected in this patient.

13 **The answer is D: Weakness of facial muscles on the lower right side of the face.** This lesion is located in the genu of the internal capsule on the left side. Corticonuclear fibers coalesce and traverse the genu, exit the hemisphere, and enter the medial area of the middle one-third of the crus cerebri. The pattern of corticonuclear fiber projections, from their position in the genu to the motor nuclei of cranial nerves, is as follows: (1) bilateral to the trigeminal motor nucleus, so there are no deficits seen from a left-sided lesion; (2) predominately contralateral to the portion of the facial nucleus that serves muscles on the lower face, so a left-sided lesion will result in weakness of the lower face on the right; (3) contralateral to the hypoglossal neurons that serve the genioglossus muscle, so a left-sided lesion results in a deviation of the tongue to the right on protrusion; (4) contralateral to the nucleus ambiguus neurons that serve the musculus uvulae, so that a left-sided lesion will, due to the arrangement of this muscle, result in a deviation of the uvula to the left on phonation; and (5) predominately ipsilateral to the accessory nucleus serving the trapezius muscle, so that a left-sided lesion produces a left-sided weakness when the shoulder is elevated against resistance. This woman experiences difficulty speaking due to the fact that she has partial weakness of the tongue, facial muscles, and uvula; when functioning normally, all of these contribute to coherent speech. This lesion does not involve the posterior limb of the internal capsule; consequently, there is no appreciable weakness of the extremities on either side of the body.

14 **The answer is D: Increased reticulospinal excitation of spinal extensor motor neurons.** In the decorticate condition, the influence of the cerebral cortex is removed; the lesion is supratentorial, but the brainstem and its descending inputs to the spinal cord (rubro-, reticulo-, vestibulospinal, etc.) are intact. The upper extremities are flexed (excessive drive on flexor lower motor neurons at cervical levels via the rubrospinal tract) and the body is extended (excessive drive on all extensor spinal motor neurons via reticulospinal and vestibulospinal fibers). In the decerebrate condition, the cone of damage has progressed through the tentorial notch and now extends into an infratentorial location. In a decerebrate situation, all corticospinal influence is removed and the red nucleus is damaged; thus, the influence of rubrospinal fibers (mainly excitatory to upper extremity flexors) is also removed. The only remaining descending brainstem pathways are vestibulospinal (medial and lateral) and reticulospinal (medial and lateral). Of these, the most important are reticulospinal fibers; the extensor rigidity is a result of excessive drive through, primarily, the descending reticulospinal fiber system. Those parts of the brainstem reticular formation that give rise to these excitatory projections receive input from a variety of sources, for example, the anterolateral system (ALS). Increased drive through the ALS, such as stimulating the skin between the toes, will increase the drive through the reticulospinal pathway and exacerbate the extensor rigidity. Since this patient has transitioned from decorticate to decerebrate rigidity, the corticospinal and rubrospinal tracts (choices B and E) are both rendered nonfunctional; reticulospinal fibers excite extensor motor neurons, not flexor motor neurons (choice C). A decrease in outflow through the cerebellothalamic pathway would have no clinical correlate/effect, since the forebrain is functionally disconnected from the brainstem.

15 **The answer is C: Inhibition of alpha motor neurons innervating the muscle associated with the activated tendon organ.** The Golgi tendon organ is located at the tendon-muscle junction, is the sensory ending on an Ib fiber (fiber diameter 12 to 20 μm, conduction velocity 8 to 120 m/s), and terminates on glycinergic interneurons in the anterior horn which, in turn, inhibits the alpha motor neurons innervating the muscle containing the activated tendon organ. Golgi tendon organs are slowly adapting receptors that sense tension within muscles and muscle force. Some tendon organs are activated by low levels of muscle tension, while others are activated by extremely high levels of muscle tension; these latter serve a protective function by reducing the likelihood of damage to the body of the muscle, damage at the muscle tendon junction, or damage at the attachment of the tendon. The activation of a Golgi tendon organ does not excite alpha or gamma motor neurons either directly or indirectly and does not inhibit gamma motor neurons. In addition, the activation of a Golgi tendon organ does not monosynaptically inhibit alpha motor neurons; rather, the Ib primary afferent fibers excite an inhibitory interneuron (glycinergic) that, in turn, inhibits the alpha motor neuron.

16 **The answer is C: Caudal part of motor decussation.** This man has the unusual observation of bilateral weakness of both lower extremities in the absence of any other obvious neurological signs or symptoms. Basically, there are only two areas where a single lesion may result in bilateral weakness of the lower extremities. The first is a lesion within the caudal part of the motor decussation; the crossing of corticospinal fibers is somatopically arranged within the decussation: fibers representing the upper extremities decussate in its rostral portion and fibers representing the lower extremities decussate in its caudal portion. A small lesion at this latter location will result in bilateral lower extremity weakness with upper extremity sparing as well as sparing of all cranial nerves and all sensory modalities. The second is a meningioma of the falx cerebri (falcine meningioma) that impinges on the anterior paracentral gyrus bilaterally; this is the somatomotor cortex for the lower extremity. However, this particular lesion may present with asymmetrical motor deficits or with motor deficits that are accompanied by some sensory deficits due to partial impingement on the posterior paracentral gyrus (the somatosensory cortex for the lower extremity). Bilateral caudal pontine damage would result in weakness of all extremities (corticospinal tract) and in a likely weakness of both lateral recti muscles (abducens nerve). A comparable lesion in the medial medulla would result in weakness of all extremities (again, corticospinal fibers), in an inability to protrude the tongue (both hypoglossal nerve roots), and in bilateral sensory deficits (medial lemniscus). Damage to the entire motor decussation would produce weakness of all extremities, but no other deficits. It is appropriate to note that this type of injury would typically be seen with atlantoaxial dislocation and would most likely have an associated XIIth nerve palsy.

17 **The answer is E (Medial crus cerebri plus oculomotor fibers on the right).** This man presents with a hemiplegia on the left upper and lower extremities and oculomotor deficits (muscle weakness, pupil dilation) on the right; this is characteristic of a lesion in the medial midbrain on the right side, a Weber syndrome (sometimes this may be called the cerebral peduncle syndrome). The best localizing sign in this case is the oculomotor deficits: it specifies a lesion in the medial midbrain encompassing the IIIrd nerve and corticospinal fibers. Recall that the exiting roots of the IIIrd nerve and the corticospinal fibers within the middle portions of the crus cerebri share a common blood supply. The lesion at A produces the same deficits, but they are on the wrong side (left oculomotor deficits, right hemiplegia). The lesion at B includes corticospinal fibers but not the oculomotor nerve, while the lesion at C involves oculomotor fibers (plus the red nucleus and cerebellothalamic fibers), but not corticospinal fibers. The lesion at D involves corticospinal fibers and some oculomotor fibers, and is on the correct side for the deficits experienced by this patient. However, this lesion also includes the medial lemniscus; deficits of discriminative touch and vibratory sense are not a complaint in this man.

18 **The answer is B: Deviation of the uvula to the right on phonation.** This right-sided lesion also encompasses the medial portion of the middle third of the crus cerebri which contains corticonuclear fibers. This particular fiber bundle contains fibers that arise from the motor cortex, descend through the crus cerebri, and pass through this particular area of the crus. These fibers influence the activity of lower motor neurons of cranial nerves, particularly those of the facial, hypoglossal, and accessory nuclei and the nucleus ambiguus. From their location in the crus cerebri, corticonuclear fibers pass to the opposite nucleus ambiguus (pharyngeal and laryngeal musculature), hypoglossal nucleus (genioglossus muscle), and the facial nucleus (that part innervating the muscles on the lower face). Corticonuclear fibers also enter the accessory nucleus on the same side, although the pattern is somewhat less consistent. In this man's case, the lesion is in the crus on the right; corticonuclear fibers travel to the left nucleus ambiguus and, because of the somewhat unique arrangement of the musculus uvulae, the uvula will deviate to the right on phonation, or when the patient says "Ah." Based on the trajectory of corticonuclear fibers, a thorough examination in this man would further show a deviation of the tongue to the left on protrusion (not the right), weakness of the lower face on the left (not the right), and, if present, difficulty elevating the right shoulder against resistance (not the left). There are no deficits of the masticatory muscle function, since these motor neurons, on each side, receive bilateral corticonuclear input.

19 **The answer is E: Weak right lower extremity and left upper extremity.** Small infarcts within the brainstem may result in deficits that may not fit the standard (or expected) plan but nonetheless reflect the realities of the structures involved. Corticospinal fibers arising from the upper extremity area of the somatomotor cortex cross in the rostral part of the motor decussation and descend to cervical levels of the spinal cord. Corticospinal fibers arising from the lower extremity region of the somatomotor cortex cross in the caudal part of the motor decussation and descend to lumbosacral levels of the spinal cord. This lesion is on the left side, is located in the rostral and lateral portion of the motor decussation and, as seen in the image, avoids the midline. In this position, the lesion damages corticospinal fibers related to the lower extremity before they decussate (left-sided lesion = right-sided weakness of the lower extremity) and corticospinal fibers related to the upper extremity after they decussate (left-sided lesion = left-sided weakness of the upper extremity). In contrast, lesions in the midline of the motor decussation result in either bilateral weakness of the upper extremities (if the lesion is in the rostral part of the decussation) or bilateral weakness of the lower extremities (if the lesion is in the caudal part of the decussation).

20 **The answer is D: Corticospinal deficits and weakness of lateral gaze to the opposite side.** There are three levels/locations in the brainstem at which damage will produce an alternating hemiplegia. These are designated based on their relative positions within the brainstem. A superior alternating hemiplegia (Weber syndrome) involves the corticospinal fibers in the crus cerebri and the exiting root of the oculomotor nerve, resulting in the deficits described in choices B and E. A middle alternating hemiplegia (Foville syndrome) involves corticospinal fibers in the basilar pons and the immediately adjacent exiting root of the abducens nerve. This results in corticospinal deficits and weakness of lateral gaze to the side opposite the extremity weakness due to the abducens involvement and the consequent weakness of the lateral rectus muscle, the condition seen in this man. An inferior alternating hemiplegia (Dejerine syndrome) involves the corticospinal fibers in the pyramid and the immediately exiting root of the hypoglossal nerve, resulting in the deficits in choice A. Corticospinal deficits accompanied by somatosensory losses on the same side of the body are the result of lesions within the hemisphere at a supratentorial location.

21 **The answer is E: Myasthenia gravis.** The key observation in this case is the observation of antibodies to acetylcholine (nicotinic) receptors and enlarged lymph nodes. On top of this are vague symptoms that may come and go. Myasthenia gravis is an autoimmune neuromuscular disease that is of unknown origin, although it likely has a genetic basis in some families. In a patient of this age range, it is not likely to be related to the presence of a thymoma; such is the case in only about 10% of patients and these are usually in the 40- to 50-year age range. In addition to the clinical picture, myasthenia gravis is characterized by muscle weakness that may wax and wane and by what is called fatigability: the patient's muscles get weaker with continued exertion or as the day wears on, but improve somewhat with rest. In general, patients will report that they feel fairly good in the morning, but are weaker and tired later in the day. Amyotrophic lateral sclerosis is a neurodegenerative disease of largely unknown origin that is rapidly fatal (2 to 5 years), is not treatable or curable, and almost exclusively affects lower motor neurons. These patients experience progressive muscle weakness, muscle wasting, and may require artificial ventilation. Foster Kennedy syndrome is optic nerve atrophy, scotoma, and papilledema caused by a tumor, usually a meningioma, of the meningeal sheath surrounding the optic nerve. Guillain-Barré syndrome is an autoimmune disorder that affects peripheral nerves (motor and sensory), is of fairly rapid onset (1 to 2 weeks), reaches its apex in about 3 to 4 weeks, and from which the patient may recover with supportive care. Multiple sclerosis is an autoimmune demyelinating disease of the central nervous system (CNS), the symptoms of which may be acute of chronic. These lesions may be "separated in time and space": the lesions may appear at different times and at different locations within the CNS and may affect almost any part of the brain; this is not a disease of the peripheral nervous system.

22 **The answer is A: Weakness of the eyelids and ocular muscles.** The general symptoms, enlargement of some lymph nodes, coupled with the presence of antibodies to acetylcholine receptors, clearly point to the likelihood of myasthenia gravis (MG). In 50% to 65% of patients, the first signs and symptoms are weakness of the levator palpebrae and/or extraocular muscles, with resultant drooping of the eyelid and diplopia; in about 85% to 90% of patients, these muscles are eventually involved and in about 10% to 15% of MG patients, these are the only muscles involved. Weakness of the oropharyngeal and/or facial muscles frequently appears after the initial weakness of ocular muscles; in about 10% to 15% of MG patients, these may be the first muscles involved or the only muscles involved. Weakness of muscles of the body (extremities, diaphragm, postural, neck extensors) may be episodic, exacerbated by exercise and/or stress, and may not affect all muscles, or muscle groups, in the same manner.

23 **The answer is D (The posterior limb of the internal capsule on the right).** The combination of motor and sensory deficits on the same side of the body, coupled with a lack of any cranial nerve deficits, is characteristic of lesions in the hemisphere. The right posterior limb of the internal capsule contains corticospinal fibers that influence spinal lower motor neurons on the left side of the body, and thalamocortical fibers from the right ventral posterolateral and posteromedial (VPL and VPM) nuclei conveying somatosensory information from the left side of the face and body. This sensory information is carried by the anterolateral and posterior column–medial lemniscus systems from the left side of the body to the right VPL and from the left spinal trigeminal nucleus to the right VPM. A lesion of these motor and sensory pathways at the level of the internal capsule on the right results in left-sided deficits of the respective modalities. The lesion of the anterior limb of the internal capsule (at A) damages anterior thalamic radiations and frontopontine fibers. The lesion at B is in the left posterior limb; the deficits of this lesion would be on the right side, not the left. The lesion at C is in the posterior thalamus, involves the VPL and VPM, and may give rise to right-sided sensory deficits. However, there may be preservation on some modalities and, if there is weakness, it is usually transient. There is no indication of a genu lesion (at E) in this man's situation.

24 **The answer is C: Inability to rotate the head to the left against resistance.** The inability to elevate her right shoulder against resistance and the location of a small defect on the right side of medulla extending into the cervical cord indicate involvement of the accessory nucleus. The accessory nucleus is located adjacent to the motor decussation in the lateral portion of the remnants of the anterior horn at the junction of the medulla with the cervical cord, and it extends into the upper cervical levels at the anterolateral border of the anterior horn. This nucleus contains lower motor neurons that innervate the trapezius and sternocleidomastoid muscles on the ipsilateral side. These muscles are frequently tested by asking the patient to attempt a movement against moderate resistance. Difficulty elevating the shoulder against resistance (the physician's hand placed on the patient's shoulder) indicates weakness of the trapezius muscle. Due to its origin, predominately from the manubrium sterni and its insertion on the mastoid process, contraction of the sternocleidomastoid muscle rotates the head toward the opposite side. Consequently, when the right sternocleidomastoid muscle is weak, the patient is not able to rotate the head to the left against resistance at the chin; the patient is able to rotate the head to the right against resistance, since the left sternocleidomastoid is intact. The motor nuclei, and the nerves, innervating the tongue, facial muscles, and muscles of the larynx, are not located at this level of the medulla or the cervical cord.

25 **The answer is D: Gamma loop.** Intrafusal muscle fibers are those small muscles located within the capsule of the muscle spindle; intrafusal muscle fibers are innervated exclusively by gamma motor neurons of the anterior horn. The gamma loop consists of the following elements: (1) supraspinal excitation of a gamma motor neuron; (2) this excitation produces contraction of intrafusal muscles with stretching of the equatorial region of the intrafusal fiber; (3) this stretching results in excitation of the Aα (Ia) fiber; and (4) the Aα afferent terminates on, and excites, the alpha motor neuron and the extrafusal muscle fibers contract. This circuit is essential to the successful completion of many motor acts. The Babinski reflex, also called the Babinski sign, is dorsiextension of the great toe with aggressive stimulation of the lateral sole of the foot; this is considered a normal response from birth to about 6 to 8 months; after that it is considered indicative of corticospinal tract damage. The crossed extensor reflex is the extension of one lower extremity in response to a noxious stimulus on the opposite foot; it may also be seen in the upper extremity. The flexor reflex, also called the withdrawal or nociceptive reflex, is the withdrawal of an extremity from a noxious stimulus applied to that extremity. Reciprocal inhibition is the inhibition of the antagonistic group of muscles to another group that is excited as part of a muscle stretch reflex; reciprocal inhibition essentially enhances the clarity of the reflex.

26 **The answer is D: 65 to 70 m/s.** The largest diameter corticospinal fibers arise from the largest pyramidal cells of the somatomotor cortex; these are called Betz cells. While it was originally believed that most corticospinal fibers arose from these large cells, it is now known that less than 10% of corticospinal fibers actually arise from large pyramidal cells of the motor cortex. The conduction velocity of these large motor fibers is in the range of 65 to 70 m/s. The vast majority of corticospinal fibers arise from smaller pyramidal cells of the motor cortex, have diameters in the range of 5 to 7 μm, and have correspondingly lower conduction velocities. Lower motor neurons have fiber diameters in the range of 12 to 20 μm and conduction velocities in the range of 70 to 110 m/s. Generally, this applies to lower motor neurons in the spinal cord as well as in the motor nuclei of cranial nerves. Visceromotor (autonomic) fibers have small diameters, are lightly myelinated, and have conduction velocities of less that 18 m/s (preganglionic) or less than 2 m/s (postganglionic).

27 **The answer is B (The approximate middle one-third of the precentral gryus).** The somatomotor cortex (also commonly called the primary somatomotor cortex, Brodmann area 4) is topographically arranged in the precentral gyrus and in the anterior paracentral gyrus. The precentral gyrus is divided into approximately thirds: the face is represented in the lateral third (C), the upper extremity with particular emphasis on the hand in the middle third (B), and the trunk in the medial third (A). The lower extremity is represented in the anterior paracentral gyrus (E) which is, functionally, the continuation of the somatomotor cortex on to the medial aspect of the hemisphere. Damage to the upper extremity area of the motor cortex on the right correlates with weakness of the upper extremity on the left. Such cortical lesions also usually involve some portions of the subcortical white matter. The lesion at D is located in the superior temporal gyrus; damage to this area would not be specifically characterized by motor deficits.

28 **The answer is B: Dejerine syndrome.** This man has the signs and symptoms that are characteristic of a medial medullary syndrome, also commonly known as the Dejerine syndrome. This is a lesion in the territory of the penetrating branches of the anterior spinal artery. The damaged structures are the corticospinal fibers in the pyramid, the immediately adjacent exiting roots of the hypoglossal nerve, and the medial lemniscus which is located immediately posterior to the pyramid. The deficits (left-sided hemiplegia and sensory losses, tongue deviation to the right) specify a lesion on the right side of the medulla involving the above structures. The hypoglossal deficit is the best localizing sign in this case. The Benedikt syndrome is a syndrome of the tegmentum of the midbrain that involves the red nucleus, oculomotor fibers, and cerebellothalamic fibers. The Dejerine-Roussy syndrome may be seen consequent to an infarct of the posterior thalamus that involves principally the sensory relay nuclei; the patient experiences sensory deficits that may include pain accompanied by a transient weakness of the extremities. The Foville syndrome results from a lesion of the caudal and basal pons that involves corticospinal fibers and the abducens root; this gives rise to an alternating or crossed deficit that may also be called a middle alternating hemiplegia. The Wallenberg syndrome, commonly called the posterior inferior cerebellar artery (PICA) syndrome (or the lateral medullary syndrome), involves the restiform body, spinal trigeminal nucleus and tract, anterolateral system, nucleus ambiguus, solitary tract and nucleus, and portions of the vestibular nuclei. A cardinal deficit in this syndrome is an alternating hemianesthesia.

29 **The answer is D: Supraspinal activation of gamma motor neurons→ intrafusal muscle fibers contract→ spindle equator stretched→ Ia fiber activity increases→ alpha motor neurons activated→ extrafusal muscle fibers contract.** Intrafusal and extrafusal muscle fibers work in concert to assure coordinated, purposeful, and effective movements of the skeletal musculature. These circuits are also essential for the maintenance of posture and the success of muscle stretch reflexes. The circuit of the gamma loop is as follows: (1) supraspinal centers such as the reticular formation excite gamma motor neurons; (2) the axons of these neurons terminate on intrafusal muscle fibers in the ends of the nuclear bag and chain fibers; (3) contraction of these muscle fibers stretches their respective equators; (4) this stretching activates the Ia endings surrounding the equator and generates an action potential in this fiber; (5) the central process of the Ia fiber monosynaptically activates the alpha motor neurons innervating the extrafusal muscle fibers from which the afferent volley arose; and (6) the extrafusal muscle fibers contract. A variation on this important circuit is what is called alpha-gamma coactivation. This is a pathway that is activated when a muscle is steadily contracting against a load: for example, repeatedly lifting dumbbells in the gym. In alpha-gamma coactivation, supraspinal centers send simultaneous messages to the alpha motor neuron, for the contraction of the extrafusal muscle fibers, and to the gamma motor neurons, for the contraction of the intrafusal muscle fibers. If the extrafsual fibers contracted independent of the intrafusal muscle fibers, the effectiveness of the muscle spindle would be rendered useless. This would have profoundly negative effects on muscle tone, reflexes, and on the ability of the muscle to contract effectively. However, in alpha-gamma coactivation, as the extrafusal muscle fibers contract, the intrafusal fibers also contract, and continual tension within the spindle is maintained throughout the range of muscle contraction. As a result of this continuous tension within the spindle, there are also continuous messages being relayed into the nervous system by the Ia fibers arising from the spindle. There is ongoing feedback on tone and tension within the muscle.

30 **The answer is E: Weakness of the extremities on the right, oculomotor palsy on the right.** In this man, the significant damage to the lower portions of the hemisphere (particularly the temporal lobe and the adjacent occipital lobe) has resulted in a herniation of the right temporal lobe into the midbrain. Rather than damaging the midbrain on the right side, the herniation has pushed the midbrain to the left and against the edge of the tentorium cerebelli. This results in deficits that reflect the nature of the damage. First, the impingement of left crus cerebri on the tentorial edge damages corticospinal fibers within the crus on the left, producing weakness of the extremities (upper and lower) on the right side of the body. Second, the shift of the midbrain to the left stretches, or results in avulsion of, the oculomotor roots on the right side; this results in loss of most eye movement (oculomotor palsy) on that side. This alternating damage, right IIIrd nerve and left corticospinal fibers, produces motor deficits that are all seen on the right side. In this particular situation, the corticospinal signs are called a false localizing sign. This clinical scenario is called the Kernohan phenomenon (more correctly called the Kernohan syndrome) and is named in recognition of James W. Kernohan who first described this clinical scenario. A Kernohan is a clinical diagnosis that is confirmed by CT or MRI. It should also be emphasized that similar deficits can be seen at other levels of the brainstem. For example, a space-taking lesion adjacent to the medulla may produce avulsion of the hypoglossal root on the same side and push the opposite pyramid against the skull, resulting in extremity weakness and hypoglossal weakness on protrusion, all on the same side.

31 **The answer is D: Right anterior paracentral gyrus.** The outlined area is the left lateral corticospinal tract. Within the spinal cord, these corticospinal fibers arise from the opposite (right in this example) cerebral cortex and are somatotopically arranged. The most lateral fibers in the tract arise from the lower extremity region of the cortex (anterior paracentral gyrus), cross in the caudal part of the motor decussation, and descend to lumbosacral cord levels. The more medial fibers arise from the upper extremity area of the precentral gyrus, cross in the rostral part of the motor decussation, and descend to cervical cord levels. Fibers to thoracic cord levels are located in-between these two parts at all levels. The lower extremity is somatotopically represented in the anterior paracentral gyrus; fibers in the spinal cord on the left arise from this gyrus on the right side of the cerebral hemisphere. The left anterior paracentral gyrus gives origin to fibers in the right lateral corticospinal tract; the left posterior paracentral gyrus is the somatosensory cortex for the right lower extremity. The right and left precentral gyri have face, upper extremity, and trunk areas of the body represented therein, but not the lower extremity.

32 **The answer is A: Acetylcholine.** All the signs and symptoms exhibited by this woman indicate that she has myasthenia gravis (MG); they are quite characteristic of this disease. These features, among others, are initial oculomotor involvement, some facial weakness, and fatigability. This is an autoimmune disease in which circulating antibodies to nicotinic acetylcholine receptors bind at receptor sites of the neuromuscular junction and not only decrease its functional integrity but also damage the morphology of the junction. About 10% of MG patients will also present with thymoma and about 60% of patients, usually those under 40 years of age, will characteristically have medullary hyperplasia of the germinal centers of lymph nodes. There are a variety of treatments for MG, and many patients, with proper treatment, have long, successful, and productive lives. Dopamine is a neurotransmitter associated with connections of the basal nuclei; a loss of dopamine in the substantia nigra, pars compacta, results in Parkinson disease. Dopamine is also found in a few other restricted areas such as tegmental areas of the midbrain. Histamine is largely restricted to the hypothalamus and is found in some hypothalamocerebellar projections. Glutamate is found at many locations in the CNS including within the efferents of the cerebral cortex such as corticospinal, corticonuclear, and corticostriate fibers. Glycine is an important inhibitory neurotransmitter in the spinal cord and, in lesser numbers, at a few other restricted regions.

33 **The answer is B: Lateral vestibular nucleus.** Fibers comprising the lateral vestibulospinal tract arise from cells within the lateral vestibular nucleus, descend on the ipsilateral side, assume a position in the ventral funiculus in the spinal cord, and terminate mainly in spinal laminae VII and VIII on interneurons that, in turn, influence alpha motor neurons in the anterior horn. The general somatotopic pattern of these projections arising from the lateral vestibular nucleus is as follows: cells in the rostral part of the nucleus terminate in cervical cord levels, cells in the middle part project to thoracic levels, and cells in the caudal portion terminate in lumbosacral cord levels. In this respect, this tract influences all levels of the spinal cord. The medial vestibulospinal tract arises from cells located mainly in the medial vestibular nucleus with some contributions from the inferior (also called the spinal) vestibular nucleus; these fibers descend bilaterally in the medial longitudinal fasciculus, but with a clear ipsilateral preponderance. This tract projects mainly to cervical and upper thoracic levels and also terminates on interneurons in laminae VII and VIII.

34 **The answer is B: Left hemiplegia, left homonymous hemianopia.** The vascular territories represented in this image are those arising from the anterior choroidal artery; this vessel is usually one of the branches of the cerebral part of the internal carotid artery. It serves lower parts of the internal capsule (some genu, posterior limb), the optic tract, and portions of the amygdaloid complex, hippocampus, and medial cortex of the temporal lobe. This lesion is on the right side damaging the corticospinal fibers as they leave the hemisphere to enter the crus cerebri (giving rise to a left hemiplegia) and damaging the right optic tract (resulting in a left homonymous hemianopia). This somewhat odd combination of deficits (hemiplegia and hemianopia) is known as the anterior choroidal artery syndrome (also called the von Monakow syndrome). Based on the extent of the vascular territory superiorly into the posterior limb of the internal capsule, the area of damage may also involve the thalamocortical radiations from the ventral posterolateral and posteromedial thalamic nuclei to the sensory cortex. In the situation where these thalamocortical radiations are involved, and assuming a lesion on the right side, in addition to the left homonymous hemianopia and left hemiplegia, the lesion will produce a left hemianesthesia. It is recognized that this lesion may also produce, to variable degrees, corticonuclear deficits, although these may be inconsistent and not the predominant signs or symptoms.

Chapter 19

The Basal Nuclei

Recall the Cautionary Tale: The images in this chapter are presented in a Clinical Orientation: this approach emphasizes how structure and function correlate with deficits in a clinical setting. MRI and CT have a universally recognized orientation and laterality. Line drawings, stained sections, and gross brain images in the Clinical Orientation are viewed here in the identical manner that one views an MRI: your right, as the physician-observer, is the patient's left, and the observer's left is the patient's right. The laterality of deficits and reference to connections follow accordingly.

QUESTIONS

Select the single best answer.

1 Which of the following nuclei is the main target of efferent fibers that have their cells of origin in the medial segment of the globus pallidus?

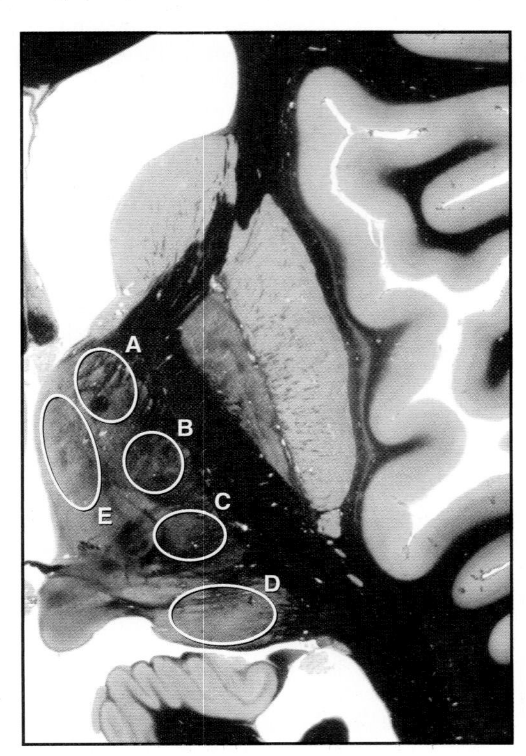

(A) A
(B) B
(C) C
(D) D
(E) E

2 A 36-year-old man presents with the initial complaint of clumsiness that he especially notices when he engages in fly fishing, his favorite sport. The history reveals that the man's father became similarly ill at 39 and died at 50 years of age. The examination reveals that the man has involuntary movements of his fingers and hands and, according to his wife, seems a bit absentminded and can easily become angry. Suspecting an inherited disorder, the physician orders a genetic analysis which reveals numerous CAG repeats on chromosome 4. An MRI of this patient would most likely reveal which of the following?
(A) Atrophy of cerebellar anterior lobe
(B) Atrophy of mammillary bodies and fornix
(C) Decrease in size of hippocampus
(D) Decrease in size of caudate nucleus
(E) Decrease in size of the temporal lobe

3 Which of the following structures, or combination of structures, constitutes the neostriatum?
(A) Globus pallidus only
(B) Putamen only
(C) Putamen + caudate nucleus
(D) Putamen + globus pallidus
(E) Putamen + globus pallidus + caudate nucleus

4 An 18-year-old man visits his family physician; his complaint is difficulty shooting in pickup basketball games with his colleagues. The examination reveals that the man has a brown-yellow ring of color at the edge of the cornea and a jerking tremor of his arms and hands with spreading of his fingers, which is seen when the man extends his upper extremities. Laboratory tests further reveal increased levels of serum copper. A diagnosis of hepatolenticular degeneration is made and a regime of treatment using penicillamine as the chelating agent is started. Which of the following might appear in this man as a consequence of his treatment?
(A) Cushing disease
(B) Guillain-Barré syndrome
(C) Lambert-Eaton syndrome
(D) Myasthenia gravis
(E) Multiple sclerosis

5 A 68-year-old man presents with a sudden onset of motor deficits. He is transported to the Emergency Department at a local hospital. An MRI reveals a lesion localized to the area outlined in the image below; this lesion is on the man's right side. Which of the following would most likely be the predominant difficulty experienced by this man?

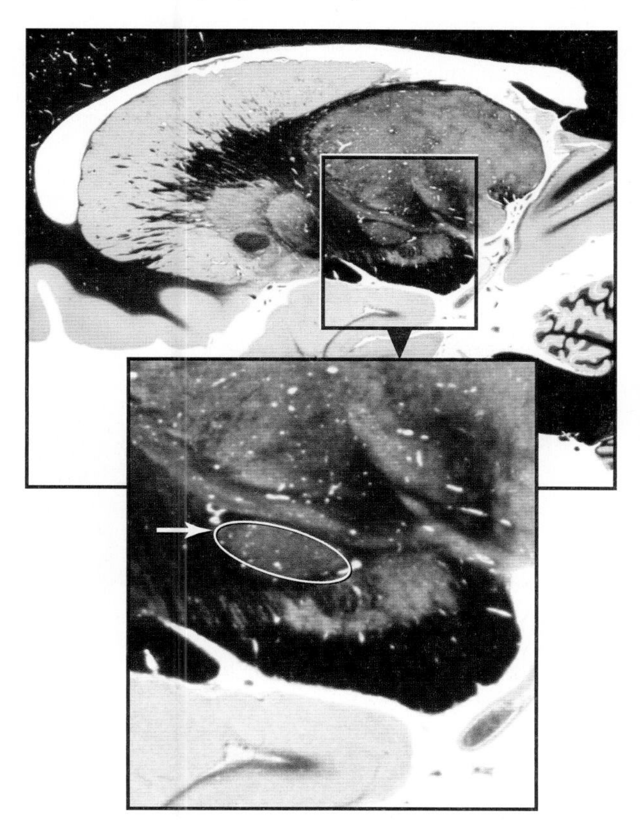

(A) Akinesia and bradykinesia
(B) Hemiballismus on the left
(C) Hemiplegia on the left
(D) Intention tremor on the right
(E) Resting tremor on the right

6 An occlusion of which of the following arteries, or group of arteries, would result in the most direct and significant damage to the head of the caudate nucleus?
(A) Anterior choroidal
(B) Anterior communicating
(C) Lenticulostriate
(D) Medial striate
(E) Thalamoperforating

7 A 64-year-old man visits his family physician as part of a routine follow-up. As this man sits quietly in his chair, he has rhythmic movements of his fingers; when he reaches to touch the physician's finger, the movement largely disappears. Which of the following describes this disorder in this man?
(A) Athetosis
(B) Dysdiadochokinesia
(C) Intention tremor
(D) Resting tremor
(E) Titubation

8 An 8-year-old girl is brought to the pediatrician by her mother. The history reveals that the girl had a serious streptococcal infection about 5 months ago that was treated with antibiotics. The examination shows that the girl has irregular rapid flowing movements of her upper extremities, movements of her facial muscles and, according to the mother, seems to get easily upset. This patient is most likely suffering which of the following?
(A) Huntington chorea
(B) Parkinson disease
(C) Sydenham chorea
(D) Tardive dyskinesia
(E) Wilson disease

9 A 63-year-old man presents with signs and symptoms clearly indicative of damage to the structure indicated at the arrow in the image below. Which of the following would be the first and most obvious deficit experienced by this patient?

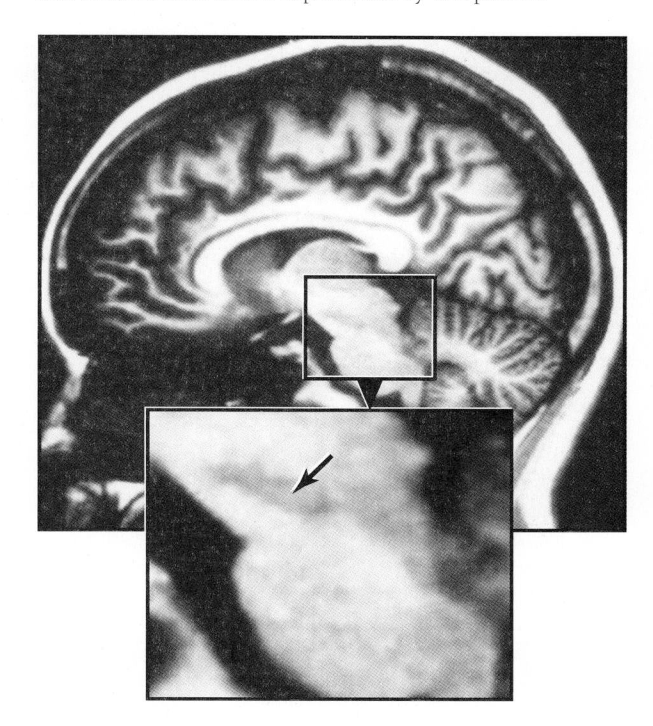

(A) Akinesia
(B) Bradyphrenia
(C) Dyssynergia
(D) Titubation
(E) Tremor

10 A 58-year-old man is diagnosed with a neurodegenerative disease of the basal nuclei. In general, his disorder results in an inability to clear glutamate from certain types of synapses within the corpus striatum. The excess glutamate results in an excess influx of calcium into the cell that initiates a cascade of cellular events resulting in the death of the cell. In which of the following does glutamate excitotoxicity play a prominent role?
(A) Hepatolenticular degeneration
(B) Huntington disease
(C) Parkinson disease
(D) Sydenham chorea
(E) Tardive dyskinesia

11 Corticostriatal excitation of striatal neurons that are projecting through the direct pathway results in which of the following?

(A) Disinhibition of thalamocortical neurons with resultant increase in activity of cerebral cortical neurons
(B) Disinhibition of thalamocortical neurons with resultant decrease in activity of cerebral cortical neurons
(C) Disinhibition of thalamocortical neurons with no change in the firing characteristics of cortical neurons
(D) Inhibition of thalamocortical neurons with resultant decrease in the firing patterns of cortical neurons
(E) Inhibition of thalamocortical neurons with resultant increase in the firing patterns of cortical neurons

12 A 59-year-old man dies following an illness of several years. His illness was characterized by irregular movements of his upper extremities and hands, tongue (resulting in dysarthria), and lower extremities. These deficits first appeared when he was 44 years old and became progressively worse in spite of medications. Along with these motor deficits, the man developed emotional lability, cognitive, and psychiatric problems; all became progressively worse. Treatment did not alter the course of his disease. The family requests an autopsy which reveals an extensive loss of medium spiny neurons within the neostriatum. This observation would suggest that this man most likely died from which of the following?

(A) Alzheimer disease
(B) Huntington disease
(C) Parkinson disease
(D) Pick disease
(E) Wilson disease

13 A 17-year-old boy undergoes an examination, which includes diagnostic laboratory tests that reveal high levels of amino acids in his urine and low levels of the α-2-globulin ceruloplasmin in his blood plasma. An MRI of this boy appears in the image below. Based on his laboratory results and on the appearance of his MRI, this boy is most likely suffering from which of the following?

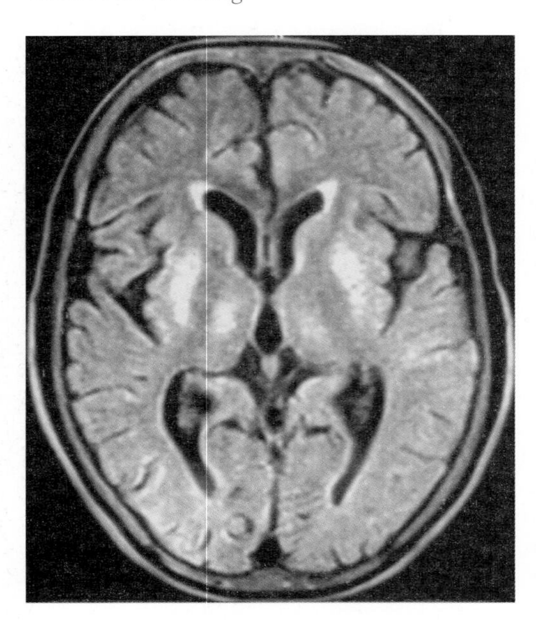

(Photograph courtesy of Professor Madhuri Behari.)

(A) Alzheimer disease
(B) Diffuse Lewy body disease
(C) Parkinson disease
(D) Pick disease
(E) Wilson disease

14 Which of the following structures in the image below contains cell bodies whose axons coalesce to form the ansa lenticularis and lenticular fasciculus?

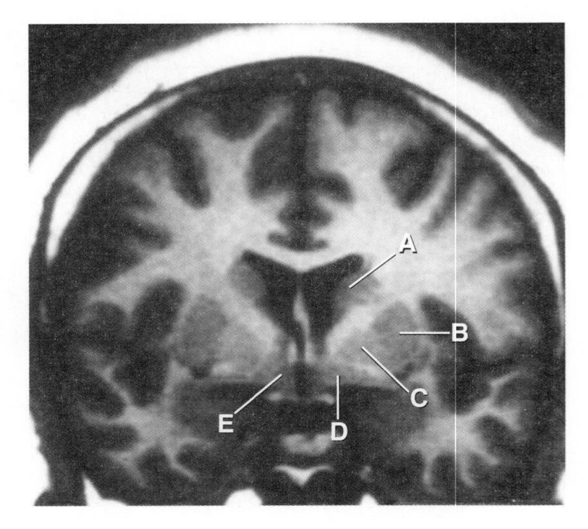

(A) A
(B) B
(C) C
(D) D
(E) E

15 A 67-year-old man has a resting tremor that was initially seen only in his left hand, but eventually involved his right hand. As this man's disease progressed, he experienced difficulty initiating voluntary movements and, once initiated, these movements were reduced in velocity and amplitude. Which of the following best describes this man's movement deficit?

(A) Akinesia
(B) Asterixis
(C) Bradykinesia
(D) Hypometria
(E) Titubation

16 A 59-year-old man presents with motor deficits that indicate profound cell loss in the shaded area in the image below. Which of the following deficits would most likely be seen in this man?

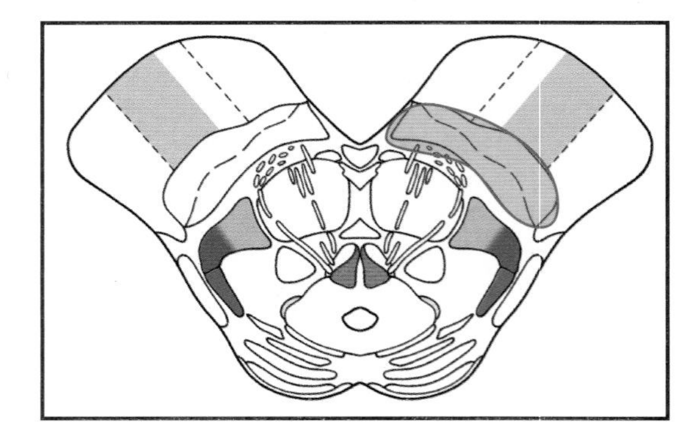

(A) Bilateral bradykinesia
(B) Hemiballismus on the right
(C) Intention tremor on the right
(D) Resting tremor on the right
(E) Resting tremor on the left

17 Which of the following neurotransmitters is specifically associated with the subthalamopallidal fibers that project to the medial segment of the globus pallidus?
(A) Acetylcholine
(B) Dopamine
(C) GABA
(D) Glutamate
(E) Glycine

18 A 21-year-old man has been diagnosed with hepatolenticular degeneration (Wilson disease). When he extends his arms, they tend to move up and down in an irregular arc and his hands flap up and down at the wrist with the fingers extended in irregular positions. Which of the following specifies this specific movement disorder?
(A) Akinesia
(B) Asterixis
(C) Athetosis
(D) Dysmetria
(E) Intention tremor

19 A 73-year-old woman is brought to the Emergency Department by her husband. The examination reveals an otherwise healthy trim woman who had a headache that responded to an OTC medication but was followed by a sudden onset of involuntary flinging movements of her right extremities, most pronounced in her right upper extremity. CT reveals a small localized hemorrhagic lesion. This lesion is most likely located in which of the following?
(A) Left caudate and putamen
(B) Left globus pallidus
(C) Left subthalamic nucleus
(D) Right globus pallidus
(E) Right subthalamic nucleus

20 A 49-year-old man visits his family physician with the complaint of "getting clumsy." The history, also provided by the man's wife, reveals that this man seems to have gotten more irritable over time: his wife indicates that he gets mad easily and it seems like "his personality has changed." The examination confirms ambulatory difficulties. The MRI of this man is shown in the given image. Which of the following is the most likely cause of this man's deficit?

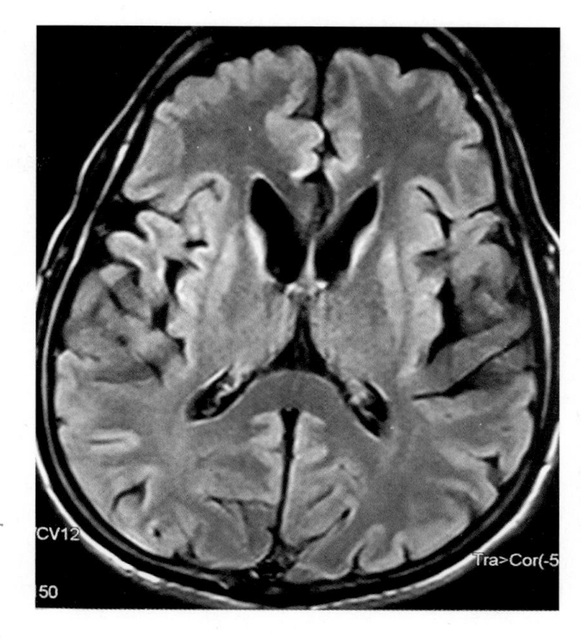

(A) Huntington disease
(B) Parkinson disease
(C) Pick disease
(D) Wilson disease
(E) von Hippel-Lindau disease

21 The laboratory results on a 19-year-old man reveal aminoaciduria. The examination shows a brown-yellow pigmentation at the corneoscleral junction and the attending physician suggests further tests to determine if this man has a metabolic disease. It is likely that these tests will show that this man has inherited difficulty metabolizing which of the following?
(A) Acetylsalicylic acid
(B) Aluminum
(C) Copper
(D) Iron
(E) Magnesium

22 An 81-year-old woman presents with symptoms (forgetfulness, anterograde and retrograde memory loss, personality changes, gait difficulties) of probable Alzheimer disease. In this disease, there is a loss of large neurons in the basal nucleus of Meynert and in the substantia innominata, and the appearance of plaques and neurofibrillary tangles in areas of the cerebral cortex. Which of the following neurotransmitters is diminished by the loss of these large cells?

(A) Acetylcholine
(B) Dopamine
(C) GABA
(D) Glutamate
(E) Substance P

23 Which of the following is the primary neurotransmitter associated with neuronal cell bodies located in the putamen and globus pallidus and with their projections?
(A) Acetylcholine
(B) Aspartate
(C) Dopamine
(D) GABA
(E) Glycine

24 Activation of the cell bodies that give rise to the fibers indicated by the arrow in the image below will result in which of the following responses?

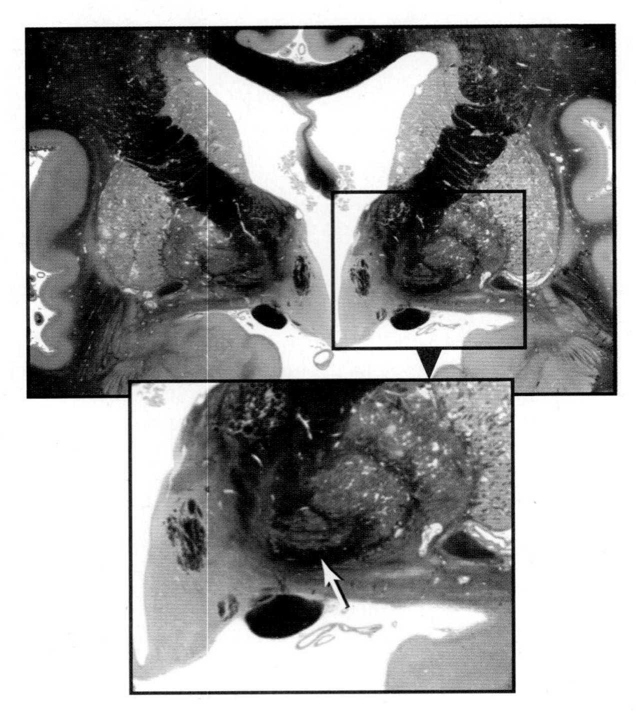

(A) Decrease in thalamic activity with corresponding increase in cortical activity
(B) Decrease in thalamic activity with corresponding decrease in cortical activity
(C) Decrease in thalamic activity with no change in cortical activity
(D) Increase in thalamic activity with corresponding increase in cortical activity
(E) Increase in thalamic activity with corresponding decrease in cortical activity

25 A 59-year-old man initially presents with a resting tremor in his right hand. Within a few weeks, this tremor also appears in his left hand. Over the next several months, he begins to assume a flexed (stooped-over) posture, has difficulty with unexpected falls (loss of righting reflex), and experiences some difficulty walking. Which of the following most likely specifies the characteristic ambulatory pattern of this man?
(A) Calcaneal gait
(B) Cerebellar gait
(C) Festinating gait
(D) Hemiplegic gait
(E) Scissors gait

26 Which of the following medical conditions, if treatment is initiated early and maintained, may have a remission of symptoms and improvement or recovery in most patients?
(A) Alzheimer disease
(B) Huntington disease
(C) Parkinson disease
(D) Pick disease
(E) Wilson disease

27 A 20-year-old-man has been seeing a neurologist for a movement disorder. One of his symptoms, for which he is being medicated, is a tremor confined largely to his hands and forearms, when his arms are extended; his fingers are spread or assume unusual postures, and the tremor is most obvious as an up-down motion at the wrist. Which of the following specifies the tremor in this man?
(A) Akinesia
(B) Asterixis
(C) Athetosis
(D) Chorea
(E) Dysmetria

28 Corticostriatal excitation of striatal neurons that are projecting through the indirect pathway of the basal nuclei results in which of the following?
(A) Disinhibition of pallidosubthalamic neurons with resultant decrease in activity of subthalamopallidal neurons, increase in pallidothalamic activity, and increase in cerebral cortical activity
(B) Disinhibition of pallidosubthalamic neurons with resultant increase in activity of subthalamopallidal neurons, decrease in pallidothalamic activity, and decrease in cerebral cortical activity
(C) Disinhibition of pallidosubthalamic neurons with resultant increase in activity of subthalamopallidal neurons, increase in pallidothalamic activity, and decrease in cerebral cortical activity
(D) Inhibition of pallidosubthalamic neurons with resultant decrease in activity of subthalamopallidal neurons, increase in pallidothalamic activity, and decrease in cerebral cortical activity
(E) Inhibition of pallidosubthalamic neurons with resultant increase in activity of subthalamopallidal neurons, increase in pallidothalamic activity, and decrease in cerebral cortical activity

29 A 68-year-old woman has a sudden onset of mild movement disorders while on a short cruise. MRI at her return revealed a hemorrhagic lesion in the area served by the lateral striate arteries. Which of the areas indicated in the image below represents the most likely location of this lesion?

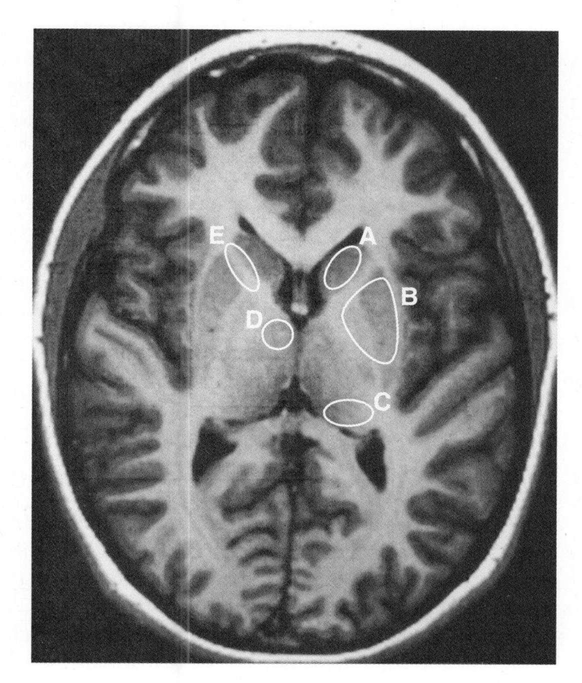

(A) A
(B) B
(C) C
(D) D
(E) E

ANSWERS

1 **The answer is B (Ventral lateral nucleus of the thalamus).** The medial segment of the globus pallidus contains cells that project to the ventral lateral nucleus of the thalamus, particularly its pars oralis portion. These pallidothalamic fibers traverse the lenticular fasciculus and the ansa lenticularis to enter the prerubral field; these two bundles coalesce in the prerubral field to enter the thalamic fasciculus for distribution to the ventral lateral nucleus, pars oralis (VLpo). These pallidothalamic fibers are GABAergic and therefore, inhibitory to their targets; they also terminate on the side ipsilateral to their origin. The VLpo projects to the motor cortex and exerts a major influence on its outflow especially via the corticospinal tract. The ventral anterior thalamic nucleus (choice A) also receives input from the medial globus pallidus and from the substantia nigra and intralaminar nuclei and projects to large areas of the frontal lobe largely avoiding the motor cortex. The dorsomedial thalamic nucleus (E) receives input from the amygdaloid complex, olfactory tubercle, various cortical areas, and the ventral pallidun and projects to the frontal and orbital cortices. The ventral postrolateral nucleus of the thalamus (C) receives input from the posterior column–medial lemniscus and anterolateral systems and projects to the somatosensory cortex (postcentral gyrus + the posterior paracentral gyrus). The pulvinar has reciprocal interconnections with the occipital extrastriate cortex and with lateral occipital and temporal cortices.

2 **The answer is D: Decrease in size of caudate nucleus.** The signs and symptoms exhibited by this man (age at the onset of his symptoms, family history, and the discovery of excessive CAG repeats) indicate that he has Huntington disease. This is a neurodegenerative disease that has movement components as well as cognitive and behavioral symptoms. As this disease progresses, additional symptoms will appear and the current ones will get worse; this disease is neither treatable nor curable. In general, patients with great numbers of CAG repeats have an earlier onset and a more severe course of the disease. The major, and most obvious, evidence in MRI is a marked decrease in the size of the caudate nucleus, specifically its head. In later stages of the disease, the ventricles may be enlarged (due to general brain atrophy) and sulci of the frontal and temporal lobes are enlarged. Atrophy of the cerebellar anterior lobe and of the mammillary bodies is seen in alcoholics, and decrease in the hippocampus and temporal lobe is seen in patients with various behavioral deficits or dementias.

3 **The answer is C: Putamen + caudate nucleus.** The putamen + caudate is the neostriatum, the putamen + globus pallidus is the lenticular nucleus, and the putamen + globus pallidus and caudate is the corpus striatum. The globus pallidus by itself is commonly called the paleostriatum or the pallidum. The putamen by itself is just called the putamen.

4 **The answer is D: Myasthenia gravis.** Use of penicillamine as a chelating agent in cases of hepatolenticular degeneration (Wilson disease) may result in the development of myasthenia gravis. This includes the development of a condition that presents with signs and symptoms similar to that seen in the adult form of myasthenia gravis: comparable muscle weakness and the presence of antibodies to acetylcholine receptors. These symptoms usually disappear after the use of penicillamine is discontinued. Cushing disease is seen in patients who have excessive corticotropin resultant to a pituitary tumor; these patients have a number of characteristic changes in body appearance. Guillain-Barré syndrome is an acute autoimmune disease of unknown etiology that advances rapidly (1 to 2 weeks), results in an ascending musculature weakness that eventually may involve respiratory, pharyngeal, and facial muscles; with supportive care the patient slowly recovers. Lambert-Eaton syndrome is a neuromuscular disease caused by faulty acetylcholine transmission that is due to the loss of voltage-gated calcium channels of the presynaptic membrane. This disease preferentially involves axial muscles, and 60% of patients with this condition also have small cell carcinoma of the lungs. Multiple sclerosis is an autoimmune demyelinating disease also of unknown etiology that has several forms, may affect any part of the CNS, is characterized by waxing and waning and fatigability, and may be successfully treated.

5 **The answer is B: Hemiballismus on the left.** This lesion involves the subthalamic nucleus, as indicated in this image, and it is on the patient's right side. The right subthalamic nucleus has interconnections with the right basal ganglia, and through these circuits regulates the right ventral lateral (VL) nucleus via pallidothalamic fibers. In turn, the right VL influences the right somatomotor cortex, which projects to the left side of the body; a lesion of the right subthalamic nucleus alters the activity of the right motor cortex and the deficits are seen on the left side of the body. In addition, these deficits (flailing, brisk movements) are usually, and predominately, seen in the upper extremity. Hemiballismus is most frequently the result of a vascular lesion. Akinesia and bradykinesia are characteristically seen in Parkinson disease, reflecting a loss of dopamine-containing cells of the substantia nigra, pars compacta. Hemiplegia may result from a variety of diseases or causes, but is a reflection of damage to corticospinal fibers. Intention tremor is seen in cases of lesions that involve cerebellar structures; a resting tremor is present in patients with diseases of the basal nuclei. None of these structures, other than the subthalamic nucleus, is damaged in this case.

6 **The answer is D: Medial striate artery.** The medial striate artery (artery of Heubner) originates from the anterior cerebral artery at, or close to, its junction with the anterior communicating artery. Clinically, it is common to refer to this vascular relationship as the ACA–ACom junction. This vessel serves the caudate head, adjacent portions of the anterior limb of the internal capsule, and smaller structures in this immediate area. The anterior choroidal artery serves the optic tract, lower portions of the internal capsule as it enters the crus cerebri, and adjacent portions of the temporal horn. Branches of the anterior communicating artery serve primarily rostral parts of the hypothalamus. The main target of lenticulostriate arteries is the lenticular nucleus and much of the posterior limb of the internal capsule (part of this structure is also served by the anterior choroidal). The thalamoperforating artery arises from P_1 and serves much of the anterior portions of the dorsal thalamus.

7 **The answer is D: Resting tremor.** A resting tremor is characteristically seen in Parkinson disease; this reflects a loss of the dopamine-containing cells of the substantia nigra, pars compacta. Tremor is the first symptom seen in about 70% of patients, and it may often appear on only one side of the body, but will become bilateral as the disease progresses. A significant number of cells must be lost for the tremor to appear. Athetosis is the slow, involuntary, and, sometimes, continuous movements seen in some neurodegenerative diseases of the basal nuclei. Dysdiadochokinesia (inability to do rapid alternating movements), intention tremor (a tremor that gets worse the closer the hand gets to the target when reaching), and titubation (a truncal or axial tremor) are all related to cerebellar lesions.

8 **The answer is C: Sydenham chorea.** The age of this patient, her prior history of a streptococcal infection, and the type of motor deficit (choreiform movements), all point to Sydenham chorea. This disease is regarded as an autoimmune condition that is a sequela to a streptococcal infection. This disease is basically a childhood occurrence, being most commonly seen in the age range of 5 to 15 years of age. Cardiac problems (endocarditis, pericarditis, myocarditis) may occur in some patients. This disease usually resolves with supportive care, unless there are other complications; some patients may experience recurrences later in their lives, and there is a 1% to 2% mortality related to cardiac complications. Huntington is seen in older patients (onset 35 to 45 years), presents initially with more jerky movements, and, while the movement disorders can be treated, there is no cure. Parkinson is usually seen in older patients (onset 45 to 65 years), although this disease does present in juvenile and young-onset forms; the symptoms can be treated but there is no cure. Wilson disease is an inherited disorder that may appear between 12 and 25 years that has a characteristic "wing-beating" tremor (asterixis); this condition can be treated. Tardive dyskinesia is a movement disorder that occurs when a patient is being treated for another disease.

9 **The answer is E: Tremor.** The structure indicated in this sagittal T1-weighted MRI is the substantia nigra. This nucleus consists of pars compacta, pars reticulata, and pars lateralis portions; the pars compacta contains dopaminergic cells whose projections participate in important pathways through the basal nuclei. A loss of these dopamine containing cells is a prime factor in the development of Parkinson disease (PD). The initial symptom seen in most patients (about 70%) with early PD is tremor. This is of the resting type, may begin on one side, but will become bilateral as the disease progresses. This resting tremor, also called a pill-rolling tremor, is one of the cardinal signs of PD. Akinesia, an inability to initiate a voluntary movement, and bradyphrenia, a slowness in processing information that results in a slow response to requests or questions, are both seen in PD, but are not the initial deficits. Akinesia, an inability to initiate a voluntary movement, and bradykinesia, a significant decrease in the ability to initiate a movement, are both features of PD, but are seen later in the disease. Dyssynergia, also called decomposition of movement, is seen in disorders of the cerebellum, and is characterized by the movement being broken down into its individual components rather than being smooth, purposeful, and accurate. Titubation, an ataxia of the trunk, is also a sign of cerebellar dysfunction.

10 **The answer is B: Huntington disease.** Normally, cortical fibers release glutamate at their synapses on striatal neurons and the result is excitation of the striatal neuron. For largely unknown reasons, in Huntington disease, the glutamate is not quickly cleared from the *N*-methyl D-aspartate receptor site. The persistence of glutamate opens many calcium channels; the resulting calcium influx leads to a cascade of events that results in cell death. In none of the other choices does glutamate excitotoxicity play a role. While cell loss in the substantia nigra, pars compacta, is a major issue in Parkinson disease, the cause is not well understood.

11 **The answer is A: Disinhibition of thalamocortical neurons with resultant increase in activity of cerebral cortical neurons.** The direct pathway through the basal nuclei consists of corticostriatal projections to the neostriatum, a striatopallidal projection to the inner segment of the globus pallidus, and a pallidothalamic projection to the ventral lateral nucleus of the thalamus. Corticostriatal projections are glutaminergic and excite the striatopallidal neurons on which they terminate. These inhibitory (GABAergic) striatopallidal neurons, in turn, inhibit the pallidothalamic neurons (these are also GABAergic) that project to the thalamus. The thalamic neurons on which these pallidothalamic fibers terminate (the ventral lateral nucleus) are now removed from their usual inhibitory (GABAergic) influence on the pallidothalamic projection; the thalamocortical neurons are disinhibited. In this scenario, the thalamocortical neurons are more active, resulting in more activity of cortical neurons. This increased activity of thalamocortical neurons reflects not the fact that the thalamopallidal inhibition has been removed but the fact that other excitatory inputs to the thalamus, such as cerebellothalamic or corticothalamic fibers, now have more influence.

12 **The answer is B: Huntington disease.** This man's disease is characterized by movement disorders of the choreiform type (described for this man), emotional, behavioral, and mental problems: all features of Huntington disease. This disease is genetic in origin: the patient has excessive CAG repeats on the 4p16.3 chromosome; the greater the number of repeats, the earlier the onset, more rapid the progression, and more severe the disease. The principal histological observation is a significant loss of the medium spinal neurons (medium, reflecting their size) in the neostriatum (caudate + putamen). The two principal cell populations in the neostriatum are these medium spiny cells (about 75% to 80%) and large multipolar cells; the former are selectively lost, and the latter are spared. Glutamate excitotoxicity is believed to be an important factor in the destruction of these neurons. The principal histological finding in Alzheimer is the presence of plaques that form around β-amyloid deposits and the presence of neurofibrillary tangles. The loss of dopamine-containing cells, and of their terminals (D1 and D2), in the neostriatum, is a characteristic feature of Parkinson disease. Since these cells also contain melanin, there is also a conspicuous loss of these dark cells. Histological changes in Pick disease (frontotemporal atrophy) are largely restricted to the appropriate areas of the cerebral cortex and consist of cell loss in mainly layers I to III, gliosis, swelling of cells (called ballooning), and the presence of fibrils (Pick bodies) in cells. Wilson disease does present with motor deficits, but of a comparatively unique type of deficit (flapping tremor). Histologically, there is involvement, up to frank cavitation, of the lenticular nucleus and other areas of the brain. However, in contrast to this man, symptoms of Wilson disease usually appear much earlier (18 to 25 years), are accompanied by a Kayser-Fleischer ring, and are treatable and curable.

13 **The answer is E: Wilson disease.** The age of this boy, laboratory data (elevated levels of amino acids in his urine, low plasma levels of ceruloplasmin), and the clarity of bilateral hyperintense areas within the putamen and, to a lesser degree, in the thalamus, indicate that he has Wilson disease. This condition is also called hepatolenticular degeneration due to the propensity for involvement of the liver and the putamen, although other brain areas, as seen in this image, may also be involved. This inherited disorder can be treated with a chelating agent to remove the copper; a diet low in copper may also be mandated. Alzheimer is a disease of the elderly, characterized by a range of behavioral, cognitive, and motor signs and symptoms; it can only be definitively determined by examination of tissue samples at autopsy. Diffuse Lewy body disease is also a disease of the elderly, and is also characterized by behavioral, cognitive, and motor deficits. Parkinson disease onset is usually seen at age 50 to 60+ and is characterized by resting tremor, bradykinesia, and typical locomotor difficulties. However, Parkinson may also be seen in patients under 20 years (juvenile Parkinson), in patients 20 to 40 years (young onset Parkinson), or as a result of repeated trauma (pugilistic Parkinson); the first two may have a genetic link, and all three are relatively rare. Pick disease, also called frontotemporal atrophy/dementia, is a middle-age+ condition (50+ years), has behavioral and cognitive consequences and loss of neurons in many areas of the brain, particularly the cerebral cortex.

14 **The answer is C (The globus pallidus).** The ansa lenticularis and the lenticular fasciculus contain pallidothalamic fibers (GABAergic) that arise from the internal segment of the globus pallidus, come together in the prerubral field, and then enter the thalamic fasciculus, from which they distribute to the ventral lateral (VL) nucleus of the thalamus (plus others). When activated, these fibers are inhibitory to the thalamic neurons on which they terminate; when these fibers are inhibited, the general activity of the VL goes up (the thalamus is disinhibited). The neostriatum (A = caudate nucleus, B = lenticular nucleus) receives a glutaminergic corticostriatal projection and gives rise to a GABAergic striatopallidal projection to the lateral pallidal segment. When activated, these fibers inhibit their targets. The ventral pallidum (D) has connections with the frontal lobe; degenerative changes of these cells result in cognitive decline. The hypothalamus (E) is a major visceral center that has connections with many nuclei throughout the brain but particularly with visceral nuclei of the brainstem and spinal cord.

15 **The answer is C: Bradykinesia.** This man has the tremor (resting), progressive (unilateral then bilateral) and locomotor features that characterize Parkinson disease. Bradykinesia is difficulty initiating a voluntary movement, for example, turning while standing and then starting to walk; the patient may have difficulty initiating the movements to start the turn, make the turn, and initiate the steps to proceed. Once underway, the patient may have difficulty stopping the forward motion; bumping into objects in the environment (table, chair, door) is commonplace. The man will also have short shuffling steps that may get more rapid as the man progresses: a festinating gait. Akinesia, also characteristic of Parkinson disease, is the complete lack of the ability to initiate a voluntary movement; asterixis, a feature of Wilson disease (or other metabolic encephalopathies), is a tremor of the arm and hand seen when the upper extremities are extended: a wing-beating or flapping tremor. Hypometria is falling short when reaching for an object: the patient reaches for a book but the reach falls short and does not get to the target. The counterpoint to hypometria is hypermetria: a reach that extends beyond the target. Titubation is a tremor of the head, neck, and axial portions of the body. Titubation and hypometria are both features of cerebellar lesions.

16 **The answer is D: Resting tremor on the right.** The lesion in this case involves the substantia nigra, partes compacta and reticularis, on the left side of the midbrain; this lesion results in deficits that are characteristic of Parkinson disease. A lesion of this structure results in a loss of the dopamine-containing cells of the nigra that project to the basal nuclei, and ultimately influences the activity of the pallidothalamic pathway through the direct and indirect pathways. In a significant number of these patients, the signs and symptoms appear on one side first, and then involve both sides; when the lesion is unilateral, the deficits are unilateral. In this case, the loss of nigral cells on the left results in a modified pallidothalamic message (on the left) and a modified thalamocortical message (also on the left). The thalamocortical fibers from the left ventral lateral nucleus of the thalamus (and their modified firing patterns) project to the left motor cortex. The left motor cortex influences motor activity on the right side of the body via the corticospinal system. Consequently, a left-sided loss of cells of the substantia nigra produces a right-sided deficit, in this case a right-sided resting tremor which is a feature of Parkinson disease. Bilateral bradykinesis would be seen following bilateral lesions involving structures and circuits of the basal nuclei. A right-sided hemiballismus would be seen following a left-sided lesion in the subthalamic nucleus, not the substantia nigra. Intention tremor is a feature of cerebellar lesions or disease (not involved in this lesion); an intention tremor on the right would most frequently be seen following a lesion of cerebellar structures on the right. For there to be a resting tremor on the left, the causative lesion would need to be on the left.

17 **The answer is D: Glutamate.** Subthalamopallidal fibers utilize the excitatory neurotransmitter glutamate; these fibers excite pallidothalamic projecting neurons. Normally, this projection is modulated by a GABAergic pallidosubthalamic projection which is inhibitory. When the pallidosubthalamic projection is active, the subthalamopallidal projection is inhibited; the activity of the pallidothalamic fibers is increased, and the activity in the thalamus and motor cortex is decreased. This is the indirect pathway; when the direct pathway is activated, the striatopallidal projection inhibits the pallidothalamic projection, and the activity of the thalamus and motor cortex increases. However, when a disease process, or lesion, upsets this balance, a motor disturbance, such as hemiballismus, may result. It should be remembered that the subthalamic nucleus also receives an excitatory glutaminergic projection from the cerebral cortex. Dopamine is found mainly in nigrostriatal

projections and may be excitatory or inhibitory depending on the receptor found on the postsynaptic membrane. Acetylcholine is found at many CNS locations; glycine is an important inhibitory neurotransmitter found mainly in the brainstem and spinal cord.

18 **The answer is B: Asterixis.** This is a tremor characteristic of Wilson disease and may also be seen in other patients with metabolic diseases; Wilson disease is an inherited defect of copper metabolism and is treatable. In addition to asterixis, these patients may have a variety of other deficits such as rigidity, hyperreflexia, a variety of behavioral problems, and an open-mouth posture. These patients also present with a green-brown pigmentation located at the corneoscleral junction called the Kayser-Fleischer ring. Akinesia (inability to initiate a voluntary movement) and athetosis (slow, writhing, vermicular movements of the hands and forearms) are seen in basal nuclei lesions/disease. Dysmetria (inability to regulate the force, distance, or speed of a movement) and intention tremor (increasingly irregular movements when reaching for a target) are commonly seen in cerebellar lesions.

19 **The answer is C: Left subthalamic nucleus.** Lesions of the subthalamic nucleus characteristically result in hemiballistic movements that are more obvious in the upper extremity on the side of the body opposite the lesion. These movements are the result of a release of the pallidothalamic neurons from the excitatory subthalamopallidal fibers and the resulting irregular bursts of thalamocortical activity that follow. The deficits are on the side opposite the lesion because of the following sequence: the left subthalamus influences the left ventral lateral nucleus of the thalamus through the left pallidum, the left thalamus influences the left motor cortex, and the left motor cortex influences the extremities on right side of the body via the corticospinal tract. Recall that the motor deficits in this patient are not the result of damage within the corticospinal system, but are due to an altered message arriving at the motor cortex and resulting altered activity of the message traveling down the corticospinal fibers. Lesions in the other choices do not result in hemiballismus.

20 **The answer is A: Huntington disease.** This man has motor difficulties, emotional and behavioral changes, coupled with distinct clear morphological changes in his MRI; collectively, these are characteristic of Huntington disease. The most obvious feature in this image is a loss of the head of the caudate nucleus, reduction in size of the putamen, and a hyperintense appearance of both of these structures. These signify a significant loss of the medium spiny neurons and resultant gliosis. Also seen are hyperintense areas in the cingulate and insular cortices; degenerative changes may also be seen in the cortex, although the neostriatial lesions are pathognomonic for Huntington disease. Genetic testing reveals CAG repeats on chromosome 4p16.3; the greater the number of repeats, the more rapid the onset, the more severe and more rapid the progress of the disease. Parkinson disease is characterized by resting tremor, bradykinesia, festinating gait, and other deficits, but the size of the neostriatum is not significantly diminished in size. In Pick disease, there is atrophy of the frontal and temporal lobes and a corresponding loss of cortical neurons with dementia, but the caudate and lenticular nuclei are unaffected. Wilson disease, hepatolenticular degeneration, is an inherited problem of copper metabolism (the metal accumulates in the basal nuclei and other areas) with motor deficits; this disease can be treated with a chelating agent that removes the copper. Von Hippel-Lindau disease is an inherited disease that results in the development of tumors comprised of vascular endothelial cells and pericytes (hemangioblastomas) localized primarily to the retina and cerebellum and seen to a lesser degree in the brainstem and spinal cord.

21 **The answer is C: Copper.** The age, results of the laboratory tests (amino acids in the urine), and the presence of a brown-green colored ring around the cornea (Kayser-Fleischer ring) clearly point to a diagnosis of Wilson disease in this man. Wilson disease, also called hepatolenticular degeneration because of involvement of the liver and the putamen, is an inherited disorder of copper metabolism. This patient will most likely eventually develop asterixis, also called a flapping or wing-beating tremor. Treatment of choice is to use a chelating agent, such as penicillamine, to remove the copper. Aluminum and magnesium salts have many medical applications and appear in many medications. Iron may be found in the basal nuclei in Hallervorden-Spatz disease, an inherited disease of iron deposition; iron chelating agents are largely ineffective and the disease is progressive. Acetylsalicylic acid is one of the most common medications in the world: aspirin. Naturally, if aluminum, magnesium, and aspirin are taken in excess they can be harmful.

22 **The answer is A: Acetylcholine.** These large cells in the basal nucleus and the innominate substance utilize acetylcholine as their neurotransmitter and have diffuse connections with the frontal and temporal lobes and with other areas. The loss of these excitatory neurons correlates, in part, with the diminution of motor and other functions in this disease. A definite diagnosis of Alzheimer disease is made only after examining postmortem tissue examples and confirming the existence of amyloid plaques and neurofibrillary tangles. Dopamine is found in the nigrostriatal projections; a loss of this neurotransmitter is seen in Parkinson disease. GABA (gamma-aminobutyric acid) is an inhibitory neurotransmitter found at many locations in the nervous system including the striatopallidal, pallidosubthalamic, and pallidothalamic projections. Glutamate is an excitatory neurotransmitter found in efferent fibers of the cerebral cortex, including corticostriatal, corticosubthalamic, corticothalamic, and corticospinal fibers, in some efferents of the hippocampus, and in granule cells of the cerebellum. Substance P is an excitatory peptide that has a long-lasting effect at its synaptic sites within the CNS. It is found within the basal nuclei; the amount of substance P is reduced in Huntington and Parkinson patients.

23 **The answer is D: GABA (Gamma-aminobutyric acid).** Gamma-aminobutyric acid, commonly abbreviated GABA, is associated with striatopallidal, pallidosubthalamic, and pallidothalamic projections, all of which are inhibitory to their targets. Dopamine is found mainly in cells of the substantia nigra, pars compacta, that project to the putamen and caudate nucleus; these projections may be excitatory (D1 receptors) or inhibitory (D2 receptors) to their target cells. Acetylcholine is found in the basal nuclei (and in many areas of the CNS), but is not the major neurotransmitter in the neostriatum. Aspartate is an excitatory transmitter found throughout the CNS that

bears similarities to glutamate. Some glycinergic cells may be present in the neostriatum and the nigra, but they are mainly found in the spinal cord; these cells are basically inhibitory interneurons, not projection neurons.

24 **The answer is B: Decrease in thalamic activity with corresponding decrease in cortical activity.** The fibers identified at the arrow are the ansa lenticularis; this bundle is comprised of pallidothalamic fibers (GABAergic) that arise from the medial segment of the globus pallidus, traverse the ansa, and enter the prerubral area, where they join the thalamic fasciculus to distribute to the ventral lateral (VL) nucleus of the thalamus on the same side. These GABAergic fibers are inhibitory to their targets in the VL; activation of these fibers inhibits (decrease the activity of) the VL cells on which they terminate, and the resulting decrease in the normal excitation of the VL thalamocortical fibers results in a corresponding decrease in cortical activity. If these pallidothalamic fibers are inhibited (not activated), the thalamus would be disinhibited; its activity would increase somewhat because of other influences, and cortical activity would also increase.

25 **The answer is C: Festinating gait.** This pattern of progression, in concert with the other deficits experienced by this man, is part of the overall clinical picture characteristic of Parkinson disease. The patient is usually flexed at the knees, and at the hip and trunk (stooped posture), and his steps are short, rapid, and may become slightly more rapid as he proceeds. Bradykinesia is a part of the characteristic appearance of the man as he attempts to start his forward movement. Calcaneal gait is so named because the patient shifts his/her weight so that progression is accomplished by walking on the heel; this may be seen in polio patients. Cerebellar gait may appear in one of two general forms. First, a general ataxia and unsteadiness to one side (based on a unilateral lesion), or second, a truncal ataxia with a wide-based stance and tremor (based on a midline lesion); this latter is called titubation. Hemiplegic gait is seen when the lower extremity is partially paralyzed; the patient proceeds in a stiff-legged manner by swinging the affected leg laterally and forward in a semicircle motion. Scissors gait, seen in cerebral palsy patients, is a situation in which the motion of each leg is toward the midline when moving forward (much like the opening and closing of scissors).

26 **The answer is E: Wilson disease.** Wilson disease, also called hepatolenticular degeneration, is an inherited (autosomal recessive, chromosome 13q14.3) disease that results in deposition of copper in all tissues of the body, but especially in the brain, liver, kidney, and at the corneoscleral junction. This gene disruption results in a reduction in the excretion of copper and its consequent accumulation in tissues. One treatment is to use a chelating agent, such as penicillamine, to remove the copper. Adverse reactions, ranging from rashes to hair loss, GI disturbances, up to a worsening of neurological deficits, have been reported with penicillamine. Consequently, other agents are sometimes used; although they may be less effective, the adverse reactions are not as severe. When properly treated, most patients (not all) improve or completely recover although they will need to restrict copper in their diet and maintain an appropriate treatment regimen. The symptoms in Alzheimer, Huntington, Parkinson, and Pick diseases can be treated with medication: these diseases are not curable, are neurodegenerative disorders, and while treatment may result in a slowing of the degenerative process, it does not result in remission or improvement.

27 **The answer is B: Asterixis.** The age of this patient and the unique type of his tremor indicate that he has Wilson disease (hepatolenticular degeneration); this is an inherited disorder of copper metabolism which can be treated. In an addition to his characteristic tremor (called asterixis), this man may have a static open-mouth appearance which results from drooping of the lower jaw. The presence of a green-brown ring at the corneoscleral junction, the Kayser-Fleisher ring, is also characteristic of this disease. Akinesia is a loss of the ability to initiate a voluntary movement (seen in Parkinson disease); athetosis is the advent of involuntary movements that are slow, vermicular, or writhing in nature (seen in a number of neurological diseases and may be in combination with other movements). Chorea is brisk, irregular, and involuntary movements that may involve not only the extremities but also the facial muscles; chorea is seen in a variety of conditions, including Huntington and Sydenham diseases. Dysmetria is the inability to judge the distance of a movement; commonly seen in cerebellar lesions, it may exist in two forms, hypometria (stopping before the target) or hypermetria (reaching beyond the target).

28 **The answer is C: Disinhibition of pallidosubthalamic neurons with resultant increase in activity of subthalamopallidal neurons, increase in pallidothalamic activity, and decrease in cerebral cortical activity.** In contrast to the direct pathway through the basal nuclei, the indirect pathway contains two extra neurons and their synaptic contacts and makes a side trip through the subthalamic nucleus. Corticostriatal fibers are glutaminergic and excite the striatopallidal fibers on which they terminate. This striatopallidal fiber is GABAergic and projects into the lateral segment of the globus pallidus; this fiber inhibits cells in the lateral segment on which it terminates. These cells of the lateral pallidal segment are also GABAergic and project to the subthalamic nucleus as pallidosubthalamic fibers, where they end in relation to subthalamic neurons. The subthalamic nucleus contains glutaminergic cells that send a subthalamopallidal projection to the medial pallidal segment, where it terminates on pallidothalamic neurons. These pallidothalamic projections are GABAergic and, when activated, will decrease the activity of the ventral lateral nucleus of the thalamus (VL) with a resultant decrease of activity of the motor cortex. Through the indirect pathway, the corticostriate fiber excites the striatopallidal fiber; since this latter fiber is GABAergic, it, in turn, inhibits the pallidosubthalamic fiber. This inhibition of the pallidosubthalamic fiber (which is itself inhibitory) disinhibits the subthalamopallidal fiber; consequently, the activity level of the subthalamopallidal fiber is increased, resulting in increased inhibitory influence of the pallidothalamic fibers on the VL. The result is a decrease in the activity of neurons within the VL and a corresponding decrease of activity of the motor cortex.

29 **The answer is B (Area including the putamen and the globus pallidus).** The lateral striate arteries, also commonly called the lenticulostriate arteries, are branches of the M_1 segment of the middle cerebral artery that generally serve large parts of the putamen, globus pallidus, and adjacent portions of the internal capsule. These parts of the basal nuclei have important functions within the overall motor system. The head of the caudate nucleus (choice A) is served by the medial striate artery, also called the artery of Heubner; this vessel usually arises at the intersection of the anterior cerebral artery (ACA) with the anterior communicating artery, or from the proximal part of the ACA. The medial striate artery also serves medial portions of the anterior limb of the internal capsule. The pulvinar of the thalamus (C) is served by the thalamogeniculate artery, which is a branch of the P_2 segment of the posterior communicating artery; other important branches of this segment include the medial and lateral posterior choroidal arteries. The anterior area of the thalamus (D) receives its blood supply from the thalamoperforating artery, a branch of P_1. The anterior limb of the internal capsule (E) receives its blood supply from both the medial and lateral striate arteries; this territory is an overlap of these two territories.

Chapter 20

The Cerebellum

Recall the Cautionary Tale: The images in this chapter are presented in a Clinical Orientation: this approach emphasizes how structure and function correlate with deficits in a clinical setting. MRI and CT have a universally recognized orientation and laterality. Line drawings, stained sections, and gross brain images in the Clinical Orientation are viewed here in the identical manner that one views an MRI: your right, as the physician-observer, is the patient's left, and the observer's left is the patient's right. The laterality of deficits and reference to connections follow accordingly.

QUESTIONS

Select the single best answer.

1 A 67-year-old man is brought to the Emergency Department after collapsing at his home. The examination reveals an intention tremor, dysmetria, and dysdiadochokinesia. A CT of this man, seen in the image below, reveals evidence of a hemorrhagic event. Damage to which of the following vessels would most likely result in this clinical picture?

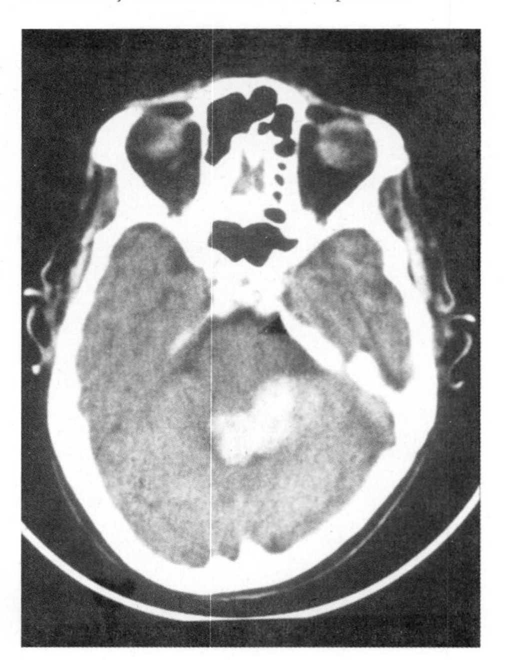

(A) Anterior inferior cerebellar artery
(B) Labyrinthine artery
(C) Posterior inferior cerebellar artery
(D) Quadrigeminal artery
(E) Superior cerebellar artery

2 During a routine physical examination, a 59-year-old woman is not able to successfully conduct a finger-to-nose test or a heel-to-shin test. These signs and symptoms would most likely indicate damage in which of the following brain regions?

(A) Basal nuclei
(B) Cerebellum
(C) Motor cortex
(D) Spinal cord
(E) Thalamus, VL nucleus

3 Which of the structures identified in the image below receives important input from the dentate, emboliform, and globose nuclei?

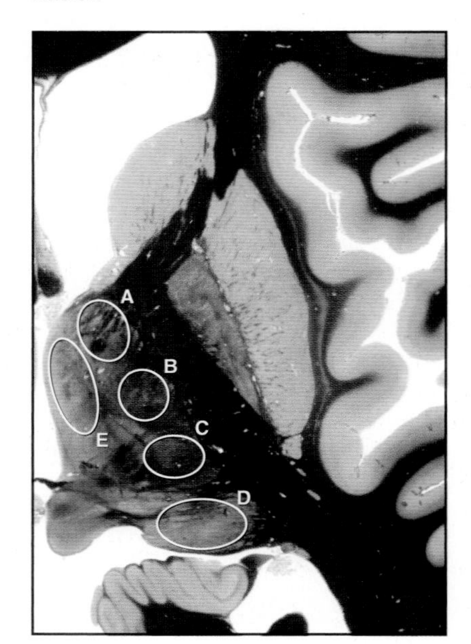

(A) A
(B) B
(C) C
(D) D
(E) E

4 A 69-year-old woman presents to her family physician with a motor complaint. The history reveals that her symptoms had a sudden onset, and the examination shows an intention tremor and dysmetria on her left side. Suspecting a hemorrhagic event, the physician orders a CT, which reveals a cerebellar stroke involving cortex and nuclei. Four months after the initial event, her deficits persisted. Which of the following would most likely be involved to produce these deficits in this woman?

(A) Left anterior inferior cerebellar artery
(B) Left superior cerebellar artery
(C) Right anterior inferior cerebellar artery
(D) Right posterior inferior cerebellar artery
(E) Right superior cerebellar artery

5 A 49-year-old man presents to the Emergency Department with motor disturbances. The examination reveals signs and symptoms consistent with a lesion involving the cerebellum. The cerebellum has no significant projections that course directly to the spinal cord to influence spinal lower motor neurons. Through which of the following does the cerebellum exert its most significant influence on spinal motor neurons?

(A) Corticospinal fibers
(B) Hypothalamospinal fibers
(C) Reticulospinal fibers
(D) Spinocerebellar fibers
(E) Vestibulospinal fibers

6 A 68-year-old hypertensive woman presents with an unsteady wide-based stance, and she is unable to walk in tandem (heel to toe). CT reveals a lesion in the cerebellum. Which of the following represents the most likely location of this lesion?

(A) Brachium conjunctivum
(B) Cortex of the anterior lobe
(C) Cortex of the posterior lobe
(D) Hemisphere plus dentate nucleus
(E) Nodulus, uvula, fastigial nucleus

7 A 59-year-old woman is brought to the Emergency Department by her daughter after the woman complained that she suddenly felt ill. The history revealed that the woman has long-standing diabetes and hypertension and that she is non-compliant with her medication regimen. The examination revealed a tremor (intention) and ataxia of her extremities on the left side and dilation of the right pupil. MRI shows a well-localized lesion. Which of the following outlined areas represents the most likely location of the lesion in this patient?

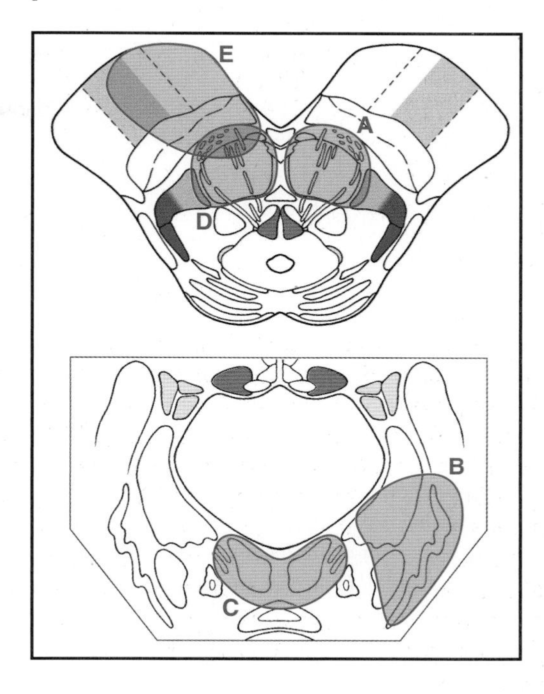

(A) A
(B) B
(C) C
(D) D
(E) E

8 A 67-year-old man is brought to the Emergency Department on a Saturday evening. The history provided by the family reveals that he has been having progressive difficulty walking and that this condition suddenly became worse over the last 3 days. In addition to his ambulatory problems, the man has hypometria and a rebound phenomenon. MRI reveals a large tumor in lateral regions of the cerebellum on the left. Which of the following generally describes the motor deficits experienced by this patient?

(A) Akinesia and bradykinesia
(B) Dyssynergia or decomposition of movement
(C) Flaccid paralysis with areflexia
(D) Increased muscle tone with hyperreflexia
(E) Resting tremor and festinating gait

9 Which of the following is an excitatory neuron within the cerebellar cortex?
(A) Basket cell
(B) Golgi cell
(C) Granule cell
(D) Purkinje cell
(E) Stellate cell

10 A 4-year-old girl is brought to the pediatrician by her mother. The history provided by the mother reveals that the girl has been getting progressively more ill over the last 8 weeks. The mother explains that over the last 10 days, her daughter has become lethargic, has vomited a number of times, and has complained that her "head hurts" (all signs of increased intracranial pressure). Suspecting a tumor, the physician orders an MRI that shows a large mass originating from the choroid plexus of the fourth ventricle; the mass contains flow voids suggesting that it is highly vascular. Which of the following is the most likely blood supply to this girl's tumor?
(A) Anterior choroidal artery
(B) Anterior inferior cerebellar artery
(C) Lateral posterior choroidal artery
(D) Medial posterior choroidal artery
(E) Posterior inferior cerebellar artery

11 Which of the following structures is indicated by the arrow in the image below?

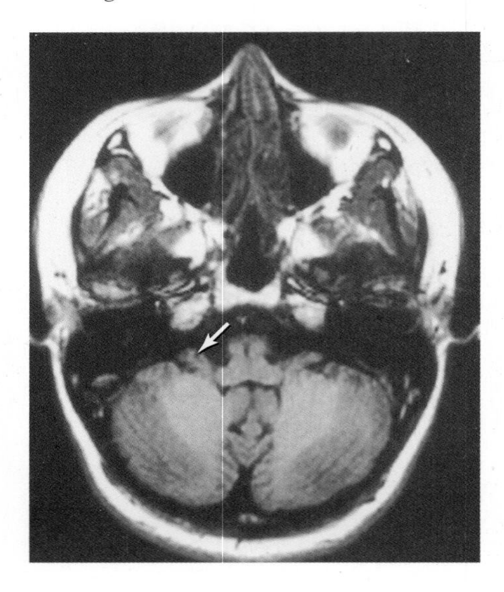

(A) Flocculus of the cerebellum
(B) Paramedian lobule of the cerebellum
(C) Roots of the vagus and glossopharyngeal nerves
(D) Roots of the vestibulocochlear and facial nerves
(E) Tonsil of the cerebellum

12 A 62-year-old man is brought to his family physician by his wife. The history reveals that the man has abused alcohol for years; this behavior has gotten significantly worse in the last 2 years. The examination reveals a slender, somewhat undernourished man with truncal instability, uncoordinated gait, and moderate ataxia of his arms. MRI shows a decrease in size of the anterior lobe of the cerebellum with widening of the fissures; this is the result of cell loss and resultant gliosis. Which of the following cells are most notably lost in this patient?
(A) Basket cells
(B) Golgi cells
(C) Granule cells
(D) Purkinje cells
(E) Stellate cells

13 A 71-year-old woman is brought to the Emergency Department by paramedics. Her son explains that she was at a family picnic, suddenly fell, could not get up without help, and complained that she "was numb on my face." The examination reveals that the woman has an alternate hemianesthesia (also called alternating hemianesthesia), nystagmus, and a pronounced ataxia. MRI reveals a medullary lesion in the territory of the posterior inferior cerebellar artery. Damage to which of the following structures within this vascular territory correlates most specifically with this woman's ambulatory difficulties?
(A) Anterolateral system
(B) Nucleus ambiguus
(C) Restiform body
(D) Reticular formation
(E) Solitary tract

14 Which of the following is the exclusive source of climbing fibers to the cerebellar cortex?
(A) Contralateral inferior olivary nuclei
(B) Contralateral pontine nuclei
(C) Contralateral reticular nuclei
(D) Ipsilateral inferior olivary nuclei
(E) Ipsilateral pontine nuclei

15 A 76-year-old man is brought to the Emergency Department (ED) from the retirement home in which he lives. The nurse who accompanied him stated that he suddenly stopped his activity for a few seconds, then fell, and was unable to get up. The neurological examination in the ED revealed several motor deficits, and an MRI was ordered, which revealed an infarct in the area outlined in the image below. Which of the following deficits would most likely be prominent in this man resultant to his lesion?

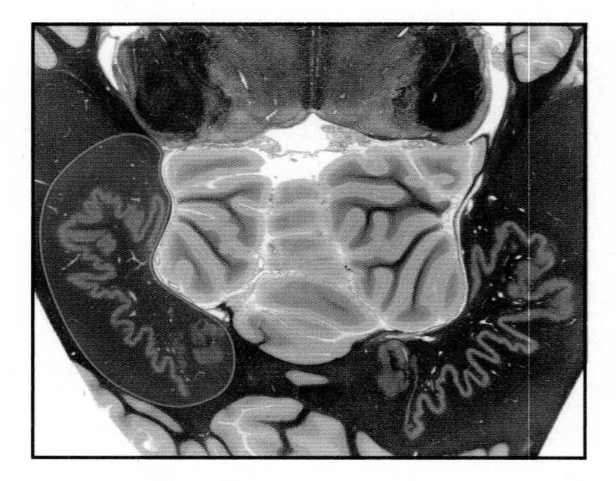

(A) Right-sided asterixis
(B) Right-sided bradykinesia
(C) Right-sided dysmetria
(D) Right-sided resting tremor
(E) Left-sided intention tremor

16 A 73-year-old man collapses at home. He is evaluated by EMSpersonnel, determined to have had a neurological event, and is transported to the Emergency Department (ED) of the nearest large hospital prepared to deal with this type of patient. The examination in the ED reveals that the man is unable to rapidly slap his hand to his knee while alternating pronation and supination of his hand with each movement. Which of the following specifies this particular deficit?

(A) Asterixis
(B) Bradykinesia
(C) Dysmetria
(D) Dysdiadochokinesia
(E) Titubation

17 The cerebellar cortex influences the action of the cerebellar nuclei through the axons of Purkinje cells that project from the cortex to the cerebellar nuclei. Direct excitation of Purkinje cells by parallel fibers would, momentarily, result in which of the following?

(A) Decrease in thalamic activity with corresponding increase in cortical activity
(B) Decrease in thalamic activity with corresponding decrease in cortical activity
(C) Increase in thalamic activity with corresponding increase in cortical activity
(D) Increase in thalamic activity with corresponding decrease in cortical activity
(E) No change in thalamic activity and no change in cortical activity

18 A 51-year-old urologist is at a hunting camp with several friends. Upon awakening in the morning, he realizes that he has tremor, has ataxia, and is unable to use his toothbrush. Recognizing the serious nature of his condition, he has one of his friends transport him, a drive of several hours, to the Emergency Department of the hospital where he practices. The examination confirms the motor deficits reported by the physician. The CT, done the following day, shows the lesion seen in the image below. Which of the following would represent the most likely outcome of this case?

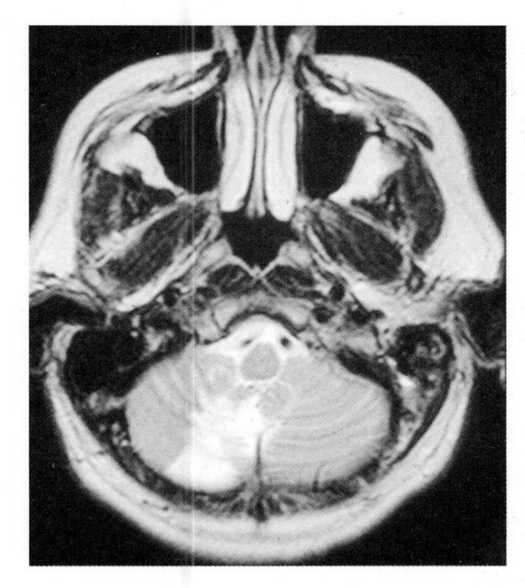

(A) Long-term deficits lasting over 12 months
(B) Permanent deficits with disability
(C) Transient deficits resolving in 4 hours or less
(D) Transient deficits resolving in 24 hours to 1 week
(E) Transient deficits resolving in about 2 to 4 weeks

19 The bundle of fibers indicated by the arrow in the image below contains cerebellar afferent fibers that preferentially end in the flocculonodular lobe, the uvula and adjacent regions of the vermis, and the fastigial nucleus. Which of the following specifies these particular fibers?

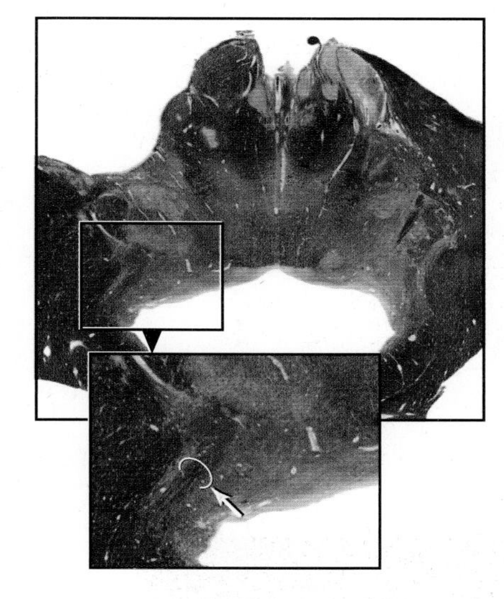

(A) Anterior spinocerebellar
(B) Pontocerebellar
(C) Posterior spinocerebellar
(D) Reticulocerebellar
(E) Vestibulocerebellar

20 A newborn infant presents with multiple developmental defects. MRI reveals several defects within the central nervous system, including the complete absence of the primary fissure. In the normal infant, this fissure separates which of the following parts of the cerebellum from each other?

(A) Anterior lobe from posterior lobe
(B) Anterior lobe vermis from paravermis
(C) Flocculus from nodulus
(D) Flocculonodular lobe from posterior lobe
(E) Posterior lobe vermis from the tonsil

Questions 21 and 22 are based on this patient:

A 69-year-old woman presents to her family physician with the complaint of difficulty walking. The history reveals that the woman's symptoms (headache, ambulatory problems) began about 6 to 8 months ago and have been getting progressively worse; her headaches have become refractory to OTC medications. The examination reveals that the woman has dysdiadochokinesia on the left, has a slurred, garbled speech, is not able to touch the physician's finger with her left index finger as he slowly passes it from left to right in front of her, and is unable to accurately slide her left heel down her right shin. MRI reveals a large tumor within the parenchyma of the cerebellum.

21 Although not a common occurrence, this woman's speech patterns may be altered. Which of the following would most likely explain these altered patterns?
(A) Paralysis of vocal musculature
(B) Paralysis of tongue and facial musculature
(C) Weakness of vocal musculature
(D) Unable to comprehend vocal commands
(E) Uncoordinated vocal musculature

22 Which of the following specifies this woman's difficulty to accurately point to a moving or stationary target?
(A) Alexia
(B) Bradykinesia
(C) Dysmetria
(D) Impaired check
(E) Resting tremor

23 A 49-year-old man presents to his family physician with the complaint of difficulty walking. The history reveals that this man first noticed his difficulties about 6 weeks ago and that they have gotten progressively worse over the intervening time frame. The examination reveals normal cranial nerve function, no sensory deficits, but an intention tremor, dysmetria, and failure of the heel-to-shin test, all on the left side. MRI shows the lesion seen in the image below. Which of the following represents the most likely location of this lesion?

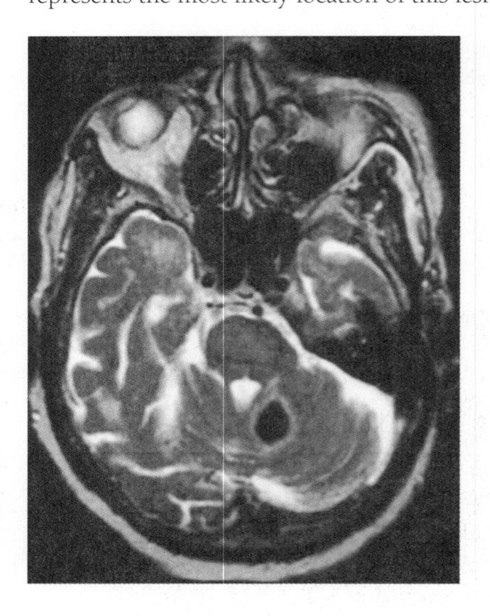

(A) Anterior lobe, lateral area
(B) Anterior lobe, medial area
(C) Posterior lobe, lateral area
(D) Posterior lobe, medial area
(E) Tonsil of the cerebellum

24 A 59-year-old morbidly obese man is brought to the Emergency Department after collapsing at his home. The history and examination reveal that he has poorly controlled hypertension and diabetes, is non-compliant with his medication, and presents with motor and sensory deficits. The examination also reveals that he has diplopia, his left eye is largely immobile (he can only look down and out), and he has a decreased perception of discriminative touch on his right upper extremity; the right lower extremity is not affected. Based on this combination of deficits, which specifies the probable location of this man's lesion, he would also most likely experience which of the following deficits?
(A) Asterixis on the right
(B) Dysdiadochokinesia on the left
(C) Intention tremor on the left
(D) Tremor and ataxia on the left
(E) Tremor and ataxia on the right

25 A 66-year-old woman is brought to the Emergency Department from a retirement community. The history of the current event reveals that her deficits appeared suddenly while she was playing cards. The examination reveals that the woman has difficulty walking; she has significant ataxia and intention tremor on her right side and is unable to slide her right heel down her left shin, but she is not weak when tested for actual muscle strength. CT reveals a focal hyperdense area. Which of the following represents the most likely location of this lesion?
(A) Cerebellar nuclei on the left side
(B) Decussation of the superior cerebellar peduncle
(C) Medial midbrain (Claude syndrome) on the left
(D) Superior cerebellar peduncle on the left side
(E) Superior cerebellar peduncle on the right side

26 Which of the following cerebellar nuclei has direct bilateral projections to brainstem nuclei that, in turn, preferentially influence spinal lower motor neurons that innervate postural (axial) musculature?
(A) Dentate
(B) Emboliform
(C) Fastigial
(D) Globose
(E) Lateral

27 A lesion in which of the outlined areas indicated in the images below would most likely produce deficits that are also seen in cerebellar lesions?

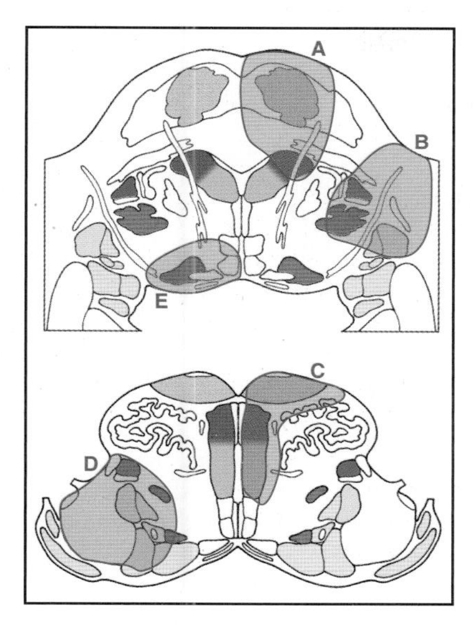

(A) A
(B) B
(C) C
(D) D
(E) E

28 The descending crossed limb of the superior cerebellar peduncle conveys cerebelloreticular and cerebelloolivary fibers from the cerebellar nuclei to the contralateral reticular formation and inferior olivary nuclei. These projections form important feedback loops for cerebelloreticular-reticulocerebellar and cerebelloolivary-olivocerebellar pathways. Which of the following represents the brainstem bundle through which these cerebelloreticular and cerebelloolivary fibers travel?

(A) Central tegmental tract
(B) Lateral lemniscus
(C) Medial longitudinal fasciculus
(D) Medial lemniscus
(E) Posterior longitudinal fasciculus

29 A 51-year-old man is brought to the Emergency Department (ED) by his wife. The man is well-known at the ED as he is a surgeon at that hospital. The history reveals that he suddenly became uncoordinated while shaving. The examination reveals that the man has an intention tremor on the finger-to-nose test, is unable to do the heel-to-shin test, and has dysmetria, all on the right side. He is also ataxic when attempting to walk. CT reveals a hemorrhagic event. Over the next 3 weeks, with physical and occupational therapies, he steadily improves, and, by 5 weeks, he is back part-time at his medical practice. Which of the following is the most likely location of this lesion?

(A) The brachium conjunctivum on the right
(B) The cerebellar cortex on the right
(C) The cerebellar nuclei on the right
(D) The midbrain on the left
(E) The midbrain on the right

30 A 13-year-old girl is brought to the pediatrician by her mother. The history provided by the girl and her mother reveals that the girl has had headaches for a long time. While her headaches earlier responded to OTC medications, more recently they have not. The examination reveals prominent temporal and forehead veins and ataxia of the cerebellar type. Suspecting a possible brain tumor, the pediatrician orders an MRI, which reveals an arteriovenous malformation involving much of the superior (tentorial) surface of the cerebellum. The venous side of this lesion appears to enter the great cerebral vein (of Galen). Which of the following is the most likely source of arterial blood to this lesion?

(A) Anterior inferior cerebellar artery
(B) Anterior choroidal artery
(C) Posterior cerebral artery
(D) Posterior inferior cerebellar artery
(E) Superior cerebellar artery

31 A 70-year-old woman presents with the complaint of difficulty walking. The history reveals that the woman suddenly had problems while she was playing with her granddaughter at the local park. The examination reveals a slender, otherwise healthy woman who has ataxia, is unable to do the finger-to-nose and the heel-to-shin tests on the right, and has slurred, garbled speech. Which of the following specifically describes this pattern of speech in this woman?

(A) Apraxia
(B) Dysarthria
(C) Dyslexia
(D) Dysmetria
(E) Dysphagia

32 Which of the following represents the largest population of fibers comprising the structure indicated at the arrow in the image below?

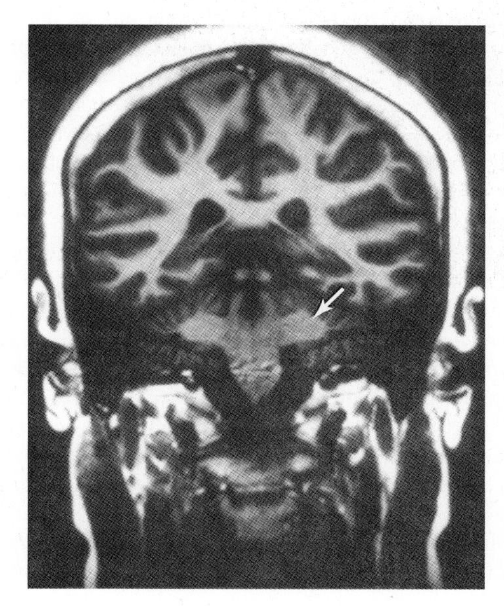

(A) Dentatocerebellar
(B) Olivocerebellar
(C) Pontocerebellar
(D) Posterior spinocerebellar
(E) Vestibulocerebellar

ANSWERS

1 **The answer is E: Superior cerebellar artery.** This axial CT reveals a hyperdense area in the territory of the superior cerebellar artery (SCA) on the left side. This artery serves the cerebellar cortex as well as most of the cerebellar nuclei on the left side. Although CT lacks anatomical detail, the characteristic shapes of the inferior portion of the temporal lobes, the superior aspect of the cerebellum, and the pons are clear. In addition, the shape of the hyperdense area reflects that of the SCA territory. The anterior inferior cerebellar artery (AICA) is a branch of the basilar artery, serves inferior and more lateral aspects of the cerebellar cortex, and gives rise to the labyrinthine artery. This latter vessel enters the internal acoustic meatus and serves the cochlea and semicircular canals. The posterior inferior cerebellar artery (PICA) originates from the vertebral artery and serves the medial and inferior portions of the cerebellar cortex and the choroid plexus within the fourth ventricle. The quadrigeminal artery is a branch of the P_1 segment of the posterior cerebral artery, courses around the midbrain within the ambient cistern, and is the primary blood supply to the mesencephalic tectum.

2 **The answer is B: Cerebellum.** These are classic tests to evaluate the integrity of the cerebellum. If there is a cerebellar lesion, the patient is not paralyzed, she does not have muscle weakness, but she cannot perform either of these movements in a coordinated manner; she may also have other deficits. In general, cerebellar lesions result in dyssynergia, a loss of the normal coordinated interaction between muscles, or muscle groups, during voluntary movements. Lesions of the motor cortex and the ventral lateral (VL) nucleus of the thalamus result in mild to profound loss of movement; damage to the spinal cord presents as paralysis in combination with various sensory losses. Patients with lesions of the basal nuclei have resting tremor, stooped posture, akinesia/bradykinesia, and a festinating gait.

3 **The answer is B (Ventral lateral [VL] nucleus of the thalamus).** The lateral three cerebellar nuclei give rise to projections that traverse the superior cerebellar peduncle, cross the midline in its decussation, and enter the VL thalamic nucleus on the opposite side. This particular population of fibers may also be called the crossed ascending limb of the brachium conjunctivum. These cerebellothalamic fibers terminate predominately in the caudal portions of the VL nucleus (VL pars caudalis, or VLpc); pallidothalamic fibers prefer the rostral, or oral, part of the VL nucleus (VL pars oralis or VLpo). Two important facts should be recalled. First, the lateral three cerebellar nuclei also give rise to a crossed descending limb of the brachium conjunctivum that projects to the brainstem reticular formation and to the inferior olivary nucleus. Second, the medial cerebellar nucleus (fastigial nucleus) projects mainly to the vestibular and reticular nuclei bilaterally but also gives rise to a small projection to the opposite VL nucleus. The ventral anterior thalamic nuclei (choice A) receive input from the pallidum, intralaminar nuclei, and substantia nigra and project to the frontal cortex; the ventral posterolateral nucleus (C) receives fibers from the posterior column and anterolateral systems and projects to the somatosensory cortex; the pulvinar nucleus (D) receives input from a variety of sources and projects to occipital, parietal, and temporal cortices. The dorsomedial nucleus (E) also receives input from a variety of sources and projects to the orbital and frontal cortices.

4 **The answer is B: Left superior cerebellar artery.** This woman has deficits on her left side that are characteristic of a cerebellar lesion, and her CT confirms a cerebellar lesion. Since her deficits are unilateral (on her left) and cortex and nuclei are involved, this lesion is proximal to the decussation of the fibers comprising the superior cerebellar peduncle and involves the area served by the superior cerebellar artery. Recall that this vessel serves cortex and most of the cerebellar nuclei. A cerebellar lesion in the left cortex and nuclei disrupts the message to the right VLpc; this thalamic nucleus, in turn, sends an altered message to the right somatomotor cortex, which, through the corticospinal tract, influences the lower motor neurons on the left side. Left-sided lesions in the cerebellum result in motor deficits on the left side of the body. Involvement of the right superior cerebellar artery would result in deficits on the right side, not the left. The anterior and posterior inferior cerebellar arteries serve, almost exclusively, only cerebellar cortex. Having said this, a lesion involving these territories will produce resting tremor and dysmetria. However, the deficits in this case are transient and will resolve within a few weeks; long-term deficits indicate damage to the cerebellar nuclei.

5 **The answer is A: Corticospinal fibers.** This influence is conveyed by the following pathway: cerebellar nuclei to the contralateral ventral lateral (VL) nucleus of the thalamus, VL nucleus of the thalamus to the motor cortex on the same side, motor cortex to the contralateral spinal motor neurons via the corticospinal fibers, and spinal motor neurons to skeletal muscles on the same side. In this respect, if there is a lesion on the right side of the cerebellum, the characteristic motor deficits are manifest in the right side of the body. The motor expression of the cerebellar lesion is seen through the corticospinal projection; there is nothing wrong with this tract, only a modified message. Spinocerebellar fibers are afferent to the cerebellum; hypothalamospinal fibers traverse the brainstem but influence visceromotor function through their terminations in brainstem visceral motor nuclei and in the intermediolateral cell column. Reticulospinal and vestibulospinal fibers receive cerebellar input and influence spinal motor neurons. However, these pathways are generally concerned with spinal motor neurons influencing postural muscles and extensor muscle tone; these pathways do not function in the circuits involved with precise or skilled movements, or tremor, that are evident in many cerebellar lesions.

6 **The answer is E: Nodulus, uvula, fastigial nucleus.** The deficits experienced by this patient are bilateral (wide-based stance, unable to walk in tandem) and affect axial body regions; there is no complaint of tremor. Lesions of midline parts of the cerebellum such as the nodulus, uvula, and fastigial nucleus result in bilateral deficits that affect axial (truncal) musculature. The patient is ataxic, is unable to walk heel-to-toe, and may walk with a wide stance; due to this way of progression, the patient may appear intoxicated. The deficits are of proximal musculature. Lesions of the cerebellar cortex of either anterior or posterior lobes are transient, may largely resolve in 3 to 6 weeks, and result in deficits on the side of the body ipsilateral to the lesion, and the deficits affect distal musculature. Damage to the brachium conjunctivum or to the hemisphere plus dentate nucleus results in deficits that are persistent and are also evident on the side ipsilateral to the injury.

7 **The answer is D (Area of red nucleus, oculomotor fibers, and cerebellothalamic fibers).** This patient has a dilation of the right pupil (the best localizing sign in this case) and tremor and ataxia on the left side. The dilated pupil caused by the damage to the parasympathetic fibers in the oculomotor nerve and the tremor and ataxia correlate with the damage to the red nucleus and the cerebellothalamic fibers (sometimes these are referred to as dentatothalamic fibers, even though this bundle contains thalamic projections from the globose and emboliform nuclei). The circuits that explain these deficits are as follows for this right-sided lesion: (1) damage to the right oculomotor nerve results in a dilation of the right pupil, (2) damage to the right red nucleus disrupts rubrospinal projections to the left side, and (3) damage to cerebellothalamic fibers on the right disrupts the message to the right ventral lateral thalamic nucleus, then to the right motor cortex, which alters the message in the corticospinal projection to the left side of the body. As one would expect, further evaluation revealed a loss of most voluntary movement of the right eye. This is the Claude syndrome. If the lesion extends into the lateral portions of the medial lemniscus, there will be a loss of proprioceptive and vibratory sense on the left lower extremity. The lesion at A produces the same deficits as does the lesion at D (as described above), but everything is on the opposite side. At B, the lesion is on the left, involves the dentate and emboliform nuclei and large portions of the brachium conjunctivum, and results in characteristic cerebellar deficits on the left. However, no portion of this lesion would account for the oculomotor deficits experienced by this patient. The lesion at C is bilateral, involves the medial cerebellar nuclei, and gives rise to bilateral deficits. Oculomotor deficits are seen in the lesion at E, but the motor deficit consists of a left-sided hemiplegia, not tremor or ataxia; this is the Weber syndrome.

8 **The answer is B: Dyssynergia or decomposition of movement.** The general sign of cerebellar disturbances is a deterioration of the smooth coordinated movements that characterize normal motor activity (reaching to pick up a glass, bringing it to the lips, and taking a drink) into the individual and uncoordinated components of the movement; this is dyssynergia/dyssynergy. Resultant to a cerebellar lesion, these movements are characterized as awkward and uncoordinated and described by clinical terms such as dysmetria (hypometria, hypermetria), intention tremor, dysdiadochokinesia (inability to perform rapid alternating movements), inability to perform heel-to-chin or finger-to-nose test, and impaired check. Akinesia (inability to initiate a movement), bradykinesia (lack of spontaneity of movement), resting tremor, and festinating gait are all signs/symptoms of basal ganglia disease. Flaccid paralysis and areflexia are seen in lower motor neuron diseases/lesions; increased muscle tone and hyperactive reflexes are seen in a variety of upper motor neuron diseases.

9 **The answer is C: Granule cell.** Granule cells are glutaminergic neurons; they are located in the granule cell that is the innermost layer of the cerebellar cortex. These neurons have small round cell bodies, dendrites that branch within the immediate vicinity of the cell body and have claw-like endings called dendritic digits, and an axon that ascends into the molecular layer where it branches parallel to the long axis of the folium to form the parallel fibers of this layer. Granule cells form excitatory synaptic contacts primarily with stellate cells, basket cells, and Purkinje cells. Granule cells that excite GABAergic stellate cells that, in turn, synapse on Purkinje cells (thereby inhibiting the Purkinje cells) will result in disinhibition of the cerebellothalamic pathway. All of the other choices (basket, Golgi, Purkinje, and stellate cells) are inhibitory cells, actually inhibitory cerebellar cortical interneurons, that utilize GABA (gamma-aminobutyric acid) as their neurotransmitter.

10 **The answer is E: Posterior inferior cerebellar artery.** This vessel originates from the vertebral artery, arches around the medulla (serving its lateral portion), makes a sharp hairpin turn through the cisterna magna (at this point giving off branches that serve the choroid plexus of the fourth ventricle), and then serves the medial portions of the posterior lobe. The anterior choroidal and lateral posterior choroidal arteries serve the choroid plexus of the lateral ventricle; the medial posterior choroidal artery serves the choroid plexus of the third ventricle. The small tuft of choroid plexus that is located in the foramen of Luschka receives its blood supply from the anterior inferior cerebellar artery. Lesions in this position comprise about 20% of all tumors in children and about 50% of all tumors in children that are located in the cerebellum and may present with a relatively rapid onset (a few weeks) that includes symptoms of increased intracranial pressure and hydrocephalus. In children, these tumors usually arise from the caudal midline of the cerebellum, while in young adults, they may arise more laterally; these tumors may metastasize.

11 **The answer is A: Flocculus of the cerebellum.** The flocculus is the hemisphere portion of the flocculonodular lobe (the nodulus being the vermis portion), which is the smallest of the three lobes (anterior, posterior, flocculonodular) of the cerebellum. It is located in the immediate vicinity of the cerebellopontine angle. In this position, it is also adjacent to the roots of the facial and vestibulocochlear nerves, to the roots of the glossopharyngeal and vagus nerves, and to the tuft of choroid plexus extending out of the foramen of Luschka. Even a cursory look at this image reveals that the flocculus (the structure at the arrow) is attached to the inferior surface of the hemisphere, not to the lateral aspect of the medulla. The roots of the facial, vestibulocochlear, vagus, and glossopharyngeal nerves attach to the lateral aspect of the medulla and the pons-medulla junction, not to the cerebellum. The paramedian lobule is part of the posterior lobe, but not visible in this plane. The tonsils of the cerebellum are visible in this image as paired structures, located immediately adjacent to the midline and dorsal to the medulla and the narrow black space, indicating the position of the fourth ventricle.

12 **The answer is D: Purkinje cells.** This man is suffering from alcoholic cerebellar degeneration. While the cause is debatable, it is most likely a combination of the effects of poor nutrition (common in alcoholic patients) with the concurrent vitamin deficiencies and the toxic effects of high blood levels of alcohol. The molecular layer contains the stellate and basket cells, the granule layer is densely populated by the granule cells and by few numbers of Golgi cells, and the Purkinje cell layer is a layer of large neurons located at the molecular-granule layers' interface. Purkinje cell dendrites extend outward into the molecular layer, and their axons descend through the granule layer. The cell loss is seen in the molecular layer (which is normally a comparatively cell-sparse layer) and in the granule layer (which is a layer of numerous, densely packed, and small neurons: a cell-rich layer); the loss of Purkinje cells is especially noticeable and prominent. There may be two types of this condition: one in which the deficits develop progressively over weeks to months, these deficits persisting, with the likelihood of structural change within the cerebellum, and a second in which the onset is acute and transient, with the likelihood of no structural change within the cerebellum.

13 **The answer is C: Restiform body.** The restiform body is frequently, and incorrectly, called the inferior cerebellar peduncle. The restiform body is the large fiber bundle located on the posterolateral aspect of the medulla; the inferior cerebellar peduncle is composed of fibers of the restiform body and fibers of the juxtarestiform body, these latter fibers representing interconnections between the vestibular and the cerebellar structures. The restiform body contains many important fibers, but most specifically related to this woman's ataxia are posterior spinocerebellar and olivocerebellar fibers. In addition to ataxia, this woman has a general asynergy on the side of the lesion, also related to the restiform damage injury. The anterolateral system conveys nociceptive information, and damage to this tract results in a loss of pain and thermal sensations on the opposite side of the body (part of the alternating hemianesthesia). The nucleus ambiguus contributes fibers to the glossopharyngeal and vagus nerves; damage to this nucleus produces dysarthria and dysphagia. The reticular formation contributes fibers to the cerebellum, and lesions in this area may result in hiccup (singultus); the solitary tract conveys visceral sensations, both general and special (taste).

14 **The answer is A: Contralateral inferior olivary nuclei.** The sole source of climbing fibers to the cerebellar cortex is from the nuclei comprising the inferior olivary complex (principal, medial accessory, dorsal accessory). These fibers are exclusively crossed: they terminate contralateral to their origin on the dendritic trees of the Purkinje cells where they form excitatory synaptic contacts (aspartate). Climbing fiber terminations are located in the molecular layer. The other nuclei give rise to projections to the cerebellar cortex that terminate as mossy fibers in the granule layer. The vast majority of mossy fibers end in the cerebellar glomeruli, as mossy fiber rosettes, where they exert an excitatory influence on the dendrites of granule cells.

15 **The answer is C: Right-sided dysmetria.** The lesion in this man encompasses the cerebellar nuclei on the right side (dentate, emboliform, and globose); the sudden onset indicates a vascular event. The superior cerebellar artery serves not only the cerebellar nuclei but also a large amount of the cerebellar cortex on its superior (tentorial) surface. Consequently, vascular lesions of the cerebellar nuclei will almost assuredly indicate damage to the overlying cortex. Also, damage to the right cerebellar nuclei, due to the doubly crossed nature of the altered message (right nuclei to left thalamus to left motor cortex to right spinal cord via corticospinal fibers), will result in motor deficits on the same side of the body as the lesion; in this case, the deficit (dysmetria) is on the right side. Dysmetria is the inability to correctly judge the distance to the target; when reaching for a target, the reach may fall short (hypometria) or pass beyond (hypermetria) the target. Symptoms of a cerebellar lesion include a general decomposition of movement, dysmetria (hypermetria, hypometria), intention tremor, dysdiadochokinesia, rebound phenomenon, and others. Asterixis is the flapping, or wing-beating, tremor seen in patients with metabolic encephalopathy, Wilson disease, an inherited metabolic disease, being one example. Bradykinesia, difficulty initiating a voluntary movement, and a resting tremor are seen in patients with Parkinson disease. Intention tremor is a characteristic of cerebellar lesions, but in this case (lesion on the right), the resting tremor would be on the right side, not on the left.

16 **The answer is D: Dysdiadochokinesia.** The inability to perform rapid alternating movements is one characteristic sign of a cerebellar lesion or of cerebellar disease. The example usually cited is alternating pats to the knee, but it can be demonstrated by one of several different examples. Dysmetria, the lack of the ability to judge the distance correctly when reaching toward a target, and titubation (axial or truncal ataxia) are both seen in cerebellar lesions, the former in lesions that involve the more lateral cerebellar areas and the latter in medial (vermis) lesions. Asterixis (flapping or wing-beating tremor) is seen in metabolic encephalopathies such as Wilson disease; bradykinesia (difficulty initiating a movement) is seen in patients with disease of the basal nuclei.

17 **The answer is B: Decrease in thalamic activity with corresponding decrease in cortical activity.** The key to this answer is to recall that parallel fibers (the axons of granule cells) are glutaminergic and, therefore, excitatory to their targets, and that Purkinje cells are GABAergic and, therefore, inhibitory to their targets. Parallel fiber excitation of Purkinje cells results in inhibition of the cerebellar nuclear cells on which these Purkinje axons terminate. Since the efferents of the cerebellar nuclei are glutaminergic and therefore are excitatory to the thalamus, inhibition of these nuclear cells by Purkinje cells will decrease their activity and, consequently, decrease their influence on the thalamus. This results in a decrease in the activity of the thalamus, in the thalamocortical pathway, and in the activity of the motor cortex. It is also important to recall that the cerebellar nuclei receive excitatory inputs from the mossy and climbing fibers that are modulated by Purkinje cells.

18 **The answer is E: Transient deficits resolving in 2 to 4 weeks.** This lesion is in the territory of the posterior inferior cerebellar artery (PICA) and involves the distal branches of this vessel; the medulla, with its two vertebral arteries, as can be seen in the image, is spared. Also, recall that the PICA serves the medial cortex on the inferior cerebellar surface but does not serve any of the cerebellar nuclei. As a general rule, lesions of the cerebellar cortex only result in transient deficits, while lesions of the cerebellar nuclei, or of the nuclei and cortex, result in permanent deficits. These characteristic cerebellar deficits usually resolve in a few weeks: this urologist was back seeing patients in 5 weeks. Deficits lasting less than 4 hours result from what is called a transient ischemic attack (TIA). In reality, about 70% of TIAs resolve in about 10 to 15 minutes and about 90% resolve within 4 to 5 hours. Deficits lasting between 24 hours and 1 week are called a reversible ischemic neurological deficit (RIND); these are relatively rare and constitute about only 2% of all cases. The extent of the lesion seen in this man is larger than what would be commonly seen in TIA or RIND, and the fact that the deficits resolve within about 1 month reflects the highly redundant nature of the cerebellar cortex. The cerebellar cortex is highly convoluted: its surface area, when spread flat, is about the same as that of the cerebral cortex when flattened.

19 **The answer is E: Vestibulocerebellar.** The structure identified by the arrow is the juxtarestiform body; the juxtarestiform body is located medial to the restiform body (seen in this image) and, in combination with the restiform body, comprises the inferior cerebellar peduncle. The juxtarestiform body contains vestibulocerebellar fibers (primary, arising from the vestibular ganglion; secondary, arising from the vestibular nuclei) and cerebellovestibular fibers arising from Purkinje cells. All of these fiber populations relate primarily to the flocculonodular lobe, the uvula, and some adjacent areas of the vermis. These function in relation to balance, equilibrium, and the modulation of certain eye movements. Although a very small part of the cerebellum, the flocculonodular lobe and its afferent sources and efferent targets have important connections and functions. Anterior spinocerebellar fibers enter the cerebellum adjacent to the superior cerebellar peduncle, and posterior spinocerebellar and reticulocerebellar fibers enter the cerebellum via the restiform body. Pontocerebellar fibers are the major fiber population found in the middle cerebellar peduncle; they arise from the pontine nuclei.

20 **The answer is A: Anterior lobe from posterior lobe.** The primary fissure is the second to appear in the developing cerebellum; this fissure appears in the corpus cerebelli and divides this structure into anterior and posterior lobes. The posterolateral fissure is the first to appear in the developing cerebellum and separates the flocculonodular lobe from the corpus cerebelli. Since the primary fissure has appeared by about 10 to 11 weeks, its absence will easily correlate with other defects. The anterior lobe vermis is continuous with the paravermal region (no division), and the separation between the posterior lobe and the tonsil does not have a designation in clinical medicine. The flocculus and nodulus collectively form the flocculonodular lobe.

21 **The answer is E: Uncoordinated vocal musculature (dyssynergia of the vocal musculature).** The most important function of the cerebellum is the coordination of purposeful and productive motor activity. The overall deficit seen in cerebellar lesions is a frank deterioration of coordinated movement (voluntary or otherwise), also called a decomposition of movement, or dyssynergia. There may be a diminution in muscle tone, but the muscles are not weak; when able to do a voluntary movement, actual muscle strength is within a normal range for the patient's age. This patient has dysarthria, resultant to her cerebellar lesion, which results from uncoordinated movements of the vocal apparatus including laryngeal muscles and the vocal folds in which the vocalis muscle is located. The vocal apparatus is not paralyzed, and its muscles are not weak; these muscles have lost their coordinated action during vocalization. This type of speech is sometimes called scanning speech. It is not a comprehension or expression (aphasia) problem, since the patient will correctly use words and grammar; it is just an altered manner and tone of speech.

22 **The answer is C: Dysmetria.** Dysmetria, one of the characteristic deficits seen in patients with cerebellar lesions, is the inability of the patient to control the speed, accuracy, and distance of a voluntary movement. Within this classification of movement disorder, hypometria is undershooting (the movement does not get to the target) and hypermetria is overshooting (the movement goes beyond the target). In this woman's case, she is not able to accurately touch the physician's finger with her finger on the afflicted side. Intention tremor, difficulty touching the finger to the nose, is a classic sign of a cerebellar lesion and is a type of dysmetria. Alexia, the inability to understand the meaning of printed words or letters, is usually indicative of a lesion of the cerebral cortex. Bradykinesia (difficulty initiating a voluntary movement) and resting tremor are both characteristic of diseases of the basal nuclei. Impaired check, also called the rebound phenomenon or the Stewart-Holmes sign, is also a clear sign of cerebellar disease; this is the inability to involuntarily stop (or check) a movement that is initiated against resistance when that resistance is suddenly removed. For example, if the upper extremity is flexed toward the chest against resistance, when that resistance is suddenly removed, the hand will slap against the chest (cerebellar disease) rather than stop before hitting the chest (individual without cerebellar disease).

23 **The answer is B: Anterior lobe, medial area.** This hypointense lesion is located in the medial area of the anterior lobe on the left side. Recall that lesions on the left side of the cerebellum will result in deficits on the left side of the body; this is due to left cerebellar projections to the right ventral lateral (VL), right VL to the right motor cortex, and right motor cortex to the left side of the body via the corticospinal fibers. This lesion may represent a rapidly growing tumor or a small hemorrhagic lesion into which rebleeding has taken place. There is a slight effacement of the midline, and the lesion imposes slightly on the superior cerebellar peduncle, pushing it into the fourth ventricle. The entire plane of the image is through the anterior portion of the cerebellum; structures that help to identify the relative level are (1) a large temporal lobe, especially on the right, (2) comparatively small fourth ventricle, (3) the exit of the trigeminal nerve seen on the right, and (4) the lower part of the eyeball (bulbus oculi) is clear on the right. A plane through the posterior lobe is characterized by the obvious large and deep midline space between the hemispheres of the posterior lobes, the vallecula cerebelli. This space is lacking in this image. The tonsils of the cerebellum are tucked in between the medulla and the posterior lobes, adjacent to each other at the midline, and immediately superior to the caudal roof of the fourth ventricle.

24 **The answer is E: Tremor and ataxia on the right.** The best localizing sign in this case is the diplopia and loss of most eye movement on the left; this indicates that the lesion involves the root of the oculomotor nerve or the oculomotor fibers as they pass through the midbrain, both on the left. These oculomotor deficits, when viewed in concert with a decrease of discriminative touch on the lower extremity on the right, place this lesion in the rostral midbrain on the left. The oculomotor fibers are damaged as they pass through the red nucleus, and the medial portion of the medial lemniscus (containing fibers conveying sensation from the opposite, right, upper extremity), immediately lateral to the red nucleus, is slightly involved. This portion of the medial lemniscus is separated from the red nucleus by cerebellothalamic fibers that are projecting, in this case, from the right cerebellar nuclei to the left ventral lateral nucleus of the thalamus, having already crossed in the decussation of the superior cerebellar peduncle (brachium conjunctivum). Since this lesion is rostral to the decussation of the brachium conjunctivum, the deficits are on the side of the body opposite the lesion (lesion left, deficits right) excepting the oculomotor deficits, which are on the left side; the oculomotor deficits represent damage to the axons of lower motor neurons and parasympathetic preganglionic fibers. The ataxia and tremor reflect damage to the red nucleus (altering the message in rubrospinal fibers) and to cerebellothalamic fibers (altering the message through the corticospinal system). Asterixis is a feature of Wilson disease; dysdiadochokinesia and the intention tremor are on the wrong side and are not the specific type of deficits seen in midbrain lesions. This is a variation on the Claude syndrome.

25 **The answer is E: Superior cerebellar peduncle on the right side.** Lesions of the superior cerebellar peduncle (or of the cerebellar nuclei) result in deficits on the same side of the body. This is due to the fact that the fibers comprising this structure on the right course to the left VLpc, from the ventral lateral nucleus (VL_{pc})to the left motor cortex, and from the left motor cortex to the right side of the body via the corticospinal tract. The motor expression of a lesion of the superior cerebellar peduncle is mediated through the corticospinal tract. The main source of fibers comprising the superior peduncle is the dentate, emboliform, and globose nuclei on the same side. This woman would also exhibit other deficits characteristic of a major cerebellar lesion. A lesion of the cerebellar nuclei on the left (choice A) would result in deficits on the left, not on the right, as in this woman. A lesion of the decussation of the superior cerebellar peduncle (B) would result in bilateral deficits that may be incapacitating. A lesion in the medial midbrain on the left (C) would result in right-sided deficits. However, if this were the case for this woman, she would have complained of double vision, and the examination would have revealed diplopia and loss of most eye movement in her left eye; none of these was reported or seen. Deficits resultant to a lesion of the left superior peduncle (D) would be seen on the left side, not the right.

26 **The answer is C: Fastigial nucleus.** The fastigial nuclei project bilaterally to the vestibular and reticular nuclei of the brainstem (among other targets); fibers to the contralateral side travel via the uncinate fasciculus of the cerebellum, also called the uncinate fasciculus of Russell. The vestibular nuclei influence postural/axial musculature, and the reticular nuclei influence extensor musculature; obviously, these muscle classes may overlap. Fastigial efferents exit to the ipsilateral side via the juxtarestiform body and to the contralateral side by crossing within the cerebellum as the uncinate fasciculus and then leaving the cerebellum in the juxtarestiform body on that side. Efferent fibers from the dentate, emboliform, and globose nuclei leave the cerebellum via the superior cerebellar nuclei, cross in its decussation, and project to the contralateral thalamus. Efferents of these three nuclei do project to the brainstem but predominately to the contralateral side via what is called the crossed descending limb of the brachium conjunctivum; this limb contains cerebelloreticular and cerebelloolivary fibers. The lateral cerebellar nucleus is an alternative name for the dentate nucleus.

27 **The answer is D (Lateral medullary lesion; lateral medullary [or Wallenberg] syndrome).** In addition to an alternating hemianesthesia, which is a cardinal feature of the lateral medullary syndrome, a lesion in the lateral medulla also results in ataxia (damage to the restiform body) and nystagmus (damage to the vestibular nuclei). This is the territory of the posterior inferior cerebellar artery (PICA). This lesion may result in a variety of deficits, and deficits in different combinations, depending on the extent of the PICA territory in individual patients. Alternating hemianesthesia (in this case right face, left body), ataxia, and nystagmus are commonly seen and, to a lesser degree, dysarthria and dysphagia (nucleus ambiguus), loss of taste (solitary tract and nucleus), and hiccup (singultus) may also be seen in a lateral medullary syndrome. The lesion at A is a Foville syndrome (corticospinal fibers, abducens root, part of medial lemniscus); at B, it is a syndrome of the lateral pons at caudal levels (motor nucleus and root of the 7th nerve, anterolateral system, spinal trigeminal tract); C is the Dejerine syndrome, also called the medial medullary syndrome (corticospinal fibers, root of the 12th nerve, medial lemniscus). The lesion at E is a one-and-a-half syndrome that produces deficits related mainly to cranial nerves (motor abducens nucleus, internal genu of the 7th nerve, abducens interneurons to the opposite medial longitudinal fasciculus and to the ipsilateral MLF from the opposite abducens nucleus). While the lesions at A and B may involve some pontine nuclei, they do not result in cerebellar deficits. For practice, review the deficits that are characteristic of these important brainstem syndromes; you may see them in the clinical years.

28 **The answer is A: Central tegmental tract.** The central tegmental tract is somewhat like a brainstem highway extending between the caudal end of the red nucleus and the inferior olivary complex. While it conveys several specific bundles of fibers, the more important are the descending crossed limb of the superior cerebellar peduncle (cerebelloreticular and cerebelloolivary fibers) and rubroolivary fibers. The former arise from the dentate, emboliform, and globose nuclei, exit via the superior cerebellar peduncle, cross the midline in its decussation, and descend to their targets in the reticular formation and inferior olive via the central tegmental tract. The latter arise mainly in the small-celled (parvocellular) portion of the red nucleus, descend in the central tegmental tract, and terminate in the principal olivary nucleus. The medial lemniscus conveys discriminative touch and vibratory sensation from the opposite side of the body, and the lateral lemniscus contains fibers of the auditory system. The medial longitudinal fasciculus contains predominately ascending fibers (from the vestibular nuclei to the nuclei of III, IV, and VI, and abducens internuclear fibers) beginning at about the level of the abducens nucleus and mainly descending fibers (medial vestibulospinal and interstitiospinal fibers) from this level caudally. The posterior longitudinal fasciculus contains descending fibers that arise in hypothalamic nuclei and descend through the brainstem distributing to a variety of targets, particularly visceral nuclei.

29 **The answer is B: The cerebellar cortex on the right.** Hemorrhagic events that involve only the cerebellar cortex, such as distal posterior inferior cerebellar artery (PICA) or anterior inferior cerebellar artery (AICA) infarcts, result in characteristic cerebellar deficits such as those experienced by this physician. However, when the lesion involves only the cortex, the deficits usually resolve relatively quickly, and the patient can eventually resume his/her normal activities. The likely explanation is twofold. First, the vascular event may leave patches of surviving cortex in the area of the lesion. Damage involving the more distal branches of the PICA and AICA results in irregular patterns of cortical lesions within their respective vascular territories. Second, and more important, is the fact that the cerebellar cortex is highly redundant. Injury to superficial cortex leaves significant areas of surviving cortex in the depths of the fissures. Also, adjacent areas of intact cortex may be recruited in the recovery process. Recall that there are two somatotopic representations of the body in the cerebellar cortex: one predominately in the anterior lobe and one in the posterior lobe (mainly in the paramedical lobule). Damage to the cortex and nuclei, to only the nuclei, or to the brachium conjunctivum (which contains cerebellar efferent axons) will result in long-term deficits. Midbrain lesions that result in cerebellar deficits will also produce diplopia (not reported in this patient).

30 **The answer is E: Superior cerebellar artery.** The superior cerebellar artery originates from the basilar artery, wraps around the pons-midbrain junction, and serves the cortex of the superior (tentorial) surface of the cerebellum and most of the cerebellar nuclei. It is the only major source of arterial blood to this region of the cerebellum. Recall that in this position, the superior cerebellar artery is located below the tentorium cerebelli while branches of the posterior cerebral artery are superior to the tentorium. The posterior cerebral artery originates at the basilar bifurcation, immediately rostral to the origin of the superior cerebellar, gives rise to branches that serve the thalamus, midbrain, and portions of the choroid plexus, and then joins the medial aspect of the temporal lobe to serve the occipital lobe above the tentorium cerebelli. The anterior choroidal artery, a branch of the internal carotid, serves the optic tract and adjacent structures in the forebrain and the choroid plexus in the temporal horn. The anterior and posterior inferior cerebellar arteries arise from the basilar and vertebral arteries, respectively, and serve the cortex comprising inferior aspects of the cerebellar hemispheres, but almost essentially none of the cerebellar nuclei. Also recall that occlusion of the superior cerebellar artery will result in long-term (read permanent) cerebellar deficits.

31 **The answer is B: Dysarthria.** As indicated by all of her signs and symptoms, this patient has a lesion in the cerebellum on the right side. Lesions within the cerebellum result in deficits on the same side of the body; recall that the right cerebellum influences the left ventral lateral nucleus and motor cortex, which influences the right side of the body via the corticospinal tract. In this case, her slurred speech is due to the fact that the musculature involved in vocalization is no longer functioning in a coordinated manner. The muscles are not paralyzed, and the patient has not lost the ability to formulate speech; the coordination of the muscles, when attempting speech, is diminished. This reflects the general dyssynergia (a loss of general muscle coordination) characteristically seen in patients with cerebellar disease/lesions. Apraxia is a loss of the ability to perform a skilled purposeful movement in spite of retention of comprehension, motor, and sensory abilities. Dyslexia is the inability to read and comprehend even though the patient's vision and ability to recognize letters and objects are intact. Dysmetria is the inability to control the speed, distance, or power of a specific movement; dysphagia is difficulty in swallowing, with the real complication of food or fluids getting into the trachea.

32 **The answer is C: Pontocerebellar.** The arrow is indicating the middle cerebellar peduncle, also called the brachium pontis; the middle peduncle arises from the cells of the pontine nuclei. This bundle of fibers is afferent to the cerebellum and is comprised primarily of pontocerebellar fibers that send collaterals to the cerebellar nuclei and continue on to end in the cerebellar cortex (granule layer) as mossy fibers; these fibers are excitatory to their targets. The middle peduncle is by far the largest of the cerebellar peduncles, and its fibers terminate in all areas of the cerebellar cortex. Dentatocerebellar fibers arise from the dentate (lateral cerebellar) nucleus, exit the cerebellum via the superior cerebellar peduncle, and end mainly in a variety of targets on the opposite side (red nucleus, thalamus, reticular formation, inferior olive). Olivocerebellar fibers arise from all portions of the inferior olive (dorsal and medial accessory nuclei, principal nucleus), cross the midline of the medulla to enter the cerebellum via the restiform body, send collaterals to the cerebellar nuclei, and end in the cortex as climbing fibers. This is the only source of climbing fibers to the cerebellar cortex. Posterior spinocerebellar fibers arise from the posterior thoracic nucleus (nucleus of Clark), ascend ipsilaterally in the spinal cord to enter the cerebellum through the restiform body, and terminate in the granule layer as mossy fibers. Vestibulocerebellar fibers arise from the vestibular nuclei, enter the cerebellum through the ipsilateral juxtarestiform body, and preferentially terminate in the flocculonodular lobe. Recall that the restiform body + the juxtarestiform body = the inferior cerebellar peduncle.

Chapter 21

The Hypothalamus

Recall the Cautionary Tale: The images in this chapter are presented in a Clinical Orientation: this approach emphasizes how structure and function correlate with deficits in a clinical setting. MRI and CT have a universally recognized orientation and laterality. Line drawings, stained sections, and gross brain images in the Clinical Orientation are viewed here in the identical manner that one views an MRI: your right, as the physician-observer, is the patient's left, and the observer's left is the patient's right. The laterality of deficits and reference to connections follow accordingly.

QUESTIONS

Select the single best answer.

1 Which of the following herniation syndromes may have diabetes insipidus as one of its complicating factors?
(A) Cingulate
(B) Transtentorial
(C) Tonsillar
(D) Uncal
(E) Upward cerebellar

2 The mother of a 16-year-old girl brings her daughter to the gynecologist concerned that the girl may be pregnant. The history reveals that the mother noticed that her daughter's bras were "wet" and the daughter acknowledged that she had fluid coming out of her breasts. The examination reveals a healthy, normally proportioned, nonpregnant girl, with a white discharge coming from her breasts; the gynecologist assures the mother that her daughter is still a virgin. Which of the following would most likely explain this condition in this girl?
(A) ACTH secreting tumor
(B) Growth hormone secreting tumor
(C) Pinealoma
(D) Pituitary apoplexy
(E) Prolactinoma

3 Which of the following nuclei is a primary target of the fiber bundle indicated by the arrow in the given image?

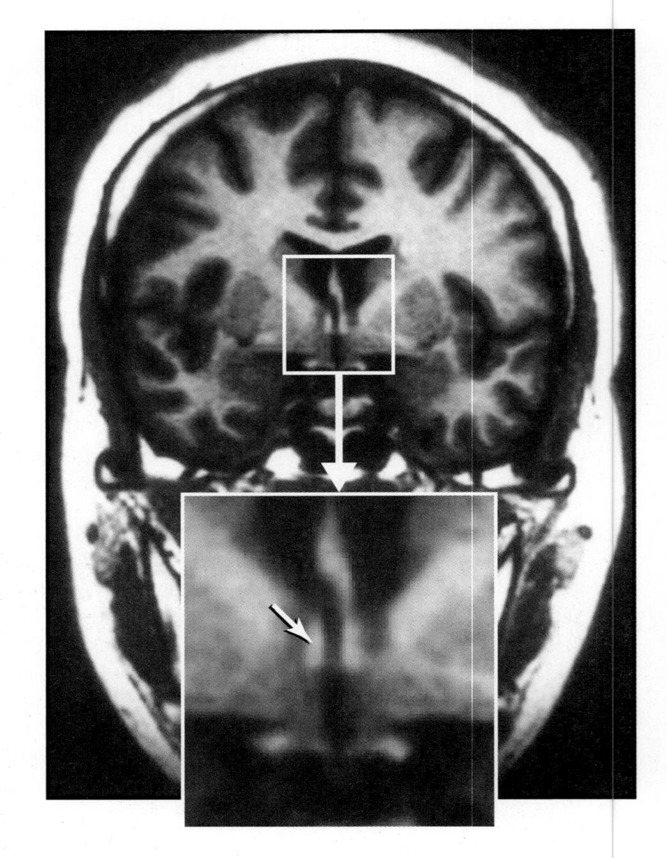

(A) Anterior nucleus of the thalamus
(B) Anterior nucleus of the hypothalamus
(C) Mammillary nuclei of the hypothalamus
(D) Nuclei of the tuberal region of the hypothalamus
(E) Suprachiasmatic nucleus of the hypothalamus

4 A 43-year-old man is brought to the Emergency Department (ED). The history, provided by the family, is known to the ED staff; the man is a serious alcoholic who suffers deficits characteristic of his problem. In addition, he has been diagnosed with diabetes insipidus. Based on the diagnosis of his diabetes, this man would most likely experience which of the following?

(A) Polycythemia
(B) Polydipsia
(C) Polyneuralgia
(D) Polymyoclonus
(E) Polyphagia

5 A 61-year-old woman presents with a complaint of persistently seeing "two of everything." The examination reveals a dilated left pupil and an inability to look up, down, or medially with that eye. MRI reveals an aneurysm of the basilar bifurcation. During surgery to clip this aneurysm, the small perforating branches of P_1 are inadvertently occluded by the clip. Which of the following structures would be most adversely affected by this disruption of blood supply?
(A) Anterior hypothalamic nucleus
(B) Mammillary nuclei
(C) Medial preoptic nuclei
(D) Paraventricular nucleus
(E) Tuberal region nuclei

6 A 46-year-old woman is brought to a primary care physician by a concerned neighbor. The history, provided largely by the neighbor, reveals that the woman lives alone on a small inheritance, does not work, and is almost constantly intoxicated. The examination reveals an undernourished woman who seems confused, cannot recall the last time she had a meal (hour or day), and makes up responses unrelated to her actual life experiences. This woman has neuron loss and petechial hemorrhages in the area indicated in the image below. This woman is most likely suffering from which of the following?

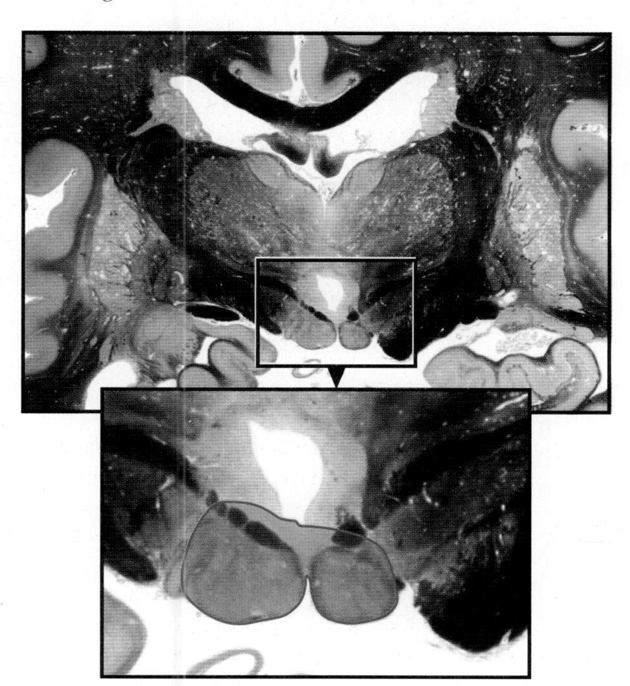

(A) Benedikt syndrome
(B) Dejerine syndrome
(C) Klüver-Bucy syndrome
(D) Korsakoff syndrome
(E) Möbius syndrome

7 A 7-year-old boy is brought to the family physician by his mother. The history reveals that the boy had a minor bicycle accident 6 weeks earlier, hit his head, but was examined and had no lesions or deficits. The mother notes that the boy has no appetite, and has not eaten normally for several weeks. The examination reveals a skinny, almost emaciated, and lethargic boy. Cranial nerves, motor and sensory systems, and reflexes are normal, considering the boy's weakened state. MRI reveals a small localized defect. Which of the following represents the most likely location of this lesion?
(A) Anterior hypothalamic nucleus
(B) Lateral hypothalamic nucleus
(C) Mammillary nuclei
(D) Suprachiasmatic nucleus
(E) Ventromedial hypothalamic nucleus

8 Which of the following fiber bundles contains bidirectional connections between the septal nuclei, lateral hypothalamic nucleus, tuberal nuclei, and brainstem areas such as the periaqueductal gray and the tegmental nuclei?
(A) Mammillotegmental tract
(B) Mammillothalamic tract
(C) Medial forebrain bundle
(D) Stria terminalis
(E) Stria medullaris thalami

9 Which of the following hypothalamic nuclei is described as being sexually dimorphic, that is being larger in males than in females?
(A) Lateral hypothalamic nucleus
(B) Medial preoptic nucleus
(C) Paraventricular nucleus
(D) Posterior hypothalamic nucleus
(E) Suprachiasmatic nucleus

10 A 48-year-old woman visits her family physician with the complaint of frequent urination. The history reveals that this symptom appeared about 3 months ago; at that time the woman remembers that she had a sudden headache that responded to OTC medications. The woman also says that she "is thirsty all the time" and has trouble sleeping because of her frequent trips to the bathroom at night. Laboratory tests reveal normal glucose levels; MRI shows a small lesion within her hypothalamus. This lesion is most likely located in which of the following hypothalamic nuclei?

(A) Arcuate
(B) Mammillary
(C) Suprachiasmatic
(D) Suparoptic
(E) Ventromedial

11 Which of the following specifies the dark area at the tip of the arrow in the image below?

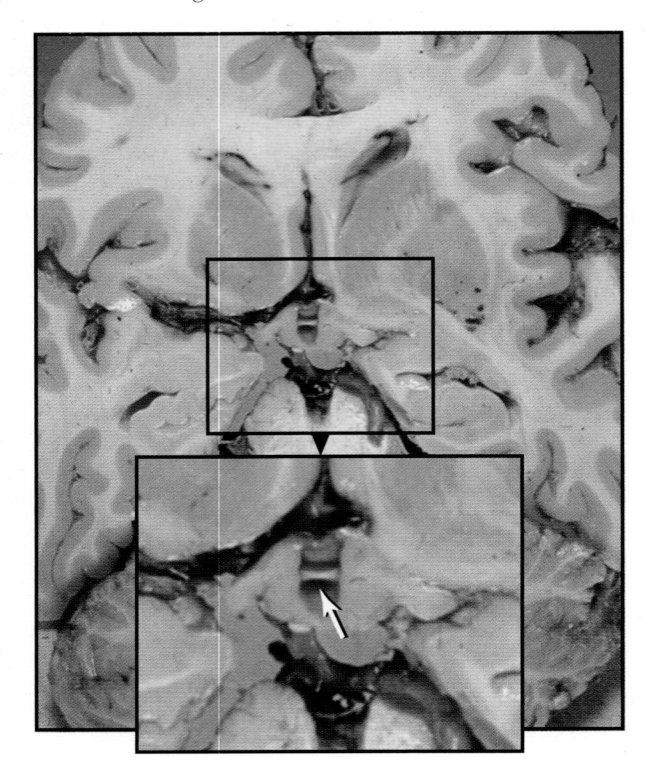

(A) Infundibular recess
(B) Interpeduncular cistern
(C) Lamina terminalis cistern
(D) Pineal recess
(E) Suprachiasmatic recess

12 A 13-year-old boy is brought to the family physician by his mother. The examination reveals that the boy is 6 ft tall, weighs 145 lb, and appears muscular, although when tested for muscle strength he is weak. His mother indicates that since he was a baby he has "been big for his age." The boy's fasting levels of growth hormone are 11 to 27 ng/mL. This boy is most likely suffering from which of the following?

(A) Acromegaly
(B) Cushing disease
(C) Gigantism
(D) Precocious puberty
(E) Wernicke syndrome

13 Which of the following areas indicated in the image below is the location of the dorsomedial nucleus of the hypothalamus in this sagittal section?

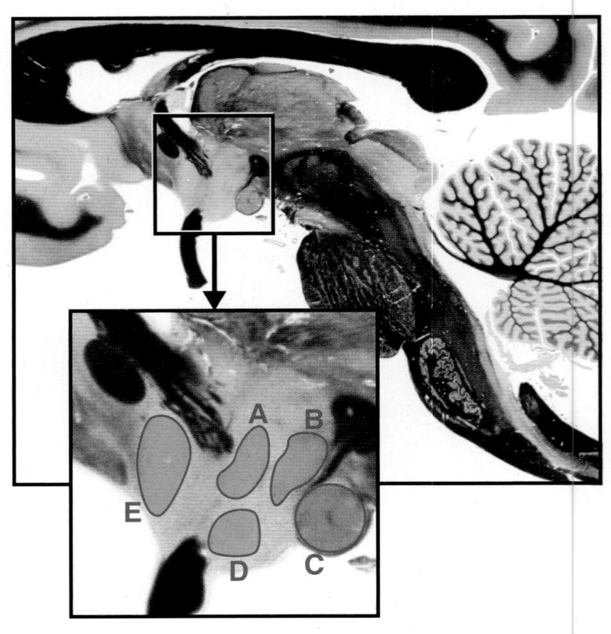

(A) A
(B) B
(C) C
(D) D
(E) E

14 A 51-year-old man is brought to the Emergency Department by the local police who state that the man was disruptive in a public place. The examination reveals an unkempt, emaciated, and intoxicated man who is somewhat uncooperative. He can recall events of past years but only states that he has been "living on the streets for sometime" when asked about more recent activities. He seems confused as to where he is, and clearly makes up answers to most questions. In addition to a nutritious diet and a restriction of alcohol intake, therapeutic doses of which of the following would most likely contribute to an improvement of this man's condition?

(A) Levodopa
(B) Penicillamine
(C) Thiamin
(D) Vitamin A
(E) Vitamin D

15 A 35-year-old man visits his family physician. He explains that over the past year or so he has experienced a progressive loss of interest in sex (decreased libido), and when aroused, he has great difficulty achieving and maintaining an erection (impotence). The examination reveals a normal-looking slender man of average height. The ophthalmological examination showed a partial loss of vision in both eyes, but not blindness; the man stated that he did not realize that this was happening. MRI reveals a pituitary tumor extending through the diaphragma sella, into the suprasellar cistern, and impinging on the hypothalamus. This man is most likely suffering from which of the following?

(A) Excessive follicle-stimulating hormone
(B) Excessive growth-stimulating hormone
(C) Excessive oxytocin production
(D) Excessive prolactin production
(E) Excessive thyrotropin production

16 A 7-year-old boy is brought to the pediatrician by his mother. The mother explains that her son fell off a jungle gym about a week ago; he was examined by a physician and did not appear to have any injuries. Since this event the mother indicates that her son seems subdued, has lost his appetite, and complains constantly about being cold. His temperature may drop to several degrees below normal, and he wears layers of clothes in an attempt to keep warm. Bilateral lesions in which of the following areas would most likely explain the symptoms in this boy?

(A) Anterior hypothalamic area
(B) Lateral hypothalamic nucleus
(C) Posterior hypothalamic area
(D) Ventromedial nucleus
(E) Suprachiasmatic nucleus

17 Which of the following fiber bundles contains axons that arise in the arcuate nucleus and pass to the median eminence and the base of the pituitary stem?

(A) Medial forebrain bundle
(B) Postcommissural fornix
(C) Stria medullaris thalami
(D) Supraopticohypophysial tract
(E) Tuberoinfundibular tract

ANSWERS

1 The answer is B: Transtentorial (or central herniation). A large, expanding mass in the central areas of the forebrain, or in the frontal lobe, may cause a shift as the mass moves downward toward the tentorial notch. Such a lesion may result in signs and symptoms characteristic of increased intracranial pressure (headache, nausea and vomiting, lethargy), visual deficits, and the mass movement downward may shear off the stalk of the pituitary. It is this latter event which will result in diabetes insipidus. There are obviously other serious consequences which are part of this particular syndrome such as oculomotor and pupil deficits, altered levels of consciousness and respiratory patterns, motor deficits, and the likelihood of decorticate and decerebrate posturing. Cingulate herniation results from a mass in one hemisphere that enlarges and forces the cingulate gyrus under the falx cerebri; this may also be called subfalcine herniation. This event may be "silent" (asymptomatic) or may involve the A_4 branches of the anterior cerebral artery, in which case the patient may have unilateral or bilateral motor and/or sensory deficits. Tonsillar herniation results from an expanding lesion in the posterior fossa that forces the cerebellar tonsils downward through the foramen magnum. If this is a sudden event, there is a possibility of compression of the medulla with damage to cardiac and respiratory centers and sudden cessation of heartbeat and respiration. Uncal herniation results from a temporal lobe lesion that forces the uncus over the edge of the tentorium cerebelli, damaging the crus cerebri (and corticospinal fibers) and the oculomotor nerve. Upward cerebellar herniation results from an infratentorial lesion that forces the cerebellum upward through the tentorial notch; this may produce hydrocephalus (occlusion of the cerebral aqueduct) or cerebellar damage by compression of the superior cerebellar artery against the tentorium.

2 The answer is E: Prolactimona. Prolactinoma (prolactin secreting adenoma) is the most common tumor of its type; in females, it causes galactorrhea (a white milklike discharge from the nipple) and amenorrhea (a cessation of menstrual cycles if they have started); both may occur in the same patient. In males, this type of tumor results in impotency and, in about 30%+, galactorrhea may also be present. In addition, this tumor may also result in infertility in both males and females and in osteoporosis. Adrenocorticotropic hormone (ACTH) secreting tumors produce the Cushing syndrome. This disease is characterized by excessive weight gain (particularly in the trunk, thorax, and back), a rounded moonlike face, facial hair growth, obvious violet stria on the trunk, and a number of other features none of which is seen in this girl. A growth hormone secreting tumor produces gigantism in patients who develop this tumor before the growth plates in their long bones close, and acromegaly (and also cardiomegaly) in adult patients; the latter have abnormally large features of the face, hands, and feet. None of these features is present in this girl. Pinealoma develop from the pineal and are characterized by hydrocephalus and signs of increased intracranial pressure (nausea and vomiting, headache, lethargy). In addition, eye movement disorders, such as Parinaud syndrome, are frequently seen. This girl has none of these. Pituitary apoplexy is seen in patients with a mass within the sella turcica: usually a hemorrhage from the mass, sudden onset, headache with visual disorders, and paralysis of some eye movements, again, not seen in this girl.

3 The answer is C: Mammillary nuclei of the hypothalamus. The structure indicated by the arrow is the column of the fornix as it arches through the rostral areas of the hypothalamus. The fornix originates in the hippocampal complex, arches around the caudal end of the thalamus (the pulvinar), passes rostrally in the medial aspect of the ventricular floor, loops around the anterior tubercle of the thalamus and forms part of the wall of the interventricular foramen, enters the hypothalamus, and courses caudally to its termination in the mammillary nuclei. In MRI, particularly in the axial plane, the fornix may be confused with the mammillothalamic tract; this latter fiber bundle arises in the mammillary nuclei and arches superiorly, and slightly rostrally, to terminate in the anterior nucleus of the thalamus. The anterior nucleus also has important projections to the cingulate gyrus of the limbic lobe. Nuclei of the tuberal region (dorsomedial, ventromedial, and arcuate) are concerned with feeding behavior (ventromedial), emotional responses including sham rage (dorsomedial), and produce releasing hormones (arcuate); these nuclei have wide connections including those into the brainstem. The suprachiasmatic nucleus receives direct input from the retina, projects to other hypothalamic nuclei, and mediates circadian rhythms.

4 The answer is B: Polydipsia. Patients with diabetes insipidus have a decreased amount of antidiuretic hormone (ADH) or may suffer from a condition of decreased output of ADH. The main clinical features of diabetes insipidus are polyuria (excessive output of large amounts of urine that has a low specific gravity) followed by polydipsia (excessive intake of water to satisfy the sense of dehydration). This is a cycle: excessive outflow, excessive intake, excessive outflow, and so on. This condition may also be exacerbated by the consumption of alcohol. Polycythemia, also called hypercythemia, is a condition of an unusually high number of erythrocytes in the vascular system. Polyneuralgia is the condition of pain and discomfort (neuralgia) associated with the distribution of several different nerves, all at the same time. Polymyoclonus is a condition of muscle contractions that are rapid, widespread in the body, and of largely unknown causes. Polyphagia is food intake far in excess of what is necessary to maintain an appropriate and healthy life.

5 The answer is B: Mammillary nuclei. The small perforators from P_1 are important sources of blood supply to the mammillary nuclei and the posterior hypothalamic nucleus (both comprise the mammillary region). The thalamoperforating branch of P_1 serves the more anterior portions of the dorsal thalamus. A few small vessels to the mammillary region also originated from the caudal portion of the posterior communicating artery. The tuberal region contains the dorsomedial, ventromedial, and arcuate nuclei; the blood supply to the nuclei constituting this region arises from the more rostral portions of the posterior communicating artery. The anterior hypothalamic nucleus, paraventricular nucleus, and the medial preoptic area receive small perforators arising from the internal carotid artery (to a lesser degree) and from the A_1 segment of the anterior cerebral artery.

6 **The answer is D: Korsakoff syndrome.** The Korsakoff syndrome is characterized by dementia, amnesia, learning and memory difficulties, and confabulation; this is commonly seen in chronic alcoholism, and is related to the accompanying malnutrition and vitamin deficiency. Treatment is cessation of alcohol intake, a healthy diet, and thiamin administration in therapeutic doses. Wernicke syndrome is a closely related condition characterized by dementia, amnesia, lethargy, significantly abnormal eye movements, ataxia, and apathy. Wernicke-Korsakoff is a condition in which patients may have a spectrum of clinical problems that includes both syndromes. Benedikt syndrome is a condition of the midbrain that includes much of the crus cerebri, oculomotor root, red nucleus, and cerebellothalamic fibers (Weber + Claude syndromes). The Dejerine syndrome, medial medullary syndrome, presents with symptoms indicating damage to corticospinal fibers, the medial lemniscus, and the hypoglossal root. The Klüver-Bucy syndrome results from bilateral damage to the temporal lobes that results in a characteristic set of behavioral problems. The Möbius syndrome is most commonly seen as a congenital bilateral paralysis of the facial muscles and of the lateral rectus muscle, although other cranial nerves may be involved. In some of these patients, the lack of a facial colliculus suggests a hypoplasia, or lack, of facial and abducens structures.

7 **The answer is B: Lateral hypothalamic nucleus.** The lateral hypothalamic zone contains the lateral hypothalamic nucleus as well as the medial forebrain bundle. Activation of the lateral hypothalamic nucleus promotes feeding behavior. However, damage to this nucleus will result in a loss of appetite, a decrease in feeding behavior, with a resultant loss of weight, as is experienced by this boy. Without appropriate intervention, there may be a catastrophic outcome. The anterior hypothalamic nucleus influences a number of visceral functions, but is particularly involved in the regulation of body temperature. The mammillary nuclei receive significant input from the hippocampal formation via the fornix and project to the anterior nucleus of the thalamus via the mammillothalamic tract. Damage to the mammillary nuclei results in memory deficits, particularly the anterograde memory deficits. The suprachiasmatic nucleus receives retinal input and modulates circadian rhythms; damage to this nucleus will abolish these natural rhythms. The ventromedial hypothalamic nucleus is a "satiety" center; activation of this nucleus is interpreted by the nervous system as "satisfied" and feeding is decreased, while damage to this nucleus is interpreted as "not satisfied" and feeding is increased, resulting in weight increase, not the case for this boy.

8 **The answer is C: Medial forebrain bundle.** The medial forebrain bundle is located in the lateral hypothalamic zone. It contains fibers that arise from the septal nuclei and nuclei located in the lateral hypothalamic zone that descend toward the brainstem, and fibers that arise in the brainstem (reticular nuclei, tegmental nuclei, etc) and ascend to the hypothalamus and septal nuclei. The mammillothalamic tract arises in the mammillary nuclei and terminates in the anterior nucleus of the dorsal thalamus, while the mammillotegmental tract is comprised of fibers that arise in the mammillary nuclei and end in the tegmental nuclei of the midbrain. The stria terminalis originates from the amygdaloid complex, follows a long circuitous course in the groove between the caudate nucleus and the thalamus (along with the terminal vein) to about the location of the interventricular foramen, then distributes to the septal nuclei and anterior regions of the hypothalamus. The stria medullaris thalami courses along the superior and medial aspect of the thalamus at the upper portion of the third ventricle between the general area of the interventricular foramen and the habenular nuclei. This bundle contains fibers arising from the septal nuclei, anterior thalamus and hypothalamus, and also has contributions from the limbic system.

9 **The answer is B: Medial preoptic nucleus.** Neurons of the medial preoptic nucleus manufacture gonadotropin-releasing hormone (GnRH). GnRH is transported to the anterior lobe of the pituitary, where it causes the release of gonadotropins (luteinizing hormone and follicle-stimulating hormone). Because gonadotropin release is continuous in males and cyclical in females, the medial preoptic nucleus tends to be more active, and consequently larger, in males than is the medial preoptic nucleus in females. Because of this gender-based variation in size, the medial preoptic nucleus is often described as being sexually dimorphic (having two forms as a function of gender). There is no evidence that any of the other hypothalamic nuclei share this characteristic feature.

10 **The answer is D: Supraoptic (nucleus).** The supraoptic nucleus produces antidiuretic hormone (ADH) that is released into the vascular system from the posterior lobe of the pituitary. ADH causes the reabsorption of water from the collecting tubules of the kidney, thus returning fluids into the vascular system. Damage to the supraoptic nucleus can decrease the amount of ADH that is available to the posterior lobe of the pituitary. Recall that ADH is produced in the supraoptic nucleus, conveyed to the posterior pituitary via the hypothalamohypophysial tract, and released by exocytosis. The decrease of ADH diminishes the amount of water that is reabsorbed in the kidneys, and greatly increases the amount of water that exits the body as copious quantities of urine. Consequently, the patient may feel unusually thirsty all the time, and compensates by drinking large amounts of water, which results in excessive urination. Lesions of the mammillary nucleus result in memory dysfunction, specifically the inability to turn immediate and short-term memory into long-term memory. The arcuate nucleus is a main location of neurons that contain releasing hormones. Lesions of the ventromedial nucleus produce increased food intake and weight gain, and damage to the suprachiasmatic nucleus can alter, or abolish, circadian rhythms.

11 **The answer is A: Infundibular recess (of the third ventricle).** In the inset showing the more detailed view, several points are clear. First, this is the lower part of the third ventricle with the lower portions of the hypothalamus, and the optic tracts just superior to the chiasm, bordering on either side of the ventricular space. Second, the lamina terminalis is seen at the rostral end of the third ventricle (in this axial plane), separating the ventricle (which is caudal to the lamina) from the cistern of the lamina terminalis (which is rostral to the lamina). Third, along this part of the floor of the third ventricle, from rostral to caudal, the following are seen: the supraoptic recess of the third ventricle (the dark area just caudal to the lamina terminalis), an elevated ridge in the floor (the light structure running across the space) formed by the optic chiasm, and a second dark area indicated by the arrow, which is the infundibular recess of the third ventricle. The infundibular recess is that portion of the third ventricle that extends into the pituitary stalk for a variable distance. The pineal recess of the third ventricle is located at the superior and caudal aspect of the third ventricle and is that small extension of the ventricular space into the stalk of the pineal. The interpeduncular cistern is located between the crus cerebri and contains the oculomotor root and the bifurcation of the basilar artery; portions of this cistern are seen in the overview portion of this figure.

12 **The answer is C: Gigantism.** Excess levels of growth hormone (normal level is less than 5 ng/mL in the fasting patient) may cause acromegaly or gigantism. In the former case, the increase in growth hormone occurs after the growth plates in long bones (the epiphyseal plates) have closed. These patients are characterized by a variety of changes that include tufting of the distal ends of the fingers, hypertension, prognathism, headache, enlargement of the tongue (macroglosia), thickening of the frontal bone (frontal bossing), thickening and pigmentation over the knuckles and heel, and others. In the latter case, the increase in growth hormone occurs before the long bones have stopped growing (they are still increasing in length) and before the epiphyseal plates have closed. Consequently, these patients have long extremities, are disproportionately tall for their age, and, while they may appear well-muscled, their muscles are infiltrated with fat and are actually weak. Cushing disease results from an overproduction of corticotropin; these patients develop a truncal obesity, violet striae on the trunk and chest, a moonlike face, and facial hirsutism. Precocious puberty may be caused by elevated levels of luteinizing hormone. The Wernicke syndrome is associated with alcoholism and characterized by ataxia, dementia, lethargy, a general apathy, and amnesia. These symptoms are closely related to those seen in the Korsakoff syndrome (the Korsakoff commonly has confabulation) and sometimes are combined into one, the Wernicke-Korsakoff, syndrome.

13 **The answer is A (Dorsomedial nucleus of the hypothalamus).** The three rostrocaudally oriented zones in the hypothalamus are the periventricular zone, the medial zone, and the lateral zone. The periventricular zone is a thin layer of neurons that is located immediately internal to the ependymal lining of the third ventricle; the medial and lateral zones are cell-rich areas that are arranged into a number of nuclei. The medial zone is divided into three regions which generally correlate with the positions of external structures; the supraoptic region is internal to the optic chiasm, the tuberal region is internal to the infundibulum, and the mammillary region is internal to the mammillay body. The tuberal region of the medial hypothalamic zone contains the dorsomedial nucleus (choice A), ventromedial nucleus (D), and the small arcuate nucleus which is not shown. The posterior hypothalamic nucleus (B) and the mammillary nucleus (medial mammillary nucleus in this plane of section (C) are both components of the mammillary region.

14 **The answer is C: Thiamin.** This man has the behavioral features and the signs and symptoms that are characteristic of the Korsakoff syndrome. On the behavioral side is the excessive consumption of alcohol and generally disheveled state that correlates with the clinical picture: confusion, amnesia and memory deficits, ataxia, and a tendency to make up answers when the patient cannot remember or does not know the answer to a question (confabulation). The Korsakoff syndrome is similar, in some respects, to the Wernicke syndrome, and when features of both are seen in the same patient, it is called the Wernicke-Korsakoff syndrome. Malnutrition and the consequent vitamin deficiency and alcoholism, in combination, are primary causes for this condition. Therapeutic doses of thiamin (vitamin B_1), with proper nutrition and cessation of alcohol intake, will result in improvement of this man's medical condition, assuming the lack of permanent structural brain lesions. Levodopa is a major drug used in the treatment of Parkinson disease. Vitamin A is essential to the synthesis of rhodopsin; a deficiency results in dryness of the cornea and conjunctiva (xerophthalmia) and night blindness (inability to perceive in reduced illumination). Vitamin D is essential to the metabolism of calcium and phosphates and promotes proper bone growth and repair and tooth growth in children. Penicillamine is a chelating agent used to remove copper in patients with Wilson disease. Some patients may develop myasthenia gravis (MG) when on a regimen of this medication; withdrawal of the medication usually results in the MG symptoms resolving over time.

15 **The answer is D: Excessive prolactin production.** A tumor that results in the overproduction of prolactin, a prolactinoma, results in a syndrome of production of a milklike fluid from the breast in the absence of pregnancy or parturition (galactorrhea), and a lack, or cessation, of menstrual periods (amenorrhea) in women. In men, a prolactinoma results in decreased libido, impotency, reduced growth of facial hair, galactorrhea, and may be accompanied by infertility. In addition, men may experience progressive weight gain. In adult males and postmenopausal females, excessive follicle-stimulating hormone causes no symptoms, unless it is from a mass effect, but in women who are menstruating it may alter their cycle, as well as give rise to deficits related to mass effect. Excessive production of growth-stimulating hormone will produce acromegaly in adults and gigantism in individuals who are still growing (their epiphyseal plates have not closed). Oxytocin produces contractions of the uterine muscles, promotes the release of milk, and, as a medication, is used to induce labor, treat postpartum bleeding, and also induce milk release. Excessive thyrotropin production affects the function of the thyroid gland and may produce cardiovascular problems, tremor, visual disturbances, and headache. In fact, headache is a common occurrence in all of these situations. In general, these types of tumors fall into the category of hypersecreting (or hormonally active) tumors.

16 **The answer is C: Posterior hypothalamic area.** The posterior hypothalamic area is concerned with mechanisms that offset conditions of decreasing body temperature. For example, peripheral vessels constrict, sweating stops, and, through visceral and somatic circuits, shivering is initiated. All of these activities serve first to conserve heat and second to produce heat through muscle activity (shivering). Lesions of the posterior hypothalamic area abolish these circuits and mechanisms, with the resultant decrease in body temperature, and other visceral (sympathetic) functions, such as a decrease in appetite. Regarding body temperature, the anterior hypothalamus is concerned with mechanisms that dissipate excess body temperature; lesions in this area result in above normal body temperatures (hyperthermia or hyperpyrexia). The lateral hypothalamic nucleus is concerned with feeding behavior; lesions in this area will produce decreased food intake. In contrast, lesions of the ventromedial nuclei result in excessive food intake and resultant weight gain. The suprachiasmatic nucleus receives input from the retina and is concerned with the regulation of circadian rhythms (light-dark cycles); lesions of this structure abolish these cycles.

17 **The answer is E: Tuberoinfundibular tract.** The tuberoinfundibular tract, also called the tuberohypophysial tract, contains axons that originate primarily in the arcuate nucleus and end in the medial and lateral portions of the median eminence. These fibers convey releasing factors. The medial forebrain bundle conveys ascending and descending fibers between preoptic and lateral hypothalamic areas and lower brainstem centers. Fibers arising in the hippocampal formation pass through the fornix, in its various named parts, and, at the level of the anterior commissure, arch caudally as the postcommisural fornix to enter the mammillary body. The stria medullaris thalami is located at the upper aspect of the dorsal thalamus, runs rostrocaudally, and conveys fibers arising in the septal nuclei, preoptic area, the anterior thalamic nuclei, and possibly from limbic structures to end in the habenula. The supraopticohypophysial tract arises in the supraoptic and paraventricular nuclei and ends in the posterior lobe of the pituitary gland; these fibers convey neurosecretory substances that are synthesized in these hypothalamic nuclei.

Chapter 22

The Limbic System

Recall the Cautionary Tale: The images in this chapter are presented in a Clinical Orientation: this approach emphasizes how structure and function correlate with deficits in a clinical setting. MRI and CT have a universally recognized orientation and laterality. Line drawings, stained sections, and gross brain images in the Clinical Orientation are viewed here in the identical manner that one views an MRI: your right, as the physician-observer, is the patient's left, and the observer's left is the patient's right. The laterality of deficits and reference to connections follow accordingly.

QUESTIONS

Select the single best answer.

1 A 23-year-old man is brought to the Emergency Department from the site of a motorcycle collision. The examination reveals that he is intoxicated, has significant facial abrasions, a broken nose, missing teeth, and possible fracture of the mandible. He also has other injuries. CT confirms several broken bones of the face and jaw, and also reveals damage and acute blood bilaterally in the area of the brain substance outlined in the image below. In the immediate future, which of the following will this man most likely experience?

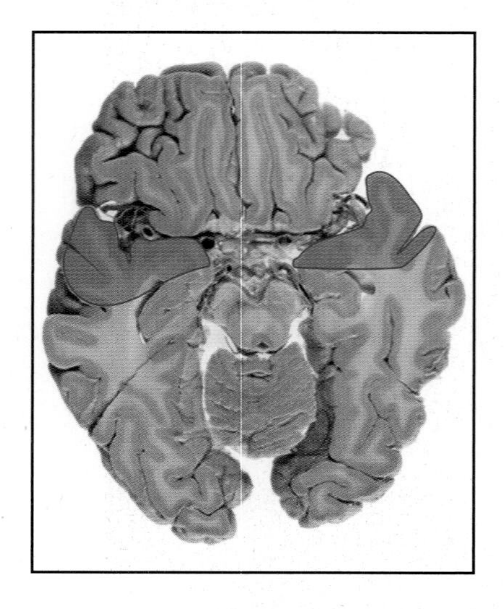

(A) Asperger disorder
(B) Broca aphasia
(C) Klüver-Bucy syndrome
(D) Wallenberg syndrome
(E) Wernicke syndrome

2 A 4-year-old girl is rushed to the Emergency Department by an emergency medical team after falling through the ice of a frozen lake; she is unconscious, cold, has a slow faint heartbeat, and she appears bluish. The history of this event, supplied by the EMS personnel, reveals that the girl fell through the ice and that it took 12 minutes to retrieve her. She is warmed and revived; her heart rate and breathing are reestablished. MRI reveals bilateral lesions in the hippocampal formation. Which of the following may this girl experience?

(A) Aphasia with agraphia
(B) Hypermetamorphosis
(C) Loss of long-term memory
(D) Loss of short-term memory
(E) Transient global amnesia

3 Which of the following is a major pathway that travels in close association with the tail and body of the caudate nucleus that conveys information from the amygdala to the hypothalamus?

(A) Amygdalofugal pathway
(B) Cingulum
(C) Medial forebrain bundle
(D) Stria medullaris thalami
(E) Stria terminalis

4 A 34-year-old woman presents with a comparatively sudden onset of inappropriate behavior. She has started eating excessively, examining objects by placing them in her mouth, and compulsively examines her environment and objects in her environment. She seems unaware that her behavior has changed. MRI of this patient would most likely reveal bilateral lesions in which of the following?

(A) Amygdaloid complex
(B) Anterior paracentral gyrus
(C) Cingulate gyrus
(D) Hippocampal formation
(E) Mammillary nuclei

5 A 43-year-old man complains to his family physician of a persistent cough with occasional red sputum. The history reveals that this has been accompanied by progressive depression, memory deficits, confusion, and general anxiety. MRI reveals bronchial tumors; biopsy shows that these are small-cell carcinomas. These tumors produce IgG antibodies that bind to the nuclear RNA of neurons within the nervous system. The combination of deficits experienced by this patient, histological confirmation of small-cell carcinoma, and the image shown below suggest which of the following diagnoses?

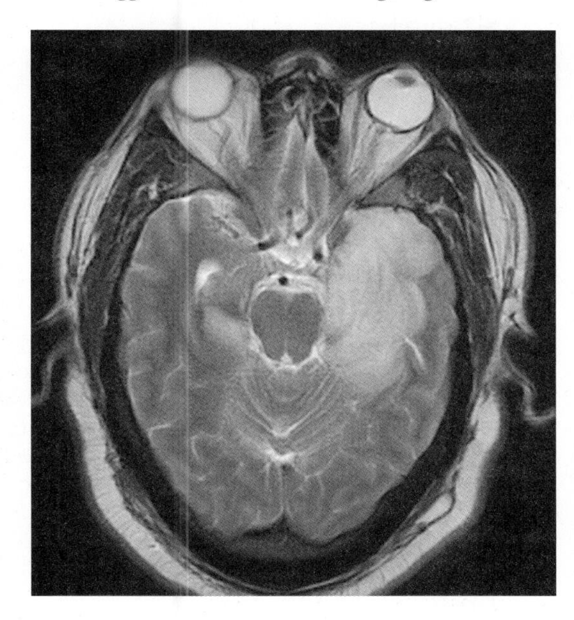

(A) Avellis syndrome
(B) Alzheimer disease
(C) Hemorrhagic lesion
(D) Lambert-Eaton syndrome
(E) Limbic encephalitis

6 A 68-year-old woman is brought to the Emergency Department by her husband. The history, provided by the woman and her husband, reveals that the woman had a sudden onset of difficulty walking. The examination reveals that the woman has a significant right-sided weakness, a decrease, but not a loss, of pain sensation and discriminative touch also on her right side, and seems confused and disoriented. CT shows a lesion involving the medial temporal cortex, large portions of the hippocampal formation and amygdaloid nucleus, and inferior portions of the internal capsule. This lesion most likely involves which of the following vascular territories?

(A) Anterior cerebral artery, A_1
(B) Anterior choroidal artery
(C) Lateral posterior choroidal artery
(D) Middle cerebral artery, M_1
(E) Posterior communicating artery

7 The circuit of Papez is a pathway within the brain that is related, in part, to emotional behaviors and responses. Which of the following schemes depicts the correct sequence of information when passing through this pathway?

(A) Cingulate gyrus to anterior nucleus of the thalamus to mammillary body to the subiculum (hippocampal formation) to cingulate gyrus
(B) Cingulate gyrus to mammillary body to anterior nucleus of the thalamus to subiculum (hippocampal formation) to cingulate gyrus
(C) Cingulate gyrus to subiculum (hippocampal formation) to anterior nucleus of the thalamus to mammillary body to cingulate gyrus
(D) Cingulate gyrus to subiculum (hippocampal formation) to mammillary body to anterior nucleus of the thalamus to cingulate gyrus
(E) Cingulate gyrus to subiculum (hippocampal formation) to septal area to anterior nucleus of the thalamus to cingulate gyrus

8 A 31-year-old man is brought to the Emergency Department from the site of a motor vehicle collision. The examination reveals a stuporous patient with extensive facial injuries, scalp lacerations, compound fracture of the left humerus, and likely fractures of the left tibia and fibula. CT confirms extensive fractures of the facial bones and mandible and of the tibia and fibula, reveals fractures of the left navicular and calcaneal bones, and shows blood in about the anterior 3 to 4 cm of the temporal lobes bilaterally. During his recovery, and when he is allowed out of bed, the man compulsively explores his hospital room, aimlessly and repeatedly examining places and objects. Which of the following describes this particular behavioral characteristic?

(A) Hypermetamorphosis
(B) Hyperkinemia
(C) Hypermetria
(D) Hyperorality
(E) Hyperphagia

9 Which of the following structures is identified between the two arrows in the image below?

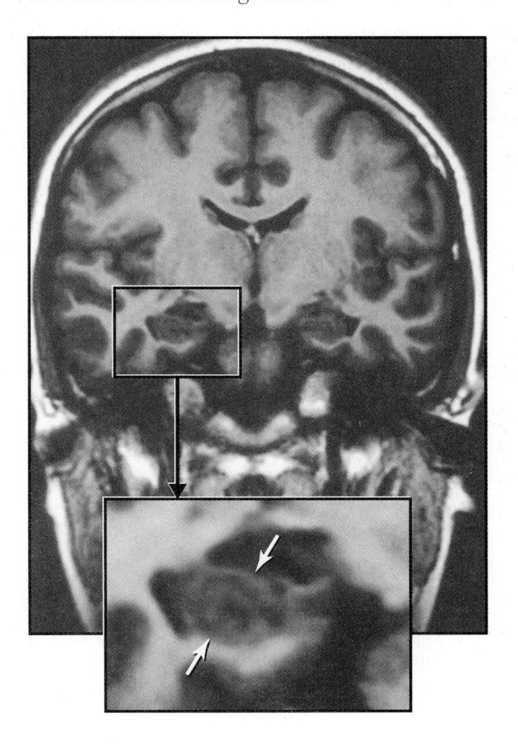

(A) Amygdaloid complex
(B) Cortex of the uncus
(C) Hippocampal commissure
(D) Hippocampal formation
(E) Tail of the caudate nucleus

10 A 20-year-old man is brought to the Emergency Department from the site of an automobile collision. CT reveals bilateral contusions of the temporal poles and of the uncus. In addition to the cortex, which of the following structures is also most likely included in the area of damage?
(A) Amygdaloid complex
(B) Hippocampal formation
(C) Nucleus accumbens
(D) Olfactory trigone
(E) Septal nuclei

11 The diagnosis of which of the following disorders is definitively confirmed only by postmortem histopathological examination of tissue samples?
(A) Charcot disease
(B) Huntington disease
(C) Parkinson disease
(D) Pick disease
(E) Wilson disease

12 Which of the following is the most likely location of the neuronal cell bodies whose axons form the structure indicated at the arrow in the given image?

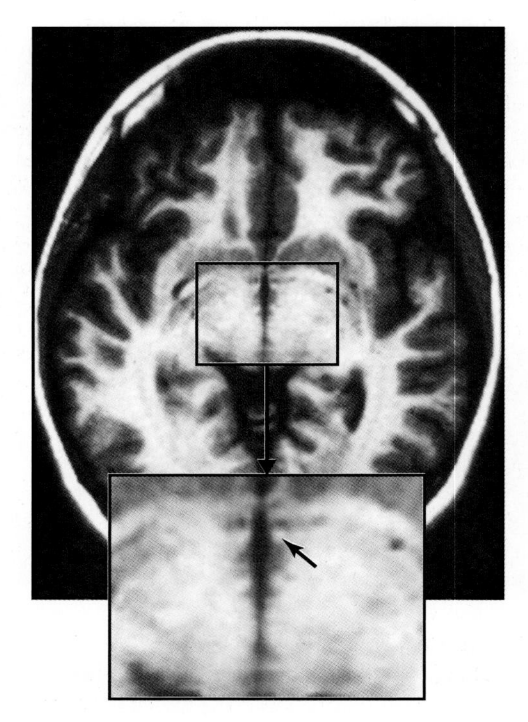

(A) Amygdaloid nucleus
(B) Anterior olfactory nucleus
(C) Habenular nuclei
(D) Hippocampal formation
(E) Mammillary nuclei

ANSWERS

1 **The answer is C: Klüver-Bucy syndrome.** The lesion in this man is localized to the rostral portions of both temporal lobes, specifically involves the amygdaloid nuclei and adjacent cortex, and can result in the Klüver-Bucy syndrome (visual agnosia, hyperphagia, hypermetamorphosis, hypersexuality, hyperorality). This particular lesion results from the temporal poles of the brain impacting into the rostral wall of the middle cranial fossa with resultant bilateral damage. Asperger syndrome is a developmental disorder resulting in a variety of impaired social skills; Broca aphasia (also called an expressive, or motor, aphasia) results from lesions of the inferior frontal gyrus. Wallenberg syndrome results from a lesion in the lateral medulla and has an alternating hemianesthesia as one of its hallmarks; Wernicke syndrome is seen in chronic alcoholics (thiamine deficiency) who have a variety of deficits such as ataxia, confusion, weakness of the extraocular muscles, diplopia, and possible degenerative changes in the mammillary bodies.

2 **The answer is D: Loss of short-term memory.** There are three basic types of memory; two are closely related and are usually lumped together. Immediate memory is the recall of events that took place seconds ago and short-term memory is of events that may be only minutes old; these two are usually collectively referred to as short-term memory. Events that are part of short-term memory (also called temporary memory), when repeatedly used and reviewed in different forms, become part of long-term memory (sometimes called remote memory). For enigmatic reasons, portions of the hippocampal formation, specifically the Sommer sector (CA1-subiculum interface), are particularly susceptible to cerebral ischemia. This type of damage interrupts the hippocampal circuits that allow the cycling of short-term memory events into long-term memory that can be successfully retrieved years later. After the insult, the patient is unable to turn new experiences into memory that may be recalled months/years later. Aphasia is an inability to comprehend sensory input or to communicate (Wernicke aphasia and Broca aphasia are common examples); agraphia is the inability to write in the absence of any deficits of the upper extremity; both of these result from cerebral lesions and may coexist in the same patient. Hypermetamorphosis is a compulsion to excessively examine items, or one's environment, or to experience rapidly changing thoughts or ideas: a type of manic excitement. Transient global amnesia is seen primarily in older individuals, occurs in the absence of trauma, stroke, or seizure activity, and results in temporary confusion and memory loss; however, the neurological exam is normal. The cause is unknown but is most likely some type of temporary ischemia.

3 **The answer is E: Stria terminalis.** The stria terminalis arises from the amygdaloid complex, mainly its basolateral areas. This bundle courses backward immediately (and medially) adjacent to the tail of the caudate nucleus, arches through the atrium of the lateral ventricle adjacent to the tail of the caudate, continues with the body of the caudate, and assumes a position in the angle between the lateral aspect of the dorsal thalamus and the body of the caudate. The stria terminalis continues rostrally to the level of the interventricular foramen, where it branches to end in a variety of targets including the preoptic area, ventromedial nucleus, and the lateral hypothalamic area. The amygdalofugal pathway arises from comparable areas of the amygdale and courses medially to the hypothalamus through the innominate substance. The cingulum is a longitudinally oriented fiber bundle located within the cingulate gyrus that conveys fibers of the limbic system including those to the hippocampus. The medial forebrain bundle contains descending and ascending fibers traveling between hypothalamic and brainstem centers. The stria medullaris thalami is located at the superior and medial aspect of the dorsal thalamus and is an important afferent path to the habenula.

4 **The answer is A: Amygdaloid complex.** This woman displays behaviors that are characteristic of the Klüver-Bucy syndrome; this syndrome is seen following bilateral lesions that involve the amygdaloid nucleus and immediately adjacent cortex and may also include inappropriate sexual behavior. Although it is sometimes stated that this syndrome has only been described in monkeys, it has, in fact, been reported in the clinical literature in humans. Lesions of the cingulate gyrus produce an improvement in symptomatic phobias, apathy, an apparent indifference to painful stimuli, and, in some instances, mutism and akinesia. The anterior paracentral gyrus is the lower extremity area of the primary motor cortex; lesions to this area produce a contralateral weakness if the lesion is on one side, or bilateral weakness if both gyri are damaged. Lesions in the hippocampus inhibit the ability of the brain to convert short-term memory events into long-term memory that can be recalled months or years later. Damage to the mammillary nuclei also results in a loss of short-term memory. Recall that the mammillary nuclei receive significant input from the hippocampal formation via the fornix, and, in turn, project to the cingulate cortex via the mammillothalamic tract.

5 **The answer is E: Limbic encephalitis.** The lesion in this man is in the medial aspect of the temporal lobe involving the amygdaloid complex, hippocampus, and the adjoining cerebral cortex. His lung tumor produces IgG antibodies (antineuronal antibodies) that react with RNA-binding proteins within the nuclei of neurons in the brain and spinal cord. This is one of the paraneoplastic syndromes. In the case of limbic encephalitis, these antibodies selectively bind to neurons located in the cortex and nuclei (amygdaloid and those of the hippocampus) of the temporal lobe and produce deficits characteristic of this condition: memory deficits, depression, anxiety, and sleep disturbances. The Avellis syndrome is a posterior fossa event that involves the vagus root at, or in, the jugular foramen and the area of the lateral medulla that contains the anterolateral system. Alzheimer disease is a disease of the aged, has an insidious onset, and is characterized by a range of behavioral features, most notably memory and learning deficits, locomotor problems, and general disorientation. In MRI, these brains have enlarged sulci with smaller gyri and hydrocephalus ex vacuo. Hemorrhagic lesions have a totally different appearance, texture, and a sudden onset. Lambert-Eaton syndrome is also one of the paraneoplastic syndromes. In this disease, there is a loss of voltage-gated calcium channels on the presynaptic membrane of the neuromuscular junction with a resultant reduction in the release of acetylcholine into the synaptic cleft; this produces a type of myasthenia gravis.

6 **The answer is B: Anterior choroidal artery.** The anterior choroidal artery arises from the cerebral part of the internal carotid artery and courses caudolaterally generally along the trajectory of the optic tract. En route, its branches serve the inferior portions of the internal capsule, the optic tract, medial temporal cortex, significant portions of the hippocampus, amygdaloid nuclei, and adjacent areas. The most obvious and most immediate deficits following occlusion of this vessel are a hemiplegia and a homonymous hemianopia, both on the side opposite the lesion; a left sided lesion results in right hemiplegia and right homonymous hemianopia. However, depending on the extent of the anterior choroidal artery territory, the patient may also experience a hemianesthesia on the same side as the weakness. This is due to the fact that, in addition to damaging corticospinal fibers, the lesion extends upward in the internal capsule far enough to include the thalamocortical fibers that arise in the ventral posterolateral nucleus and travel to the sensory cortex. Long-term deficits will also include memory deficits, forgetfulness, confusion, and behavior/personality changes indicative of medial temporal lobe damage. The A_1 segment of the anterior cerebral artery serves rostral hypothalamic regions, the septal area, and may be the origin of the medial striate artery (of Heubner). The lateral posterior choroidal artery arises from the P_2 segment of the posterior cerebral artery and enters the choroidal fissure to serve the choroid plexus in the lateral ventricle; it also serves brain structures en route. The major branches of the M_1 segment of the middle cerebral artery are uncal, polar temporal, and the lateral striate arteries (also called lenticulostriate). Branches of the posterior communicating artery serve mainly portions of the hypothalamus.

7 **The answer is D: Cingulate gyrus to subiculum (hippocampal formation) to mammillary body to anterior nucleus of the thalamus to cingulate gyrus.** The complete circuit of Papez is generally regarded as beginning in the cortex of the cingulate gyrus. From the cingulate cortex these axons project either directly or indirectly (via relays in the entorhinal cortex), as part of the subcortical bundle called the cingulum, to end in the subiculum of the hippocampal formation. Fibers originating in the subiculum, and in the adjacent Ammon horn, coalesce to form the fimbria of the hippocampus and the various part of the fornix (crus, body, and column) as it arches around the thalamus to form, at the anterior commissure, the postcommissural fornix. This portion of the fornix passes through the hypothalamus to end in the mammillary nuclei. Cells of both the medial and lateral mammillary nuclei contribute fibers to the mammillothalamic tract that distributes to the various subdivisions of the anterior nucleus of the thalamus. In turn, the anterior thalamic nucleus projects to the cortex of the cingulate gyrus, thus completing the basic features of the Papez circuit.

8 **The answer is A: Hypermetamorphosis.** This man's head injury not only includes extensive bony damage to his face but also has resulted in bilateral injuries that involve the cortex of the rostral temporal lobe as well as the amygdala and possibly the rostral aspect of the hippocampus. This man's compulsive and repeated exploration of his room is called hypermetamorphosis and is one on the features seen in the Klüver-Bucy syndrome; this behavior indicates rapidly, and random, changing ideas and thoughts that are not organized and represent a type of manic excitement. Hyperorality, the placing of inappropriate objects into the mouth or examining such objects with the mouth or tongue, and hyperphagia, excessive eating (also called gluttony), are also features of the Klüver-Bucy syndrome. Hyperkinemia is the condition under which there is excessively high cardiac output, not increased heart rate, but increased volume of blood, a supranormal outflow. Hypermetria is a motor deficit seen in individuals with cerebellar lesions; this is overreaching of an intended target, reaching beyond the target and bringing the upper extremity back to acquire the target.

9 **The answer is D: Hippocampal formation.** The hippocampus is characterized, in coronal plane, by alternating layers of cell bodies and fibers. This feature shows in this T1-weighted MRI; the layers of fibers appear whiter (these are hyperintense since it is a shift toward the appearance of fat, more white, in a T1 image), and the intervening layers of cells appear darker (these are hypointense since it is a shift toward the appearance of air, more black, in a T1 image). This is much like the appearance of a jelly roll where the jelly alternates with the roll. The hippocampus is located in the ventromedial wall of the temporal horn of the lateral ventricle; the space of the ventricle appears black above and lateral to the hippocampus. The amygdaloid nucleus, which internally appears homogenous in MRI, is located in the anterior wall of the temporal horn just internal to the uncus and is more rostral than the plane of this image. The cortex of the uncus is located at the rostral and medial aspect of the temporal lobe, basically at the medial end of the parahippocampal gyrus, and is not related to the temporal horn of the lateral ventricle. The hippocampal commissure is the crossing of fibers from the fornix on one side to the fornix on the other side; it is located on the inferior surface of the splenium of the corpus callosum. The tail of the caudate nucleus is the part of this structure that is found in the lateral wall of the temporal horn between the atrium and to where it ends just before reaching the amygdala.

10 **The answer is A: Amygdaloid complex.** The uncus, an elevation located at the rostral and medial aspect of the parahippocampal gyrus, overlies the nuclei that collectively form the amygdaloid complex. Because of its proximity and common features of blood supply, damage of this type to the temporal pole and uncus will spread to encompass the subjacent amygdala. The hippocampal formation is located in the medial wall of the temporal horn, internal to the main part of the parahippocampal gyrus, and caudal (in the long axis of the temporal lobe) to the amygdala. The nucleus acumbens is located at the junction of the head of the caudate with rostral and ventral portions of the lenticular nucleus; in this location it is adjacent to the septum pellucidum and the septal nuclei; this is why it is sometimes called the nucleus accumbens septi. The olfactory trigone is located in the area between the medial and lateral olfactory stria. Contusion of the brain is basically a bruising of the brain usually without damage to the pia and arachnoid mater, but with extravasated blood in the brain substance.

11 **The answer is D: Pick disease.** Pick disease is one of a group of conditions in which atrophy of the frontal and temporal lobes is a major factor; this is sometimes called a lobar atrophy. The signs and symptoms include atrophy of frontal and temporal lobes (clearly seen in MRI) of the cerebral hemispheres (atrophy of these cortical regions is seen in a variety of other conditions and is, therefore, not diagnostic on its own), dementia, apathy and lethargy, changes in personality, and occasionally, speech disorders. A definitive diagnosis is revealed following a histopathological examination which reveals neuron loss (predominately in cortical laminae 1 to 3, gliosis, swelling (called ballooning) of surviving neurons, and the presence of Pick bodies (agyrophilic fibrils) in some neurons that may be ubiquitin- and tau-positive. Charcot disease (amyotrophic lateral sclerosis) is a neurodegenerative disease that affects corticospinal and corticonuclear systems, lower motor neurons of the spinal cord, and has an insidious onset and comparatively rapid course of about 3 to 6 years. Huntington disease is an inherited disease that can be diagnosed based on its onset, characteristic loss of the caudate head in MRI, and genetic testing. Parkinson disease is a disease of largely unknown etiology that can be diagnosed by the older age of onset, characteristic signs and symptoms, progression, and response to medication (levodopa). Wilson disease is an inherited disorder of copper metabolism that can also be diagnosed by characteristic signs and symptoms: flapping tremor (asterixis), early age of onset, cell loss or small cavitations in the basal nuclei (and other areas), and positive response to a chelating agent to remove the copper.

12 **The answer is D: Hippocampal formation.** The arrow is pointing to the column of the fornix in this axial T1-weighted MRI; the fornix arises in cells of the hippocampal formation, most specifically in the subiculum. In the detail, note that the anterior commissure is clearly evident on the right and left sides of the image as the large white structures and that the position of this commissure at the midline can also been identified. The column of the fornix (also called the postcommissural fornix) is immediately caudal to the anterior commissure and terminates in the mammillary nuclei; the precommissural fornix passes rostral to the anterior commissure to end in the preoptic nucleus and septal area. Fibers from the amygdaloid nucleus form the stria terminalis; the habenular nuclei give rise to the habenulointerpeduncular tract, both of which are not seen in this image. Some of the fibers that arise in the anterior olfactory nucleus cross the midline within the anterior commissure to distribute to olfactory structures on the opposite side of the brain. The mammillary nuclei give rise to the mammillothalamic tract which, in this image, appears to be about the same size as the fornix and is located 3 to 4 mm caudal to the fornix. The mammillothalamic tract projects to the anterior nucleus of the dorsal thalamus.

Chapter 23

General Review and Identification Questions

Recall the Cautionary Tale: The images in this chapter are presented in a Clinical Orientation: this approach emphasizes how structure and function correlate with deficits in a clinical setting. MRI and CT have a universally recognized orientation and laterality. Line drawings, stained sections, and gross brain images in the Clinical Orientation are viewed here in the identical manner that one views an MRI: your right, as the physician-observer, is the patient's left, and the observer's left is the patient's right. The laterality of deficits and reference to connections follow accordingly.

This chapter is designed to provide an opportunity to identify structures primarily in MRI and CT and in a few examples of stained slices and gross brain slices. In the clinical years, and in clinical settings, it is absolutely essential to be able to recognize brain structures in MRI and CT for two reasons. First, if the shape or relationship of some portion of the brain is altered from the normal pattern, it is important to be able to recognize the altered pattern. Being able to recognize the abnormal pattern depends on knowing, understanding, and being able to visualize the normal pattern in your mind's eye. Second, if there is a space-taking lesion either within the substance of the central nervous system or external to the central nervous system and impinging thereon, it is essential to understand what specific structure is affected (structure and function) and how the deficits experienced by the patient correlate with the image.

*In most of the questions in this chapter, the images are of normal brains; three show clinical conditions with the opportunity to identify specifics within the image. The clinical situations posed in all questions are meant to stimulate one's thinking as to **what structures may relate to the deficits in the example.** The items labeled in the image include one or more structures, damage to which would correlate with deficits described in the example. This allows a functional correlation and, at the same time, an opportunity to identify central nervous system structures in a variety of images.*

QUESTIONS

Select the single best answer.

1 A 67-year-old man experiences a sudden explosive headache and he is transported to the Emergency Department. CT reveals that he has suffered a subarachnoid hemorrhage (SAH). The most common cause of blood in the subarachnoid space is trauma (traumatic SAH); the second most common cause is rupture of an intracranial aneurysm (non-traumatic or spontaneous SAH). The sudden, severe (some patients describe it as a "thunderclap") headache is characteristic of aneurysm rupture with blood being present in many portions of the subarachnoid space. Name the specific parts, identified in the image below, of the subarachnoid space that contain blood.

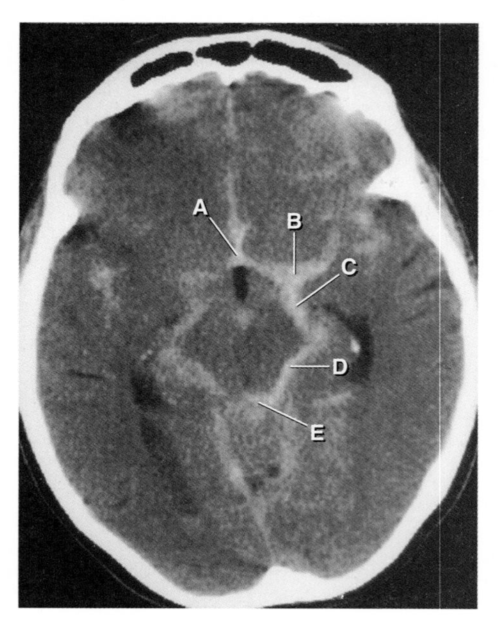

(A) ______________________
(B) ______________________
(C) ______________________
(D) ______________________
(E) ______________________

2 A 38-year-old obese woman presents with persistent headache and mild papilledema. Suspecting idiopathic intracranial hypertension, the physician orders an MRI to assess any degree of ventricular enlargement or shift of brain structures. This procedure will expedite this evaluation. Identify the structures indicated in the image below; changes in their configuration may signal a disease process.

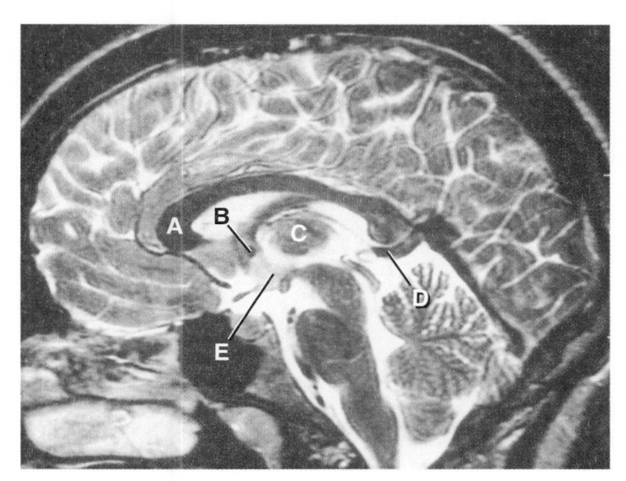

(A) ______________________
(B) ______________________
(C) ______________________
(D) ______________________
(E) ______________________

3 A 39-year-old man presents to his physician's office with a complaint of clumsiness. The history reveals that the man's father died at 52 years of age with a severe motor disease and dementia. The examination reveals that, in addition to his motor problems, the man is forgetful, somewhat depressed, and displays choreiform movements. The man states the he did not marry because his mother told him that he would probably get the same disease that his father had. An MRI was ordered to see if there is any evidence of a neurodegenerative process. Identify the structures indicated in the given image, some of which may be of concern regarding this man's case.

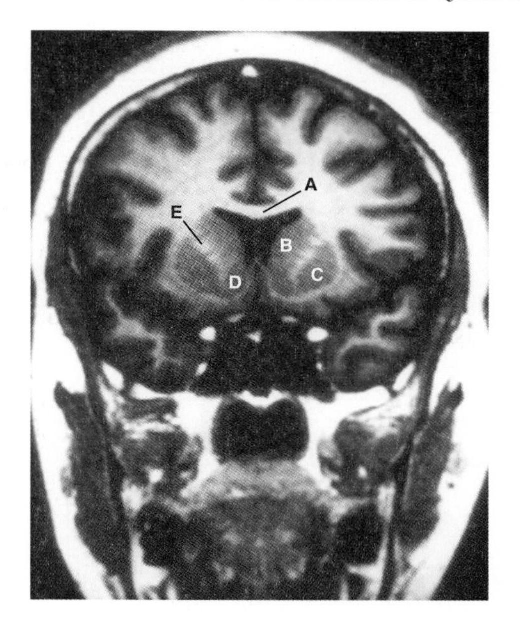

(A) ______________________
(B) ______________________
(C) ______________________
(D) ______________________
(E) ______________________

4 A 16-year-old girl is brought to the pediatrician by her mother. The history of this visit is that the girl has a white substance coming out of her breasts. The mother contends, and the examination confirms, that the girl is not pregnant. In addition, the physician discovers that the girl has a partial visual loss in both eyes, of which she was unaware. An MRI is ordered. Identify the structures indicated in the image below, some of which may be of particular interest in this girl's case.

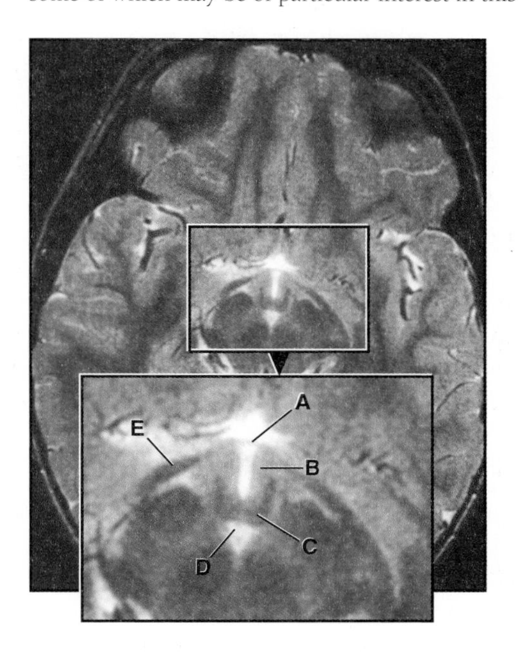

(A) ____________________
(B) ____________________
(C) ____________________
(D) ____________________
(E) ____________________

5 A 71-year-old woman is brought to the Emergency Department from the retirement home in which she is residing. The history of the current event was a sudden onset of left-sided weakness. The examination also reveals that she is somnolent and has hyperactive reflexes on her left side. MRI reveals a hemorrhagic lesion in the right hemisphere. Identify the structures in the image below, some of which may be within the vascular territory involved in this woman's case.

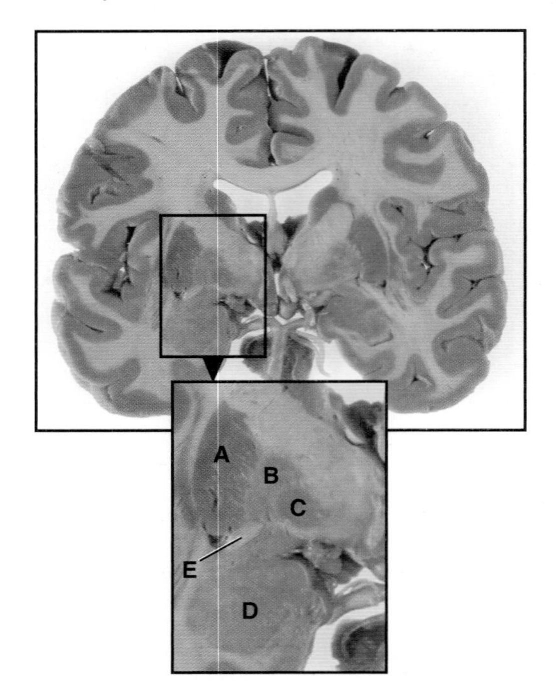

(A) ____________________
(B) ____________________
(C) ____________________
(D) ____________________
(E) ____________________

6 A 23-year-old man is involved in a motorcycle collision and is transported to the Emergency Department. CT reveals that he has a traumatic brain injury (TBI) to portions of the brain located in the posterior fossa; within a few hours, a portion of his cerebellum herniates downward. This constitutes a medical emergency that must be addressed immediately, or the patient will almost certainly die. Identify the structures in the given image, some of which are involved in this man's case.

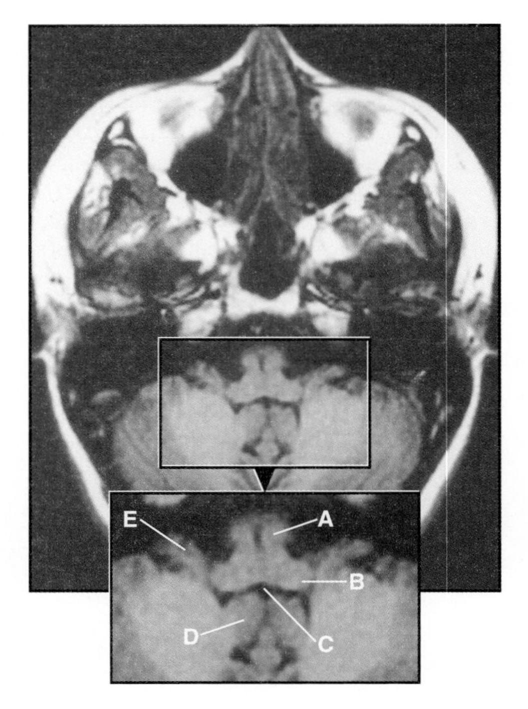

(A) ____________________
(B) ____________________
(C) ____________________
(D) ____________________
(E) ____________________

7 A 19-year-old man is brought to the Emergency Department from the site of a fight outside a bar. The history reveals that the man was involved in an altercation and was stabbed in the back of his neck with a knife with a 6-in blade. The examination reveals that the man has a Brown-Séquard syndrome consistent with an injury at the C5-6 interspace. CT and MRI confirm that the blade entered the vertebral column between C5 and C6, was deflected slightly upward, and passed through the dural sac hemisecting the spinal cord. Identify the structures in the image below, most of which would be affected by this trauma.

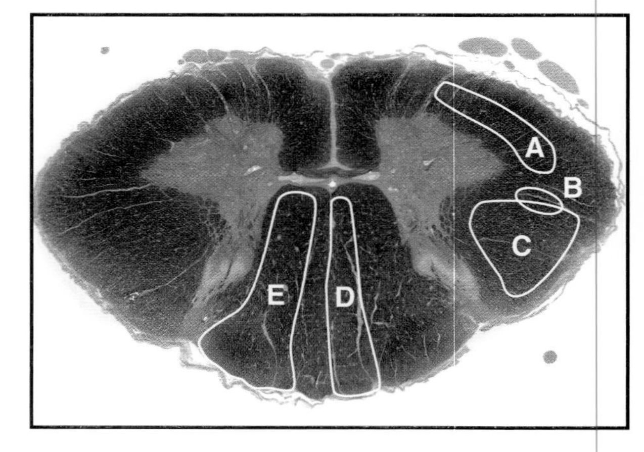

(A) ____________________
(B) ____________________
(C) ____________________
(D) ____________________
(E) ____________________

8 A 53-year-old woman presents with a persistent headache. The history reveals that this malady has persisted for several weeks and seems to be getting slightly worse; it is largely refractory to OTC pain medication. An MRI is done to see if there is evidence of a mass lesion or ventricular enlargement. Identify the spaces in the image below, the shapes and sizes of which may be affected in the case of a variety of intracranial lesions or conditions.

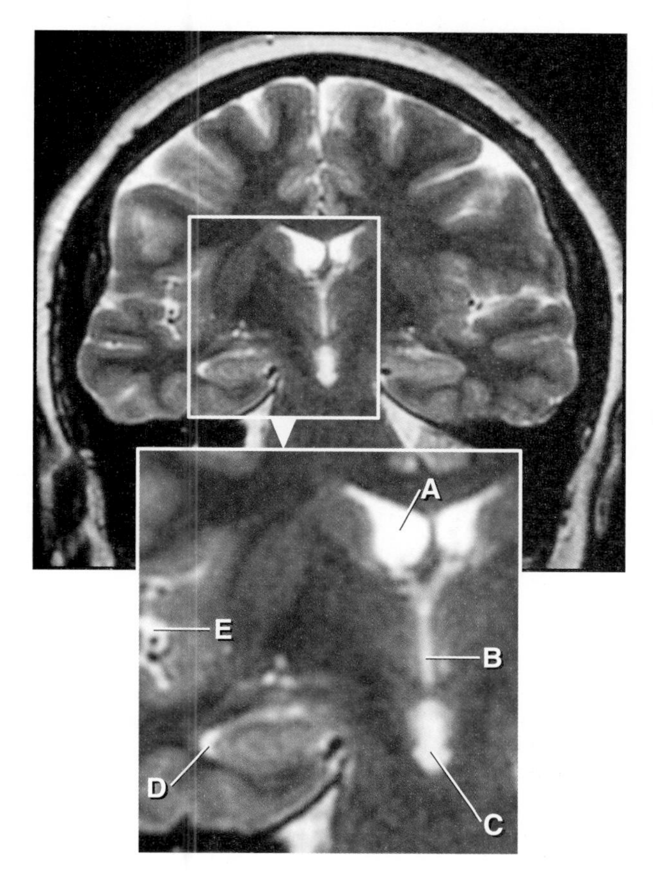

(A) ______

(B) ______

(C) ______

(D) ______

(E) ______

9 A 29-year-old woman presents with atypical pain that waxes and wanes, at times is severe, and can be initiated by putting lipstick on her lips and makeup on her cheek. An MRI is done to ascertain if there might be an organic reason for this pain. Identify the structures in the given image, some of which might be related to this women's clinical condition.

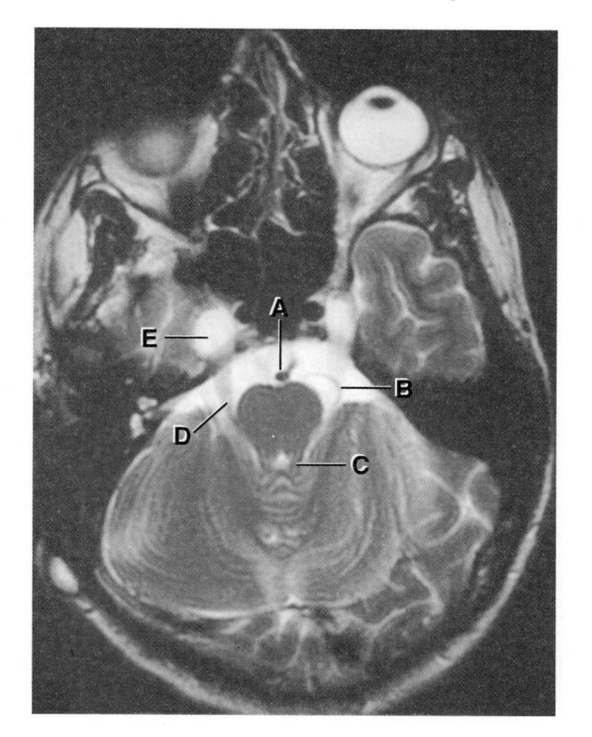

(A) ______

(B) ______

(C) ______

(D) ______

(E) ______

10 During a Clinical Pathology Conference, the brain of a 54-year-old man is sectioned in the axial plane. The students in attendance are quizzed regarding their general knowledge of brain structures in one of these slices. Identify the structures in the image below; some of these were asked by the Pathology Faculty member.

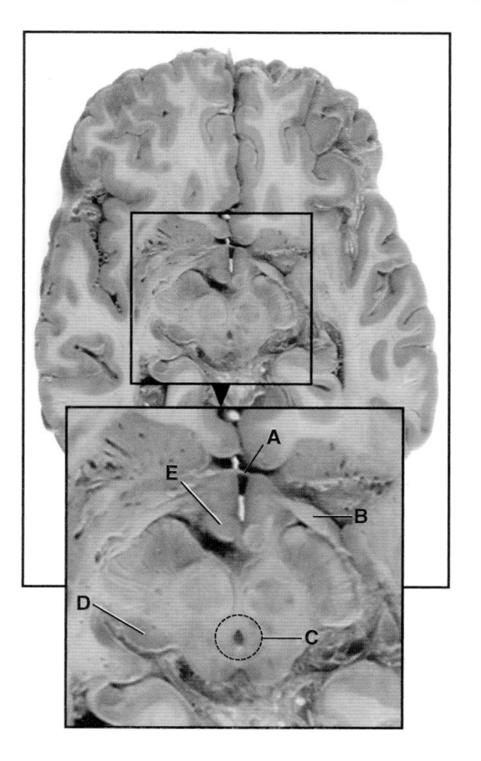

(A) ______________________________
(B) ______________________________
(C) ______________________________
(D) ______________________________
(E) ______________________________

11 A 57-year-old man presents with tremor of unknown origin. The examination reveals a right-sided tremor, a slightly ataxic gait, but the sensory examination is normal. An MRI is done in an attempt to determine if any organic cause can be identified. Identify the structures in the image below, some of which may be compromised in this man's case.

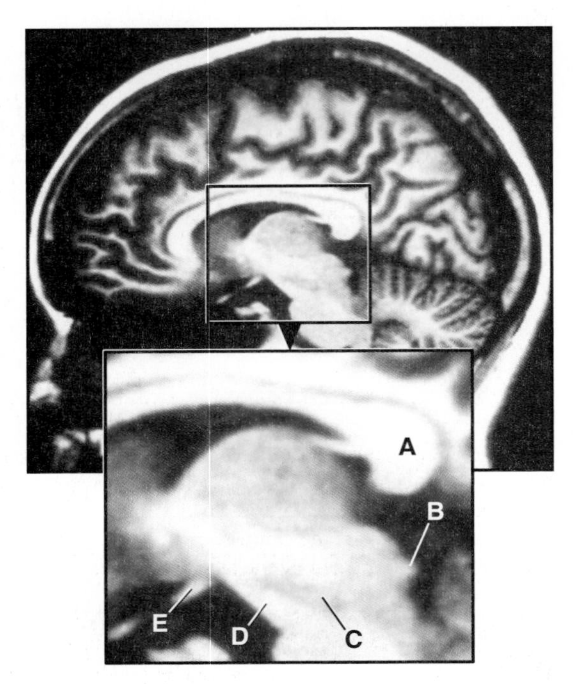

(A) ______________________________
(B) ______________________________
(C) ______________________________
(D) ______________________________
(E) ______________________________

12 A 65-year-old man presents at his physician's office with neck pain. The examination reveals neck pain on the right (when the man moves his neck) that is centered at the base of his neck and over the left shoulder. A CT myelogram, focusing on levels C4-7, is performed as part of the evaluation of this man's case. Identify the structures and spaces in the given image that would be seen in this man's CT.

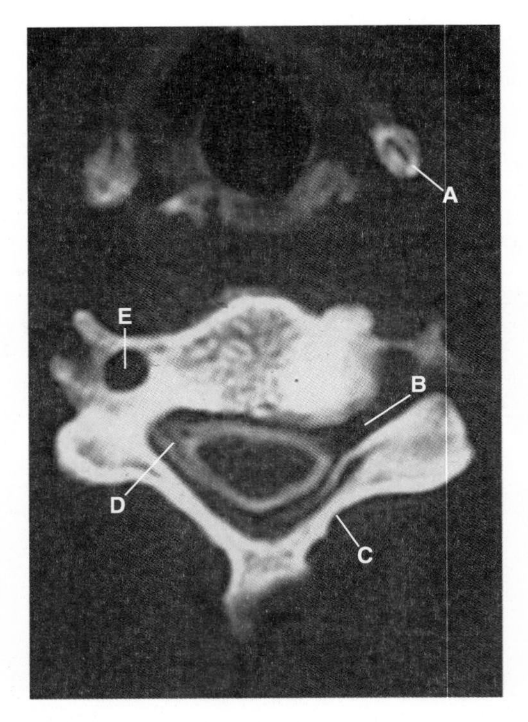

(A) ______________________________
(B) ______________________________
(C) ______________________________
(D) ______________________________
(E) ______________________________

13 A 61-year-old man has a sudden onset of difficulty speaking. The examination in the Emergency Department reveals that the man is not weak, clearly understands commands, and can respond to these commands, but has great difficulty formulating speech. It is not a problem of weakness or paralysis of the vocal apparatus but rather a problem of turning thoughts or concepts into speech. MRI reveals a lesion in the cerebral cortex. Identify the structures in the image below; some of them may be involved in a lesion of the type experienced by this man.

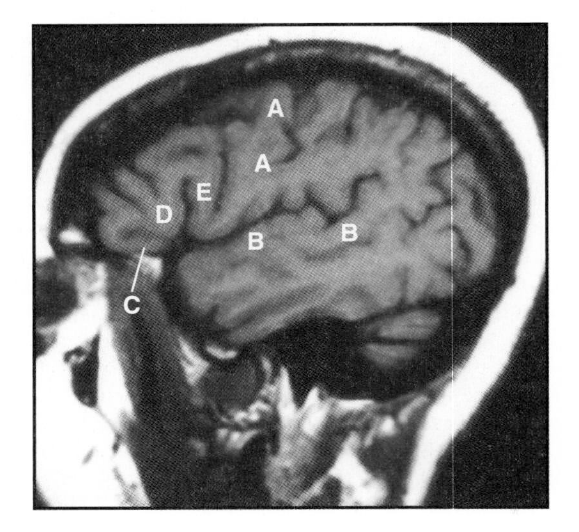

(A) ______________________________
(B) ______________________________
(C) ______________________________
(D) ______________________________
(E) ______________________________

14 The medulla oblongata (commonly just called the medulla) is rostrally continuous with the pons and caudally continuous with the upper levels of the cervical spinal cord. Several pathways that are essential in clinical diagnoses, the nuclei of several cranial nerves, and respiratory and cardiac centers are located in the medulla. In fact, alternating, or crossed, deficits (long tract deficit on one side, cranial nerve deficit on the opposite side) are characteristic of brainstem lesions. Identify the structures in the image below; these are important medullary structures damage to which would result in testable deficits.

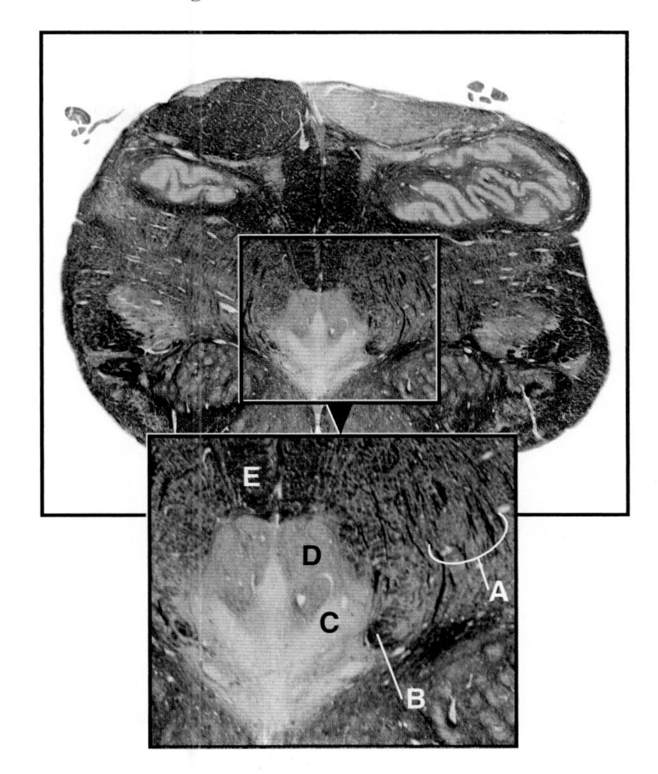

(A) ______________________________
(B) ______________________________
(C) ______________________________
(D) ______________________________
(E) ______________________________

15 A 71-year-old woman experiences a sudden onset of a right-sided weakness. The examination reveals weakness of upper and lower extremities, right-sided sensory disturbances, and a decrease in her level of consciousness. An MRI was done to determine if there is evidence of a lesion. Identify the structures and space in the given image; in a lesion of the type experienced by this woman, at least one of these structures would most likely be involved.

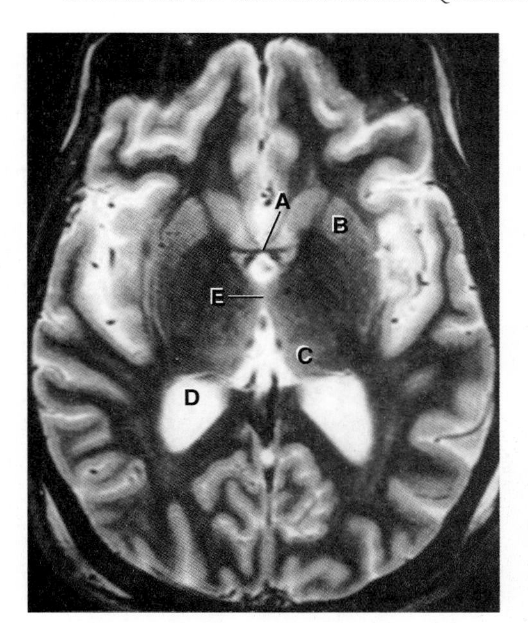

(A) ______________________________
(B) ______________________________
(C) ______________________________
(D) ______________________________
(E) ______________________________

16 A 36-year-old woman is brought to the Emergency Department from the site of a motor vehicle collision. She is not conscious and has extensive injuries to her head and body. CT reveals the injuries to the brain shown in the image below. Identify what is indicated by the labels in this image.

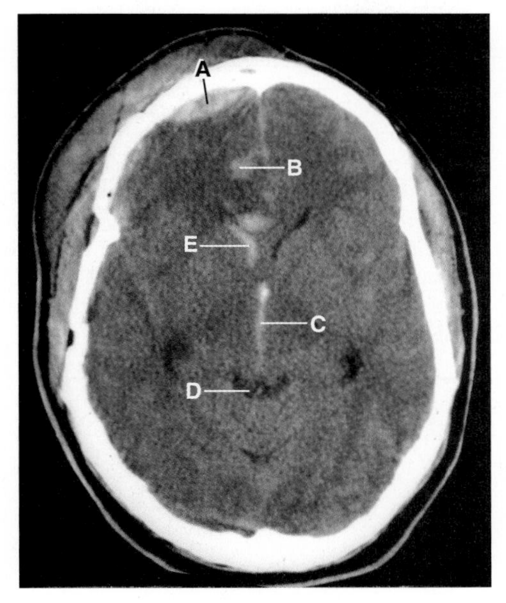

(A) ______________________________
(B) ______________________________
(C) ______________________________
(D) ______________________________
(E) ______________________________

17 A 16-year-old boy is brought to the family physician by his mother. He complains of difficulty with his eyes. The examination reveals that the boy is not blind in either eye or in any hemifield or quadrant, that eye movements in the horizontal plane or downward are essentially normal, that his pupillary light reflex is intact, but that he has a paralysis of upward gaze in both eyes. Identify the structures and space in the image below; this boy's condition may relate to at least one of these labeled entities.

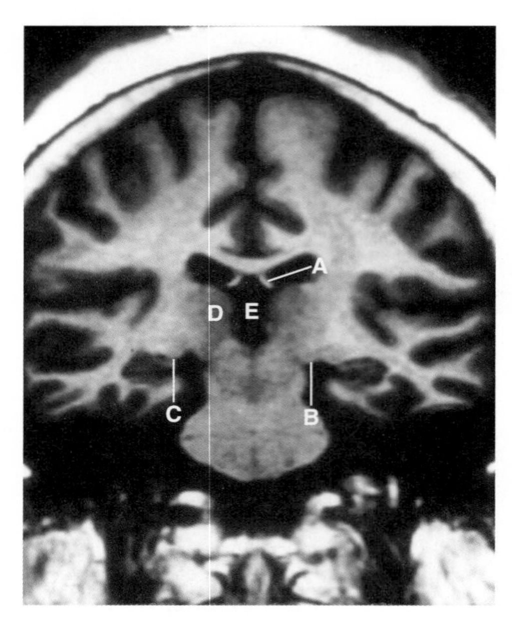

(A) ______________________
(B) ______________________
(C) ______________________
(D) ______________________
(E) ______________________

18 A 23-year-old man is brought to the Emergency Department from the site of a motorcycle collision. He was wearing leather pants and jacket but still has abrasions. He particularly complains of lower back pain. CT reveals a ruptured disc at L4-5 protruding into a stenotic vertebral canal and impinging on the dural sac. An image representing this lesion is shown below. If not treated, what are the main deficits that this man would most likely experience?

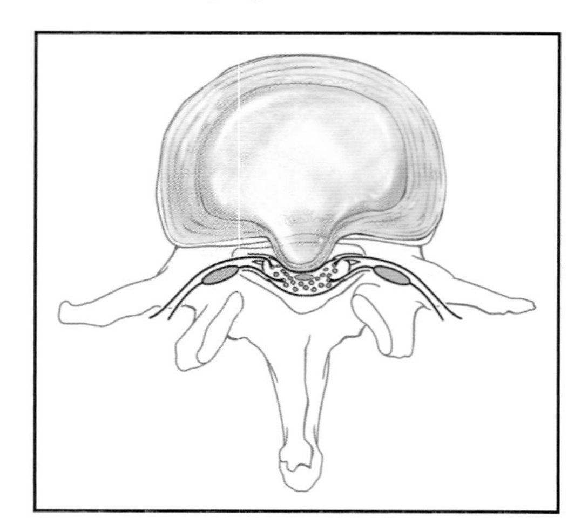

(A) ______________________
(B) ______________________
(C) ______________________
(D) ______________________
(E) ______________________

19 A 72-year-old woman is brought to the Emergency Department from her home. The history, provided by her son, reveals that the woman experienced a sudden excruciating headache followed by nausea and vomiting. CT reveals blood in the cisterns at the base of the brain, strongly suggesting that she suffered a ruptured aneurysm. Identify the structures indicated in the image below, several of which could be the source of the type of vascular defect present in this woman.

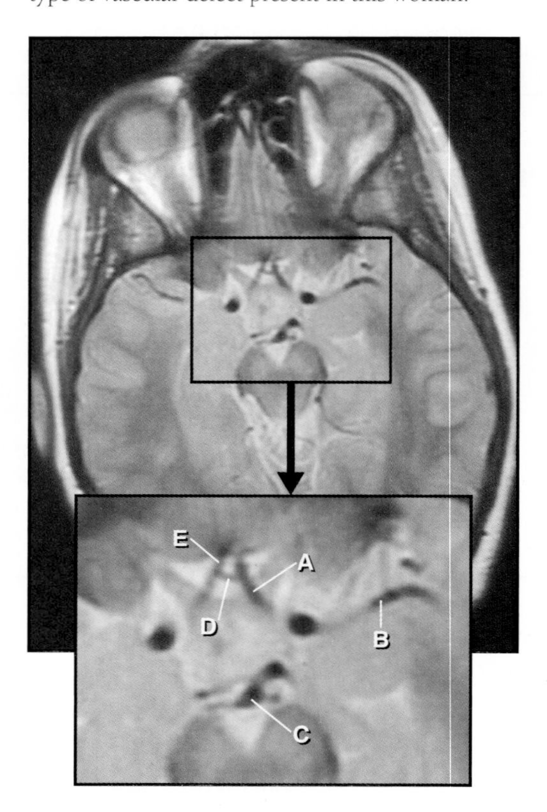

(A) ______________________
(B) ______________________
(C) ______________________
(D) ______________________
(E) ______________________

20 A 31-year-old man visits his ophthalmologist. He thinks that he needs to have the prescription for his glasses changed, since he has experienced progressively more difficulty seeing things over the last 2 months. The examination reveals that the man does not have visual field deficits (he is not blind), but does have diplopia. An MRI is done to investigate possible organic causes for this man's deficit. Identify the structures in the given image, damage to at least one of which would produce the deficit seen in this man.

(A) ______________________________
(B) ______________________________
(C) ______________________________
(D) ______________________________
(E) ______________________________

21 The dorsal thalamus, commonly just called the thalamus, is the major synaptic station between all types of ascending information arising at lower levels and traveling to the cerebral cortex. Only olfactory information does not synapse in the thalamus en route to the cerebral cortex. Identify the thalamic nuclei in the axial image below.

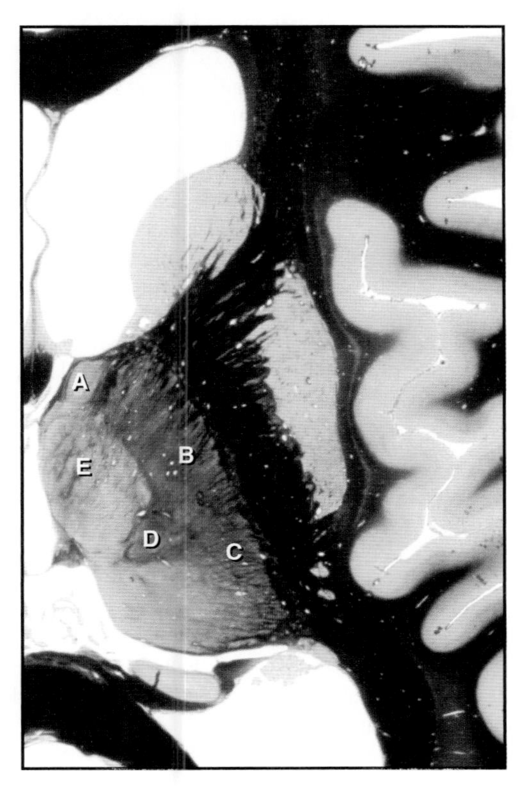

(A) ______________________________
(B) ______________________________
(C) ______________________________
(D) ______________________________
(E) ______________________________

22 A 39-year-old woman presents with progressive signs and symptoms suggesting a probable pituitary tumor. Laboratory tests indicate that she has elevated blood corticotropin levels that produce high levels of adrenal cortisol secretion. This type of lesion results in deficits, and bodily changes, that are characteristic of Cushing disease. Identify the structures indicated in the image below; in Cushing disease some of these structures may be damaged.

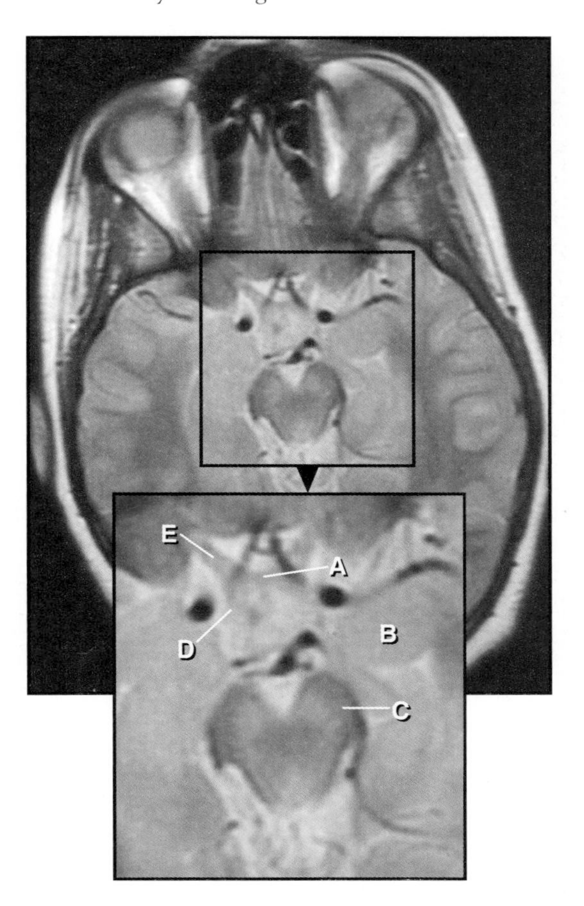

(A) ______________________________
(B) ______________________________
(C) ______________________________
(D) ______________________________
(E) ______________________________

ANSWERS

1 **The answers are:** The occurrence of a sudden severe headache (commonly described as a "thunderclap") is characteristic of subarachnoid hemorrhage (SAH). Depending on the severity of the hemorrhage, blood may occupy few or many of the subarachnoid cisterns. A = Cistern of the lamina terminalis (contains A_1 and A_2 segments, artery of Heubner, anterior communicating artery, and other major branches); B = Sylvian cistern (contains the M_1 segment and its branches); C = Crural cistern (contains the anterior choroidal artery and the basal vein of Rosenthal); D = Ambient cistern (contains P_2 and P_3 segments and the quadrigeminal, superior cerebellar, and lateral posterior choroidal arteries, basal vein, and oculomotor nerve); E = Quadrigeminal cistern (contains the P_4 segment, quadrigeminal artery, vein of Galen, and root of the trochlear nerve).

2 **The answers are:** The most likely diagnosis in this woman is idiopathic intracranial hypertension (pseudotumor cerebri). This condition is seen in obese women of child-bearing age. A = Genu of the corpus callosum; B = Anterior commissure; C = Dorsal thalamus (in this sagittal view and plane, the dorsomedial nucleus is the main thalamic structure present); D = Vein of Galen; E = Hypothalamus (in this sagittal view, the plane is through the medial hypothalamic zone: for example, the mammillary body is obvious).

3 **The answers are:** This man has all the signs and symptoms of Huntington disease; this can be definitively confirmed by genetic testing. In an MRI, the physician will be especially attentive to the condition of the neostriatum. A = Body of the corpus callosum; B = Head of the caudate nucleus (this structure becomes progressively diminished in size in Huntington disease, the head of the caudate becomes flattened); C = Lenticular nucleus (this nucleus also diminishes in size in Huntington disease); D = Nucleus accumbens (also involved in a number of neurodegenerative diseases); E = Anterior limb of the internal capsule.

4 **The answers are:** This girl has a hypersecreting tumor of the pituitary; her symptoms are highly suggestive of a prolactinoma. These tumors cause characteristic endocrine disorders and may damage visual structures at the base of the hypothalamus resulting in visual field deficits. A = Lamina terminalis (this thin structure forms the rostral wall of the third ventricle, separating this space from the cistern of the lamina terminalis that is immediately rostral to it); B = Hypothalamus; C = Mammillary body (a distinct caudal portion of the hypothalamus); D = Interpeduncular fossa (contains the basilar bifurcation and oculomotor root); E = Optic tract (on the surface of the crus cerebri).

5 **The answers are:** The sudden onset of a hemiplegia and the confirmation of a hemisphere lesion in MRI suggest that this hemorrhagic event arises in the territory of the lenticulostriate branches of the M_1 segment. One cardinal sign is the sudden onset of weakness on one side of the body; other deficits usually appear with time. A = Putamen nucleus (along with the caudate forms the neostriatum); B = Lateral, or external, segment of the globus pallidus (some of its cells project to the subthalamic nucleus); C = Medial, or internal, segment of the globus pallidus (cells of this structure give rise to the pallidothalamic projection); D = Amygdaloid complex (important structure of the limbic system, bilateral lesions produce Klüver-Bucy syndrome); E = Anterior commissure (after crossing the midline, these fibers course caudolaterally immediately inferior to the lenticular nucleus).

6 **The answers are:** This is tonsillar herniation: a situation in which increased pressure within the posterior fossa forces the tonsil of the cerebellum downward through the foramen magnum. As it does so, there is the very real possibility of pressure being exerted on the medulla, with resultant damage to respiratory and cardiac centers located therein producing sudden respiratory and cardiac arrest. A = Medullary pyramid (contains corticospinal fibers); B = Restiform body (contains many cerebellar afferent fibers such as olivocerebellar and posterior spinocerebellar); C = Fourth ventricle; D = Cerebellar tonsil (notice its proximity to the medulla); E = Flocculus of the cerebellum (the lateral part of the flocculonodular lobe).

7 **The answers are:** Spinal cord hemisection (a Brown-Séquard syndrome) may be the result of trauma and presents as a classic set of deficits or, as is more commonly the case, may result from vertebral fracture or tumor, and may be described as the "functional hemisection." In this latter case, the deficits may present as a cord hemisection, but the injury may be a compression or distortion of the cord with damage, rather than a sharp clean lesion that would be seen in a knife wound. A = Anterolateral system (conveys pain and thermal sense from the right side beginning about two levels below the damage); B = Rubrospinal tract (predominant influence on flexor motor neurons at cervical levels on the left side); C = Lateral corticospinal tract (influences lower motor neurons on the left side, somatotopically arranged, from lateral to medial lower extremity, trunk, upper extremity); D = Gracile fasciculus (conveys proprioception, discriminative touch, and vibratory sense from left lower extremity); E = Cuneate fasciculus (conveys proprioception, discriminative touch, and vibratory sense from right upper extremity).

8 **The answers are:** Recurring headache may signal an organic cause such as a lesion, aneurysm, or increase in intracranial pressure, or it may exist in the absence of any organic cause; the latter is usually the case. In this woman, the T2-weighted MRI appears essentially normal. A = Body of the lateral ventricle (note the clarity of the caudate nucleus; based on the angle of this image, this is close to the junction of the head of the caudate with the body); B = Third ventricle (characteristically narrow and located on the midline); C = Interventricular fossa (contains the basilar bifurcation and oculomotor root); D = Temporal horn of the lateral ventricle (also called inferior horn); E = Subarachnoid space on the insular cortex (contains M_2 and M_3 branches and the deep middle cerebral vein).

9 **The answers are:** This woman's symptoms are characteristic of trigeminal neuralgia (also called tic douloureux or Fothergill neuralgia); intermittent and severe pain that may be initiated by stimulating the corner of the mouth or cheek. There may be a variety of causes, one being aberrant branches of the superior cerebellar artery impinging on the trigeminal root. A = Basilar artery; B = Branch of the superior cerebellar artery passing adjacent to the trigeminal root; C = Superior cerebellar peduncle (also called brachium conjunctivum); D = Root of the trigeminal nerve; E = Subarachnoid space around the trigeminal ganglion.

10 **The answers are:** The topics covered in a brain-cutting in a clinical pathological conference (CPC) vary widely. The faculty may point out normal structures, small lesions that represent incidental findings, areas of color or textural differences, and, of course, the larger obvious lesions. This type of conference can be a valuable learning experience. A = Lamina terminalis (thin membrane forming the separation between the third ventricle and the cistern of the lamina terminalis); B = Optic tract (characteristically located on the surface of the crus cerebri); C = Periaqueductal gray (receives input from the anterolateral system and makes important contributions to pathways for the descending inhibition of pain); D = Medial geniculate nucleus (relay for auditory information); E = Mammillary nuclei (receives information via the fornix, projects to anterior thalamic nucleus; lesions result in disrupted memory).

11 **The answers are:** Tremor can result from many different types of clinical problems. Tremor is commonly seen in cerebellar lesions (intention tremor), in diseases of the basal nuclei (such as Parkinson, where it is a resting tremor), and in neurodegenerative diseases (such as Huntington disease). The type of motor disorder may be characteristic of the disease (such as Wilson disease and asterixis). A = Splenium of the corpus callosum; B = Inferior colliculus (receives a variety of inputs particularly auditory); C = Substantia nigra (loss of the dopamine-containing cells results in Parkinson disease and a resting tremor); D = Crus cerebri (contains corticopontine, corticospinal, and corticonuclear fibers); E = Optic tract (the optic nerve is attached to the base of the diencephalon and courses over the crus cerebri).

12 **The answers are:** This man has an intervertebral disc problem at the C5-6 interspace which impacts on the C6 roots. Pain indicative of a cervical disc problem is commonly experienced upon awakening; assuming a nontraumatic cause, the majority of these resolve without surgical intervention. About 20% of cervical disc problems are seen at the C5-6 level, while the majority (65% to 70%) are at the C6-7 level. A = Inferior cornu of the thyroid cartilage (note the opening of the trachea); B = Intervertebral foramen (conveys anterior and posterior roots and small arteries and veins); C = Lamina of the cervical vertebra; D = Posterior root (along with the anterior root forms the spinal nerve); E = Foramen transversarium (contains the vertebral artery).

13 **The answers are:** This man is suffering an expressive aphasia (also called a Broca aphasia) that is seen in lesions that are located in the inferior frontal gyrus of the dominant hemisphere. In fact, the inferior frontal gyrus is sometimes called the Broca convolution. These are usually vascular in origin, although tumors may also produce such deficits. A = Precentral gyrus (motor cortex; this gyrus plus the anterior paracentral gyrus form the primary somatomotor cortex, Brodmann area 4); B = Superior temporal gyrus (Brodmann area 22); C = Pars orbitalis of the inferior frontal gyrus (Brodmann area 47); D = Pars triangularis of the inferior frontal gyrus (Brodmann area 45); E = Pars opercularis of the inferior frontal gyrus (Brodmann area 44). Lesions that cause an expressive aphasia are most commonly located in areas 44 + 45. The pars orbitalis is separated from the pars triangularis by the anterior horizontal ramus of the lateral (Sylvian) fissure; the pars triangularis is separated from the pars opercularis by the anterior ascending ramus of the lateral (Sylvian) fissure.

14 **The answers are:** A = Internal arcuate fibers (cells of origin are in the posterior column nuclei; fibers cross the midline as the sensory decussation to enter the medial lemniscus on the opposite side); B = Solitary tract (the cells immediately adjacent to the tract collectively constitute the solitary nucleus; at this level [caudal medulla] it contains primarily general visceral afferent fibers; the solitary nucleus at this level is called the cardiorespiratory nucleus); C = Dorsal motor nucleus of the vagus (contains general visceral efferent preganglionic parasympathetic cells whose axons distribute within the vagus nerve); D = Hypoglossal nucleus (motor to the ipsilateral side of the tongue [left side in this example]; paralysis of the genioglossus muscle results in deviation of the tongue to the weak side); E = Medial longitudinal fasciculus (at this level contains primarily descending fibers including the medial vestibulospinal tract).

15 **The answers are:** This woman most likely experienced a hemorrhagic lesion in the hemisphere on the left side; her motor and sensory deficits are on the right, and no cranial nerve deficits are reported. A large internal region of the hemisphere, including much of the lenticular nucleus and the posterior limb of the internal capsule, will result in the type of signs and symptoms seen in this woman. This vascular territory is usually that of the lenticulostriate (lateral striate) branches of the M_1 segment. A = Anterior commissure (immediately caudal to the commissure, the dark structure is the column of the fornix; the mammillothalamic tract appears as a black spot in the anterior and medial thalamus); B = Putamen, the lateral part of the lenticular nucleus (the globus pallidus is represented by the light structures immediately medial to the putamen); C = Pulvinar (the large caudal portion of the thalamus); D = Atrium of the lateral ventricle (the body, posterior horn, and temporal horn, all coalesce to form the atrium); E = Massa intermedia (present in about 80% of brains, not present in about 20% of brains; no deficit associated with its absence).

16 **The answers are:** As can be seen on the right side of this image, this patient has soft tissue damage, a fractured skull, and traumatic brain injuries. A = Small epidural hematoma (these lesions occur when the dura mater is stripped from the inner table of the skull and blood collects in the space that is created; patients with large epidural lesions may experience a "talk-and-die" episode; there is not a preexisting epidural space around the brain); B = Parenchymatous hemorrhage, or bleeding into, and collecting within, the tissues of the brain (note that much of the right frontal lobe is hypodense; the appearance of this area is shifted toward the appearance of CSF, a shift toward more black, indicating ischemia of the tissue); C = Blood in the third ventricle (blood within the ventricle forms a cast of that space, clearly showing its dimensions); D = Quadrigeminal cistern (in this patient, it contains no blood); E = Blood in the anterior horn of the lateral ventricle (again, blood within the ventricle forms a cast of that space clearly showing its dimensions; note the characteristic shape of the head of the caudate nucleus and compare this area with the corresponding ventricular space on the left side).

17 **The answers are:** This boy has a dorsal midbrain syndrome, commonly called the Parinaud syndrome or Parinaud ophthalmoplegia. This is a paralysis of upward gaze of both eyes (it is conjugate) when the patient attempts to look up without moving his head. This may result from damage to the mesencephalic tectum, particularly the superior colliculi, or from tumors or vascular lesions that are located in the superior cistern and impinge on the tectum. A = Body of the fornix (it is continuous with the crus of the fornix, then with the hippocampal formation); B = Medial geniculate body and nucleus (the elevation on the surface is the body; the nucleus is the cells internal to the elevation, auditory function); C = Lateral geniculate body and nucleus (the elevation on the surface is the body; the nucleus is the cells internal to the elevation, visual in function); D = Pulvinar nucleus (located on either side of the superior cistern); E = Superior (quadrigeminal) cistern (this cistern is located on the midline between the two pulvinar, is wide in contrast to the very narrow third ventricle, and is located caudal to the posterior commisure [which is the landmark stipulating the caudal end of the third ventricle]).

18 **The answers are:** This man has a cauda equina syndrome (CES). This particular example may be classified as a Group I CES (sudden onset with no previous symptoms, traumatic in this case). Group II CES is the situation in which there is a history of recurrent low back pain and/or sciatica with the onset of CES which can be specifically related to a recent event. CES III is the situation where there is an onset of low back pain and/or sciatica that evolves into CES without a specific precipitating event. CES may also result from tumors, epidural hematoma, or from tissue debris following removal of a disc. Recall that the spinal cord ends (at the conus medullaris) at about L1-2 and the dural sac ends at about S2. Between these levels, the dural sac contains the anterior and posterior roots that collectively form the cauda equina, which is located within the lumbar cistern. Probable symptoms include A = Urinary retention (commonly seen, the bladder is only partially emptied); B = Saddle anesthesia (commonly seen, sensory loss over the buttocks, anus, perineum, medial thigh); C = Weakness of lower extremity (may involve single or multiple roots, may become paraplegic); D = Decreased or absent reflexes (lower extremity reflexes, such as the Achilles, may be affected); E = Sexual deficits (while these may not be initially seen, they may be experienced as progressive deficits). Lower back pain (as reported by this man) and/or sciatica may be present as ipsilateral or bilateral deficits; in some situations, it may be absent.

19 **The answers are:** The sudden occurrence of a severe headache (these are described by the patient as "like a thunderclap," "the worst of my life," or "like a bomb going off in the head") is a characteristic feature of a rupture of an aneurysm with bleeding into the subarachnoid space. Recall that the subarachnoid cisterns are simply enlarged, and named, portions of the subarachnoid space. Aneurysm rupture is the second most common cause of blood in the subarachnoid space, trauma being the first. The most common location of aneurysm in the internal carotid circulation is in the immediate vicinity of the anterior communicating artery and its junction with the anterior cerebral artery (ACA). The most common location of aneurysm in the vertebrobasilar system is at the bifurcation of the basilar artery. A = A_1 segment of the ACA (located between the internal carotid and anterior communicating arteries); B = M_1 segment of the middle cerebral artery; C = Basilar artery at its bifurcation (the branches of the superior cerebellar arteries and the larger [in this individual] initial segment of the posterior cerebral artery [P_1] on the left is seen); D = Anterior communicating artery (separates A_1 from A_2); E = A_2 segment of the ACA (also called the infracallosal segment).

20 **The answers are:** Diplopia, double vision or "seeing two of everything," results from damage to any of the cranial nerves (oculomotor, trochlear, or abducens) that innervate the extraocular muscles. The patient does not experience visual loss but is unable to bring the visual axes of both eyes to bear on the same point in space. Additional deficits frequently accompany the diplopia, regardless of which nerve is involved. A = Optic tract (it is attached to the inferior aspect of the brain immediately adjacent to the crus cerebri [not labeled]); B = Posterior cerebral artery; C = Superior cerebellar artery (note that the oculomotor nerve is passing between these two vessels at this particular point); D = Oculomotor nerve (note the nerve exiting from the midbrain immediately in front of the basilar pons, passing in between the posterior cerebral and superior cerebellar arteries, and coursing toward the superior orbital fissure through which it exits the cranial cavity to enter the orbit); E = Internal carotid artery (cerebral part).

21 **The answers are:** A = Anterior nucleus of the thalamus (receives input from the mammillary body via the mammillothalamic tract, projects to the cingulate gyrus); B = Ventral lateral nucleus of the thalamus (receives input from the left basal nuclei and the right cerebellar nuclei and projects to the left motor cortex); C = Ventral posterolateral nucleus of the thalamus (receives input via the anterolateral system and the posterior column–medial lemniscus system, projects to the somatosensory cortex, conveying information from the right side of the body sans the head); D = Centromedian nucleus of the thalamus (receives input from the left basal nuclei and lesser projections from the spinal cord and reticular formation, projects to the left motor and sensory cortex, globus pallidus, subthalamic nucleus, and substantia nigra); E = Dorsomedial nucleus of the thalamus (receives input from the amygdala, orbitofrontal and temporal cortex, and several olfactory structures, projects to all areas of the cortex of the frontal lobe excepting the primary motor cortex).

22 **The answers are:** Hypersecreting pituitary tumors that produce excessive corticotropin result in a complex of body changes and deficits characteristic of Cushing disease (truncal obesity, facial hair, violaceous striae [these are clearly different from the stria of pregnancy], hypertension, cervical hump, and emotional lability). Cushing disease is more common in women. In 10% of cases, the tumor is sufficiently large to produce visual field deficits. A = Optic chiasm (damage to this structure results in a bitemporal hemianopsia); B = Amygdala (an important part of the limbic system); C = Crus cerebri (contains corticospinal and corticonuclear fibers, and fronto-, parieto-, occipito- and temporopontine fibers); D = Optic tract (damage to this structure will result in a homonymous hemianopsia in the opposite visual fields, in this case, a left homonymous hemianopsia); E = Optic nerve (damage to this structure will result in blindness in the eye on the same side, in this case, blindness in the right eye; there will also be a loss of the direct and consensual pupillary light reflex when a light is shined in the right eye).

Index

Page numbers in *italics* denote figures; those followed by a t denote tables

D

E

F

G

H

N

O

P

V

W